Preface

This volume consists of a selection of papers concerning a new approach to the problem of Wiener-Hopf factorization for rational and analytic matrix-valued (or operator-valued) functions. It is a result of developments which took place during the past ten years. The main advantage of this new approach is that it allows one to get the Wiener-Hopf factorization explicitly in terms of the original function. The starting point is a special representation of the function which is taken from Mathematical Systems Theory where it is known as a realization. For the case of rational matrix-valued functions the final theorems express the factors in the factorization and the indices in terms of the three matrices which appear in the realization.

This book consists of two parts. Part I concerns canonical and, more generally, minimal factorization. Part II is dedicated to non-canonical Wiener-Hopf factorization (i.e., the factorization indices are not all zero). Each part starts with an editorial introduction which contains short descriptions of each of the papers.

This book is a result of research which for a large part was done at the Vrije Universiteit at Amsterdam and was started about ten years ago. It is a pleasure to thank the department of Mathematics and Computer Science of the Vrije Universiteit for its support and understanding during all those years. We also like to thank the Econometrics Institute of the Erasmus Universiteit at Rotterdam for its technical assistance with the preparations of this volume.

Amsterdam, June 1986 I. Gohberg, M.A. Kaashoek

TABLE OF CONTENTS

OT 21:
Operator Theory: Advances and Applications
Vol. 21

Birkhäuser Verlag
Basel · Boston · Stuttgart

Constructive Methods of Wiener-Hopf Factorization

Edited by

I. Gohberg
M. A. Kaashoek

1986

Birkhäuser Verlag
Basel · Boston · Stuttgart

Volume Editorial Office

Department of Mathematics and Computer Science
Vrije Universiteit
P. O. Box 7161
1007 MC Amsterdam
The Netherlands

Library of Congress Cataloging in Publication Data

Constructive methods of Wiener-Hopf factorization.
(Operator theory, advances and applications ;
vol. 21)
Includes bibliographies and index.
1. Wiener-Hopf operators. 2. Factorization of
operators. I. Gohberg, I. (Israel), 1928– .
II. Kaashoek, M. A. III. Series: Operator theory,
advances and applications ; v. 21.
QA329.2.C665 1986 515.7'246 86–21587

CIP-Kurztitelaufnahme der Deutschen Bibliothek

Constructive methods of Wiener-Hopf factorization
/ ed. by I. Gohberg ; M. A. Kaashoek. – Basel ;
Boston ; Stuttgart : Birkhäuser, 1986.
(Operator theory ; Vol. 21)
ISBN 3-7643-1826-0
NE: Gohberg, Israel [Hrsg.]; GT

Printed in Germany
ISBN 3-7643-1826-0
ISBN 0-8176-1826-0

PART I

CANONICAL AND MINIMAL FACTORIZATION

EDITORIAL INTRODUCTION

The problem of canonical Wiener-Hopf factorization appears in different mathematical fields, theoretical as well as applied. To define this type of factorization consider the matrix-valued function

$$(1) \qquad W(\lambda) = I_m - \int_{-\infty}^{\infty} e^{i\lambda t} k(t)dt, \qquad -\infty < \lambda < \infty,$$

where k is an $m \times m$ matrix function of which the entries are in $L_1(-\infty,\infty)$ and I_m is the $m \times m$ identity matrix. A *(right) canonical (Wiener-Hopf) factorization* of W relative to the real line is a multiplicative decomposition:

$$(2) \qquad W(\lambda) = W_-(\lambda)W_+(\lambda), \qquad -\infty < \lambda < \infty,$$

in which the factors W_- and W_+ are of the form

$$W_-(\lambda) = I_m - \int_{-\infty}^{0} e^{i\lambda t} k_1(t)dt, \qquad \operatorname{Im}\lambda \le 0,$$

$$W_+(\lambda) = I_m - \int_{0}^{\infty} e^{i\lambda t} k_2(t)dt, \qquad \operatorname{Im}\lambda \ge 0,$$

where k_1 and k_2 are $m \times m$ matrix functions with entries in $L_1(-\infty,0]$ and $L_1[0,\infty)$, respectively, and

$$\det W_-(\lambda) \ne 0 \quad (\operatorname{Im}\lambda \le 0), \qquad \det W_+(\lambda) \ne 0 \quad (\operatorname{Im}\lambda \ge 0).$$

Such a factorization does not always exist, but if $W(\lambda)$ (or, more generally, its real part) is positive definite for all real λ, then the matrix function W admits a canonical factorization (see [8], [7]). Sometimes iterative methods can be used to construct a canonical factorization. In the special case when $W(\lambda)$ is a rational matrix function there is an algorithm of elementary row and column operations which leads in a finite number of steps to a canonical factorization provided such a factorization exists.

In the late seventies a new method has been developed to deal with factorization problems for rational matrix functions. This method is based on a special representation of the function, namely in the form

(3) $$W(\lambda) = I_m + C(\lambda I_n - A)^{-1}B, \quad -\infty < \lambda < \infty,$$

where $A : \mathbb{C}^n \to \mathbb{C}^n$, $B : \mathbb{C}^m \to \mathbb{C}^n$ and $C : \mathbb{C}^n \to \mathbb{C}^m$ are linear operators. The representation (3), which comes from Mathematical Systems Theory, is called a *realization*. The smallest possible number n in (3) is called the *degree* of W (notation: $\delta(W)$), and if $n = \delta(W)$, then (2) is said to be a *minimal realization* of W. The factorization method based on (3) has led to the following canonical factorization theorem ([1], Section 4.5):

THEOREM 1. *Let* $W(\lambda) = I_m + C(\lambda I_n - A)^{-1}B$ *be a minimal realization, and assume* A *has no real eigenvalue. Then* W *admits a canonical Wiener-Hopf factorization relative to the real line if and only if the following two conditions are fulfilled:*

(i) $A^\times := A - BC$ *has no real eigenvalue,*

(ii) $\mathbb{C}^n = M \oplus M^\times$,

where M *(resp.* $M^\times$*) is the space spanned by the eigenvectors and generalized eigenvectors corresponding to the eigenvalues of* A *(resp.* $A^\times$*) in the upper (resp. lower) half plane. Furthermore, in that case* W *admits a canonical factorization* $W(\lambda) = W_-(\lambda)W_+(\lambda)$, $\lambda \in \mathbb{R}$, *with*

$$W_-(\lambda) = I_m + C(\lambda I_n - A)^{-1}(I - \Pi)B,$$

$$W_+(\lambda) = I_m + C\Pi(\lambda I_n - A)^{-1}B,$$

$$W_-(\lambda)^{-1} = I_m - C(I - \Pi)(\lambda I_n - A^\times)^{-1}B,$$

$$W_+(\lambda)^{-1} = I_m - C(\lambda I_n - A^\times)^{-1}\Pi B,$$

where Π *is the projection of* $\mathbb{C}^n$ *along* M *onto* $M^\times$.

In [1] this theorem was obtained as a corollary of a more general theorem about minimal factorization. Consider a factorization $W(\lambda) = W_1(\lambda)W_2(\lambda)$, where W_1 and W_2 are rational $m \times m$ matrix functions which are analytic at infinity and have the value I_m at infinity (i.e., both W_1 and W_2 can be represented in the form (3)). Then $\delta(W) \le \delta(W_1) + \delta(W_2)$, and the factorization $W(\lambda) = W_1(\lambda)W_2(\lambda)$ is called *minimal* if $\delta(W) = \delta(W_1) + \delta(W_2)$.

THEOREM 2. *Let* $W(\lambda) = I_m + C(\lambda I_n - A)^{-1}B$ *be a minimal realization, and let* $W(\lambda) = W_1(\lambda)W_2(\lambda)$ *be a minimal factorization (with* W_1 *and* W_2 *analytic at infinity and* $W_1(\infty) = W_2(\infty) = I_m$). *Then there exists a unique pair* $M, M^{\times}$ *of subspaces of* $\mathbb{C}^n$ *such that*

(i) $AM \subset M, \quad A^{\times}M^{\times} \subset M^{\times}$,

(ii) $\mathbb{C}^n = M \oplus M^{\times}$,

and

$$W_1(\lambda) = I_m + C(\lambda I_n - A)^{-1}(I - \Pi)B, \tag{4}$$

$$W_2(\lambda) = I_m + C\Pi(\lambda I_n - A)^{-1}B, \tag{5}$$

where Π *is the projection of* $\mathbb{C}^n$ *along* M *onto* $M^{\times}$. *Conversely, if* M *and* $M^{\times}$ *are subspaces of* $\mathbb{C}^n$ *such that* (i) *and* (ii) *hold, then* $W(\lambda) = W_1(\lambda)W_2(\lambda)$ *with* W_1 *and* W_2 *given by* (4) *and* (5) *and this factorization is minimal.*

Part I of the present book concerns a number of different further developments connected with Theorems 1 and 2. In what follows we shall briefly characterize each of the papers of this part. (We conclude the introduction with a few historical remarks.)

The first paper "*Left versus right canonical Wiener-Hopf factorization*", by J.A. Ball and A.C.M. Ran, concerns connections between left (i.e., the order of the factors in (2) is interchanged) and right canonical factorization. Starting from a left canonical factorization with the factors given in realized form, the authors determine the conditions for the existence of a right canonical factorization and they construct explicitly the factors in terms of the data of the original left factorization. Both symmetric and non-symmetric factorizations are discussed. Several application are made.

The second paper "*Wiener-Hopf equations with symbols analytic in a strip*", by H. Bart, I. Gohberg and M.A. Kaashoek, extends Theorem 1 to the case when W is the Fourier transform of an $m \times m$ matrix function k with entries in $e^{\omega|t|}L_1(-\infty,\infty)$, where ω is some negative constant. First for such functions a realization is constructed. It turns out that in this case one has to replace the space $\mathbb{C}^n$ in the realization by an infinite dimensional Banach space and one has to allow that the operators A and C are unbounded. To prove the analogue of Theorem 1 for this class of matrix functions requires to overcome a number of difficulties connected with semigroup theory. Applications concern the Riemann boundary value problem and inversion and Fredholm properties of Wiener-Hopf integral operators.

The third paper "*On Toeplitz and Wiener-Hopf operators with contour-*

wise rational matrix and operator symbols", by I. Gohberg, M.A. Kaashoek, L. Lerer and L. Rodman, generalizes Theorem 1 for the case when the domain of definition of the function W is not the real line but a curve consisting of several disjoint closed contours. On each of the contours the function W is represented in the form (3), but the operators A, B and C in the realizations depend on the contour. The paper contains applications to different classes of integral equations of convolution type.

The fourth paper "*Canonical pseudo-spectral factorization and Wiener-Hopf integral equations*", by L. Roozemond, deals with Theorem 1 for the case when det W(·) has zeros on the real line. This case concerns realizations for which the operator $A^{\times} = A - BC$ has real eigenvalues. An appropriate generalization of the factorization theorem is given. Applications are made to Wiener-Hopf integral equations of so-called non-normal type.

The last paper of the present part "*Minimal factorization of integral operators and cascade decompositions of systems*", by I. Gohberg and M.A. Kaashoek, extends the concepts of canonical and minimal factorization into the direction of integral operators. Note that Theorems 1 and 2 can be expressed as factorization theorems for integral operators of the type $I + V$, where

$$(V\varphi)(t) = \varphi(t) + \int_0^t Ce^{(t-s)A}B\varphi(s)ds, \quad 0 \le t \le \tau.$$

This observation has led the authors to far reaching generalizations of Theorems 1 and 2 for different classes of integral operators. For example, in this way lower/upper factorization of integral operators appears as a special kind of minimal factorization. Also the connections with Mathematical Systems Theory, which formed one of the starting points for Theorems 1 and 2, are developed further here and concern now a cascade decomposition theory for time varying linear systems with well-posed boundary conditions.

Since there seems to be some confusion about references concerning Theorem 2 (see, e.g., [10], page 345 and [13], pages 13-16) we conclude this introduction with some historical remarks. Theorem 2 is due to H. Bart, I. Gohberg, M.A. Kaashoek and P. Van Dooren; it was produced at a mini-conference held at Amsterdam and Delft in February 1978 and published in [3]. A first predecessor of Theorem 2 is the famous Brodskii-Livsic division theorem for characteristic operator functions (see [4] and the references given there). In the theorem of Brodskii-Livsic the operator $A - BC$ is just the adjoint A^* of the main operator A and the factors are also required to have this property. Consequentially, in the Brodskii-Livsic theorem the space M is the important space and $M^\times$ appears only implicitly as the orthogonal complement of M. Another predecessor of Theorem 2 connected with the theory of characteristic operator functions can be found in the work of V.M. Brodskii [6], where the operator $A - BC$ appears as $(A^*)^{-1}$ (see also the review [5] by M.S. Brodskii).

A next predecessor of Theorem 2 is a factorization theorem due to L.A. Sahnovic. The main new feature in the Sahnovic theorem is the appearance of a second main operator and a second invariant subspace. Let us describe in more detail the result. Let $W(\cdot)$ and $W(\cdot)^{-1}$ be given in realized form:

$$(6) \qquad W(\lambda) = I_m + C_1(\lambda I_n - A_1)^{-1}B_1, \qquad W(\lambda)^{-1} = I_m - C_2(\lambda I_n - A_2)^{-1}B_2.$$

These two realizations are assumed to be minimal, and hence there is an invertible operator $S : \mathbb{C}^n \to \mathbb{C}^n$ such that

$$A_1S - SA_2 = B_1C_2, \qquad SB_2 = B_1, \qquad C_1S = C_2.$$

Now suppose that there exists an orthogonal projection P of $\mathbb{C}^n$ such that

$$A_1(I - P) = (I - P)A_1(I - P), \qquad A_2P = PA_2P$$

and $S_{11} = PSP$ is invertible on Im P. Then the Sahnovic theorem states that $T_{22} = (I - P)S^{-1}(I - P)$ is invertible on Im (I - P) and W factors as $W(\lambda) = W_1(\lambda)W_2(\lambda)$ with

$$W_1(\lambda) = I_m + C_2PS_{11}^{-1}(\lambda - A_{11})^{-1}PB_1,$$

$$W_2(\lambda) = I_m + C_1(I - P)(\lambda - A_{22})^{-1}T_{22}^{-1}(I - P)B_2,$$

where $A_{11} = PAP$ and $A_{22} = (I - P)A(I - P)$.

This factorization theorem was announced in [12] and its proof appeared recently in [13]. (In [12] and [13] also a weaker version appears with the spaces $\mathbb{C}^n$ and $\mathbb{C}^m$ in the realizations (6) replaced by arbitrary Hilbert spaces.) In 1978 N.M. Kostenko [11] proved a theorem (which is reproduced in [13]) which showed that under certain extra conditions on the location of the zeros and the poles of the factors the conditions in the Sahnovic theorem are not only sufficient but also necessary for factorization.

Unfortunately, the proof in [11] (and also in [13]) contains a gap, and the Kostenko theorem is not correct (in fact, this theorem does not take into account the possibility of pole-zero cancellation between the factors).

As a distant relative preceding Theorem 2 we also mention the Gohberg-Lancaster-Rodman factorization theorem for monic matrix polynomials which describes factorization in the class of monic matrix polynomials in terms of certain invariant subspaces of the companion matrix (see [9] and the references given there; also [2]).

REFERENCES

1. Bart, H., Gohberg, I. and Kaashoek, M.A.: *Minimal factorization of matrix and operator functions*, Operator Theory: Advances and Applications, Vol. 1, Birkhäuser Verlag, Basel etc., 1979.

2. Bart, H., Gohberg, I. and Kaashoek, M.A.: Operator polynomials as inverses of characteristic functions, *Integral Equations and Operator Theory* 1 (1978), 1-12.

3. Bart, H., Gohberg, I., Kaashoek, M.A. and Van Dooren, P.: Factorizations of transfer functions, *SIAM J. Control Opt.* 18 (6) (1980), 675-696.

4. Brodskii, M.S.: *Triangular and Jordan representations of linear operators*, Transl. Math. Monographs, Vol.32, Amer. Math. Soc. Providence, R.I., 1970.

5. Brodskii, M.S.: Unitary operator colligations and their characteristic functions, *Uspehi Mat. Nauk* 33 (1978), 141-168 (Russian) = *Russian Math. Surveys* 33 (1978), 159-191.

6. Brodskii, V.M.: Some theorems on knots and their characteristic functions, *Funktsional. Anal. i Prilozhen.* 4 (1970), 95-96 (Russian) = *Functional Analysis and Applications* 4 (1970), 250-251.

7. Clancey, K. and Gohberg, I.: *Factorization of matrix functions and singular integral operators*, Operator Theory: Advances and Applications, Vol. 3, Birkhäuser Verlag, Basel etc, 1981.

8. Gohberg, I. and Krein, M.G.: Systems of integral equations on a half line with kernels depending on the difference of arguments, *Uspehi Mat. Nauk* 13 (1958), no. 2 (80), 3-72 (Russian) = *Amer. Math. Soc. Transl.* (2) 14 (1960), 217-287.

9. Gohberg, I., Lancaster, P. and Rodman, L.: *Matrix polynomials*, Academic Press. Inc., London, 1982.

10. Helton, J.W. and Ball, J.A.: The cascade decompositions of a given system vs the linear fractional decompositions of its transfer

function, *Integral Equations and Operator Theory* 5 (1982), 341-385.

11. Kostenko, K.: A necessary and sufficient condition for factorization of a rational operator-function, *Funktsional. Anal. i Prilozhen.* 12 (1978), 87-88 (Russian) = *Functional Analysis and Applications* 12 (1978), 315-317.

12. Sahnovic, L.A.: On the factorization of an operator-valued transfer function, *Dokl. Akad. Nauk SSR* 226 (1976), 781-784 (Russian) = *Soviet Math. Dokl.* 17 (1976), 203-207.

13. Sahnovic, L.A.: Problems of factorization and operator identities, *Uspehi Mat. Nauk* 41 (1986), 3-55.

Operator Theory:
Advances and Applications, Vol. 21

LEFT VERSUS RIGHT CANONICAL WIENER-HOPF FACTORIZATION

Joseph A. Ball[1)] and André C. M. Ran[2)]

In this paper the existence of a right canonical Wiener-Hopf factorization for a rational matrix function is characterized in terms of a left canonical Wiener-Hopf factorization. Formulas for the factors in a right factorization are given in terms of the formulas for the factors in a given left factorization. Both symmetric and nonsymmetric factorizations are discussed.

1. INTRODUCTION

It is well known that Wiener-Hopf factorization of matrix and operator valued functions has wide applications in analysis and electrical engineering. Applications include convolution integral equations (see eg. [GF]), singular integral equations (see eg. [CG]), Toeplitz operators, the study of Riccati equations (see eg. [W] and [H]) and, more recently, the model reduction for linear systems. The latter application, presented in detail by the authors in [BR 1,2] both for continuous time and discrete time systems, lies at the basis of the present paper. We mention that Glover [Gl] first solved the model reduction problem for continuous time systems without using Wiener-Hopf factorization.

The model reduction problem for discrete time systems as presented in [BR 2] leads to the following question concerning Wiener-Hopf factorization. We are given a pxq rational matrix function $K(\lambda)=C(\lambda I-A)^{-1}B$ with all its poles in the open unit disk, and a number $\sigma>0$. Construct the function

1) The first author was partially supported by a grant from the National Science Foundation.

2) The second author was supported by a grant from the Niels Stensen Stichting at Amsterdam.

$$(1.1) \qquad W(\lambda) = \begin{bmatrix} I_p & 0 \\ K(\bar{\lambda}^{-1})^* & \sigma I_q \end{bmatrix} \begin{bmatrix} I_p & 0 \\ 0 & -I_q \end{bmatrix} \begin{bmatrix} I_p & K(\lambda) \\ 0 & \sigma I_q \end{bmatrix}.$$

One needs a factorization of $W(\lambda)$ of the form

$$(1.2) \qquad W(\lambda) = X_+(\bar{\lambda}^{-1})^* \begin{bmatrix} I_p & 0 \\ 0 & -I_q \end{bmatrix} X_+(\lambda)$$

where X_+ and its inverse is analytic on the disk. Note that (1.1) and (1.2) constitute symmetric left and right Wiener–Hopf factorization of $W(\lambda)$, respectively. This problem was solved in [BR 2] using the geometric factorization approach given in [BGK 1].

In this paper we shall discuss the following more general problem. Given a left canonical Wiener–Hopf factorization

$$W(\lambda) = Y_+(\lambda)Y_-(\lambda)$$

with respect to a contour Γ, where Y_+ and Y_- are given in realization form, give necessary and sufficient conditions for the existence of a right canonical Wiener–Hopf factorization

$$W(\lambda) = X_-(\lambda)X_+(\lambda),$$

and provide formulas for X_- and X_+ in realization form. This is discussed in Section 2.

In Section 3 we give applications of the main result of Section 2 to the invertibility of singular integral operators. As is well known the invertibility of the singular integral operator on $L^p(\Gamma)$, where Γ is a contour in the complex plane $\mathbb{C}$, with symbol W is equivalent to the existence of a right canonical Wiener–Hopf factorization for W on Γ. Theorem 2.1 then gives necessary and sufficient conditions for the invertibility of the singular integral operator with symbol W under the assumption that the singular integral operator with symbol W^{-1} (or with symbol W^T) is invertible and the right canonical factorization of W^{-1} (respectively W^T) is known. We thank Kevin Clancey for pointing out to us this application of our main result.

In Sections 4 and 5 we indicate how symmetrized versions of the factorization problem can be handled also as a direct application of Theorem 2.1. Section 4 deals with the situation where the symmetry is with respect to the unit circle; this is the case which comes up for discrete time systems. In Section 5 the symmetry is with respect to the imaginary axis; this is germane to model reduction for continuous time systems. In this way we may view some factorization formulas from [BR 1,2] needed for the model reduction problem as essentially specializations of the more general formulas in Theorem 2.1.

2. LEFT AND RIGHT CANONICAL WIENER-HOPF FACTORIZATION

In this section we shall analyze the existence of a right canonical Wiener-Hopf factorization for a given rational matrix function in terms of a given left canonical Wiener-Hopf factorization. The factorizations are with respect to some fixed but general contour Γ in the complex plane. The analysis is built on the geometric approach to factorization given in [BGK 1]. We first establish some notation and terminology from [BGK 1].

By a Cauchy contour Γ we mean the positively oriented boundary of a bounded Cauchy domain in the complex plane $\mathbb{C}$; such a contour consists of finitely many nonintersecting closed rectifiable Jordan curves. We denote by F_+ the interior domains of Γ and by F_- the complement of its closure $\overline{F}_+$ in the extended complex plane $\mathbb{C}\cup\{\infty\}$.

Now suppose that W is a rational mxm matrix function invertible at ∞ and with no poles or zeros on the contour Γ. By a <u>right canonical (spectral) factorization</u> of W with respect to Γ we mean a factorization

$$W(\lambda) = W_-(\lambda)W_+(\lambda) \qquad (\lambda\in\Gamma) \tag{2.1}$$

where W_- and W_+ are also rational mxm matrix functions such that W_- has no poles or zeros on $\bar{F}_-$ (including ∞) and W_+ has no poles or zeros on $\bar{F}_+$. If the factors W_- and W_+ are interchanged in (2.1), we speak of a <u>left canonical (spectral) factorization</u>.

We assume throughout that all matrix functions are analytic and invertible at ∞; without loss of generality we may then assume that the value at ∞ is the identity matrix I_m. By a <u>realization</u> for the matrix function W we mean a representation for W of the form

$$W(\lambda) = I_m + C(\lambda I_n - A)^{-1}B.$$

Here A is an nxn matrix (for some n) while C and B are mxn and nxm matrices respectively.

Left and right canonical factorizations are discussed at length in the book [BGK 1]. There it is shown how to compute realizations for the factors W_-, W_+ for a right canonical factorization $W=W_-W_+$ if one knows a realization $W(\lambda)=I_m+C(\lambda I_n-A)^{-1}B$ for W. We shall suppose that we know the factors Y_+ and Y_- of a left canonical factorizations $W(\lambda)=Y_+(\lambda)Y_-(\lambda)$, say

$$Y_+(\lambda) = I + C_+(\lambda I - A_+)^{-1}B_+$$

and

$$Y_-(\lambda) = I + C_-(\lambda I - A_-)^{-1}B_-.$$

We then give a necessary and sufficient condition for a right canonical factorization to exist, and in that case we compute the factors W_- and W_+ of a right canonical factorization $W(\lambda)=W_-(\lambda)W_+(\lambda)$ in terms of the realizations of Y_+ and Y_-. The analysis is a straightforward application of the geometric factorization principle in Chapter I of [BGK 1] (see also [BGKvD]). The result is as follows.

THEOREM 2.1. *Suppose that the rational* mxm *matrix function* $W(\lambda)$ *has left canonical factorization* $W(\lambda)=Y_+(\lambda)Y_-(\lambda)$ *where*

$$Y_+(\lambda) = I_m + C_+(\lambda I_{n_+} - A_+)^{-1}B_+ \tag{2.2}$$

and

$$Y_-(\lambda) = I_m + C_-(\lambda I_{n_-} - A_-)^{-1}B_-. \tag{2.3}$$

We may assume that A_- *and* $A_-^x := A_- - B_-C_-$ *are* $n_- x n_-$ *matrices with spectra in* F_-, *and that* A_+ *and* $A_+^x := A_+ - B_+C_+$ *are* $n_+ x n_+$ *matrices with spectra in* F_+. *Let* P *and* Q *denote the unique solutions of the Lyapunov equations*

$$A_-^x P - P A_+^x = B_- C_+ \tag{2.4}$$

and

$$A_+ Q - Q A_- = -B_+ C_-. \tag{2.5}$$

Then W *has a right canonical factorization if and only if the* $n_+ x n_+$ *matrix* $I_{n_+} - QP$ *is invertible, or equivalently, if and only if the* $n_- x n_-$ *matrix* $I_{n_-} - PQ$ *is invertible. When this is the case, the factors* $W_-(\lambda)$ *and* $W_+(\lambda)$ *for a right canonical factorization* $W(\lambda)=W_-(\lambda)W_+(\lambda)$ *are given by the formulas*

$$(2.6)\qquad W_-(\lambda) = I+(C_+Q+C_-)(\lambda I_{n_-}-A_-)^{-1}(I-PQ)^{-1}(-PB_++B_-)$$

and

$$(2.7)\qquad W_+(\lambda) = I+(C_++C_-P)(I-QP)^{-1}(\lambda I_{n_+}-A_+)^{-1}(B_+-QB_-)$$

with inverses given by

$$(2.8)\qquad W_-(\lambda)^{-1}=I-(C_+Q+C_-)(I-PQ)^{-1}(\lambda I_{n_-}-A_-+B_-C_-)^{-1}(-PB_++B_-)$$

and

$$(2.9)\qquad W_+(\lambda)^{-1}=I-(C_++C_-P)(\lambda I_{n_+}-A_++B_+C_+)^{-1}(I-QP)^{-1}(B_+-QB_-).$$

PROOF. From the realizations (2.2) and (2.3) for the functions $Y_+(\lambda)$ and $Y_-(\lambda)$ we compute a realization for their product $W(\lambda)=Y_+(\lambda)Y_-(\lambda)$ as

$$W(\lambda) = I + C(\lambda I-A)^{-1}B$$

where

$$(2.10)\qquad A = \begin{bmatrix} A_+ & B_+C_- \\ 0 & A_- \end{bmatrix},\ B = \begin{bmatrix} B_+ \\ B_- \end{bmatrix},\ C = [C_+\ C_-]$$

(see p.6 of [BGK 1]). The matrix $A^\times:=A-BC$ equals

$$A^\times = \begin{bmatrix} A_+^\times & 0 \\ -B_-C_+ & A_-^\times \end{bmatrix}$$

where $A_+^\times:=A_+-B_+C_+$ and $A_-^\times:=A_--B_-C_-$. Now by assumption the spectrum $\sigma(A_+)$ of A_+ is contained in F_-, while that of A_- is contained in F_+. From the triangular form of A we see that $\sigma(A)=\sigma(A_+)\cup\sigma(A_-)$ and that the spectral subspace for A associated with F_- must be $\text{Im}\begin{bmatrix} I_{n_-} \\ 0 \end{bmatrix}$. Now the spectral subspace $\mathcal{M}$ for A corresponding to F_+ is determined by the fact that it must be complementary to the spectral subspace $\text{Im}\begin{bmatrix} I_{n_-} \\ 0 \end{bmatrix}$ for F_-, and that it must be invariant for A. The first condition forces $\mathcal{M}$ to have the form $\mathcal{M}=\text{Im}\begin{bmatrix} Q \\ I_{n_-} \end{bmatrix}$ for some $n_+ \times n_-$ matrix Q (the "angle operator" for $\mathcal{M}$). The second condition ($A\mathcal{M}\subset\mathcal{M}$) requires that

$$\begin{bmatrix} A_+ & B_+C_- \\ 0 & A_- \end{bmatrix}\begin{bmatrix} Q \\ I_{n_-} \end{bmatrix} = \begin{bmatrix} Q \\ I_{n_-} \end{bmatrix}X$$

for some $n_- \times n_-$ matrix X. From the second row in this identity we see that $X=A_-$ and then from the first row we see that

$$A_+Q + B_+C_- = QA_-.$$

Thus the angle operator Q must be a solution of the Lyapunov equation (2.5). By our assumption that the spectra of A_+ and A_- are disjoint, it follows directly from the known theory of Lyapunov equations that there is a unique solution Q. We have thus identified the spectral subspace $\mathcal{M}$ of A for F_+ as $\mathcal{M}=\text{Im}\begin{bmatrix} Q \\ I_{n_-} \end{bmatrix}$ where Q is the unique solution of (2.5).

Since by assumption $A_+^\times$ has its spectrum in F_- while $A_-^\times$ has its spectrum in F_+, the same analysis applies to $A^\times$. We see that the spectral subspace of $A^\times$ for F_+ is the coordinate space $\text{Im}\begin{bmatrix} 0 \\ I_{n_+} \end{bmatrix}$ while the spectral subspace $\mathcal{M}^\times$ of $A^\times$ for F_- is the space

$$\mathcal{M}^{\times} = \operatorname{Im}\begin{bmatrix} I_{n_+} \\ P \end{bmatrix}$$

where P is the unique solution of the Lyapunov equation (2.4). Again, since the spectra of $A_+^{\times}$ and $A_-^{\times}$ are disjoint, we also see directly that the solution P exists and is unique.

Now we apply Theorem 1.5 from [BGK 1]. One concludes that the function W has a right canonical factorization $W(\lambda)=W_-(\lambda)W_+(\lambda)$ if and only if $\mathbb{C}^{n_++n_-}=\mathcal{M}\dot{+}\mathcal{M}^{\times}$, that is, if and only if

$$\mathbb{C}^{n_++n_-} = \operatorname{Im}\begin{bmatrix} Q \\ I_{n_-} \end{bmatrix} \dot{+} \operatorname{Im}\begin{bmatrix} I_{n_-} \\ P \end{bmatrix}.$$

(Here $\dot{+}$ indicates a direct sum decomposition.)

One easily checks that this direct sum decomposition holds if and only if the square matrix $\begin{bmatrix} I_{n_+} & Q \\ P & I_{n_-} \end{bmatrix}$ is invertible. By standard row and column operations this matrix can be diagonalized in either of two ways:

$$\begin{bmatrix} I_{n_+} & Q \\ P & I_{n_-} \end{bmatrix} = \begin{bmatrix} I & Q \\ 0 & I \end{bmatrix}\begin{bmatrix} I-QP & 0 \\ 0 & I \end{bmatrix}\begin{bmatrix} I & 0 \\ P & I \end{bmatrix}$$

$$= \begin{bmatrix} I & 0 \\ P & I \end{bmatrix}\begin{bmatrix} I & 0 \\ 0 & I-PQ \end{bmatrix}\begin{bmatrix} I & Q \\ 0 & I \end{bmatrix}.$$

Thus we see that the invertibility of $\begin{bmatrix} I & Q \\ P & I \end{bmatrix}$ is equivalent to the invertibility of I−QP and also to the invertibility of I−PQ.

Now suppose this condition holds. Let π be the projection of $\mathbb{C}^{n_++n_-}$ onto $\mathcal{M}^{\times}=\operatorname{Im}\begin{bmatrix} I_{n_+} \\ P \end{bmatrix}$ along $\mathcal{M}=\operatorname{Im}\begin{bmatrix} Q \\ I_{n_-} \end{bmatrix}$. It is straightforward to compute that

$$\pi = \begin{bmatrix} I_{n_+} \\ P \end{bmatrix}(I_{n_+} - QP)^{-1}[I_{n_+}, -Q]$$

and that

$$I - \pi = \begin{bmatrix} Q \\ I_{n_-} \end{bmatrix}(I_{n_-} - PQ)^{-1}[-P, I_{n_-}].$$

From Theorem 1.5 [BGK 1] one obtains the formulas for the right canonical spectral factors of W:

$$W_-(\lambda) = I + C(I-\pi)(\lambda I - A(I-\pi))^{-1}(I-\pi)B$$

and

$$W_+(\lambda) = I + C\pi(\lambda I - \pi A\pi)^{-1}\pi B$$

Let $S: \mathbb{C}^{n_-} \to \operatorname{Im}(I-\pi)$ be the operator $S = \begin{bmatrix} Q \\ I_{n_-} \end{bmatrix}$ with inverse $S^{-1} = [0 \;\; I_{n_-}]\big|_{\operatorname{Im}(I-\pi)}$. Similarly, let $T: \mathbb{C}^{n_+} \to \operatorname{Im}\pi$ be the operator $T = \begin{bmatrix} I_{n_+} \\ P \end{bmatrix}$ with inverse $T^{-1} = [I_{n_+} \;\; 0]\big|_{\operatorname{Im}\pi}$. The above formulas for W_- and W_+ may be rewritten as

$$W_-(\lambda) = I + C(I-\pi)S(\lambda I_{n_-} - S^{-1}A(I-\pi)S)^{-1}S^{-1}(I-\pi)B \tag{2.11}$$

and

$$W_+(\lambda) = I + C\pi T(\lambda I_{n_+} - T^{-1}\pi A\pi T)^{-1}T^{-1}\pi B. \tag{2.12}$$

Now one computes from formulas (2.10) and the Lyapunov equations (2.4) and (2.5) that

$$S^{-1}AS = [0 \;\; I_{n_-}]\begin{bmatrix} A_+ & B_+C_- \\ 0 & A_- \end{bmatrix}\begin{bmatrix} Q \\ I_{n_-} \end{bmatrix} = [0 \;\; I_{n_-}]\begin{bmatrix} Q \\ I_{n_-} \end{bmatrix}A_- = A_-$$

as well as

$$C(I-\pi)S = CS = C_+Q + C_-$$

and

$$S^{-1}(I-\pi)B = (I_{n_-}-PQ)^{-1}(-PB_+ + B_-).$$

Similarly we compute

$$T^{-1}\pi AT = (I_{n_+}-QP)^{-1}[I_{n_+},-Q]\begin{bmatrix} A_+ & B_+C_- \\ 0 & A_- \end{bmatrix}\begin{bmatrix} I_{n_+} \\ P \end{bmatrix}$$

$$= (I_{n_+}-QP)^{-1}[A_+, B_+C_- - QA_-]\begin{bmatrix} I_{n_+} \\ P \end{bmatrix} = (I_{n_+}-QP)^{-1}A_+(I_{n_+}-QP)$$

as well as

$$C\pi T = CT = C_+ + C_-P$$

and

$$T^{-1}\pi B = (I-QP)^{-1}(B_+ - QB_-).$$

Substituting these expressions into formulas (2.11) and (2.12) yields the expressions for $W_-(\lambda)$ and $W_+(\lambda)$ in the statement of the Theorem. The formulas for $W_-(\lambda)^{-1}$ and $W_+(\lambda)^{-1}$ follow immediately from these and the general formula for the inverse of a transfer function

$$(2.13) \qquad (I+C(\lambda I-A)^{-1}B)^{-1} = I - C(\lambda I-A+BC)^{-1}B$$

(see [BGK 1] p.7) once the associate operators

$$\widetilde{A}_-^{x} := A_- - (I-PQ)^{-1}(-PB_+ + B_-)(C_+Q + C_-)$$

and

$$\widetilde{A}_+^{x} := A_+ - (B_+ - QB_-)(C_+ + C_-P)(I-QP)^{-1}$$

are computed. Again use the Lyapunov equations to deduce that

$$(-PB_+ + B_-)(C_+Q + C_-)$$

$$= -PB_+C_+Q + B_-C_+Q - PB_+C_- + B_-C_-$$

$$= -PB_+C_+Q + (A_-^{x}P - PA_+^{x})Q - P(QA_- - A_+Q) + B_-C_-$$

$$- PB_+C_+Q + (A_- - B_-C_-)PQ - P(A_+ - B_+C_+)Q - PQA_-$$

$$+ PA_+Q + B_-C_- = A_-PQ - PQA_- + B_-C_-(I-PQ)$$

and thus

$$\widetilde{A}_-^{\times} = (I-PQ)^{-1}[(I-PQ)A_- - A_-PQ + PQA_- - B_-C_-(I-PQ)]$$

$$= (I-PQ)^{-1}A_-^{\times}(I-PQ).$$

A completely analogous computation gives

$$\widetilde{A}_+^{\times} = (I-QP)(A_+ - B_+C_+)(I-QP)^{-1}$$

$$= (I-QP)A_+^{\times}(I-QP)^{-1}.$$

Now apply formula (2.13) to the representations for $W_-(\lambda)$ and $W_+(\lambda)$ in the Theorem together with the above expressions for $\widetilde{A}_+^{\times}$ and $\widetilde{A}_-^{\times}$ to derive the desired expressions for $W_-(\lambda)^{-1}$ and $W_+(\lambda)^{-1}$.

REMARK. Theorem 2.1 actually holds in greater generality than that stated here. Specifically the matrix functions Y_- and Y_+ may be irrational as long as they have (possibly infinite dimensional) realizations as in the Theorem.

3. APPLICATION TO SINGULAR INTEGRAL OPERATORS

For Γ a contour as above, introduce the operator of singular integration S_Γ: $L_p^n(\Gamma) \to L_p^n(\Gamma)$ on Γ by

$$(S_\Gamma \varphi)(\lambda) = \frac{1}{\pi i}\int_\Gamma \frac{\varphi(\tau)}{\tau-\lambda}\, d\tau$$

where integration over Γ is in the Cauchy principal value sense. Introduce $P_\Gamma = \frac{1}{2}(I+S_\Gamma)$, $Q_\Gamma = \frac{1}{2}(I-S_\Gamma)$; then P_Γ and Q_Γ are projections on $L_p^n(\Gamma)$. We consider the singular integral operator S: $L_p^n(\Gamma) \to L_p^n(\Gamma)$

(3.1) $(S\varphi)(\lambda)=A(\lambda)(P_\Gamma\varphi)(\lambda)+B(\lambda)(Q_\Gamma\varphi)(\lambda),$

where $A(\lambda)$ and $B(\lambda)$ are rational matrix functions with poles and zeros off Γ. The symbol of S is the function $W(\lambda)=B(\lambda)^{-1}A(\lambda)$. It is well known (see eg. [CG], [GK]) that S is invertible if and only if $W(\lambda)$ admits a right canonical factorization

(3.2) $W(\lambda)=W_-(\lambda)W_+(\lambda)$

in which case

(3.3) $(S^{-1}\varphi)(\lambda)$

$$= W_+^{-1}(\lambda)(P_\Gamma W_-^{-1}B^{-1}\varphi)(\lambda)+W_-(\lambda)(Q_\Gamma W_-^{-1}B^{-1}\varphi)(\lambda).$$

Theorem 2.1 can be used to study the invertibility of S in terms of the invertibility of either one of the following operators

$$(S_1\varphi)(\lambda)=B(\lambda)(P_\Gamma\varphi)(\lambda)+A(\lambda)(Q_\Gamma\varphi)(\lambda)$$

$$(S_2\varphi)(\lambda)=\{B(\lambda)^{-1}\}^T(P_\Gamma\varphi)(\lambda)+\{A(\lambda)^{-1}\}^T(Q_\Gamma\varphi)(\lambda).$$

Note that the symbol of S_1 is $W(\lambda)^{-1}$ and the symbol of S_2 is $W(\lambda)^T$. More precisely we have the following theorems, the proofs of which are immediate by combining the above remarks with Theorem 2.1.

THEOREM 3.1 Assume that S_1 is invertible and let the right factorization of the symbol of S_1 be given by

$$W(\lambda)^{-1}=A(\lambda)^{-1}B(\lambda)=Y_-(\lambda)^{-1}Y_+(\lambda)^{-1}$$

where

$$Y_-(\lambda)^{-1}=I_m-C_-(\lambda I-A_-^x)^{-1}B_-,$$

$$Y_+(\lambda)^{-1}=I_m-C_+(\lambda I-A_+^x)^{-1}B_+.$$

Set $A_-=A_-^x+B_-C_-$ and $A_+=A_+^x+B_+C_+$. Let P and Q denote the unique solutions of the Lyapunov equations (2.4) and (2.5), respectively. Then S is invertible if and only if I−QP is invertible, or equivalently, if and only if I−PQ is invertible.

THEOREM 3.2 Assume that S_2 is invertible and let the right factorization of the symbol of S_2 be given by

$$W(\lambda)^T = B(\lambda)^T A^{-1}(\lambda)^T = Y_-(\lambda)^T Y_+(\lambda)^T$$

where

$$Y_-(\lambda)^T = I_m + B_-^T(\lambda I - A_-^T)^{-1} C_-^T$$

$$Y_+(\lambda)^T = I_m + B_+^T(\lambda I - A_+^T)^{-1} C_+^T.$$

Set $A_-^x = A_- - B_- C_-$ and $A_+^x = A_+ - B_+ C_+$. Let P and Q denote the unique solutions of the Lyapunov equations (2.4) and (2.5), respectively. Then S is invertible if and only if I−QP is invertible, or equivalently, if and only if I−PQ is invertible.

In both cases the formulas for the factors $W_-(\lambda)$, $W_+(\lambda)$ in the factorization (3.2) of the symbol of S as well as the formulas for their inverses are given by (2.6)−(2.9). Then (3.3) gives an explicit formula for S^{-1}.

The two theorems above can be reformulated of course completely in terms of S and it symbol $W(\lambda)$. Actually if $W(\lambda)$ admits a left canonical factorization $W(\lambda) = Y_+(\lambda) Y_-(\lambda)$ with factors Y_+, Y_- as in (2.2), (2.3) then invertibility of S is equivalent to invertibility of I−PQ, where P and Q are the unique solutions of (2.4) and (2.5), respectively. In fact, in terms of [BGK 2] I−PQ is an indicator for the operator S, as well as for the Toeplitz operator with symbol W. Indeed, according to

[BGK 2], Theorem III.2.2 an indicator for S is given by the operator $\hat{P}^{x}\big|_{Im\hat{P}}: Im\hat{P}\to Im\hat{P}^{x}$, where $\hat{P}$ (resp. $\hat{P}^{x}$) is the spectral projection of A(resp. A^{x}) corresponding to F_{+} (here A,A^{x} come from a realization of W). From the proof of Theorem 2.1 one sees easily that $Im\hat{P}=Im\begin{bmatrix} Q \\ I \end{bmatrix}$ and $\hat{P}^{x} = \begin{bmatrix} 0 & 0 \\ -P & I \end{bmatrix}$. Hence $\hat{P}^{x}\big|_{Im\hat{P}}$ is actually given by I−PQ.

4. SPECTRAL AND ANTISPECTRAL FACTORIZATION ON THE UNIT CIRCLE

Suppose that $W(\lambda)$ is a rational mxm matrix function analytic and invertible on the unit circle $\{|\lambda|=1\}$ such that $W(\frac{1}{\bar{\lambda}})^{*} = W(\lambda)$. For convenience in the sequel, in general in this section we shall use W^{*} to designate the function $W^{*}(\lambda)=W(\frac{1}{\bar{\lambda}})^{*}$. Note that $W=W^{*}$ for a rational matrix function W if and only if $W(\lambda)$ is self-adjoint for $|\lambda|=1$. Since $W(\lambda)$ by assumption is also invertible on the unit circle, $W(e^{i\tau})$ must have a constant number (say p) of positive eigenvalues and q=m−p of negative eigenvalues for all real τ. By a <u>signed antispectral factorization</u> of W (with respect to the unit circle) we mean a factorization of the form

$$W(\lambda) = Y_{-}^{*}(\lambda)\begin{bmatrix} I_{p} & 0 \\ 0 & -I_{q} \end{bmatrix}Y_{-}(\lambda)$$

where $Y_{-}(\lambda)$ is analytic and invertible on the exterior of the unit disk

$\bar{D}_e=\{|\lambda|\geq 1\}$. By a <u>signed spectral factorization</u> of W (with respect to the unit circle) we mean a factorization of the form

$$W(\lambda) = X_+^*(\lambda)\begin{bmatrix} I_p & 0 \\ 0 & -I_q \end{bmatrix}X_+(\lambda)$$

where $X_+(\lambda)$ is analytic and invertible on the closed unit disk $\bar{D}=\{|\lambda|\leq 1\}$. The problem which we wish to analyze in this section is a symmetrized version of that considered in Section 2: namely, given a signed antispectral factorization $W(\lambda)=Y_-^*(\lambda)\begin{bmatrix} I_p & 0 \\ 0 & -I_q \end{bmatrix}Y_-(\lambda)$, give necessary and sufficient conditions for the existence of a signed spectral factorization, and, for the case where these are satisfied, give an explicit formula for a spectral factor $X_+(\lambda)$.

We first remark that a function $W=W^*$ (invertible on the unit circle) has a signed spectral factorization if and only if it has a canonical right Wiener-Hopf factorization with respect to the unit circle. Indeed, if $W(\lambda)=X_+^*(\lambda)\begin{bmatrix} I_p & 0 \\ 0 & -I_q \end{bmatrix}X_+(\lambda)$ is a signed spectral factorization, then $W(\lambda)=W_-(\lambda)W_+(\lambda)$ where $W_-(\lambda):=X_+^*(\lambda)\begin{bmatrix} I_p & 0 \\ 0 & -I_q \end{bmatrix}$ and $W_+(\lambda):=X_+(z)$ is a right canonical factorization as discussed in §2 (with the contour Γ chosen to be the unit circle $\{|\lambda|=1\}$). Note that here we do not insist on normalizing the value at infinity to be I_m. Conversely, suppose $W(\lambda)=W_-(\lambda)W_+(\lambda)$ is a right canonical factorization with respect to the unit circle. Then $W(\lambda)=W^*(\lambda)=W_+^*(\lambda)W_-^*(\lambda)$ is another. But it is known that such (nonnormalized) factorizations are unique up to a constant invertible factor; thus $W_-(\lambda)=W_+^*(\lambda)c$ for some nonsingular mxm matrix c, and $W(\lambda)=W_+^*(\lambda)cW_+(\lambda)$. Plugging in $\lambda=1$ we see that $c=W_+(1)^{*-1}W(1)W_+(1)^{-1}$ is self-adjoint with p positive and q negative eigenvalues. We then may factor c as $c=d^*\begin{bmatrix} I_p & 0 \\ 0 & -I_q \end{bmatrix}d$ and $W(\lambda)=X_+^*(\lambda)\begin{bmatrix} I_p & 0 \\ O & -I_q \end{bmatrix}X_+(\lambda)$ is a signed spectral factorization, where $X_+(\lambda)=dW_+(\lambda)$.

It remains only to use this connection and the work of Section 2 on Wiener-Hopf factorization to get an analogous result for signed spectral factorization.

THEOREM 4.1. *Suppose that the rational* mxm *matrix function* $W(\lambda)=W^*(\lambda)$ *has a signed antispectral factorization*

$$W(\lambda) = Y_-^*(\lambda)\begin{bmatrix} I_p & 0 \\ 0 & -I_q \end{bmatrix}Y_-(\lambda)$$

where

$$Y_-(\lambda) = Y_-(\infty)[I_m + C_-(\lambda I-A_-)^{-1}B_-]$$

We may assume that A_- *and* $A_-^x:=A_--B_-C_-$ *have their spectra in the open unit disk* D. *We also assume that* $Y_-(\infty)$ *and* $Y_-^*(\infty) = Y_-(0)^*$ *are invertible, so* $W(\infty)$ *and* $W(0)=W(\infty)^*$ *are invertible. We denote by* Ψ *the Hermitian matrix*

$$\Psi = Y_-(\infty)^*\begin{bmatrix} I_p & 0 \\ O & -I_q \end{bmatrix}Y_-(\infty).$$

Let P *and* Q *denote the unique solutions of the Lyapunov equations*

$$(4.1)\qquad A_{-}^{\times}P(A_{-}^{\times})^{*} - P = -B_{-}\Psi^{-1}B_{-}^{*}$$

and

$$(4.2)\qquad A_{-}^{*}QA_{-} - Q = C_{-}^{*}\Psi C_{-}.$$

Then W *has a signed spectral factorization if and only the matrix* I–QP *is invertible.*

Suppose now that this is the case, so I–QP *is invertible. Let* $Z=(I-QP)^{-1}$. *Then the Hermitian matrix,*

$$(4.3)\qquad c=\Psi - \Psi C_{-}A_{-}^{-1}Z^{*}B_{-} - B_{-}^{*}ZA_{-}^{*-1}C_{-}^{*}\Psi + B_{-}^{*}ZQB_{-}$$
$$+ \Psi C_{-}A_{-}^{-1}PZA_{-}^{*-1}C_{-}^{*}\Psi$$

is invertible and has p *positive and* q *negative eigenvalues. Thus* c *has a factorization*

$$c = d^{*}\begin{bmatrix} I_p & 0 \\ 0 & -I_q \end{bmatrix} d$$

for an invertible matrix d. *Then*

$$W(\lambda) = X_{+}^{*}(\lambda)\begin{bmatrix} I_p & 0 \\ 0 & -I_q \end{bmatrix} X_{+}(\lambda)$$

is a signed spectral factorization of W(λ) *where*

$$(4.4)\qquad X_{+}(\lambda) = d\{I + (-\Psi^{-1}B_{-}^{*}(A_{-}^{\times})^{*-1} + C_{-}P)$$
$$\cdot Z(\lambda I - A_{-}^{*-1})^{-1}(A_{-}^{*-1}C_{-}^{*}\Psi - QB_{-})\}$$

with

$$(4.5)\qquad X_{+}(\lambda)^{-1} = \{I - (-\Psi^{-1}B_{-}^{*}(A_{-}^{\times})^{*-1} + C_{-}P)$$
$$\cdot(\lambda I - (A_{-}^{\times})^{*-1})^{-1}Z(A_{-}^{*-1}C_{-}^{*}\Psi - QB_{-})\}d^{-1},$$

PROOF. In Theorem 2.1 it was assumed that $W(\infty) = I_m$ and that $W(\lambda) = Y_+(\lambda)Y_-(\lambda)$ where $Y_+(\infty)=Y_-(\infty)=I_m$. We thus consider here the matrix function $W(\infty)^{-1}W(\lambda)$ and its left canonical factorization (with respect to the unit circle)

$$W(\infty)^{-1}W(\lambda) = Y_+(\lambda)(Y_-(\infty)^{-1}Y_-(\lambda))$$

where

$$Y_+(\lambda) = W(\infty)^{-1}Y_-^*(\lambda)\begin{bmatrix} I_p & 0 \\ 0 & -I_q \end{bmatrix}Y_-(\infty).$$

By assumption

$$Y_-(\infty)^{-1}Y_-(\lambda) = I_m + C_-(\lambda I-A_-)^{-1}B_-$$

where A_-, B_-, C_- are given. Note that then

$$\begin{aligned} Y_-^*(\lambda) &= Y_-(\tfrac{1}{\bar{\lambda}})^* \\ &= [I_m + B_-^*(\lambda^{-1}I-A_-^*)^{-1}C_-^*]Y_-(\infty)^* \\ &= [I_m - B_-^*A_-^{*-1}(\lambda I-A_-^{*-1})^{-1}\lambda C_-^*]Y_-(\infty)^* \\ &= [I_m - B_-^*A_-^{*-1}C_-^* - B_-^*A_-^{*-1} \\ &\quad \cdot(\lambda I-A_-^{*-1})^{-1}A_-^{*-1}C_-^*]Y_-(\infty)^* \end{aligned}$$

and thus

$$\begin{aligned} W(\infty) &= Y_-^*(\infty)\begin{bmatrix} I_p & 0 \\ 0 & -I_q \end{bmatrix}Y_-(\infty) \\ &= (I_m - B_-^*A_-^{*-1}C_-^*)\Psi \end{aligned}$$

Thus $Y_+(\lambda)$ has the form

$$\begin{aligned} Y_+(\lambda) &= \Psi^{-1}(I_m - B_-^*A_-^{*-1}C_-^*)^{-1} \\ &\quad \cdot\{(I_m - B_-^*A_-^{*-1}C_-^*) - B_-^*A_-^{*-1}(\lambda I-A_-^{*-1})^{-1}A_-^{*-1}C_-^*\}\Psi \end{aligned}$$

$$= I_m - \Psi^{-1}(I_m - B_-^* A_-^{*-1} C_-^*)^{-1} B_-^* A_-^{*-1}$$

$$\cdot (\lambda I - A_-^{*-1})^{-1} A_-^{*-1} C_-^* \Psi.$$

Certainly, $W(\infty)^{-1}W(\lambda)$ has a right canonical factorization if and only if $W(\lambda)$ does, and by the remarks above, this in turn is equivalent to the existence of a signed spectral factorization for W. To get conditions for a right canonical factorization for $W(\infty)^{-1}W(\lambda)$, we apply Theorem 2.1 with A_-, B_-, C_- as given here, but with A_+, B_+, C_+ given by

$$A_+ = A_-^{*-1}$$

$$B_+ = A_-^{*-1} C_-^* \Psi$$

and

$$C_+ = -\Psi^{-1}(I_m - B_-^* A_-^{*-1} C_-^*)^{-1} B_-^* A_-^{*-1}$$

$$= -\Psi^{-1} B_-^* (I_m - A_-^{*-1} C_-^* B_-^*)^{-1} A_-^{*-1}$$

$$= -\Psi^{-1} B_-^* (A_-^* - C_-^* B_-^*)^{-1}$$

so

$$C_+ = -\Psi^{-1} B_-^* (A_-^x)^{*-1}.$$

We next compute

$$A_+^x := A_+ - B_+ C_+$$

$$= A_-^{*-1} + A_-^{*-1} C_-^* \Psi \Psi^{-1} B_-^* (A_-^x)^{*-1}$$

$$= A_-^{*-1} [(A_-^x)^* + C_-^* B_-^*] (A_-^x)^{*-1}$$

$$= (A_-^x)^{*-1}$$

Thus the Lyapunov equation (2.4) for this setting becomes

$$A_-^x P - P(A_-^x)^{*-1} = -B_-\Psi C_-^*(A_-^x)^{*-1}$$

which we prefer to write in the equivalent form (4.1). Similarly the Lyapunov equation (2.5) becomes upon substituting the above expressions for A_+ and B_+

$$A_-^{*-1}Q - QA_- = -A_-^{*-1}C_-^*\Psi C_-$$

which is equivalent to (4.2). Thus the invertibility of I−QP, where Q and P are the unique solutions of the Lyapunov equations (4.1) and (4.2) is a necessary and sufficient condition for the existence of a signed spectral factorization of $W(\lambda)$. Note that P^* is a solution of (4.1) whenever P is. By our assumptions on the spectrum of A_-^x, the solution of (4.1) is unique, and hence $P=P^*$. Similarly $Q=Q^*$ for the solution Q of (4.2).

Now suppose I−QP is invertible and set $Z=(I-QP)^{-1}$. In computations to follow, we shall use that

$$Z^* = (I-PQ)^{-1},\ PZ = Z^*P,\ ZQ = QZ^*.$$

By the formulas (2.6)–(2.9) in Theorem 2.1, we see that $W(\infty)^{-1}W(\lambda)$ has the right canonical factorization $W(\infty)^{-1}W(\lambda)=W_-(\lambda)W_+(\lambda)$ where

$$W_+(\lambda) = I + (-\Psi^{-1}B_-^*(A_-^x)^{*-1} + C_-P)Z \cdot(\lambda I-A_-^{*-1})^{-1}(A_-^{*-1}C_-^*\Psi-QB_-) \tag{4.6}$$

and

$$W_+(z)^{-1} = I - (-\Psi^{-1}B_-^*(A_-^x)^{*-1} + C_-P) \cdot(\lambda I - (A_-^x)^{*-1})^{-1}Z(A_-^{*-1}C_-^*\Psi - QB_-). \tag{4.7}$$

In particular $W(\lambda)=W(\infty)W_-(\lambda)\cdot W_+(\lambda)$ is a right canonical factorization of

W, as is also $W(\lambda) = W^*(\lambda) = W_+^*(\lambda)\cdot W_-^*(\lambda)W(\infty)^*$. By the uniqueness of the right canonical factorization, we know that there is a (constant) invertible matrix c such that $W(\infty)W_-(\lambda)=W_+^*(\lambda)c$. Thus

(4.8) $$W(\lambda) = W_+^*(\lambda)cW_+(\lambda).$$

By evaluating both sides of (4.8) at a point λ on the unit circle and using the original signed antispectral factored form for W, we see that c is invertible with p positive and q negative eigenvalues . Thus c can be factored as $c=d^*\begin{bmatrix} I_p & 0 \\ 0 & -I_q \end{bmatrix}d$ for an invertible matrix d. Then (4.8) becomes

$$\begin{aligned} W(\lambda) &= W_+^*(\lambda)d^*\begin{bmatrix} I_p & 0 \\ 0 & -I_q \end{bmatrix}dW_+(\lambda) \\ &= X_+^*(\lambda)\begin{bmatrix} I_p & 0 \\ 0 & -I_q \end{bmatrix}X_+(\lambda), \end{aligned}$$

a signed spectral factorization of $W(\lambda)$, where $X_+(\lambda)=dW_+(\lambda)$. Using formulas (4.6) and (4.7), we get the desired formulas (4.4) and (4.5) for $X_+(\lambda)$ and $X_+(\lambda)^{-1}$ once we verify that the constant c in (4.8) is given by formula (4.3).

To evaluate c, we set $\lambda=\infty$ in (4.8) to get

$$\begin{aligned} c &= W_+^*(\infty)^{-1}W(\infty) \\ &= W_+(0)^{*-1}W(\infty). \end{aligned}$$

From (4.7) we see that

$$\begin{aligned} W_+(0)^{*-1} &= (W_+(0)^{-1})^* \\ &= I + (\Psi C_-A_-^{-1}-B_-^*Q)Z^*A_-^x(-(A_-^x)^{-1}B_-\Psi^{-1}+PC_-^*) \end{aligned}$$

while we have already observed that $W(\infty)=(I-B_-^*A_-^{*-1}C_-^*)\Psi$. To compute the product $c=W_+(0)^{*-1}W(\infty)$, we first simplify the expression $(-(A_-^x)^{-1}B_-\Psi^{-1}+PC_-^*)W(\infty)$ as follows:

$$(-(A_-^x)^{-1}B_-\Psi^{-1}+PC_-^*)(\Psi-B_-^*A_-^{*-1}C_-^*\Psi)$$

$$= -(A_-^x)^{-1}B_- + PC_-^*\Psi + (A_-^x)^{-1}B_-\Psi^{-1}B_-^*A_-^{*-1}C_-^*\Psi$$

$$- PC_-^*B_-^*A_-^{*-1}C_-^*\Psi$$

$$= -(A_-^x)^{-1}B_- + PC_-^*\Psi + (A_-^x)^{-1}[P-A_-^xP(A_-^x)^*]A_-^{*-1}C_-^*\Psi$$

$$- PC_-^*B_-^*A_-^{*-1}C_-^*\Psi$$

(from the Lyapunov equation (4.1))

$$= -(A_-^x)^{-1}B_- + PC_-^*\Psi + (A_-^x)^{-1}PA_-^{*-1}C_-^*\Psi$$

$$- P(A_-^*-C_-^*B_-^*)A_-^{*-1}C_-^*\Psi - PC_-^*B_-^*A_-^{*-1}C_-^*\Psi$$

$$= (A_-^x)^{-1}[-B_- + PA_-^{*-1}C_-^*\Psi].$$

Thus

$$c = W_+(0)^{*-1}W(\infty)$$

$$= (I-B_-^*A_-^{*-1}C_-^*)\Psi + (\Psi C_-A_-^{-1}-B_-^*Q)Z^*(-B_-+PA_-^{*-1}C_-^*\Psi)$$

$$= \Psi-B_-^*A_-^{*-1}C_-^*\Psi - \Psi C_-A_-^{-1}Z^*B_-$$

$$+ \Psi C_-A_-^{-1}Z^*PA_-^{*-1}C_-^*\Psi$$

$$+ B_-^*QZ^*B_- - B_-^*QZ^*PA_-^{*-1}C_-^*\Psi.$$

Now use that $QZ^*P = ZQP = -I+Z$ to get

$$c = \Psi - B_-^* A_-^{*-1} C_-^* \Psi - \Psi C_- A_-^{-1} Z^* B_-$$

$$+ \Psi C_- A_-^{-1} Z^* P A_-^{*-1} C_-^* \Psi$$

$$+ B_-^* Q Z^* B_- - B_-^*(-I+Z) A_-^{*-1} C_-^* \Psi$$

$$= \Psi - \Psi C_- A_-^{-1} Z^* B_-$$

$$+ \Psi C_- A_-^{-1} P Z A_-^{*-1} C_-^* \Psi$$

$$+ B_-^* Z Q B_- - B_-^* Z A_-^{*-1} C_-^* \Psi$$

which agrees with (4.3). This completes the proof of Theorem 4.1.

The model reduction problem for discrete time systems from [BR 2] involves the application of Theorem 4.1 to a function $Y_-(\lambda)$ of a special form.

COROLLARY 4.2. Suppose $K(z)=C(\lambda I-A)^{-1}B$ is a pxq rational matrix function of McMillan degree n such that all poles of K are in the open unit disk D. Thus we may assume that $\sigma(A) \subset D$. For σ a positive real number, define the matrix function W(z) by

$$W(z) = \begin{bmatrix} I_p & 0 \\ K^*(\lambda) & \sigma I_q \end{bmatrix} \begin{bmatrix} I_p & 0 \\ 0 & -I_q \end{bmatrix} \begin{bmatrix} I_p & K(\lambda) \\ 0 & \sigma I_q \end{bmatrix}$$

and let P and Q be the unique solutions of the Lyapunov equations

(4.9) $$A(\sigma^2 P)A^* - (\sigma^2 P) = BB^*$$

and

(4.10) $$A^* Q A - Q = C^* C.$$

Then $W(\lambda)$ has a signed spectral factorization if and only if the matrix I-QP is invertible.

When this is the case, the factor $X_+(\lambda)$ for a signed spectral factorization $W(\lambda)=X_+^*(\lambda)\begin{bmatrix} I_p & 0 \\ 0 & -I_q \end{bmatrix}X_+(\lambda)$ is computed as follows. Set $Z=(I-QP)^{-1}$ and let c be the (p+q)x(p+q) matrix

$$(4.11)\qquad c = \begin{bmatrix} I+CA^{-1}PZA^{*-1}C^* & -CA^{-1}Z^*B \\ -B^*ZA^{-1*}C^* & -\sigma^2 I+B^*ZQB \end{bmatrix}.$$

Then c is Hermitian with p positive and q negative eigenvalues, and so has a factorization

$$(4.12)\qquad c = d^*\begin{bmatrix} I_p & 0 \\ 0 & -I_q \end{bmatrix}d$$

for an invertible (p+q)x(p+q) matrix d. Then the spectral factor $X_+(\lambda)$ for $W(\lambda)$ in this case is given by

$$(4.13)\qquad X_+(\lambda)=d\left\{\begin{bmatrix} I_p & 0 \\ 0 & I_q \end{bmatrix} + \begin{bmatrix} CP \\ \sigma^{-2}B^*A^{*-1} \end{bmatrix}Z(\lambda I-A^{*-1})^{-1}[A^{*-1}C^*,-QB]\right\}$$

with inverse given by

$$(4.14)\qquad X_+(\lambda)^{-1} =$$

$$\left\{\begin{bmatrix} I_p & 0 \\ 0 & I_q \end{bmatrix}-\begin{bmatrix} CP \\ \sigma^{-2}B^*A^{*-1} \end{bmatrix}(\lambda I-A^{*-1})^{-1}Z[A^{*-1}C^*,-QB]\right\}d^{-1}$$

PROOF. The result follows immediately from Theorem 4.1 upon taking

$$Y_-(\lambda) = \begin{bmatrix} I_p & 0 \\ 0 & \sigma I_q \end{bmatrix}\begin{bmatrix} I_p & K(\lambda) \\ 0 & I_q \end{bmatrix}$$

$$= \begin{bmatrix} I_p & 0 \\ 0 & \sigma I_q \end{bmatrix}\left\{\begin{bmatrix} I_p & 0 \\ 0 & I_q \end{bmatrix} + \begin{bmatrix} C \\ 0 \end{bmatrix}(\lambda I-A)^{-1}[0,B]\right\}.$$

Note that both $Y_-(\lambda)$ and

$$Y_-(\lambda)^{-1} = \begin{bmatrix} I_p & -K(\lambda) \\ 0 & I_q \end{bmatrix}\begin{bmatrix} I_p & 0 \\ 0 & \sigma^{-1}I_q \end{bmatrix}$$

are analytic in the complement of the unit disk D (including ∞) since all poles of $K(\lambda)$ are assumed to be in D.

5. SYMMETRIZED LEFT AND RIGHT CANONICAL SPECTRAL FACTORIZATION ON THE IMAGINARY AXIS

Suppose that $W(\lambda)$ is a rational mxm matrix function analytic and invertible on the iw-axis (including ∞) which enjoys the additional symmetry property $W(-\bar{\lambda})^{*}=W(\lambda)$. For convenience we shall denote $W(-\bar{\lambda})^{*}$ by $W^{*}(\lambda)$ in the sequel in this section. Thus on the iw-axis $W(\lambda)$ is Hermitian. Since $W(\lambda)$ is also invertible on the iw-axis, $W(iw)$ must have a constant number (say p) of positive eigenvalues and $q=m-p$ negative eigenvalues for all real w. By a <u>left spectral factorization</u> of W (with respect to the imaginary axis) we mean a factorization of the form

$$W(\lambda) = Y_{+}^{*}(\lambda) \begin{bmatrix} I_p & 0 \\ 0 & -I_q \end{bmatrix} Y_{+}(\lambda)$$

where $Y_{+}(\lambda)$ is analytic and invertible on the closed right half plane $\{\mathrm{Re}\lambda\geq 0\}$. By a <u>right spectral</u> factorization of W (with respect to the iw axis) we mean a factorization of the form

$$W(\lambda) = X_{-}^{*}(\lambda) \begin{bmatrix} I_p & 0 \\ 0 & -I_q \end{bmatrix} X_{-}(\lambda)$$

where $X_{-}(\lambda)$ is analytic and invertible on the closed left half plane $\{\mathrm{Re}\lambda\leq 0\}$. The problem we wish to analyze in this section is the half plane version of that considered in the previous section: namely given a left spectral factorization $W(\lambda) = Y_{+}^{*}(\lambda) \begin{bmatrix} I_p & 0 \\ 0 & -I_q \end{bmatrix} Y_{+}(\lambda)$, compute a right spectral factorization $W(\lambda) = X_{-}^{*}(\lambda) \begin{bmatrix} I_p & 0 \\ 0 & -I_q \end{bmatrix} X_{-}(\lambda)$. The result is the following.

THEOREM 5.1. _Suppose the rational mxm matrix function_ $W(\lambda)=W^*(\lambda)$ _has a left spectral factorization_

$$W(\lambda) = Y_+^*(\lambda)\begin{bmatrix} I_p & 0 \\ 0 & -I_q \end{bmatrix} Y_+(\lambda)$$

where

$$Y_+(\lambda) = Y_+(\infty)[I+C_+(\lambda I-A_+)^{-1}B_+].$$

We may assume that A_+ _and_ $A_+^x := A_+-B_+C_+$ _have their spectra in the open left half plane_ $\{\mathrm{Re}\lambda<0\}$. _Let_ P _and_ Q _denote the unique solutions of the Lyapunov equations_

$$A_+^xP+P(A_+^x)^* = -B_+W(\infty)^{-1}B_+^* \tag{5.1}$$

and

$$A_+^*Q+QA_+=C_+^*W(\infty)C_+. \tag{5.2}$$

Then W _has a right spectral factorization if and only if the matrix_ I−QP _is invertible, or equivalently, if and only if the matrix_ I−PQ _is invertible. When this is the case, the factor_ $X_-(\lambda)$ _for a right spectral factorization_

$$W(\lambda) = X_-^*(\lambda)\begin{bmatrix} I_p & 0 \\ 0 & -I_q \end{bmatrix} X_-(\lambda)$$

_of_ W _can be taken to be_

$$X_-(\lambda)$$

$$=Y_+(\infty)\{I+(-W(\infty)^{-1}B_+^*+C_+P)(I-QP)^{-1}(\lambda I+A_+^*)^{-1}(C_+^*W(\infty)-QB_+)\}$$

with inverse

$$X_-(\lambda)^{-1} = \{I-(-W(\infty)^{-1}B_+^* + C_+P)$$

$$\cdot(\lambda I+A_+^*-C_+^*B_+^*)^{-1}(I-QP)^{-1}(C_+^*W(\infty)-QB_+)\}Y_+(\infty)^{-1}.$$

PROOF. In Theorem 2.1 it was assumed that $W(\infty)=I_m$ and that $W(\lambda)=Y_-(\lambda)Y_+(\lambda)$ where $Y_-(\infty)=Y_+(\infty) = I_m$. We thus consider here $W(\infty)^{-1}W(\lambda)$ and its left Wiener-Hopf factorization

$$W(\infty)^{-1}W(\lambda) = Y_-(\lambda)\ (Y_+(\infty)^{-1}Y_+(\lambda))$$

where

$$Y_-(\lambda):= W(\infty)^{-1}Y_+(-\bar{\lambda})^* \begin{bmatrix} I_p & 0 \\ 0 & -I_q \end{bmatrix} Y_+(\infty).$$

From

$$Y_+(\infty)^{-1}Y_+(\lambda) = I+C_+(\lambda I-A_+)^{-1}B_+$$

we get

$$Y_-(\lambda) = I-W(\infty)^{-1}B_+^*(\lambda I+A_+^*)^{-1}C_+^*W(\infty).$$

We thus define

$$A_- = -A_+^*$$

$$B_- = C_+^*W(\infty)$$

and

$$C_- = -W(\infty)^{-1}B_+^*$$

and apply the results of Theorem 2.1 with the roles of + and − interchanged. The Lyapunov equations (2.4) and (2.5) specialize to (5.1) and (5.2). Thus $W(\infty)^{-1}W(\lambda)$ has a right Wiener-Hopf factorization if and only if $I-QP$ is invertible, where P and Q are the solutions of (5.1) and (5.2). When this is the case then $W(\infty)^{-1}W(\lambda)=W_+(\lambda)W_-(\lambda)$ where W_+ and W_- can be computed as in Theorem 2.1 (interchanging + and −), where $W_+(\infty) = W_-(\infty) = I_m$. One easily sees that $X_-(\lambda):=$

$Y_+(\infty)W_-(\lambda)$ is the factor for a right spectral factorization for $W(\lambda)$. This choice of $X_-(\lambda)$ then produces the formulas in Theorem 5.1.

For the application to the model reduction problem for continuous time systems (see [BR 1] and [Gl]), one needs to apply Theorem 5.1 to a function $Y_+(\lambda)$ having a special form.

COROLLARY 5.2. *Suppose* $G(\lambda) = C(\lambda I-A)^{-1}B$ *is a stable rational* pxq *matrix function of McMillan degree* n. *Thus we may assume that the spectrum of the* nxn *matrix* A *is in the open left half plane* $\{\mathrm{Re}\lambda<0\}$. *For* σ *a positive real number, let* $W(\lambda)$ *be defined by*

$$W(\lambda) = \begin{bmatrix} I_p & 0 \\ G^*(\lambda) & \sigma I_q \end{bmatrix} \begin{bmatrix} I_p & 0 \\ 0 & -I_q \end{bmatrix} \begin{bmatrix} I_p & G(\lambda) \\ 0 & \sigma I_q \end{bmatrix}$$

and let the nxn *matrices* P *and* Q *be the unique solutions of the Lyapunov equations*

$$A(\sigma^2 P) + (\sigma^2 P) A^* = BB^* \tag{5.3}$$

and

$$A^*Q + QA = C^*C. \tag{5.4}$$

Then $W(\lambda)$ *has a right spectral factorization if and only if the matrix* I−QP *(or equivalently* I−PQ) *is invertible. When this is the case, the factor* $X_-(\lambda)$ *for a symmetrized right canonical factorization*

$$W(\lambda) = X_-^*(\lambda) \begin{bmatrix} I_p & 0 \\ 0 & -I_q \end{bmatrix} X_-(\lambda)$$

of W *can be taken to be*

$$X_-(\lambda) = \begin{bmatrix} I_p & 0 \\ 0 & \sigma I_q \end{bmatrix} + \begin{bmatrix} CP \\ \sigma^{-1}B^* \end{bmatrix} (I-QP)^{-1}(\lambda I_n+A^*)^{-1}[C^*,-QB]$$

with inverse given by

$$X_-(\lambda)^{-1} = \begin{bmatrix} I_p & 0 \\ 0 & \sigma^{-1} I_q \end{bmatrix} - \begin{bmatrix} CP \\ \sigma^{-2} B^* \end{bmatrix}$$

$$\cdot (\lambda I_n + A^*)^{-1} (I - QP)^{-1} [C^*, -\sigma^{-1} QB]$$

PROOF. The result follows immediately from Theorem 3.1 upon taking

$$Y_+(\lambda) = \begin{bmatrix} I_p & G(\lambda) \\ 0 & \sigma I_q \end{bmatrix}$$

$$= \begin{bmatrix} I_p & 0 \\ 0 & \sigma I_q \end{bmatrix} \left\{ \begin{bmatrix} I_p & 0 \\ 0 & I_q \end{bmatrix} + \begin{bmatrix} C \\ 0 \end{bmatrix} (\lambda I_n - A)^{-1} [0, B] \right\}.$$

Note that both $Y_+(\lambda)$ and $Y_+(\lambda)^{-1} = \begin{bmatrix} I_p & -G(\lambda) \\ 0 & I_q \end{bmatrix} \begin{bmatrix} I_p & 0 \\ 0 & \sigma^{-1} I_q \end{bmatrix}$ are analytic in the closed right half plane since all poles of $G(\lambda)$ are by assumption in the open left half plane.

REFERENCES

[BR.1] Ball, J.A. and Ran, A.C.M., Hankel norm approximation of a rational matrix function in terms of its realization, in Proceedings of 1985 Symposium on the Mathematical Theory of Networks and Systems (Stockholm), to appear.

[BR.2] Ball, J.A. and Ran, A.C.M., Optimal Hankel norm model reductions and Wiener-Hopf Factorization I: The canonical case, SIAM J. Control and Opt., to appear.

[BGK.1] Bart, H.; Gohberg, I. and Kaashoek, M.A., Minimal Factorization of Matrix and Operator Functions, OT1 Birkhäuser, Basel, 1979.

[BGK.2] Bart, H.; Gohberg, I. and Kaashoek, M.A., The coupling method for solving integral equations, in Topics in Operator Theory Systems and Networks (ed. H. Dym and I. Gohberg), OT 12 Birkhäuser, Basel, 1983, 39-73.

[BGKvD] Bart, H.; Gohberg, I.; Kaashoek, M.A. and van Dooren, P., Factorization of transfer functions, SIAM J. Control and Opt. 18 (1980), 675-696.

[CG] Clancey, K. and Gohberg, I., Factorization of Matrix Functions and Singular Integral Operators, OT3 Birkhäuser, Basel, 1981.

[Gl] Glover, K., All optimal Hankel-norm approximations of linear multivariable systems and their L^∞-error bounds, Int. J. Control 39 (1984), 1115-1193.

[GF] Gohberg, I.C. and Feldman, I.A., Convolution Equations and Projection Methods for their Solutions, Amer. Math. Soc. (Providence), 1974.

[GK] Gohberg, I.C. and Krupnik, N.Ja., Einfuhrung in die Theorie der eindimensionalen singulären Integraloperatoren, Birkhäuser, Basel,1979.

[H] Helton, J.W., A spectral factorization approach to the distributed stable regulator problem: the algebraic Riccati equation, SIAM J. Control and Opt. 14 (1976), 639-661.

[W] Willems, J., Least squares stationary optimal control and the algebraic Riccati equation, IEEE Trans. Aut. Control AC-16 (1971), 621-634.

J.A. Ball
Department of Mathematics
Virginia Tech
Blacksburg, VA 24061 USA

A.C.M. Ran
Subfaculteit der Wiskunde
en Informatica
Vrije Universiteit
1007 MC Amsterdam
The Netherlands

Operator Theory:
Advances and Applications, Vol. 21

WIENER-HOPF EQUATIONS WITH SYMBOLS ANALYTIC IN A STRIP

H. Bart, I. Gohberg, M.A. Kaashoek

The explicit method of factorization and inversion developed in [BGK1], [BGK5] and [BGK6] is extended to a larger class of Wiener-Hopf integral equations, namely those with m×m matrix symbols of the form $I - \hat{k}(\lambda)$, where $\hat{k}$ is the Fourier tranform of a function k from the class $e^{\omega|t|}L_1^{m\times m}(\mathbb{R})$, $\omega < 0$.

0. INTRODUCTION

Let $W(\lambda) = I - \hat{k}(\lambda)$, where $\hat{k}$ is the Fourier transform of an m×m matrix function k, the entries of which are integrable on the real line. When $\hat{k}$ has rational entries, the matrix function W may be represented in the form

$$(0.1) \qquad W(\lambda) = I + C(\lambda-A)^{-1}B, \qquad -\infty < \lambda < \infty.$$

Here A is a square matrix of order n (where n may be larger than m), A does not have eigenvalues on the real line, and B and C are matrices of sizes n×m and m×n, respectively. The representation (0.1), which is called a <u>realization</u> of W, has been used (see [BGK1], Ch.IV) to give explicit formulas for a Wiener-Hopf factorization of W in terms of the matrices A, B and C. Also in this way the inverse of the Wiener-Hopf integral operator with symbol W, its Fredholm characteristics and other related properties may be expressed explicitly in terms of the three matrices A, B and C (see [BGK1], Section IV.5; [BGK2]). In the case when $\hat{k}$ is analytic on the extended real line (including infinity), the same ideas can be applied. The only difference is that in the realization (0.1) for A, B and C one has to use bounded linear operators acting between (possibly infinite dimensional) Banach spaces (see [BGK2], [BGK3] and

[BK]. The next step was taken in [BGK4], [BGK5] and [BGK6] for certain matrix functions that are analytic in a strip around the real line but not at infinity.

In the present paper we treat m×m matrix functions $\hat{k}$ such that k belongs to the class $e^{\omega|t|}L_1^{m\times m}(\mathbb{R})$ for some $\omega < 0$. Again the starting point is the representation (0.1), but now the operators A,B and C have the following properties. The operator -iA, which acts in a Banach space X, is a (possibly unbounded) exponentially dichotomous operator of exponential type ω (see Section I.1 for the definition of these notions), the operator B from $\mathbb{C}^m$ into X is a (bounded) linear operator and C is an A-bounded linear operator between X and $\mathbb{C}^m$ such that the linear tranformation

$$(0.2) \qquad x \to \begin{cases} iCe^{-itA}(I-P)x, & t > 0, \\ -iCe^{-itA}Px, & t < 0, \end{cases}$$

maps the domain of A into the space of all functions in $e^{\omega|t|}L_1^m(\mathbb{R})$ with a derivative in $L_1^m(\mathbb{R})$, and extends to a bounded linear operator defined on X with values in $e^{\omega|t|}L_1^m(\mathbb{R})$. The operator P appearing in (0.2) is the separating projection for the exponentially dichotomous operator -iA (see Section I.1).

To use the method of factorization and inversion referred to above for the class of functions considered in the present paper, the main difficulty is to prove the following result. If $\det W(\lambda) \neq 0$ for $-\infty < \lambda < \infty$, then

$$W(\lambda)^{-1} = I - C(\lambda - A^{\times})^{-1}B, \qquad -\infty < \lambda < \infty,$$

where $A^{\times} = A - BC$, the operator $-iA^{\times}$ is an exponentially dichotomous operator, the operator C is $A^{\times}$-bounded, the map (0.2) remains bounded (for ω sufficiently close to zero) if A and P are replaced by $A^{\times}$ and the separating projection $P^{\times}$ of $-iA^{\times}$, respectively, and, finally, the difference $P-P^{\times}$ is a compact operator.

The theorem which gives the special representation (0.1) for the matrix functions considered in this paper, and the results mentioned in the previous paragraph, are proved in the first chapter of the paper. In the second chapter we give applications to inverse Fourier transforms, Wiener-Hopf factorization, the Riemann-Hilbert boundary problem, and the inversion and Fredholm properties of Wiener-Hopf integral operators. The applications are established along the same lines of reasoning as in [BGK4], [BGK5] and [BGK6]. However, in the present context also it was necessary to overcome several new technical difficulties.

I. REALIZATION

I.1. Preliminaries

Let X be a complex Banach space and let S be a linear operator defined on a linear subspace $\mathcal{D}(S)$ of X with values in X, written $S(X \to X)$. We say that S is exponentially dichotomous if S is densely defined (i.e., $\mathcal{D}(S)$ is dense in X) and X admits a topological direct sum decomposition

$$(1.1) \qquad X = X_- \oplus X_+$$

with the following properties: the decomposition reduces S, the restriction of $-S$ to X_- is the infinitesimal generator of an exponentially decaying strongly continuous semigroup, and the same is true for the restriction of S to X_+. These requirements determine the decomposition (1.1) uniquely and the projection of X onto X_- along X_+ is called the separating projection for S. In case S is bounded, S is exponentially dichotomous if and only if the spectrum $\sigma(S)$ of S does not meet the imaginary axis and then the separating projection is just the Riesz projection corresponding to the part of $\sigma(S)$ lying in the right half plane. In general, the condition that S is exponentially diochotomous involves a more complicated spectral splitting of the (possibly connected) extended spectrum of S. The details (including a characterization of exponentially dichotomous operators in terms of two-sided Laplace transforms) may be found in [BGK4] or [BGK6].

Suppose $S(X \to X)$ is exponentially dichotomous, and let (1.1) be the decomposition having the properties described above. With respect to this decomposition, we write

$$S = \begin{pmatrix} S_- & 0 \\ 0 & S_+ \end{pmatrix}.$$

The _bisemigroup_ $E(.;S)$ _generated by_ S is then defined as follows:

$$(1.2) \qquad E(t;S)x = \begin{cases} -e^{tS_-}Px, & t < 0, \\ e^{tS_+}(I-P)x, & t > 0. \end{cases}$$

Here P is the separating projection for S. The operator S will sometimes be referred to as the _bigenerator_ of $E(.;S)$. Note that the function $E(.;S)$ takes its values in $L(X)$, the Banach space of all bounded linear operators on X.

From standard semigroup theory (cf. [HP], [P]) it is known that there exists a real constant ω such that

$$(1.3) \qquad \sup_{t \leqq 0} e^{\omega t} \| e^{tS_-} \| < \infty, \qquad \sup_{t \geqq 0} e^{-\omega t} \| e^{tS_+} \| < \infty.$$

In the present situation (exponential decay), we may take the constant ω negative. If (1.3) is fulfilled, we say that S (or the bisemigroup generated by S) is of _exponential type_ ω. Note that (1.3) is equivalent to

$$(1.4) \qquad \sup_{t \neq 0} e^{-\omega|t|} \| E(t;S) \| < \infty.$$

For later use we recall a few simple facts about bisemigroups. Suppose $S(X \to X)$ is exponentially dichotomous, and let $E(.;S)$ be the corresponding bisemigroup given by (1.2). Take $x \in X$. The function $E(.;S)x$ is continuous on $\mathbb{R}\setminus\{0\}$ and exponentially decaying (in both directions). It also has a jump (discontinuity) at the origin and in fact

$$\lim_{t\to 0-} E(t;S)x = -Px, \qquad \lim_{t\to 0+} E(t;S)x = (I-P)x,$$

where P is the separating projection for S. If x belongs to the domain $\mathcal{D}(S)$ of S, then $E(.;S)x$ is differentiable on $\mathbb{R}\backslash\{0\}$ and

$$\frac{d}{dt} E(t;S)x = E(t;S)Sx = SE(t;S)x, \qquad t \neq 0.$$

Obviously, the derivative of $E(.;S)x$ is continuous on $\mathbb{R}\backslash\{0\}$, exponentially decaying (in both directions) and has a jump at the origin. From (1.2) it is clear that

$$E(t;S)P = PE(t;S) = E(t;S), \qquad t < 0,$$

$$E(t;S)(I-P) = (I-P)E(t;S) = E(t;S), \qquad t > 0.$$

Moreover, the following semigroup properties hold:

$$E(t+s;S) = -E(t;S)E(s;S), \qquad t,s < 0,$$

$$E(t+s;S) = E(t;S)E(s;S), \qquad t,s > 0.$$

I.2. Realization triples

Let m be a positive integer. By $\mathcal{D}_1^m(\mathbb{R})$ we denote the linear subspace of $L_1^m(\mathbb{R}) = L_1(\mathbb{R};\mathbb{C}^m)$ consisting of all $f \in L_1^m(\mathbb{R})$ for which there exists $g \in L_1^m(\mathbb{R})$ such that

$$(2.1) \qquad f(t) = \begin{cases} \int_{-\infty}^{t} g(s)ds, & \text{a.e. on } (-\infty,0], \\ -\int_{t}^{\infty} g(s)ds, & \text{a.e. on } [0,\infty). \end{cases}$$

If $f \in \mathcal{D}_1^m(\mathbb{R})$, then there is only one $g \in L_1^m(\mathbb{R})$ satisfying (2.1). This g is called the <u>derivative</u> of f and denoted by f'. We also stipulate

$$f_-(0) = \int_{-\infty}^{0} g(s)ds, \qquad f_+(0) = -\int_{0}^{\infty} g(s)ds,$$

which implies that

$$(2.2) \qquad f_-(0) - f_+(0) = \int_{-\infty}^{\infty} g(s)ds = \int_{-\infty}^{\infty} f'(s)ds.$$

We shall encounter this identity in Section 4 below.

Let X be a complex Banach space, let $A(X \to X)$, $B: \mathbb{C}^m \to X$ and $C(X \to \mathbb{C}^m)$ be linear operators and let ω be a negative constant. We call $\Theta = (A,B,C)$ a realization triple of exponential type ω if the following conditions are satisfied:

(1) $-iA$ is exponentially dichotomous of exponential type ω,

(2) $\mathcal{D}(C) \supset \mathcal{D}(A)$ and C is A-bounded,

(3) there exists a linear operator $\Lambda_\Theta: X \to L_1^m(\mathbb{R})$ such that

(i) $\sup_{\|x\| \leqq 1} \int_{-\infty}^{\infty} e^{-\omega|t|} \|\Lambda_\Theta x(t)\| dt < \infty$,

(ii) Λ_Θ maps $\mathcal{D}(A)$ into $\mathcal{D}_1^m(\mathbb{R})$ and

$$\Lambda_\Theta x = iCE(.;-iA)x, \qquad x \in \mathcal{D}(A).$$

Note that B, being a linear operator from $\mathbb{C}^m$ into X, is automatically bounded. Observe also that (i) implies that Λ_Θ is bounded and maps X into $L_{1,\omega}^m(\mathbb{R})$, where

$$(2.3) \qquad L_{1,\omega}^m(\mathbb{R}) = \{f \in L_1^m(\mathbb{R}) \mid e^{-\omega|.|}f(.) \in L_1^m(\mathbb{R})\}.$$

Taking into account (ii) and the fact that $\mathcal{D}(A)$ is dense in X, one sees that Λ_Θ is determined uniquely. Since ω is negative, $L_{1,\omega}^m(\mathbb{R})$ given by (2.3) is a linear subspace of $L_1^m(\mathbb{R})$. The space X is called the state space, the space $\mathbb{C}^m$ the input/output space of the triple Θ.

Suppose Θ is a realization triple of exponential type ω and $\omega \leqq \omega_1 < 0$. Then Θ is a realization triple of exponential type ω_1 too. To see this, note that (1) and (i) are fulfilled with ω replaced by ω_1. When the actual value of ω is irrelevant, we simply call Θ a realization triple. So $\Theta = (A,B,C)$ a <u>realization triple</u> (without further qualification) if Θ is a realization triple of exponential type ω for some $\omega < 0$. As we saw, the operator Λ_Θ does not depend on the value of ω, and the same is true with regard to the separating projection for $-iA$. This projection will be denoted by P_Θ.

In [BGK4], [BGK5] and [BGK6] we have been dealing with the situation where $-iA(X \to X)$ is exponentially dichotomous, $B: \mathbb{C}^m \to X$ is a (bounded) linear operator and $C: X \to \mathbb{C}^m$ is a bounded linear operator too. It is of interest to note that under these circumstances $\Theta = (A,B,C)$ is a realization triple. In fact, if $-iA$ is of exponential type ω, then Θ is a realization triple of exponential type ω_1 whenever $\omega \leqq \omega_1 < 0$. To see this, define $\Lambda_\Theta x$ for each $x \in X$ by $\Lambda_\Theta x = iCE(.;-iA)x$ and use the inequality (1.4).

Let $\Theta = (A,B,C)$ be a realization triple with state space X. The projection P_Θ of X associated with Θ is defined in terms of A alone, the operator Λ_Θ from X into $L_1^m(\mathbb{R})$ is completely determined by A and C. Next we introduce another operator, namely $\Gamma_\Theta: L_1^m(\mathbb{R}) \to X$, which depends only on A and B. The definition is

$$\Gamma_\Theta \phi = \int_{-\infty}^{\infty} E(-t;-iA)B\phi(t)dt.$$

It is easy to see that Γ_Θ is well-defined, linear and bounded. Note in this context that because of the finite dimensionality of $\mathbb{C}^m$, the operator function $E(.;-iA)B$ is continuous on $\mathbb{R}\backslash\{0\}$ with a (possible) jump at the origin, all with respect to the operator norm topology.

PROPOSITION 2.1. <u>Let</u> Θ <u>be a realization triple of exponential type</u> ω, <u>and let</u> Γ_Θ <u>be as above. Then</u> Γ_Θ <u>is compact and maps</u> $\mathcal{D}_1^m(\mathbb{R})$ <u>into</u> $\mathcal{D}(A)$.

PROOF. The compactness of Γ_Θ has already been established in [BGK5], Lemma 3.2. So we shall concentrate on the second assertion of the proposition.

Take ϕ in $\mathcal{D}_1^m(\mathbb{R})$. We need to show that $\Gamma_\Theta\phi \in \mathcal{D}(A)$. For simplicity we restrict ourselves to the case when ϕ vanishes almost everywhere on $(-\infty,0]$. Write

$$\phi(t) = -\int_t^\infty \psi(s)ds, \qquad t > 0,$$

where $\psi \in L_1^m(\mathbb{R})$. Then

$$\Gamma_\Theta\phi = -\int_0^\infty\Big(\int_t^\infty E(-t;-iA)B\psi(s)ds\Big)dt$$

and applying Fubini's theorem we get

$$\Gamma_\Theta\phi = -\int_0^\infty\Big(\int_0^s E(-t;-iA)B\psi(s)dt\Big)ds.$$

Since A is exponentially dichotomous, the origin belongs to the resolvent set of A. So it makes sense to consider the operator function $iE(-t;-iA)A^{-1}B$. This function is differentiable on $[0,\infty)$ and its derivative is the continuous operator function $-E(-t;-iA)B$. (Pointwise this is clear from the results collected together in Section I.1; next use the finite dimensionality of $\mathbb{C}^m$). Thus

$$-\int_0^s E(-t;-iA)Bdt = iE(-s;-iA)A^{-1}B - iP_\Theta A^{-1}B$$

and consequently

$$\Gamma_\Theta\phi = \int_0^\infty \Big(iE(-s;-iA)A^{-1}B - iP_\Theta A^{-1}B\Big)\psi(s)ds$$

$$= A^{-1}\Big[\int_0^\infty\Big(iE(-s;-iA)B - iP_\Theta B\Big)\psi(s)ds\Big].$$

But then $\Gamma_\Theta\phi \in \operatorname{Im}A^{-1} = \mathcal{D}(A)$. □

We conclude this section with the following observation concerning Γ_Θ. Let Q be the projection of $L_1^m(\mathbb{R})$ onto $L_1^m[0,\infty)$ along $L_1^m(-\infty,0]$, where $L_1^m(\mathbb{R})$ is identified in the usual way with $L_1^m(-\infty,0] \oplus L_1^m[0,\infty)$. Then it is clear from the properties of bisemigroups mentioned in the last paragraph of Section I.1 that

$$(2.4) \qquad (I-P)\Gamma_\Theta Q = 0, \qquad P\Gamma_\Theta(I-Q) = 0.$$

In other words, Γ_Θ maps $L_1^m[0,\infty)$ into ImP and $L_1^m(-\infty,0]$. into KerP.

I.3. <u>The realization theorem</u>

Suppose $\Theta = (A,B,C)$ is a realization triple of (negative) exponential type ω. For fixed y in $\mathbb{C}^m$ (the input/output space of Θ), we have that $\Lambda_\Theta By \in L_{1,\omega}^m(\mathbb{R})$. Thus the expression

$$(3.1) \qquad k_\Theta(t)y = \Lambda_\Theta By(t), \qquad \text{a.e. on } \mathbb{R}$$

determines a unique element k_Θ of $L_{1,\omega}^{m\times m}(\mathbb{R})$, the linear subspace of $L_1^{m\times m}(\mathbb{R}) = L_1(\mathbb{R};\mathbb{C}^{m\times m})$ consisting of all $h \in L_1^{m\times m}(\mathbb{R})$ for which $e^{-\omega|\cdot|}h(.) \in L_1^{m\times m}(\mathbb{R})$. We call k_Θ the <u>kernel associated with</u> Θ.

Since $k_\Theta \in L_{1,\omega}^{m\times m}(\mathbb{R}) \subset L_1^{m\times m}(\mathbb{R})$, the Fourier transform $\hat{k}_\Theta$ of k_Θ is an analytic m×m matrix function on the strip $|\mathrm{Im}\lambda| < -\omega$. More explicitly, the following can be said.

THEOREM 3.1. <u>Let</u> Θ <u>be a realization triple of exponential type</u> ω. <u>Then</u>

$$(3.2) \qquad \hat{k}_\Theta(\lambda) = -C(\lambda-A)^{-1}B, \qquad |\mathrm{Im}\lambda| < -\omega.$$

The A-boundedness of C implies that $C(\lambda-A)^{-1}$ is a well-defined bounded linear operator depending analytically on λ in the strip $|\mathrm{Im}\lambda| < -\omega$.

PROOF. It suffices to show that for $x \in X$ and $|\mathrm{Im}\lambda| < -\omega$

$$(3.3) \qquad C(\lambda-A)^{-1}x = -\int_{-\infty}^{\infty} e^{i\lambda t}\Lambda_\Theta x(t)dt,$$

i.e., $-C(\lambda-A)^{-1}x$ is equal to the Fourier transform $\hat{\Lambda}_\Theta x$ of $\Lambda_\Theta x$.

Take $|\mathrm{Im}\lambda| < -\omega$. As was observed already, $C(\lambda-A)^{-1}$ is a bounded linear operator. Now consider the mapping $x \to \hat{\Lambda}_\Theta x(\lambda)$ from X into $\mathbb{C}^m$. This mapping is linear and bounded too. Linearity is obvious and boundedness follows from the estimate

$$\|\hat{\Lambda}_\Theta x(\lambda)\| \leqq \int_{-\infty}^{\infty} e^{-\omega|t|}\|\Lambda_\Theta x(t)\|dt,$$

together with condition (i) in Section I.2. So it is enough to check the identity (3.3) for vectors x belonging to the dense linear subspace $\mathcal{D}(A)$ of X.

Take $x \in \mathcal{D}(A)$ and put $z = Ax$. According to formula (1.5) in [BGK6]

$$(\lambda-A)^{-1}z = -i\int_{-\infty}^{\infty} e^{i\lambda t}E(t;-iA)zdt.$$

Recall that CA^{-1} is a bounded linear operator. It follows that

$$\begin{aligned} C(\lambda-A)^{-1}x &= CA^{-1}(\lambda-A)^{-1}z \\ &= -\,i\int_{-\infty}^{\infty} e^{i\lambda t}CA^{-1}E(t;-iA)zdt \\ &= -i\int_{-\infty}^{\infty} e^{i\lambda t}CE(t;-iA)xdt \\ &= -\int_{-\infty}^{\infty} e^{i\lambda t}\Lambda_\Theta x(t)dt, \end{aligned}$$

the latter equality holding by virtue of condition (ii) in Section I.2. □

Representations of the type (3.1) will play an important role in this paper. They are called <u>(spectral) exponential representations</u>. Instead of (3.1) we shall also write

(3.4) $\qquad k_\Theta(t) = iCE(t;-iA)B.$

Note, however, that the latter expression has to be understood in the right way involving condition (ii) in Section 2. Functions of the type (3.4) with B and C bounded were considered in [BGK4], [BGK5] and [BGK6].

An identity of the type (3.2) is called a _realization_. This notion is taken from systems theory (cf. [K], KFA], [Ka] and [BGK1]). The realizations appearing in the present paper feature not only a (possibly) unbounded state space operator A but also a (possibly) unbounded output operator C. However, C and A, are related in the way described in Section I.2.

I.4. Construction of realization triples

The kernel associated with a realization triple of exponential type ω belongs to $L_{1,\omega}^{m\times m}(\mathbb{R})$. In this section we shall see that, conversely, each $k \in L_{1,\omega}^{m\times m}(\mathbb{R})$ appears as such a kernel. An explicit construction is contained in the next theorem.

THEOREM 4.1. _Let_ $k \in L_{1,\omega}^{m\times m}(\mathbb{R})$, _where_ m _is a positive integer and_ ω _is a negative constant. Then_ k _is the kernel associated with a realization triple of exponential type_ ω. _In fact_ $k = k_\Theta$, _where the realization triple_ $\Theta = (A,B,C)$ _with state space_ $X = L_1^m(\mathbb{R})$ _and input/output space_ $\mathbb{C}^m$ _is defined as follows_:

$$\mathcal{D}(A) = \mathcal{D}(C) = \mathcal{D}_1^m(\mathbb{R}),$$

$$Af(t) = \begin{cases} -i\omega f(t) + if'(t), & \underline{\text{a.e. on}}\ (-\infty,0], \\ i\omega f(t) + if'(t), & \underline{\text{a.e. on}}\ [0,\infty), \end{cases}$$

$$By(t) = e^{-\omega|t|}k(t)y, \qquad \underline{\text{a.e. on}}\ \mathbb{R},$$

$$Cf = if_-(0) - if_+(0).$$

Recall from Section I.2 that for $f \in \mathcal{D}(A) = \mathcal{D}_1^m(\mathbb{R})$ the left and right evaluations $f_-(0)$ and $f_+(0)$ at the origin are well-defined. In view of (2.2) one can rewrite the expression for Cf as

$$(4.1)\qquad Cf = i\int_{-\infty}^{\infty} f'(t)dt.$$

PROOF. As is well-known, the backward translation semigroup on $L_1^m[0,\infty)$ is strongly continuous. The infinitesimal generator of this semigroup has $\mathcal{D}_1^m[0,\infty)$ as its domain and its action amounts to taking the derivative (cf. the first paragraph in Section I.2). Using this one sees that $-iA$ an exponentially dichotomous operator of exponential type ω and that the bisemigroup associated with $-iA$ acts as follows: For $t < 0$

$$E(t;-iA)f(s) = \begin{cases} -e^{-\omega t}f(t+s), & \text{a.e. on } (-\infty,0], \\ 0, & \text{a.e. on } [0,\infty); \end{cases}$$

for $t > 0$

$$E(t;-iA)f(s) = \begin{cases} 0 & \text{a.e. on } (-\infty,0], \\ e^{\omega t}f(t+s), & \text{a.e. on } [0,\infty). \end{cases}$$

The separating projection for $-iA$ is the projection of $L_1^m(\mathbb{R})$ onto $L_1^m(-\infty,0]$ along $L_1^m[0,\infty)$.

Define $\Lambda: X \to L_1^m(\mathbb{R})$ by

$$(4.2)\qquad \Lambda f(t) = e^{\omega|t|}f(t), \qquad \text{a.e. on } \mathbb{R}.$$

Then Λ satisfies the conditions (i) and (ii) of Section I.2 with Λ_Θ replaced by Λ. For (i) this is obvious. As to the first part of (ii), observe that $f \in \mathcal{D}_1^m(\mathbb{R})$ and $\omega < 0$ implies that $e^{\omega|\cdot|}f(\cdot) \in \mathcal{D}_1^m(\mathbb{R})$ too. To check the second part of (ii), one uses the above description of the bisemigroup $E(t;-iA)$ and the definition of C or (4.1). From the latter expression it is also clear that $\|Cf\| \leqq -\omega\|f\|+\|Af\|$, and so C is A-bounded.

Since $k \in L_{1,\omega}^{m\times m}(\mathbb{R})$, the operator B is well-defined, linear (and bounded). Thus $\Theta = (A,B,C)$ is a realization triple of exponential type ω. We claim that the kernel k_Θ associated with Θ coincides with k. Indeed, for $y \in \mathbb{C}^m$ the following identities hold a.e. on $\mathbb{R}$:

$$k_\Theta(t)y = \Lambda By(t) = e^{\omega|t|}By(t) = k(t)y.$$

Since $\mathbb{C}^m$ is finite dimensional, it follows that $k_\Theta(t) = k(t)$ a.e. on $\mathbb{R}$. In other words k_Θ and k coincide as elements of $L^{m\times m}_{1,\omega}(\mathbb{R})$. □

REMARK 4.2. Let $\Theta = (A,B,C)$ be the realization triple constructed in the preceding theorem. As we have already seen, the projection P_Θ associated with Θ is the projection of $L^m_1(\mathbb{R})$ onto $L^m_1(-\infty,0]$ along $L^m_1[0,\infty)$, i.e., $P_\Theta = I-Q$, where Q is as in the last paragraph of Section 1.2. Also Λ_Θ is the bounded linear operator acting on $L^m_1(\mathbb{R})$ defined by (4.2). Finally Γ_Θ is the bounded linear operator on $L^m_1(\mathbb{R})$ given by

$$\Gamma_\Theta\phi(f) = \begin{cases} -e^{\omega t}\int_{-\infty}^{0} k(t+s)\phi(-s)ds, & \text{a.e. on } (-\infty,0], \\ e^{-\omega t}\int_{-\infty}^{\infty} k(t+s)\phi(-s)ds, & \text{a.e. on } [0,\infty), \end{cases}$$

(cf. formula (2.4) in Section I.2 and formulas (2.8) and (2.9) in Section II.2 below).

Theorem 4.1 is related to certain results on infinite dimensional realization in systems theory (see, for example, [F], Ch.III, Section 6 and [CG]).

I.5. Basic properties of realization triples

It is convenient to introduce the following notation. Let $\Theta = (A,B,C)$ be a realization triple. The linear operator $A-BC$, having $\mathcal{D}(A)$ as its domain, will be denoted by $A^\times$. Note that $A^\times$ does not only depend on A, but also on B and C. By $\Theta^\times$ we indicate the triple $(A^\times,B,-C)$. For reasons that will become clear later, $\Theta^\times$ is sometimes called the _inverse_ of Θ. For what follows, it is essential to know under what circumstances $\Theta^\times$ is again a realization triple.

THEOREM 5.1. _Let $\Theta = (A,B,C)$ be a realization triple. The following statements are equivalent_:

(i) $\Theta^\times$ _is a realization triple_,

(ii) $\det\left(I + C(\lambda-A)^{-1}B\right)$ _does not vanish on the real line_,

(iii) $A^{\times}$ <u>has no spectrum on the real line</u>,

(iv) $-iA^{\times}$ <u>is exponentially dichotomous</u>.

Condition (iii) simply means that for each real λ, the linear operator $\lambda - A^{\times}$ maps $\mathcal{D}(A^{\times}) = \mathcal{D}(A)$ in a one-one way onto X.

PROOF. Suppose $A^{\times}$ has no spectrum on the real line. Then for each real λ, the linear operator $I - C(\lambda - A^{\times})^{-1}B$ acting on $\mathbb{C}^m$ is well-defined. A straightforward computation shows that it is the inverse of $I + C(\lambda - A)^{-1}B$. Conversely, if $W(\lambda) = I + C(\lambda - A)^{-1}B$ is invertible, then

$$(5.1) \qquad (\lambda - A^{\times})^{-1} = (\lambda - A)^{-1} - (\lambda - A)^{-1}BW(\lambda)^{-1}C(\lambda - A)^{-1}.$$

This proves the equivalence of (ii) and (iii). By definition (i) implies (iv) and it is obvious that (iv) implies (iii). It remains prove that (ii) and (iii) together give (i).

Suppose (ii) and (iii) are satisfied. Taking $\lambda = 0$ in (5.1), one gets $(A^{\times})^{-1} = A^{-1} + A^{-1}BW(0)^{-1}CA^{-1}$, where $W(0) = I - CA^{-1}B$. Since C is A-bounded, the operator CA^{-1} is bounded, and hence $(A^{\times})^{-1}$ is bounded. It follows that $A^{\times}$ is closed. Also, $C(A^{\times})^{-1} = CA^{-1} + CA^{-1}BW(0)^{-1}CA^{-1}$ is bounded, and from this the $A^{\times}$-boundedness of C is immediate.

With $W(\lambda)$ as in the first paragraph of this proof, we have $W(\lambda) = I - \hat{k}_{\Theta}(\lambda)$, where $k_{\Theta} \in L_{1,\omega}^{m\times m}(\mathbb{R})$. By Wiener's theorem there exists $k^{\times} \in L_1^{m\times m}(\mathbb{R})$ such that $W(\lambda)^{-1} = I - \hat{k}^{\times}(\lambda)$. Taking $|\omega|$ smaller (if necessary), we may assume that $k^{\times} \in L_{1,\omega}^{m\times m}(\mathbb{R})$ too (cf. [GRS]). But then $-iA^{\times}$ is exponentially dichotomous of exponential type ω. The proof of this is based on Theorem 4.1 in [BGK6] and (5.1). The argument goes along the lines indicated in Part III of the proof of Theorem 7.1 in [BGK6]. Analogously to what we have there, the bisemigroup generated by $-iA^{\times}$ is given by

$$E(t;-iA^{\times})x = E(t;-iA)x + \int_{-\infty}^{\infty} E(t-s;-iA)B\Lambda_{\Theta}x(s)ds$$

$$- \int_{-\infty}^{\infty}\Big(\int_{-\infty}^{\infty} E(t-s-u)Bk^{\times}(u)_{\Theta}\Lambda\ x(s)du\Big)ds.$$

We leave the details to the reader.

The negative constant ω having been taken sufficiently close to zero, one has that Θ is of exponential type ω and $k^{\times} \in L_{1,\omega}^{m\times m}(\mathbb{R})$. A standard reasoning now shows that the convolution product $k^{\times}*\Lambda_{\Theta}x$,

$$(k*\Lambda_{\Theta}x)(t) = \int_{-\infty}^{\infty} k^{\times}(t-s)\Lambda_{\Theta}x(s)ds, \quad \text{a.e. on } \mathbb{R},$$

determines a (bounded) linear operator from X into $L_1^m(\mathbb{R})$ such that

$$\sup_{\|x\| \leqq 1} \int_{-\infty}^{\infty} e^{-\omega|t|}\|(k*\Lambda_{\Theta}x)(t)\|dt < \infty.$$

But then the expression

$$(5.2) \qquad \Lambda^{\times}x = -\Lambda_{\Theta}x + k^{\times}*\Lambda_{\Theta}x$$

defines a (bounded) linear operator $\Lambda^{\times}$: $X \to L_1^m(\mathbb{R})$ for which condition (i) in Section I.2, with Λ_{Θ} replaced by $\Lambda^{\times}$, is satisfied.

Next, take $x \in X$ and consider the Fourier transform of $\Lambda^{\times}x$. Clearly, $\hat{\Lambda}^{\times}(\lambda) = -(I - \hat{k}^{\times}(\lambda))\hat{\Lambda}_{\Theta}x(\lambda) = -W(\lambda)^{-1}\hat{\Lambda}_{\Theta}x(\lambda)$. As we have seen above $W(\lambda)^{-1} = I - C(\lambda-A^{\times})^{-1}B$. Also we know from (3.3) that $\hat{\Lambda}_{\Theta}x(\lambda) = -C(\lambda-A)^{-1}x$. Hence $\hat{\Lambda}^{\times}x(\lambda) = (I - C(\lambda-A^{\times})^{-1}B)C(\lambda-A)^{-1}x$. It follows that

$$(5.3) \qquad \hat{\Lambda}^{\times}x(\lambda) = C(\lambda-A^{\times})^{-1}x.$$

Here λ is real (or, more generally, $|\mathrm{Im}\lambda| < -\omega$).

Take $x \in \mathcal{D}(A^{\times}) = \mathcal{D}(A)$, and write $x = (A^{\times})^{-1}z$. Then $C(\lambda-A^{\times})^{-1}x = C(A^{\times})^{-1}(\lambda-A^{\times})^{-1}z$, where $C(A^{\times})^{-1}$ is a bounded linear operator from X into $\mathbb{C}^m$. Using (5.3) and formula (1.5) in [BGK6], one gets

$$\hat{\Lambda}^{\times}x(\lambda) = -iC(A^{\times})^{-1}\int_{-\infty}^{\infty} e^{i\lambda t}E(t;-iA^{\times})zdt$$

$$= -\, i\int_{-\infty}^{\infty} e^{i\lambda t}CE(t;-iA^{\times})xdt.$$

But then we may conclude that

$$\Lambda^{\times}x(t) = -iCE(t;-iA^{\times})x, \quad \text{a.e. on } \mathbb{R}$$

It remains to prove that $\Lambda^{\times}x \in \mathcal{D}_1^m(\mathbb{R})$.

In view of the properties of Λ_Θ and the identity (5.2), it suffices to show that $k^{\times}*\Lambda_\Theta x$ belongs to $\mathcal{D}_1^m(\mathbb{R})$. Considering $\Lambda_\Theta x$ as a function, we may write

$$\Lambda_\Theta x(t) = \begin{cases} \int_{-\infty}^{t} g(s)ds, & t < 0, \\ -\int_{k}^{\infty} g(s)ds, & t > 0, \end{cases}$$

where $g \in L_1^m(\mathbb{R})$. Put

$$h = k^{\times}*g - \left(\int_{-\infty}^{\infty} g(s)ds\right)k.$$

Then $h \in L_1^m(\mathbb{R})$ and

$$(k^{\times}*\Lambda_\Theta x)(t) = \begin{cases} \int_{-\infty}^{t} h(s)ds, & \text{a.e. on } (-\infty,0], \\ -\int_{t}^{\infty} h(s)ds, & \text{a.e. on } [0,\infty). \end{cases}$$

So $k^{\times} * \Lambda_\Theta x \in \mathcal{D}_1^m(\mathbb{R})$ indeed. □

REMARK 5.2. Suppose $\Theta = (A,B,C)$ and $\Theta^{\times} = (A^{\times},B,-C)$ are both realization triples. Then it is clear from the proof of Theorem 5.1 that $\Lambda_\Theta^{\times}$ and Λ_Θ are related by

$$\Lambda_\Theta^{\times}x(t) = -\Lambda_\Theta x(t) + \int_{-\infty}^{\infty} k^{\times}(t-s)\Lambda_\Theta x(s)ds, \quad \text{a.e. on } \mathbb{R},$$

where $(I - \hat{k}_\Theta(\lambda))^{-1} = I - \hat{k}^\times(\lambda)$. For typographical reasons we wrote $\Lambda_\Theta^\times$ instead of $\Lambda_\Theta^\times$. Similar notations (such as $P_\Theta^\times$ and $k_\Theta^\times$) will be used below.

THEOREM 5.3. *Suppose* $\Theta = (A,B,C)$ *and* $\Theta^\times = (A^\times,B,-C)$ *are realization triples. Then the associated projections* P_Θ *and* $P_\Theta^\times$ *are related by*

$$(5.4) \qquad P_\Theta^\times = P_\Theta + \Gamma_\Theta\Lambda_\Theta^\times,$$

and $P_\Theta^\times - P_\Theta$ *is a compact linear operator.*

PROOF. By Proposition 2.1, the operator Γ_Θ is compact. So it suffices to establish (5.4), and for this it is enough to show that the identity $\Gamma_\Theta\Lambda_\Theta^\times x = P_\Theta^\times x - P_\Theta x$ holds on the domain of A. For $x \in \mathcal{D}(A)$, we have $\Lambda_\Theta^\times x = -iCE(.;-iA^\times)x$. Using this, the desired result is obtained along the lines indicated in the proof of [BGK5], Lemma 3.2. □

II. APPLICATIONS

II.1. Inverse Fourier transforms

In this section we shall give an explicit formula for the inverse Fourier transform of a function of the type $I - (I-\hat{k}(\lambda))^{-1}$, where $k \in L_{1,\omega}^{m\times m}(\mathbb{R})$ with $\omega < 0$. Recall from Section I.4 that $L_{1,\omega}^{m\times m}(\mathbb{R})$ coincides with the class of all kernels of realization triples of exponential type ω.

THEOREM 1.1. *Let* $\Theta = (A,B,C)$ *be a realization triple. Then* $\det(I-\hat{k}_\Theta(\lambda))$ *does not vanish on the real line if and only if* $\Theta^\times = (A^\times,B,-C)$ *is a realization triple, and in that case*

$$(I - \hat{k}_\Theta(\lambda))^{-1} = I - k_\Theta^\times(\lambda), \qquad \lambda \in \mathbb{R},$$

where $k_\Theta^\times$ *is the kernel associated with* $\Theta^\times$, *i.e.*,

$$k_\Theta^\times(t)y = \Lambda_\Theta^\times By(t), \qquad \text{a.e. on } \mathbb{R}.$$

Less precise, the latter identity can be written as

$k_\Theta^\times(t) = -iCE(t;-iA^\times)B$. The condition that $\Theta^\times$ is a realization triple can be replaced by any of the equivalent conditions in Theorem 5.1 of Chapter I.

PROOF. The first part of the theorem is immediate from Theorems 3.1 and 5.1 in Ch. I. To prove the second part, assume that $\det(I-\hat{k}_\Theta(\lambda))$ does not vanish on the real line, and hence $\Theta^\times = (A^\times,B,-C)$ is a realization triple. Let $k^\times \in L_1^m(\mathbb{R})$ be such that $(I - \hat{k}_\Theta(\lambda))^{-1} = I - k^\times(\lambda)$, $\lambda \in \mathbb{R}$. The existence of $k^\times$ is guaranteed by Wiener's theorem. Now, for λ on the real line,

$$\begin{aligned} I - k^\times(\lambda) &= (I + C(\lambda-A)^{-1}B)^{-1} \\ &= I - C(\lambda-A^\times)^{-1}B = I - k_\Theta^\times(\lambda), \end{aligned}$$

and hence $k^\times(t) = k_\Theta^\times(t)$ a.e. on $\mathbb{R}$. □

Theorem 1.1 can be reformulated in terms of full line convolution integral operators.

THEOREM 1.2. *Consider the convolution integral operator on* $L_1^m(\mathbb{R})$ *defined by*

$$L\phi(t) = \int_{-\infty}^{\infty} k_\Theta(t-s)\phi(s)ds, \quad \text{a.e. on } \mathbb{R},$$

where k_Θ *is the kernel of the realization triple* $\Theta = (A,B,C)$. *Then* $I-L$ *is invertible if and only if* $\Theta^\times = (A^\times,B,-C)$ *is a realization triple and the inverse of* $I-L$ *is given by*

$$(I-L)^{-1}\psi(t) = \psi(t) - \int_{-\infty}^{\infty} k_\Theta^\times(t-s)\psi(s)ds, \quad \text{a.e. on } \mathbb{R},$$

$$k_\Theta^\times(t)y = \Lambda_\Theta^\times By(t), \quad \text{a.e. on } \mathbb{R}.$$

The condition that $\Theta^\times$ is a realization triple can be replaced by any of the equivalent conditions in Theorem 5.1 of Ch. I. In a concise manner, the conclusion of Theorem 1.2 may be phrased as $(I-L)^{-1} = I-L^\times$, where $L^\times$ stands of course for the (full line) convolution integral operator associated with $\Theta^\times$.

II.2. Coupling

In Sections II.4 and II.5 below we want to apply the coupling method developed in [BGK3]. The next result contains the key step in this direction.

THEOREM 2.1. *Suppose* $\Theta = (A,B,C)$ *and* $\Theta^{\times} = (A^{\times},B,-C)$ *are realization triples, and introduce*

$$K : L_1^m[0,\infty) \to L_1^m[0,\infty),\ K\phi(t) = \int_0^\infty k_\Theta(t-s)\phi(s)ds, \quad \text{a.e. on } [0,\infty),$$

$$K^{\times}: L_1^m[0,\infty) \to L_1^m[0,\infty),\ K^{\times}\phi(t) = \int_0^\infty k_\Theta^{\times}(t-s)\phi(s)ds, \quad \text{a.e. on } [0,\infty),$$

$$U : \operatorname{Im}P_\Theta^{\times} \to L_1^m[0,\infty),\ Ux(t) = \Lambda_\Theta x(t), \quad \text{a.e. on } [0,\infty),$$

$$U^{\times}: \operatorname{Im}P_\Theta \to L_1^m[0,\infty),\ U^{\times}x(t) = -\Lambda_\Theta^{\times}x(t), \quad \text{a.e. on } [0,\infty),$$

$$R : L_1^m[0,\infty) \to \operatorname{Im}P_\Theta,\ R\phi = \int_0^\infty E(-t;-iA)B\phi(t)dt,$$

$$R^{\times}: L_1^m[0,\infty) \to \operatorname{Im}P_\Theta^{\times},\ R^{\times}\phi = -\int_0^\infty E(-t;-iA^{\times})B\phi(t)dt,$$

$$J : \operatorname{Im}P_\Theta^{\times} \to \operatorname{Im}P_\Theta, \quad Jx = P_\Theta x,$$

$$J^{\times}: \operatorname{Im}P_\Theta \to \operatorname{Im}P_\Theta^{\times}, \quad J^{\times}x = P_\Theta^{\times}x,$$

where m *is the dimension of the (common) input/output space* $\mathbb{C}^m$ *of* Θ *and* $\Theta^{\times}$. *Then all these operators are well-defined, linear and bounded. Moreover*

$$\begin{pmatrix} I-K & U \\ R & J \end{pmatrix} : L_p^m[0,\infty) \oplus \operatorname{Im}P_\Theta^{\times} \to L_p^m[0,\infty) \oplus \operatorname{Im}P_\Theta$$

is invertible with inverse

$$\begin{pmatrix} I-K^{\times} & U^{\times} \\ R^{\times} & J^{\times} \end{pmatrix} : L_p^m[0,\infty) \oplus \operatorname{Im}P_\Theta \to L_p^m[0,\infty) \oplus \operatorname{Im}P_\Theta^{\times}.$$

In terms of [BGK3], the theorem says that the operators I-K and $J^{\times}$ are *matricially coupled* with coupling relation

$$(2.1)\qquad \begin{pmatrix} I-K & U \\ R & J \end{pmatrix}^{-1} = \begin{pmatrix} I-K^{\times} & U^{\times} \\ R^{\times} & J^{\times} \end{pmatrix}.$$

The operator $J^{\times}$ is also called an <u>indicator</u> for I-K. Note that K is the Wiener-Hopf integral operator with kernel k_{Θ}. Analogously $K^{\times}$ is the Wiener-Hopf integral operator with kernel $k_{\Theta}^{\times}$.

PROOF. All operators appearing in Theorem 3.1 are well-defined, linear and bounded, and acting between the indicated spaces. In this context three observations should be made. First, let Q be the projection of $L_1^m(\mathbb{R})$ onto $L_1^m[0,\infty)$ along $L_1^m(-\infty,0]$. Then $U = Q\Lambda_{\Theta}|\mathrm{Im}P_{\Theta}^{\times}: \mathrm{Im}P_{\Theta}^{\times} \to L_1^m[0,\infty)$ and $U^{\times} = -Q\Lambda_{\Theta}^{\times}|\mathrm{Im}P_{\Theta}: \mathrm{Im}P_{\Theta} \to L_1^m[0,\infty)$. Second, viewing P_{Θ} as an operator from X into $\mathrm{Im}P_{\Theta}$, we have

$$R = P_{\Theta}\Gamma_{\Theta}|L_1^m[0,\infty): L_1^m[0,\infty) \to \mathrm{Im}P_{\Theta},$$

and, similarly,

$$R^{\times} = -P_{\Theta}^{\times}\Gamma_{\Theta}^{\times}|L_1^m[0,\infty): L_1^m[0,\infty) \to \mathrm{Im}P_{\Theta}^{\times}.$$

Proving Theorem 2.1 that is checking the coupling relation (2.1), amounts to verifying eight identities. Pairwise these identities have analogous proofs. So actually only four identities have to be taken care of. These will be dealt with below.

First we shall prove that

$$(2.2)\qquad R(I-K^{\times}) + JR^{\times} = 0.$$

Take ϕ in $L_1^m[0,\infty)$. We need to show that $RK^{\times}\phi = P_{\Theta}R^{\times}\phi + R\phi$. Whenever this is convenient, it may be assumed that ϕ is a continuous function with compact support in $(0,\infty)$.

Applying Fubini's theorem, one gets

$$RK^{\times}\phi = \int_0^\infty\Big(\int_0^\infty E(-t;-iA)Bk_\Theta^{\times}(t-s)\phi(s)ds\Big)dt$$
$$= \int_0^\infty\Big(\int_0^\infty E(-t;-iA)Bk_\Theta^{\times}(t-s)\phi(s)dt\Big)ds.$$

For $s > 0$, consider the identity

$$(2.3)\qquad \int_0^\infty E(-t;-iA)B\Lambda_\Theta^{\times}x(t-s)dt = E(-s;-iA)x - P_\Theta E(-s;-iA^{\times})x.$$

To begin with, take $x \in \mathcal{D}(A) = \mathcal{D}(A^{\times})$. Then, for $t \neq 0,s$,

$$\frac{d}{dt}\Big(E(-t;-iA)E(t-s;-iA^{\times})x\Big) = iE(-t;-iA)BCE(t-s;-iA^{\times})x$$
$$= iE(-t;-iA)BC(A^{\times})^{-1}E(t-s;-iA^{\times})A^{\times}x.$$

Because $C(A^{\times})^{-1}$ is bounded, the last expression is a continuous function of t on the intervals $[0,s]$ and $[s,\infty)$. It follows that (2.3) holds for $x \in \mathcal{D}(A)$. The validity of (2.3) for arbitrary $x \in X$ can now be obtained by a standard approximation argument based on the fact that $\mathcal{D}(A)$ is dense in X and the continuity of the operators involved. Substituting (2.3) in the expression for $RK^{\times}\phi$, one immediately gets (2.2).

Next we deal with the identity

$$(2.4)\qquad RU^{\times} + JJ^{\times} = I_{\mathrm{Im}P_\Theta}.$$

Take x in $\mathrm{Im}P_\Theta$. Then

$$(2.5)\qquad RU^{\times}x = -\int_0^\infty E(-t;-iA)B\Lambda_\Theta^{\times}x(t)dt.$$

Apart from the minus sign, the right hand side of (2.5) is exactly the same as the left hand side of (2.3) for $s = 0$. It is easy to check that (2.3) also holds for $s = 0$, provided that the right hand side is interpreted as $-P_\Theta x + P_\Theta P_\Theta^{\times}x$. Thus $RU^{\times}x = P_\Theta^{\times}x - P_\Theta P_\Theta^{\times}x = x - P_\Theta P_\Theta^{\times}x$, and (2.4) is proved.

In the third place, we shall establish

$$(2.6)\qquad (I-K)U^{\times} + UJ^{\times} = 0.$$

Take $x \in \operatorname{Im}P_\Theta$. Then $U^\times x = -Q\Lambda_\Theta^\times x$, where Q is the projection of $L_1^m(\mathbb{R})$ onto $L_1^m[0,\infty)$ along $L_1^m(-\infty,0]$. Here the latter two spaces are considered as subspaces of $L_1^m(\mathbb{R})$. Observe now that $Q\Lambda_\Theta^\times x = \Lambda_\Theta^\times(I-P_\Theta^\times)x$. For $x \in \mathcal{D}(A) = \mathcal{D}(A^\times)$ this is evident, and for arbitrary x one can use an approximation argument. Hence $KU^\times x = Qh$, where $-h = k_\Theta * \Lambda_\Theta^\times(I-P_\Theta^\times)x$ is the (full line) convolution product of k_Θ and $\Lambda_\Theta^\times(I-P_\Theta^\times)x$. Taking Fourier transforms, one gets

$$\begin{aligned}\hat{h}(\lambda) &= C(\lambda-A)^{-1}BC(\lambda-A^\times)^{-1}(I-P_\Theta^\times)x\\ &= C(\lambda-A)^{-1}(I-P_\Theta^\times)x - C(\lambda-A^\times)^{-1}(I-P_\Theta^\times)x.\end{aligned}$$

Put $g = U^\times x + UP_\Theta^\times x$. Then $g = Qg$. Also $g = -\Lambda_\Theta^\times(I-P_\Theta^\times)x + \Lambda_\Theta(I-P_\Theta)P_\Theta^\times x$, and hence

$$\hat{g}(\lambda) = -C(\lambda-A^\times)^{-1}(I-P_\Theta^\times)x - C(\lambda-A)^{-1}(I-P_\Theta)P_\Theta^\times x.$$

Since $x \in \operatorname{Im}P_\Theta$, it follows that

$$\hat{h}(\lambda) - \hat{g}(\lambda) = C(\lambda-A)^{-1}P_\Theta(I-P_\Theta^\times)x.$$

So $\hat{h}(\lambda)-\hat{g}(\lambda)$ is the Fourier transform of $-\Lambda_\Theta P_\Theta(I-P_\Theta^\times)x$. But then $h-g = -\Lambda_\Theta P_\Theta(I-P_\Theta^\times)x = -(I-Q)\Lambda_\Theta(I-P_\Theta^\times)x$. Applying Q, we now get $Qh = Qg = g$. In other words, $KU^\times x = U^\times x + UP_\Theta^\times x$ for all $x \in X$, which is nothing else than (2.6).

Finally, we prove

$$(2.7)\qquad (I-K)(I-K^\times) + UR^\times = I.$$

Let L be the (full line) convolution integral operator associated with Θ, featuring in Theorem 1.2. Since Θ and $\Theta^\times$ are both realization triples, the operator I-L is invertible and $(I-L)^{-1} = I-L^\times$, where $L^\times$ is the convolution integral operator associated with $\Theta^\times$. With respect to the decomposition $L_1^m(\mathbb{R}) = L_1^m[0,\infty) \oplus L_1^m(-\infty,0]$, we write I-L and its inverse in the form

$$I-L = \begin{pmatrix} I-K & L_- \\ * & * \end{pmatrix}, \qquad I-L^\times = \begin{pmatrix} I-K^\times & * \\ L_+^\times & * \end{pmatrix}.$$

Clearly $(I-K)(I-K^\times) + L_-L_+^\times = I$. So, in order to prove (2.7), it suffices to show that $L_-L_+^\times = UR^\times$.

Suppose, for the time being, that

$$(2.8) \qquad L_-\phi_- = -Q\Lambda_\Theta\Gamma_\Theta\phi_-, \qquad \phi_- \in L_1^m(-\infty,0],$$

$$(2.9) \qquad L_+^\times\phi_+ = (I-Q)\Lambda_\Theta^\times\Gamma_\Theta^\times\phi_+, \qquad \phi_+ \in L_1^m[0,\infty).$$

As was observed in the fourth paragraph of the present proof, (2.3) also holds for $s = 0$, that is

$$\int_0^\infty E(-t;-iA)B\Lambda_\Theta^\times x(t)dt = P_\Theta(I-P_\Theta^\times)x.$$

Analogously, one has

$$\int_{-\infty}^0 E(-t;-iA)B\Lambda_\Theta^\times x(t)dt = (I-P_\Theta)P_\Theta^\times x.$$

Hence

$$\begin{aligned} L_-L_+^\times\phi_+ &= -Q\Lambda_\Theta\Gamma_\Theta(I-Q)\Lambda_\Theta^\times\Gamma_\Theta^\times\phi_+ \\ &= -Q\Lambda_\Theta(I-P_\Theta)P_\Theta^\times\Gamma_\Theta^\times\phi_+ \\ &= UR^\times\phi_+. \end{aligned}$$

It remains to verify (2.8) and (2.9).

Let us first prove (2.8) for the case when $\operatorname{Im}B \subset \mathcal{D}(A)$. Then we can write B in the form $B = A^{-1}B_1$, where $B_1\colon \mathbb{C}^m \to X$ is a (bounded) linear operator. Write $C_1 = CA^{-1}$. Then $C_1\colon X \to \mathbb{C}^m$ is a bounded linear operator too. Also, for each $y \in \mathbb{C}^m$,

$$k_\Theta(t)y = iCE(t;-iA)By = iC_1E(t;-iA)B_1y, \qquad \text{a.e. on } \mathbb{R}.$$

Since $\mathbb{C}^m$ is finite dimensional, we may assume that

$$k_\Theta(t) = iC_1E(t;-iA)B_1, \quad t \neq 0.$$

Take $\phi_- \in L_1^m(-\infty,0]$. Then $L_-\phi_- \in L_1^m[0,\infty)$, and almost everywhere on $[0,\infty)$

$$L_-\phi_-(t) = -\int_{-\infty}^{0} k_\Theta(t-s)\phi_-(s)ds$$

$$= -\int_{-\infty}^{0} iC_1E(t-s;-iA)B_1\phi_-(s)ds.$$

Next, use the semigroup properties of the bisemigroup $E(.;-iA)$ mentioned in Section I.1. It follows that almost everywhere on $[0,\infty)$

$$L_-\phi_-(t) = -iC_1E(t;-iA)\int_{-\infty}^{0} E(-s;-iA)B_1\phi_-(s)ds$$

$$= -iCE(t;-iA)\int_{-\infty}^{0} E(-s;-iA)B\phi_-(s)ds.$$

Thus $L_-\phi_- = QL_-\phi_- = -Q\Lambda_\Theta\Gamma_\Theta\phi_-$.

We have proved (2.8) now for the case when $\operatorname{Im}B \subset \mathcal{D}(A)$. The general situation, where $\operatorname{Im}B$ need not be contained in $\mathcal{D}(A)$ can be treated with an approximation argument based on the fact that B can be approximated (in norm) by (bounded) linear operators from $\mathbb{C}^m$ into X having their range inside $\mathcal{D}(A)$. This is true because $\mathcal{D}(A)$ is dense in X and $\mathbb{C}^m$ is finite dimensional. The proof of (2.9) is similar. □

II.3. Inversion and Fredholm properties

In this section we study inversion and Fredholm properties of the Wiener-Hopf integral operator K,

$$(3.1) \qquad K\phi(t) = \int_{0}^{\infty} k(t-s)\phi(s)ds, \quad \text{a.e. on } [0,\infty).$$

It will be assumed that the m×m matrix kernel k admits a spectral exponential representation. This implies that $k \in L_1^m(\mathbb{R})$, and so K is a well-defined bounded linear operator on $L_1^m[0,\infty)$.

THEOREM 3.1. *Assume the kernel* k *of the integral operator* K *given by* (3.1) *is the kernel associated with the realization triple* $\Theta = (A,B,C)$, *i.e.,* $k = k_\Theta$. *Then* $I-K$ *is a Fredholm operator if and only if* $\Theta^\times = (A^\times,B,-C)$ *is a realization triple, and in that case the following statements hold true:*

(i) $\mathrm{Im}P_\Theta \cap \mathrm{Ker}P_\Theta^\times$ *is finite dimensional,* $\mathrm{Im}P_\Theta + \mathrm{Ker}P_\Theta^\times$ *is closed with finite codimension in the (common) state space of* Θ *and* $\Theta^\times$, *and*

$$\mathrm{ind}(I-K) = \dim(\mathrm{Im}P_\Theta \cap \mathrm{Ker}P_\Theta^\times) - \mathrm{codim}(\mathrm{Im}P_\Theta + \mathrm{Ker}P_\Theta^\times),$$

(ii) *a function* ϕ *belongs to* $\mathrm{Ker}(I-K)$ *if and only if there exists a (unique)* $x \in \mathrm{Im}P_\Theta \cap \mathrm{Ker}P_\Theta^\times$ *such that*

$$\phi(t) = \Lambda_\Theta x(t) \quad \textit{a.e. on } [0,\infty),$$

(iii) $\dim \mathrm{Ker}(I-K) = \dim(\mathrm{Im}P_\Theta \cap \mathrm{Ker}P_\Theta^\times)$,

(iv) *a function* ψ *in* $L_1^m[0,\infty)$ *belongs to* $\mathrm{Im}(I-K)$ *if and only if*

$$\int_0^\infty E(-t;-iA^\times)B\psi(t)dt \in \mathrm{Im}P_\Theta + \mathrm{Ker}P_\Theta^\times,$$

(v) $\mathrm{codim}\ \mathrm{Im}(I-K) = \mathrm{codim}(\mathrm{Im}P_\Theta + \mathrm{Ker}P_\Theta^\times)$.

The condition that $\Theta^\times = (A^\times,B,-C)$ is a realization triple may be replaced by any of the equivalent conditions in Theorem 5.1 of Ch.I or by the condition that $\det\big(I-\hat{k}(\lambda)\big)$ does not vanish on the real line.

PROOF. The proof of the if part, including that of (i) - (v), amounts to combining Theorem 2.1 of the previous section, Theorem 5.3 in Ch.I and the results obtained in [BGK3], Section I.2. For details, see [BGK5], first part of the proof of Theorem 2.1.

To prove the only if part of the theorem, one may reason as follows. From [GK] it is known that I-K is Fredholm if and only if $\det(I-\hat{k}(\lambda))$ does not vanish on the real line. By Theorems 3.1 and 5.1 in Ch.I, the latter condition is equivalent to the requirement that $\Theta^{\times}$ is a realization triple. It is also possible to use a perturbation argument as in [BGK5], second part of the proof of Theorem 2.1. □

Next we consider the special case when I-K is invertible.

THEOREM 3.2. _Assume the kernel_ k _of the integral operator_ K _defined by_ (3.1) _is the kernel associated with the realization triple_ $\Theta = (A,B,C)$, _i.e.,_ $k = k_{\Theta}$. _Then_ I-K _is invertible if and only if the folowing two conditions are satisfied_:

(1) $\Theta^{\times} = (A^{\times},B,-C)$ _is a realization triple,_

(2) $X = \operatorname{Im}P_{\Theta} \oplus \ker P_{\Theta}^{\times}.$

Here X _is the_ (_common_) _state space of_ Θ _and_ $\Theta^{\times}$. _If_ (1) _and_ (2) _hold, then the inverse of_ I-K _is given by_

$$(I-K)^{-1}\psi t) = \psi(t) - \int_0^\infty k_{\Theta}^{\times}(t-s)\psi(s)ds - [\int_0^\infty \Lambda_{\Theta}^{\times}(I-\Pi)E(-s;-iA^{\times})B\psi(s)ds](t)$$

(_a.e. on_ $[0,\infty)$), _where_ Π _is the projection of_ X _onto_ $\operatorname{Ker}P_{\Theta}^{\times}$ _along_ $\operatorname{Im}P_{\Theta}$.

A somewhat different expression for the inverse of I-K will be given at the end of Section 4 below. Analogous results can be obtained for left inverses, right inverses and generalized inverves (cf. [BGK5], Theorem 2.2 and the discussion thereafter). Also here, the condition that $\Theta^{\times}$ is a realization triple may be replaced by any of the equivalent conditions in Theorem 5.1 of Ch.I or by the condition that $\det(I-\hat{k}(\lambda))$ does not vanish on the real line.

PROOF. The first part of the theorem is immediate from

Theorem 3.1. With regard to the second part, we argue as follows.

Suppose (1) and (2) are satisfied. Then I-K is invertible and by virtue of Theorem 2.1 in Ch.I of [BGK3], its inverse is given by

$$(I-K)^{-1} = I - K^{\times} - U^{\times}(J^{\times})^{-1}R^{\times}.$$

Here $K^{\times}$, $U^{\times}$ $J^{\times}$, and $R^{\times}$ are as in Theorem 2.1. So, for $\psi \in L_1^m[0,\infty)$,

$$(I-K)^{-1}\psi = \psi - Q(k_{\Theta}^{\times}*\psi) - Q\Lambda_{\Theta}^{\times}(J^{\times})^{-1}P_{\Theta}^{\times}\Gamma_{\Theta}^{\times}\psi,$$

where Q is the projection of $L_1^m(\mathbb{R})$ onto $L_1^m[0,\infty)$ along $L_1^m(-\infty,0]$. The desired result is now clear from the fact that $(J^{\times})^{-1}P_{\Theta}^{\times}$ is the projection of X onto $\operatorname{Im}P_{\Theta}$ along $\operatorname{Ker}P_{\Theta}^{\times}$. □

The projection Π appearing in Theorem 3.2 maps $\mathcal{D}(A)$ into itself. This fact is interesting in its own right and will also be used in the next section. Therefore we formally state it as a proposition.

PROPOSITION 3.3. <u>Suppose</u> $\Theta = (A,B,C)$ <u>and</u> $\Theta^{\times} = (A^{\times},B,-C)$ <u>are realization triples, and assume in addition that</u> $X = \operatorname{Im}P_{\Theta} \oplus \operatorname{Ker}P_{\Theta}^{\times}$. <u>Here</u> X <u>is the</u> (<u>common</u>) <u>state space of</u> Θ <u>and</u> $\Theta^{\times}$. <u>Let</u> Π <u>be the projection of</u> X <u>onto</u> $\operatorname{Im}P_{\Theta}$ <u>along</u> $\operatorname{Ker}P_{\Theta}^{\times}$. <u>Then</u> Π <u>maps the domain of</u> A <u>into itself</u>.

PROOF. With the notation of Theorem 2.1, we have $\Pi = I - (J^{\times})^{-1}P^{\times}$. So it suffices to show that $(J^{\times})^{-1}$ maps $\mathcal{D}(A)\cap\operatorname{Im}P^{\times}$ into $\mathcal{D}(A)\cap\operatorname{Im}P$.

According to Theorem 2.1 in Ch.I of [BGK3] and Theorem 2.1 in the present paper, the inverse of $J^{\times}$ is given by $(J^{\times})^{-1} = J - R(I-K)^{-1}U$. Take $x \in \mathcal{D}(A)\cap\operatorname{Im}P^{\times}$. Then $Ux = Q\Lambda_{\Theta}x \in \mathcal{D}_1^m(\mathbb{R})\cap L_1^m[0,\infty)$, where Q is the projection of $L_1^m(\mathbb{R})$ onto $L_1^m[0,\infty)$ along $L_1^m(-\infty,0]$. From [GK] we know that $(I-K)^{-1}$ is the product of two operators of the type $Q(I+F)|L_1^m[0,\infty)$, where F is a convolution integral operator on $L_1^m(\mathbb{R})$ with an L_1-kernel (see also [BGK1], Section 4.5). Hence

$(I-K)^{-1}$ maps $\mathcal{D}_1^m(\mathbb{R})\cap L_1^m[0,\infty)$ into itself. Note in this context that if $k\in L_1^{m\times m}(\mathbb{R})$ and $f\in\mathcal{D}_1^m(\mathbb{R})$, then $k*f\in\mathcal{D}_1^m(\mathbb{R})$ too and $(k*f)' = k*f' + [f_+(0)-f_-(0)]k$. Thus $(I-K)^{-1}Ux\in\mathcal{D}_1^m(\mathbb{R})$ and Proposition 2.1 in Ch.I tells us that we end up in $\mathcal{D}(A)$ by applying Γ_Θ. We conclude that $R(I-K)^{-1}U$ maps $\mathcal{D}(A)\cap \operatorname{Im}P_\Theta^\times$ into $\mathcal{D}(A)\cap \operatorname{Im}P_\Theta$. The same is true for J and (consequently) for $(J^\times)^{-1}$. □

II.4. Canonical Wiener-Hopf factorization

We begin by recalling the definition of canonical Wiener-Hopf factorization. Let

$$W(\lambda) = I - \hat{k}(\lambda), \tag{4.1}$$

where $k\in L_1^{m\times m}(\mathbb{R})$. So $W(\lambda)$ belongs to the ($m\times m$ matrix) Wiener algebra with respect to the real line. A factorization

$$W(\lambda) = W_-(\lambda)W_+(\lambda),\qquad \lambda\in\mathbb{R} \tag{4.2}$$

will be called a <u>canonical Wiener-Hopf factorization</u> if the following conditions are satisfied:

(a) there exist $k_-\in L_1^{m\times m}(-\infty,0]$ and $k_+\in L_1^{m\times m}[0,\infty)$ such that $W_-(\lambda) = I-\hat{k}_-(\lambda)$ and $W_+(\lambda) = I-\hat{k}_+(\lambda)$,

(b) det $W_-(\lambda)$ does not vanish on the closed lower half plane $\operatorname{Im}\lambda\leqq 0$, and det $W_+(\lambda)$ does not vanish on the closed upper half plane $\operatorname{Im}\lambda\geqq 0$.

If (4.2) is a canonical Wiener-Hopf factorization, then $W_-(\lambda)$ is continuous on the closed lower half plane, analytic in the open lower half plane and

$$\lim_{\substack{\lambda\to\infty\\ \operatorname{Im}\lambda\leqq 0}} W_-(\lambda) = I.$$

Analogous observations can be made for $W_+(\lambda)$, with the understanding that the lower half plane has to be replaced by the upper half plane. By Wiener's theorem, there exist $k_-^\times \in L_1^{m\times m}(-\infty,0]$ and $k_+^\times \in L_1^{m\times m}[0,\infty)$ such that $W_-(\lambda)^{-1} = I - \hat{k}_-^\times(\lambda)$ and $W_+(\lambda)^{-1} = I - \hat{k}_+^\times(\lambda)$. Also canonical Wiener-Hopf factorizations as introduced above are unique (provided of course that they exist).

In this section we discuss canonical Wiener-Hopf factorization of m×m matrix functions $W(\lambda)$ of the form (4.1) under the additional assumption that $k \in L_{1,\omega}^{m\times m}(\mathbb{R})$ for some negative constant ω. In particular $W(\lambda)$ is analytic in a strip around the real axis (but not necessarily at ∞). Recall from Section I.4 that $L_{1,\omega}^{m\times m}(\mathbb{R})$ coincides with the class of all kernels of realization triples of exponential type ω.

THEOREM 4.1. *Consider the function* $W(\lambda) = I - \hat{k}(\lambda)$, *where* $k = k_\Theta$ *is the kernel associated with the realization triple* $\Theta = (A,B,C)$, *i.e.*,

$$W(\lambda) = I + C(\lambda-A)^{-1}B.$$

Then $W(\lambda)$ *admits a canonical Wiener-Hopf factorization if and only if*

(1) $\Theta^\times = (A^\times,B,-C)$ *is a realization triple*,

(2) $X = \operatorname{Im}P_\Theta \oplus \operatorname{Ker} P_\Theta^\times$.

Here X *is the (common) state space of* Θ *and* $\Theta^\times$. *Suppose* (1) *and* (2) *are satisfied, and let* Π *be the projection of* X *onto* $\operatorname{Ker}P_\Theta^\times$ *along* $\operatorname{Im}P_\Theta$. *Then the canonical Wiener-Hopf factorization of* $W(\lambda)$ *has the form* (4.2) *with*

$$W_-(\lambda) = I + C(\lambda-A)^{-1}(I-\Pi)B, \tag{4.3}$$

$$W_+(\lambda) = I + C\Pi(\lambda-A)^{-1}B, \tag{4.4}$$

$$W_-(\lambda)^{-1} = I - C(I-\Pi)(\lambda-A^\times)^{-1}B, \tag{4.5}$$

$$W_+(\lambda)^{-1} = I - C(\lambda-A^\times)^{-1}\Pi B. \tag{4.6}$$

As we have seen at the end of the previous section, the projection Π maps $\mathcal{D}(A)$ into $\mathcal{D}(A) \subset \mathcal{D}(C)$. It follows that the right hand sides of (4.4) and (4.5) are well-defined for the appropriate values of λ. This fact is obvious for the right hand sides of (4.3) and (4.6). Without proof we note that the operator $C\Pi$ featuring in (4.4) is A-bounded. Analogously $C(I-\Pi)$ appearing in (4.5) is $A^{\times}$-bounded.

PROOF. Suppose (1) and (2) are satisfied, and let Π be the projection of X onto $\mathrm{Ker}P_{\Theta}^{\times}$ along $\mathrm{Im}P_{\Theta}$. Then Π maps $\mathcal{D}(A) = \mathcal{D}(A^{\times})$ into itself. Hence, along with

$$(4.7) \qquad X = \mathrm{Im}P_{\Theta} \oplus \mathrm{Ker}P_{\Theta}^{\times},$$

we have the decomposition

$$(4.8) \qquad \mathcal{D}(A) = [\,\mathcal{D}(A) \cap \mathrm{Im}P_{\Theta}] \oplus [\,\mathcal{D}(A) \cap \mathrm{Ker}P_{\Theta}^{\times}].$$

This enables us to apply a generalized version (involving unbounded operators) of the Factorization Principle introduced and used in [BGKVD] and [BGK1]. In fact, the proof of the second part of the theorem is a straightforward modification of the proof of the first part of Theorem 1.5 in [BGK1]. The details are as follows.

With respect to the decomposition (4.7) and (4.8),we write

$$A = \begin{pmatrix} A_1 & A_0 \\ 0 & A_2 \end{pmatrix}, \quad B = \begin{pmatrix} B_1 \\ B_2 \end{pmatrix}, \quad C = \begin{pmatrix} C_1 & C_2 \end{pmatrix}.$$

Put $\Theta_1 = (A_1, B_1, C_1)$. Then Θ_1 is a realization triple. This is clear from the fact that $A_1(\mathrm{Im}P_{\Theta} \to \mathrm{Im}P_{\Theta})$ and $C_1(\mathrm{Im}P_{\Theta} \to \mathbb{C}^m)$ are the restrictions of A and C to $\mathcal{D}(A) \cap \mathrm{Im}P_{\Theta}$, respectively. Here $\mathbb{C}^m$ is of course the (common) input/output space of Θ and $\Theta^{\times}$. Note that iA_1 is the infinitesimal generator of a C_0-semigroup of negative exponential type. Hence the kernel k_1 associated with Θ_1 has its support in $(-\infty, 0]$ and

$$W_1(\lambda) = I - \hat{k}_1(\lambda) = I + C_1(\lambda-A_1)^{-1}B_1$$

is defined and analytic on an open half plane of the type $\mathrm{Im}\lambda < -\omega$ with ω strictly negative.

Next we consider $\Theta_1^\times = (A_1^\times, B_1, -C_1)$, the inverse of Θ_1. As in Sections II.2 and II.3, let $J^\times = P_\Theta^\times|\mathrm{Im}P_\Theta: \mathrm{Im}P_\Theta \to \mathrm{Im}P_\Theta^\times$. Then $J^\times$ is invertible and maps $\mathcal{D}(A)\cap\mathrm{Im}P_\Theta$ onto $\mathcal{D}(A)\cap\mathrm{Im}P_\Theta^\times$. It is easy to check that $J^\times$ provides a similarity between the operator $A_1^\times$ and the restriction of $A^\times$ to $\mathcal{D}(A^\times)\cap\mathrm{Im}P_\Theta^\times$. Hence $iA_1^\times$ is the infinitesimal generator of a C_0-semigroup of negative exponential type. But then Theorem 5.1 in Ch.I guarantees that $\Theta_1^\times$ is a realization triple. Further, the kernel $k_1^\times$ associated with $\Theta_1^\times$ has its support in $(-\infty,0]$ and

$$W_1(\lambda)^{-1} = I - \hat{k}_1^\times(\lambda) = I - C_1(\lambda-A_1^\times)^{-1}B_1$$

for all λ with $\mathrm{Im}\lambda < -\omega$. Here it is assumed that the negative constant ω has been taken sufficiently close to zero.

Put $\Theta_2 = (A_2,B_2,C_2)$ and $\Theta_2^\times = (A_2^\times,B_2,-C_2)$, where $A_2^\times = A_2-B_2C_2$. So Θ_2 and $\Theta_2^\times$ are each others inverse. Obviously $\Theta_2^\times$ is a realization triple, and a similarity argument of the type presented above yields that the same is true for Θ_2. The operators $-iA_2$ and $-iA_2^\times$ are infinitesimal generators of C_0-semigroups of negative exponential type. Hence the kernels k_2 and $k_2^\times$ associated with Θ_2 and $\Theta_2^\times$, respectively, have their support in $[0,\infty)$. Finally, taking $|\omega|$ smaller if necessary, we have that

$$W_2(\lambda) = I - \hat{k}_2(\lambda) = I + C_2(\lambda-A_2)^{-1}B_2$$

and

$$W_2(\lambda)^{-1} = I - \hat{k}_2^\times(\lambda) = I - C_2(\lambda-A_2^\times)^{-1}B_2$$

are defined and analytic on $\mathrm{Im}\lambda > -\omega$.

We may assume that both Θ and $\Theta^\times$ are of exponential type ω. For $|\mathrm{Im}\lambda| < -\omega$, one then has

$$W(\lambda) = I + C_1(\lambda - A_1)B_1^{-1} + C_2(\lambda - A_2)^{-1}B_2$$
$$+ C_1(\lambda - A_1)^{-1}A_0(\lambda - A_2)^{-1}B_2.$$

Now $\operatorname{Ker}P_{\Theta}^{\times}$ is an invariant subspace for

$$A^{\times} = \begin{pmatrix} A_1^{\times} & A_0 - B_1C_2 \\ -B_2C_1 & A_2^{\times} \end{pmatrix},$$

and so $A_0 = B_1C_2$. Substituting this in the above expression for $W(\lambda)$, we get $W(\lambda) = W_1(\lambda)W_2(\lambda)$. Clearly this is a canonical Wiener-Hopf factorization. One verifies without difficulty that $W_1(\lambda) = W_-(\lambda)$ and $W_2(\lambda) = W_+(\lambda)$, where $W_-(\lambda)$ and $W_+(\lambda)$ are as defined in the theorem.

This settles the second part of the theorem. In order to establish the first, we recall from [GK] that $W(\lambda)$ admits a canonical Wiener-Hopf factorization (if and) only if I-K is invertible, where K is as in Sections II.2 and II.3. The desired result is now clear from Theorem 3.2 above. □

Let us return to the second part of Theorem 4.1 and its proof. Adopting (or rather extending) the terminology and notation of Section 1.1 in [BGK1], we say that Θ_1 is the <u>projection of</u> Θ <u>associated with</u> $I-\Pi$ and write $\Theta_1 = pr_{I-\Pi}(\Theta)$. Similarly, $\Theta_2 = pr_{\Pi}(\Theta)$, $\Theta_1^{\times} = pr_{I-\Pi}(\Theta^{\times})$ and $\Theta_2^{\times} = pr_{\Pi}(\Theta^{\times})$. What we have got then is a canonical Wiener-Hopf factorization $W(\lambda) = W_-(\lambda)W_+(\lambda)$ with

$$W_-(\lambda) = I - \hat{k}_-(\lambda), \quad W_-(\lambda)^{-1} = I - \hat{k}_-^{\times}(\lambda),$$
$$W_+(\lambda) = I - \hat{k}_+(\lambda), \quad W_+(\lambda)^{-1} = I - \hat{k}_1^{\times}(\lambda),$$

where k_-, k_+, $k_-^{\times}$ and $k_+^{\times}$ are the kernels associated with $pr_{I-\Pi}(\Theta)$, $pr_{\Pi}(\Theta)$, $pr_{I-\Pi}(\Theta^{\times})$ and $pr_{\Pi}(\Theta^{\times})$, respectively. For $\omega < 0$ sufficiently close to zero, $k_- \in L_{1,\omega}^{m\times m}(-\infty,0]$, $k_+ \in L_{1,\omega}^{m\times m}[0,\infty)$, $k_-^{\times} \in L_{1,\omega}^{m\times m}(-\infty,0]$ and $k_+ \in L_{1,\omega}^{m\times m}[0,\infty)$.

Suppose conditions (1) and (2) of Theorem 4.1 are

satisfied, and let K be the Wiener-Hopf integral operator defined by (3.1) with $k = k_\Theta$. Then I-K is invertible by Theorem 3.2. For the inverse of I-K one can now write

$$(I-K)^{-1}\psi(t) = \psi(t) - \int_0^\infty \gamma(t,s)\psi(s)ds,$$

where (almost everywhere)

$$\gamma(t,s) = k_+^\times(t-s) + k_-^\times(t-s) - \int_0^{\min(t,s)} k_+^\times(t-r)k_-^\times(r-s)dr$$

and $k_-^\times, k_+^\times$ are as above (see [GK]; cf. also [BGK1], Section 4.5).

II.5. The Riemann-Hilbert boundary value problem

In this section we deal with the Riemann-Hilbert boundary value problem (on the real line)

$$(5.1) \qquad W(\lambda)\Phi_+(\lambda) = \Phi_-(\lambda), \quad -\infty < \lambda < \infty,$$

the precise formulation of which reads as follows: Given an $m\times m$ matrix function $W(\lambda)$, $-\infty < \lambda < \infty$, with continuous entries, describe all pairs Φ_+, Φ_- of $\mathbb{C}^m$-valued functions such that (5.1) is satisfied while, in addition, Φ_+ and Φ_- are the Fourier transforms of integrable $\mathbb{C}^m$-valued functions with support in $[0,\infty)$ and $(-\infty,0]$, respectively. For such a pair of functions, we have that Φ_+ (resp. Φ_-) is continuous on the closed upper (resp. lower) half plane, analytic in the open upper (resp. lower) half plane and vanishes at infinity.

The functions $W(\lambda)$ that we shall deal with are of the type considered in the previous section. So $W(\lambda)$ is of the form (4.1) with $k \in L_{1,\omega}^{m\times m}(\mathbb{R})$ for some negative constant ω. In particular $W(\lambda)$ is analytic in a strip around the real axis.

THEOREM 5.1. <u>Consider the function</u> $W(\lambda) = I - \hat{k}(\lambda)$, <u>where</u> $k = k_\Theta$ <u>is the kernel associated with the realization triple</u> $\Theta = (A,B,C)$, <u>i.e.,</u>

$$W(\lambda) = I + C(\lambda-A)^{-1}B.$$

Assume $\Theta^{\times} = (A^{\times}, B, -C)$ _is a realization triple too_ (_or equivalently_, $\det W(\lambda) \neq 0$ _for all_ $\lambda \in \mathbb{R}$). _Then the pair of functions_ Φ_+, Φ_- _is a solution of the Riemann-Hilbert boundary value problem_ (5.1) _if and only if there exists a_ (_unique_) _vector_ x _in_ $\operatorname{Im} P_\Theta \cap \operatorname{Ker} P_\Theta^{\times}$ _such that_

$$\Phi_+(\lambda) = C(\lambda - A^{\times})^{-1}x = \int_0^{\infty} e^{i\lambda t}\Lambda_\Theta^{\times}x(t)dt,$$

$$\Phi_-(\lambda) = C(\lambda - A)^{-1}x = -\int_{-\infty}^{0} e^{i\lambda t}\Lambda_\Theta x(t)dt.$$

PROOF. The proof is analogous to that of Theorem 5.1 in [BGK5]. For the if part, use formula (3.3) in Ch.I; for the only if part employ Theorem 3.1. □

REFERENCES

[BGK1] Bart. H., Gohberg, I. and Kaashoek, M.A.: _Minimal factorization of matrix and operator functions_. Operator Theory: Advances and Applications, Vol. 1, Birkhäuser Verlag, Basel etc. 1979.

[BGK2] Bart, H., Gohberg, I. and Kaashoek, M.A.: _Wiener-Hopf integral equations, Toeplitz matrices and linear systems_. In: Toeplitz Centennial, Operator Theory: Advances and Applications, Vol. 4 (Ed. I. Gohberg), Birkhäuser Verlag, Basel etc., 1982, pp. 85-135.

[BGK3] Bart, H., Gohberg, I. and Kaashoek, M.A.: _The coupling method for solving integral equations_. In: Topics in Operator Theory, Systems and Network, The Rehovot Workshop, Operator Theory: Advances and Applications, Vol. 12 (Eds. H. Dym, I. Gohberg), Birkhäuser Verlag, Basel etc., 1984, pp. 39-73.

[BGK4] Bart, H., Gohberg, I. and Kaashoek, M.A.: _Exponentially dichotomous operators and inverse Fourier transforms_. Report 8511/M, Econometric Institute, Erasmus University Rotterdam, The Netherlands, 1985.

[BGK5] Bart, H., Gohberg, I. and Kaashoek, M.A.: _Fredholm theory of Wiener-Hopf equations in terms of realization of their symbols. Integral Equations and Operator Theory 8, 590-613 (1985).

[BGK6] Bart, H., Gohberg, I. and Kaashoek, M.A.: Wiener-Hopf factorization, inverse Fourier transforms and exponentially dichotomous operators. J. Functional Analysis, 1986 (to appear).

[BGKVD] Bart, H., Gohberg, I., Kaashoek, M.A. and Van Dooren, P.: Factorizations of transfer functions. SIAM J. Control Optim. 18, 675-696 (1980).

[BK] Bart, H. and Kroon, L.G., An indicator for Wiener-Hopf integral equations with invertible analytic symbol. Integral Equations and Operator Theory 6, 1-20 (1983). See also the addendum to this paper: Integral Equations and Operator Theory 6, 903-904 (1983).

[CG] Curtain, R.F. and Glover, K: Balanced realisations for infinite dimensional systems. In: Operator Theory and Systems, Proceedings Workshop Amsterdam, June 1985, Operator Theory: Advances and Applications, Vol.19(Eds. H. Bart, I. Gohberg, M.A. Kaashoek), Birkhäuser Verlag, Basel etc., 1986, pp. 86-103.

[F] Fuhrmann, P.A.: Linear systems and operator theory in Hilbert space. McGraw-Hill, New York, 1981.

[GRS] Gelfand, I., Raikov, D. and Shilov, G.: Commutative normed rings. Chelsea Publishing Company, Bronx, New York, 1964.

[GK] Gohberg, I. and Krein, M.G.: Systems of integral equations on a half line with kernels depending on the difference of argument. Uspehi Mat. Nauk 13, no.2 (80), 3-72 (1958). Translated as: Amer. Math. Soc. Trans. (2) 14, 217-287 (1960).

[HP] Hille, E. and Phillips, R.S.: Functional analysis, and semigroups. Amer. Math. Soc., Providence R.I., 1957.

[Ka] Kailath, T.: Linear systems. Prentice Hall Inc., Englewood Cliffs N.J., 1980.

[K] Kalman, R.E.: Mathematical description of linear dynamical systems., SIAM J. Control 1, 152-192 (1963).

[KFA] Kalman, R.E., Falb, P. and Arbib, M.A.: Topics in

mathematical system theory. McGraw-Hill, New York etc., 1960.

[P] Pazy, A.: _Semigroups of linear operators and applications to partial differential equations_. Applied Mathematical Sciences, Vol. 44, Springer-Verlag, New York, etc. 1983.

H. Bart
Econometric Institute
Erasmus Universiteit
Postbus 1738
3000 DR Rotterdam
The Netherlands

I. Gohberg
Dept. of Mathematical Sciences
The Raymond and Beverly Sackler
Faculty of Exact Sciences
Tel-Aviv University
Ramat-Aviv. Israel

M.A. Kaashoek
Subfaculteit Wiskunde en
Informatica
Vrije Universiteit
Postbus 7161
1007 MC Amsterdam
The Netherlands

Operator Theory:
Advances and Applications, Vol. 21

ON TOEPLITZ AND WIENER-HOPF OPERATORS WITH CONTOURWISE RATIONAL MATRIX AND OPERATOR SYMBOLS

I. Gohberg, M.A. Kaashoek, L.Lerer, L. Rodman*)

Dedicated to the memory of David Milman

Explicit formulas for the (generalized) inverse and criteria of invertibility are given for block Toeplitz and Wiener-Hopf type operators. We consider operators with symbols defined on a curve composed of several non-intersecting simple closed contours. Also criteria and explicit formulas for canonical factorization of matrix functions relative to a compound contour are presented. The matrix functions we work with are rational on each of the compounding contours but the rational expressions may vary from contour to contour. We use realizations for each of the rational expressions and the final results are stated in terms of invertibility properties of a certain finite matrix called indicator, which is built from the realizations. The analysis does not depend on finite dimensionality and is carried out for operator valued symbols.

TABLE OF CONTENTS

*) The work of this author partially supported by the Fund for Basic Research administrated by the Israel Academy for Sciences and Humanities.

INTRODUCTION

The main part of this paper concerns Toeplitz operators of which the symbol W is an $m \times m$ matrix function defined on a disconnected curve Γ. The curve Γ is assumed to be the union of $s+1$ nonintersecting simple smooth closed contours $\Gamma_0, \Gamma_1, \dots, \Gamma_s$ which form the positively oriented boundary of a finitely connected bounded domain in $\mathbb{C}$. Our main requirement on the symbol W is that on each contour Γ_j the function W is the restriction of a rational matrix function W_j which does not have poles and zeros on Γ_j and at infinity. Using the realization theorem from system theory (see, e.g., [1], Chapter 2) the rational matrix function W_j (which differs from contour to contour) may be written in the form

$$(0.1) \qquad W_j(\lambda) = I + C_j(\lambda - A_j)^{-1}B_j \quad , \quad \lambda \in \Gamma_j \quad ,$$

where A_j is a square matrix of size $n_j \times n_j$, say, B_j and C_j are matrices of sizes $n_j \times m$ and $m \times n_j$, respectively, and the matrices A_j and $A_j^\times = A_j - B_jC_j$ have no eigenvalues on Γ_j. (In (0.1) the functions W_j are normalized to I at infinity.) Our aim is to get the inverse of the Toeplitz operator with symbol W and the canonical factorization of W in terms of the matrices A_j, B_j, C_j $(j = 0,1,\dots,s)$. When the rational matrix function W_j do not depend on j, our results contain those of [1], Sections 1.2 and 4.4 and of [3], Section III.2. The case when $W_0, W_1, \dots, W_s$ are (possibly different) matrix polynomials has been treated in [9].

In achieving our goal an important role is played by the (left) indicator, an $(s+1) \times (s+1)$ block matrix $S = [S_{jk}]_{j,k=0}^{s}$ of which the entries are defined as follows. For $j = 0,1,\dots,s$ let P_j (resp., $P_j^\times$) be the Riesz projection corresponding to the eigenvalues of A_j (resp., $A_j^\times$) in the outer domain of Γ_j and for $j,k = 0,1,\dots,s$ define $S_{jk} : \operatorname{Im} P_k^\times \to \operatorname{Im} P_j$ by setting

$$S_{jj}x = P_jx \qquad (x \in \operatorname{Im} P_j^\times) \ ;$$

$$S_{jk}x = -\frac{1}{2\pi i}\int_{\Gamma_j} (\lambda - A_j)^{-1}B_jC_k(\lambda - A_k^\times)^{-1}x \, d\lambda \ (x \in \operatorname{Im} P_k^\times \ , \ j \neq k).$$

Note that for $x \in \operatorname{Im} P_j^\times$ the term $(\lambda - A_j^\times)^{-1}$ makes sense for each λ in the inner domain of Γ_k, and hence S_{jk} is well-defined.

The Toeplitz operator we study is defined in the following way. Let $\mathcal{B}$ be the Banach space of all $\mathbb{C}^m$-valued functions on Γ that are Hölder continuous with a fixed exponent $\alpha \in (0,1)$, and let $P_\Gamma = \frac{1}{2}(I + S_\Gamma)$, where $S_\Gamma : \mathcal{B} \to \mathcal{B}$ is the operator of singular integration on Γ. The Toeplitz operator T with symbol W on Γ is now defined on $\mathcal{B}^+ = P_\Gamma(\mathcal{B})$ by setting

$$T\varphi = P_\Gamma(W(\cdot)\varphi(\cdot)) \ ,\varphi \in \mathcal{B}^+ .$$

The following is a sample of our main results. We prove that the Toeplitz operator T with symbol W is invertible if and only if the indicator S corresponding to W has a nonzero determinant and we express the inverse of T in terms of S^{-1} and the matrices A_j, B_j, C_j $(j = 0, 1,\ldots,s)$. This result is a by-product of a more general theorem which gives explicit formulas for a generalized inverse of T and for the kernel and image of T. Also through the indicator we obtain criteria and explicit formulas for canonical factorization of the symbol. Our results about the canonical factorization are also valid for unbounded domains, which allows us to obtain inversion formulas for Wiener-Hopf pair equations and equations with two kernels.

This paper consists of nine sections. In Section 1 we introduce the left indicator and also its dual version. The main results on Toeplitz operators are stated in Section 2 and proved in Section 3. In the fourth section we give applications to the barrier problem. Canonical factorization is treated in Section 5 for bounded domains and in Section 6 for unbounded domains. Sections 7 and 8 contain the applications to operators of Wiener-Hopf type. The discrete case is discussed in the last section.

We remark that our analysis does not depend on finite dimensionality and is carried out for operator-valued symbols. So in the representation (0.1) we allow the operators A_j, B_j and C_j to act between infinite dimensional spaces.

In this paper the following notation is used. Given Banach space operators $K_i : X_1 \to X_i$, $i = 1,\ldots,m$, we denote by $\mathrm{diag}[K_i]_{i=1}^m$ or $K_1 \oplus\ldots\oplus K_m$ the operator K acting on the direct sum $X_1 \oplus X_2 \oplus\ldots\oplus X_m$ as follows: $Kx = K_i x$, $x \in X_i$. If $K_i : X_i \to Y$, $i = 1, \ldots ,m$ are operators, the notation row $[K_i]_{i=1}^m$ stands for the naturally defined operator $[K_1K_2 \dots K_m]: X_1\oplus\dots\oplus X_m \to Y$. Analogously, for operators $K_i : Y \to X_i$ we denote

$$\mathrm{col}[K_i]_{i=1}^m = \begin{bmatrix} K_1 \\ K_2 \\ \cdot \\ \cdot \\ \cdot \\ K_m \end{bmatrix} : Y \to X_1 \oplus X_2 \oplus\dots\oplus X_m.$$

The notation $\mathrm{col}[x_i]_{i=1}^m$ is also used to designate vectors $x \in X_1\oplus\dots\oplus X_m$ whose i-th coordinate is x_i, $i = 1,\dots,m$. The identity operator on the Banach space X is denoted I_X.

1. Indicator

Let $\Gamma_0,\dots,\Gamma_s$ be a system of simple closed rectifiable non-intersecting smooth contours in the complex plane which form the positively oriented boundary Γ of a finitely connected bounded open set Ω^+. Denote by $\Omega_0^-,\Omega_1^-,\dots,\Omega_s^-$ the outer domains of the curves $\Gamma_0, \Gamma_1,\dots,\Gamma_s$ respectively. We assume that $\Omega_1^-,\dots,\Omega_s^-$ are in the inner domain of Γ_0 and Ω_0^- is unbounded. Put $\Omega^- = \bigcup_{j=0}^{s} \Omega_j^-$. The notation $L(Y)$ is used for the Banach algebra of all bounded linear operators acting on a (complex) Banach space Y.

Consider an $L(Y)$-valued function $W(\lambda)$ on Γ which on each $\Gamma_j (j=0,\dots,s)$ admits the representation

$$W(\lambda) = W_j(\lambda) = I + C_j(\lambda - A_j)^{-1}B_j,\ \lambda \in \Gamma_j, \tag{1.1}$$

where $C_j:X_j\to Y$, $A_j:X_j\to X_j$, $B_j : Y \to X_j$ are (bounded linear)operators and X_j is a Banach space.

The representation of the form (1.1) is called a <u>realization</u> of W_j. An important particular case appears when $\dim Y < \infty$ and the functions $W_0, W_1,\dots, W_s$ are rational with value I at infinity. It is well-known that for such functions realizations (1.1) exist always with finite dimensional X_j, $j = 0,1,\dots,s$. More generally (see Chapter 2 in [1]), a realization (1.1) exists if and only if the $L(Y)$-valued function W_j is analytic in a neighborhood of Γ_j.

Given $X(\lambda)$ realized as in (1.1), assume that the operators A_j and $A_j^{\times} \overset{\text{def}}{=} A_j - B_jC_j$ have no spectrum on Γ_j. Note that this implies (two-sided bounded) invertibility of $W(\lambda)$ for each $\lambda \in \Gamma$. In fact, there is a realization

$$W(\lambda)^{-1} = W_j(\lambda)^{-1} = I - C_j(\lambda - A_j^{\times})^{-1}B_j \ , \ \lambda \in \Gamma_j .$$

Conversely, if $\dim Y < \infty$ and the rational matrix function $W_i(\lambda)$ together with its inverse $W_i(\lambda)^{-1}$ has no poles on Γ_i $(i = 0,\dots,s)$, then one can take the realization (1.1) for $W_0,\dots,W_s$ to be minimal (see, e.g., [1], Chapter 2) which ensures that $\sigma(A_j)\cap\Gamma_j = \sigma(A_j^{\times})\cap\Gamma_j = \emptyset$, $j = 0,\dots,s$.

We introduce now the notion of indicator for the realizations (1.1) of the function $W(\lambda)$. By definition, P_j is the spectral projection of A_j corresponding to the part of $\sigma(A_j)$ in the domain Ω_j^- $(j = 0,\dots,s)$. Similarly, $P_j^{\times}$ is the spectral projection of $A_j^{\times}$ corresponding to the part of $\sigma(A_j^{\times})$ in Ω_j^-. Introduce the spaces

$$Z = \overset{s}{\underset{j=0}{\oplus}} \operatorname{Im} P_j \ , \ Z^{\times} = \overset{s}{\underset{j=0}{\oplus}} \operatorname{Im} P_j^{\times} .$$

Observe that in these direct sum representations of the spaces Z and $Z^{\times}$ some of the summands $\operatorname{Im} P_j$ and $\operatorname{Im} P_j^{\times}$ may be zeros. However, this does not affect the formalism that follows.

Consider the operator matrix $R = [R_{ij}]_{i,j=0}^{s} : Z \to Z^{\times}$,

where

$$R_{jj}\, x = P_j^{\times} x \qquad (x \in \operatorname{Im} P_j) \ ,$$

$$R_{kj}\, x = \frac{1}{2\pi i}\int_{\Gamma_k} (\lambda - A_k^{\times})^{-1}B_k\, C_j(\lambda - A_j)^{-1}\, x\, d\lambda$$

$$(x \in \operatorname{Im} P_j \ , \ k \neq j).$$

Note that for $x \in \operatorname{Im} P_j$ the term $(\lambda - A_j)^{-1}x$ is well-defined for each $\lambda \notin \Omega_j^-$. Since for $k \neq j$ the curve Γ_k is outside Ω_j^-, it is clear that for $x \in \operatorname{Im} P_j$ the function

$$\psi(\lambda) = (\lambda - A_k^{\times})^{-1}\,(I - P_k^{\times})B_k\, C_j(\lambda - A_j)^{-1}x$$

is analytic in Ω_k^- (and has a zero of order ≥ 2 at ∞ if $k = 0$), and thus $\operatorname{Im} R_{kj} \subset \operatorname{Im} P_k^{\times}$ for $k \neq j$. We shall refer to R as the right indicator of the function $W(\lambda)$ relative to the realization (1.1) (and to the multiple contour Γ). We emphasize that the right indicator depends not only on $W(\lambda)$ and Γ but also on the realizations (1.1) for $W(\lambda)$. If $(A_j, B_j, C_j) = (A,B,C)$ for each j, then $R_{kj}\, x = P_k^{\times} x$ for $x \in \operatorname{Im} P_j$ and R is the usual indicator (see [3]).

The notion of left indicator is introduced analogously. Namely, the left indicator of $W(\lambda)$ relative to realizations (1.1) is the operator $S = [S_{ij}]_{i,j=0}^{s} : Z^{\times} \to Z$,

where

$$S_{jj}\, x = P_j\, x \quad (x \in \operatorname{Im} P_j^{\times})\ ;$$

$$S_{jk}\, x = \frac{-1}{2\pi i} \int_{\Gamma_j} (\lambda - A_j)^{-1} B_j C_k (\lambda - A_k^{\times})^{-1}\, x\, d\lambda \qquad (x \in \operatorname{Im} P_k^{\times}\ ,\ j \neq k).$$

Again, one checks easily that S_{jk} maps $\operatorname{Im} P_k^{\times}$ into $\operatorname{Im} P_j$, so S is well-defined.

We shall see in section 5 that invertibility of the right (left) indicator is equivalent to existence of right (left) canonical factorization of $W(\lambda)$. Here we mention only that the indicators satisfy certain Lyapunov equations. In fact,

$$SM^{\times} - MS = -BC^{\times}\ , \tag{1.2}$$

where

$$M = \operatorname{diag}\,[A_i | \operatorname{Im} P_i]_{i=0}^{s} \ :\ Z \to Z, \tag{1.3}$$

$$M^{\times} = \operatorname{diag}\,[A_i^{\times} | \operatorname{Im} P_i^{\times}]_{i=0}^{s} \ :\ Z^{\times} \to Z^{\times}, \tag{1.4}$$

$$C^{\times} = \operatorname{row}[C_i | \operatorname{Im} P_i^{\times}]_{i=0}^{s} \ :\ Z^{\times} \to Y\ , \tag{1.5}$$

$$B = \operatorname{col}[P_i B_i]_{i=0}^{s} \ :\ Y \to Z. \tag{1.6}$$

The equality (1.2) is easily verified; indeed, one has

$$S_{jk}\, A_k^{\times}\, x - A_j\, S_{jk}\, x = -\, P_j B_j C_k\, x \quad (x \in \operatorname{Im} P_k^{\times}\ ,\ j \neq k)$$

(ref. Section 1.4 in [5]). Analogously,

$$RM - M^{\times} R = -B^{\times} C\ , \tag{1.7}$$

where

(1.8) $\qquad C = \mathrm{row}[C_i | \mathrm{Im} P_i]_{i=0}^{s} \ : \ Z \to Y ,$

and

(1.9) $\qquad B^{\times} = - \mathrm{col}[P_i^{\times} B_i]_{i=0}^{s} \ : \ Y \to Z^{\times} .$

2. Toeplitz operators on compounded countours

In this section we shall introduce and study Toeplitz operators whose symbol is an operator valued function $W(\lambda)$ admiting representation (1.1).

Let $\mathcal{B}$ be an admissible Banach space of Y-valued functions (e.g., $\mathcal{B} = H^{\alpha}(\Gamma,Y)$, the Banach space of all functions from Γ into Y that are Hölder continuous with a fixed Hölder exponent α , $0 < \alpha < 1$). Define

$$P_{\Gamma}:\mathcal{B} \to \mathcal{B}, (P_{\Gamma}\varphi)(\lambda)=\frac{1}{2}(\varphi(\lambda) + \frac{1}{\pi i} \int_{\Gamma} \frac{\varphi(\tau)}{\tau-\lambda} d\tau) \ ,\lambda \in \Gamma ,$$

where the integral is understood in the Cauchy principal value sense. Then P_{Γ} is a (bounded) projection, and we have the direct sum decomposition $\mathcal{B} = \mathcal{B}^{+} \oplus \mathcal{B}^{-}$ with $\mathcal{B}^{+} = \mathrm{Im}\, P_{\Gamma}$, $\mathcal{B}^{-} = \mathrm{Ker}\, P_{\Gamma}$. Here $\mathcal{B}^{+}$ consists of all functions from $\mathcal{B}$ which admit analytic extensions on Ω^{+}, while $\mathcal{B}^{-}$ consists of all functions from $\mathcal{B}$ which admit analytic extensions on Ω^{-} and take value 0 at infinity. Let M_W be the operator of multiplication by W on $\mathcal{B}$, that is,

$$(M_W\varphi)\ (\lambda) = W(\lambda)\varphi(\lambda) \ (\lambda\in\Gamma \ ,\varphi\in\mathcal{B}) ,$$

and write M_W as a 2×2 operator matrix with respect to the direct sum decomposition $\mathcal{B} = \mathcal{B}^{+} \oplus \mathcal{B}^{-}$:

$$M_W = \begin{bmatrix} T & M_{12} \\ M_{21} & M_{22} \end{bmatrix} \ : \ \mathcal{B}^{+} \oplus \mathcal{B}^{-} \to \mathcal{B}^{+} \oplus \mathcal{B}^{-} .$$

We shall refer to T as the *Toeplitz operator with symbol* W. For the 2×2 operator matrix representing the operator of multiplication by $W(\lambda)^{-1}$ we shall use notation

$$M_{W^{-1}} = \begin{bmatrix} T^{\times} & M_{12}^{\times} \\ M_{21}^{\times} & M_{22}^{\times} \end{bmatrix} \ : \ \mathcal{B}^{+} \oplus \mathcal{B}^{-} \to \mathcal{B}^{+} \oplus \mathcal{B}^{-} \ ;$$

so $T^{\times}$ is the Toeplitz operator with symbol W^{-1}.

THEOREM 2.1. *Let* T *be a Toeplitz operator whose symbol* $W(\lambda)$ *admits the representation* (1.1) *and let* S *be the left indicator of* $W(\lambda)$ (*with respect to the realizations* (1.1) *and contour* Γ). *Put* $Z = \bigoplus_{j=0}^{s} \operatorname{Im} P_j$ *and* $Z^{\times} = \bigoplus_{j=0}^{s} \operatorname{Im} P_j^{\times}$. *Then the operators* $T \oplus I_{\mathcal{B}^-} \oplus I_Z : \mathcal{B}^+ \oplus \mathcal{B}^- \oplus Z \to \mathcal{B}^+ \oplus \mathcal{B}^- \oplus Z$ *and* $S \oplus I_{\mathcal{B}^+} \oplus I_{\mathcal{B}^-} : Z^{\times} \oplus \mathcal{B}^+ \oplus \mathcal{B}^- \to Z \oplus \mathcal{B}^+ \oplus \mathcal{B}^-$ *are equivalent. More precisely,*

$$(T \oplus I_{\mathcal{B}^-} \oplus I_Z)Q_1 = Q_2(S \oplus I_{\mathcal{B}^+} \oplus I_{\mathcal{B}^-}),$$

where

$$Q_1 = \begin{bmatrix} M_{12}^{\times} M_{21} U^{\times} & T^{\times} & M_{12}^{\times} M_{22} \\ M_{22}^{\times} M_{21} U^{\times} & M_{21}^{\times} & M_{22}^{\times} M_{22} \\ S & 0 & N \end{bmatrix} \quad : \quad Z^{\times} \oplus \mathcal{B}^+ \oplus \mathcal{B}^- \to \mathcal{B}^+ \oplus \mathcal{B}^- \oplus Z$$

and

$$Q_2 = \begin{bmatrix} -M_{12} M_{21}^{\times} U & TT^{\times} & -M_{12} M_{22}^{\times} M_{22} \\ M_{21}^{\times} U & M_{21}^{\times} & M_{22}^{\times} M_{22} \\ I_Z & 0 & N \end{bmatrix} : Z \oplus \mathcal{B}^+ \oplus \mathcal{B}^- \to \mathcal{B}^+ \oplus \mathcal{B}^- \oplus Z$$

are invertible operators with inverses given by

$$Q_1^{-1} = \begin{bmatrix} -N^{\times} M_{21} & -N^{\times} M_{22} & R \\ T & M_{12} & 0 \\ M_{22}^{\times} M_{21} & M_{22}^{\times} M_{22} & -M_{21}^{\times} U \end{bmatrix}$$

and

$$Q_2^{-1} = \begin{bmatrix} NM_{21}^{\times} & -NM_{22}^{\times}M_{22} & SR \\ I_{\mathcal{B}^+} & M_{12} & 0 \\ -M_{21}^{\times} & M_{22}^{\times}M_{22} & -M_{21}^{\times}U \end{bmatrix} .$$

Here R *is the right indicator of* $W(\lambda)$, *and*

$$N : \mathcal{B}^- \to Z \ , \ N\varphi = \mathrm{col}\Big(\frac{1}{2\pi i}\int_{\Gamma_j} (\tau-A_j)^{-1}P_jB_j\varphi(\tau)d\tau\Big)_{j=0}^{s} \ , \varphi \in \mathcal{B}^- ;$$

$$U : Z \to \mathcal{B}^+ \ , \ U(\mathrm{col}(x_j)_{j=0}^{s}) = \sum_{j=0}^{s} C_j(\lambda-A_j)^{-1} P_jx_j \ ,$$

$$x_j \in \mathrm{Im}\, P_j \quad , \quad j = 0,\dots,s ;$$

$$N^{\times} : \mathcal{B}^- \to Z^{\times} \ , \ N^{\times}\varphi = \mathrm{col}\Big(\frac{1}{2\pi i}\int_{\Gamma_j} (\tau-A_j^{\times})^{-1}P_j^{\times} B_j\varphi(\tau)d\tau\Big)_{j=0}^{s} \ , \ \varphi \in \mathcal{B}^- \ ;$$

$$U^{\times} : Z^{\times} \to \mathcal{B}^+ \ , \ U^{\times}(\mathrm{col}(x_j)_{j=0}^{s}) = -\sum_{j=0}^{s} C_j(\lambda-A_j^{\times})^{-1} P_j^{\times} x_j \ ,$$

$$x_j \in \mathrm{Im}\, P_j^{\times} \quad , \quad j = 0,\dots,s .$$

Theorem 2.1 will allow us to write down explicitly the connections between the kernels and the images of T and S as well as express the generalized inverses of T and S in terms of each other. Recall that an operator V is said to have a generalized inverse V^I whenever $V = VV^IV$.

THEOREM 2.2. *Let* T *be a Toeplitz operator whose symbol* $W(\lambda)$ *admits the representation* (1.1), *and let* S *be the left indicator of* $W(\lambda)$ (*with respect to the realizations* (1.1)).
Then

$$(2.1) \quad \mathrm{Ker}\, T = \{\varphi : \Gamma \to Y \mid \varphi(\lambda) \doteq U^{\times}x \overset{\mathrm{def}}{=} -\sum_{j=0}^{s} C_j(\lambda-A_j^{\times})^{-1} P_j^{\times} x_j ,$$
$$\mathrm{col}(x_j)_{j=0}^{s} \in \mathrm{Ker}\, S\}$$

and $\mathrm{Im}\, T$ *consists of all functions* $\varphi \in \mathcal{B}^+$ *such that*

$$\Lambda\varphi \overset{\text{def}}{=} \operatorname{col}\Big(\frac{1}{2\pi i}\int_{\Gamma_j} P_j(\lambda - A_j)^{-1} B_j\, W(\lambda)^{-1}\varphi(\lambda) +$$

$$- \frac{1}{2\pi i}\int_{\Gamma_j} P_j P_j^{\times}(\lambda - A_j^{\times})^{-1} B_j \varphi(\lambda) d\lambda\Big)_{j=0}^{s} \in \operatorname{Im} S. \tag{2.2}$$

Further, if S^I *is a generalized inverse of* S, *then*

$$T^I = T^{\times} + U^{\times}[R(I - SS^I) + S^I]\,\Lambda \tag{2.3}$$

is a generalized inverse of T. *In* (2.3) R *is the right indicator of* $W(\lambda)$ *and the operators* $U^{\times}$ *and* Λ *are as in* (2.1) *and* (2.2).

The proofs of Theorems 2.1 and 2.2 will be given in the next section.

3. Proof of the main theorems.

The proof will be given in several steps. Firstly, we shall establish a general equivalence theorem.

Consider the following spaces and (linear bounded) operators:

$$\mathcal{B} = \mathcal{B}_1 \oplus \mathcal{B}_2 \quad , \quad \mathcal{B}^{\times} = \mathcal{B}_1^{\times} \oplus \mathcal{B}_2^{\times} \quad ,$$

$$Z = Z_1 \oplus Z_2 \quad , \quad Z^{\times} = Z_1^{\times} \oplus Z_2^{\times} \quad ,$$

$$U : Z_1 \to \mathcal{B}_2^{\times} \quad , \quad U^{\times} : Z_1^{\times} \to \mathcal{B}_2 \quad ,$$

$$\rho = \begin{bmatrix} N & G \\ 0 & H \end{bmatrix} : \mathcal{B}_1 \oplus \mathcal{B}_2 \to Z_1 \oplus Z_2 \quad ,$$

$$\rho^{\times} = \begin{bmatrix} N^{\times} & G^{\times} \\ 0 & H^{\times} \end{bmatrix} : \mathcal{B}_1^{\times} \oplus \mathcal{B}_2^{\times} \to Z_1^{\times} \oplus Z_2^{\times} \quad ,$$

$$D = \begin{bmatrix} D_{11} & D_{12} \\ -UN & D_{22} \end{bmatrix} : \mathcal{B}_1 \oplus \mathcal{B}_2 \to \mathcal{B}_1^{\times} \oplus \mathcal{B}_2^{\times} \quad ,$$

$$D^{\times} = \begin{bmatrix} D_{11}^{\times} & D_{12}^{\times} \\ -U^{\times}N^{\times} & D_{22}^{\times} \end{bmatrix} : \mathcal{B}_1^{\times} \oplus \mathcal{B}_2^{\times} \to \mathcal{B}_1 \oplus \mathcal{B}_2 \quad ,$$

$$K = \begin{bmatrix} R-G^{\times}U & K_{12} \\ -H^{\times}U & K_{22} \end{bmatrix} \quad : \quad Z_1 \oplus Z_2 \to Z_1^{\times} \oplus Z_2^{\times} \quad ,$$

$$K^{\times} = \begin{bmatrix} R^{\times}-GU^{\times} & K_{12}^{\times} \\ -HU^{\times} & K_{22}^{\times} \end{bmatrix} \quad : \quad Z_1^{\times} \oplus Z_2^{\times} \to Z_1 \oplus Z_2 \quad .$$

THEOREM 3.1 *Assume*

(i) $\rho^{\times} D = K\rho$, $U^{\times} R = -D_{22}^{\times} U$, $UR^{\times} = -D_{22} U^{\times}$;

(ii) D *is invertible* , $D^{-1} = D^{\times}$;

(iii) K *is invertible* , $K^{-1} = K^{\times}$.

Then the operators $D_{11} \oplus I_{Z_1^{\times}}$ *and* $R \oplus I_{\mathcal{B}_1^{\times}}$ *are equivalent. More precisely:*

$$E \begin{bmatrix} R & 0 \\ 0 & I_{\mathcal{B}_1^{\times}} \end{bmatrix} = \begin{bmatrix} D_{11} & 0 \\ 0 & I_{Z_1^{\times}} \end{bmatrix} F \quad , \tag{3.1}$$

where the operators

$$E = \begin{bmatrix} D_{12}U^{\times} & D_{11}D_{11}^{\times} \\ I_{Z_1^{\times}} & N^{\times} \end{bmatrix} \quad : \quad Z_1^{\times} \oplus \mathcal{B}_1^{\times} \to \mathcal{B}_1^{\times} \oplus Z_1^{\times} \; ,$$

$$F = \begin{bmatrix} D_{12}^{\times}U & D_{11}^{\times} \\ R & N^{\times} \end{bmatrix} \quad : \quad Z_1 \oplus \mathcal{B}_1^{\times} \to \mathcal{B}_1 \oplus Z_1^{\times}$$

are invertible operators with inverses given by

$$E^{-1} = \begin{bmatrix} -N^{\times} & RR^{\times} \\ I_{\mathcal{B}_1^{\times}} & -D_{12}U^{\times} \end{bmatrix} \; , \quad F^{-1} = \begin{bmatrix} -N & R^{\times} \\ D_{11} & -D_{12}U^{\times} \end{bmatrix} \quad .$$

Proof. We shall derive the following identities:

(3.2a) $D_{11}D_{12}^{\times}U - D_{12}U^{\times}R = 0$,

(3.2b) $D_{11}^{\times}D_{12}U^{\times} - D_{12}^{\times}UR^{\times} = 0$,

(3.3a) $D_{11}D_{11}^{\times} - D_{12}U^{\times}N^{\times} = I_{\mathcal{B}_1^{\times}}$,

(3.3b) $D_{11}^{\times}D_{11} - D_{12}^{\times}UN = I_{\mathcal{B}_1}$,

(3.4a) $NT^{\times} = R^{\times} N^{\times}$,

(3.4b) $N^{\times} T = R N$,

(3.5a) $R^{\times} R - N D_{12}^{\times} U = I_{Z_1}$,

(3.5b) $R R^{\times} - N^{\times} D_{12} U^{\times} = I_{Z_1^{\times}}$.

Since our hypotheses are symmetric with respect to the upper index $\times$, it suffices to prove the first identity from each pair.

From the second identity in (i) we know that $U^{\times} R = -D_{22}^{\times} U$. Thus

(3.6) $$D_{11} D_{12}^{\times} U - D_{12} U^{\times} R = (D_{11} D_{12}^{\times} + D_{12} D_{22}^{\times})U .$$

Now observe that $D_{11} D_{12}^{\times} + D_{12} D_{22}^{\times} = 0$, because $D D^{\times} = I$. Thus (3.2a) holds. The identity (3.3a) also follows from $D D^{\times} = I$. To prove (3.4a) equate the left upper entries on both sides of the equality $\rho M^{\times} = K^{\times}\rho^{\times}$. Equating the right upper entries of this equality gives

$$N D_{12}^{\times} + G D_{22}^{\times} = (R^{\times} - G U^{\times})G^{\times} + K_{12}^{\times} H^{\times} .$$

So

$$R^{\times} R - N D_{12}^{\times} U =$$

$$= R^{\times} R + G D_{22}^{\times} U - (R^{\times} - G U^{\times})G^{\times} U - K_{12}^{\times}H^{\times} U$$

$$= (R^{\times} - G U^{\times}) (R - G^{\times} U) + K_{12}^{\times}(-H^{\times} U) +$$

$$+ G D_{22}^{\times} U + G U^{\times} R.$$

Now use $D^{\times}_{22}\, U = -U^{\times} R$ and $K^{\times} K = I$. We obtain $R^{\times} R - N\, D^{\times}_{12}\, U =$ $= I_{Z_1}$, and (3.5a) is proved.

From (3.2a) it is clear that formula (3.1) holds true. Formula (3.3a) allows us to write

$$E = \begin{bmatrix} D_{12}U^{\times} & I_{\mathcal{B}_1^{\times}} \\ I_{Z_1^{\times}} & 0 \end{bmatrix} \begin{bmatrix} I_{Z_1^{\times}} & N^{\times} \\ 0 & I_{\mathcal{B}_1^{\times}} \end{bmatrix} .$$

This implies that E is invertible, and

$$E^{-1} = \begin{bmatrix} -N^{\times} & I_{Z_1^{\times}} + N^{\times} D_{12}U^{\times} \\ I_{\mathcal{B}_1^{\times}} & -D_{12}\, U^{\times} \end{bmatrix} .$$

Using (3.5b) we see that E^{-1} has the desired form. By computing the matrix products $F \circ F^{-1}$ and $F^{-1} \circ F$ one easily sees (using the formulas (3.2) - (3.5)) that F is invertible with inverse of the desired form. □

Observe that equalities (3.2) - (3.5) amount to

$$\text{(3.7)} \qquad \begin{bmatrix} D_{11} & -D_{12}U^{\times} \\ -N & R^{\times} \end{bmatrix}^{-1} = \begin{bmatrix} D^{\times}_{11} & D^{\times}_{12}\, U \\ N^{\times} & R \end{bmatrix} .$$

In the terminology of [3] the operators D_{11} and R are *matricially coupled*. Once (3.7) is established, the equality (3.1) together with the formulas for $E^{\pm 1}$ and $F^{\pm 1}$ follow also from a general result on matricially coupled operators (Theorem I.1.1 in [3]).

Further, we shall prove three lemmas. In the rest of this section we use the notations introduced in Section 2. The function $W(\lambda)$ is assumed to admit the realizations (1.1).

LEMMA 3.2. *For* $\varphi_- \in \mathcal{B}^-$ *and* $\lambda \in \Gamma$ *we have*

$$\text{(3.8)} \qquad \frac{1}{2\pi i} \int_{\Gamma_j} \frac{W(\tau)-W(\lambda)}{\tau-\lambda}\, \varphi_-(\tau)\, d\tau = (-U_j N_j \varphi_-)(\lambda)\ , \quad j = 0,\ldots,s,$$

where

(3.9) $N_j : \mathcal{B}^- \to \operatorname{Im}P_j \ , \ N_j\varphi_- = \frac{1}{2\pi i}\int_{\Gamma_j} (\tau - A_j)^{-1} P_jB_j\varphi_-(\tau)d\tau \ ;$

$U_j : \operatorname{Im}P_j \to \mathcal{B}^+ \ , \quad (U_jx)(\lambda) = C_j(\lambda - A_j)^{-1} P_jx \ , \quad x \in \operatorname{Im}P_j \ .$

Proof. Take $\lambda \in \Gamma_k$. First assume $k \neq j$. Since λ is outside Γ_j, we know that $(\cdot - \lambda)^{-1}\varphi_-(\cdot)$ has an analytic continuation in Ω_j^- (and a zero of order ≥ 2 at ∞ for $j = 0$). Hence

$$\frac{1}{2\pi i}\int_{\Gamma_j} \frac{\varphi_-(\tau)}{\tau - \lambda} d\tau = 0 \ . \tag{3.10}$$

Also $C_j(\cdot - A_j)^{-1}(I - P_j)B_j\varphi_-(\cdot)$ has an analytic continuation in Ω_j^- (and a zero of order ≥ 2 at ∞ for $j = 0$). It follows that

$$\frac{1}{2\pi i}\int_{\Gamma_j} \frac{W(\tau) - W(\lambda)}{\tau - \lambda}\varphi_-(\tau)d\tau =$$

$$= \frac{1}{2\pi i}\int_{\Gamma_j} \frac{C_j(\tau - A_j)^{-1}P_jB_j}{\tau - \lambda}\varphi_-(\tau)d\tau =$$

$$= C_j(\lambda - A_j)^{-1}P_j\left(\frac{1}{2\pi i}\int_{\Gamma_j} \frac{(\lambda - A_j)(\tau - A_j)^{-1}P_jB_j}{\tau - \lambda}\varphi_-(\tau)d\tau\right) =$$

$$= C_j(\lambda - A_j)^{-1}P_j\left(-\frac{1}{2\pi i}\int_{\Gamma_j} (\tau - A_j)^{-1}P_jB_j\varphi_-(\tau)d\tau\right) +$$

$$+ C_j(\lambda - A_j)^{-1}P_j\ B_j\left(\frac{1}{2\pi i}\int_{\Gamma_j} \frac{\varphi_-(\tau)}{\tau - \lambda}d\tau\right) \ .$$

Note that the last term is zero because of (3.10). So we have proved (3.8) for $\lambda \in P_k$, $k \neq j$.

Next, take $\lambda \in P_j$. Using the resolvent equation we see that

$$\frac{1}{2\pi i}\int_{\Gamma_j} \frac{W(\lambda) - W(\lambda)}{\tau - \lambda}\varphi_-(\tau)d\tau =$$

$$= C_j(\lambda - A_j)^{-1}\left(-\frac{1}{2\pi i}\int_{\Gamma_j} (\tau - A_j)^{-1}\ B_j\varphi_-(\tau)d\tau\right) \ .$$

Since $(\cdot - A_j)^{-1}(I - P_j)B_j\varphi_-(\cdot)$ has an analytic continuation in Ω_j^- (and a zero of order ≥ 2 for $j = 0$), we may conclude that

$$-\frac{1}{2\pi i}\int_{\Gamma_j}(\tau - A_j)^{-1}(I-P_j)B_j\varphi_-(\tau)d\tau = 0 ,$$

and thus (3.8) holds. □

LEMMA 3.3. *Let* $\psi \in B$ *have an analytic continuation outside* Ω_j^- *which is zero at* ∞ *when* $j \neq 0$ *. Then*

$$\frac{1}{2\pi i}\int_{\Gamma_\nu}\frac{W(\tau)-W(\lambda)}{\tau - \lambda}\psi(\tau)d\tau = -(U_\nu N_\nu\psi)\ (\lambda) \quad (\nu \neq j), \lambda \in \Gamma,$$

where U_ν *is as in Lemma 3.2, and* N_ν *is given by the formula (3.9) with* φ_- *replaced by* $\varphi \in B$ *(so,* N_ν *is considered as an operator* $B \to \operatorname{Im} P_j$ *).*

Proof. Take $\lambda \in \Gamma_k$ with $k \neq \nu$. Then $(\cdot - \lambda)^{-1}\psi(\cdot)$ has an analytic continuation in Ω_ν^- (and a zero of order ≥ 2 at ∞ for $\nu = 0$). It follows that

$$\frac{1}{2\pi i}\int_{\Gamma_\nu}\frac{W(\tau)-W(\lambda)}{\tau - \lambda}\psi(\tau)d\tau =$$

$$= \frac{1}{2\pi i}\int_{\Gamma_\nu}\frac{C_\nu(\tau-A_\nu)^{-1}B_\nu}{\tau - \lambda}\psi(\tau)d\tau \ +$$

$$+ C_k(\lambda-A_k)^{-1}B_k\left(-\frac{1}{2\pi i}\int_{\Gamma_\nu}\frac{\psi(\tau)}{\tau-\lambda}d\tau\right) =$$

$$= \frac{1}{2\pi i}\int_{\Gamma_\nu}\frac{C_\nu(\tau-A_\nu)^{-1}B_\nu}{\tau - \lambda}\psi(\tau)d\tau \ .$$

Note that $C_\nu(\cdot - A_\nu)^{-1}(I - P_\nu)B_\nu\psi(\cdot)$ has an analytic continuation in Ω_ν^- (and a zero of order ≥ 2 at ∞ for $\nu = 0$). Since $\lambda \in \Gamma_k$ with $k \neq \nu$, it follows that

$$\frac{1}{2\pi i}\int_{\Gamma_\nu}\frac{C_\nu(\tau-A_\nu)^{-1}B_\nu}{\tau - \lambda}\psi(\tau)d\tau =$$

$$= \frac{1}{2\pi i}\int_{\Gamma_\nu} \frac{C_\nu(\tau-A_\nu)^{-1}P_\nu B_\nu}{\tau - \lambda} \psi(\tau)d\tau$$

$$= C_\nu(\lambda-A_\nu)^{-1}P_\nu\Big(- \frac{1}{2\pi i} \int_{\Gamma_\nu} (\tau-A_\nu)^{-1}P_\nu B_\nu\psi(\tau)d\tau \;+$$

$$+ \frac{1}{2\pi i}\int_{\Gamma_\nu} P_\nu B_\nu \frac{\psi(\tau)}{\tau - \lambda} d\tau\Big) =$$

$$= -(U_\nu N_\nu\psi)\ (\lambda).$$

Next, take $\lambda \in \Gamma_\nu$. Then

$$\frac{1}{2\pi i}\int_{\Gamma_\nu} \frac{W(\tau)-W(\lambda)}{\tau - \lambda} \psi(\tau)d\tau =$$

$$= C_\nu(\lambda-A_\nu)^{-1} \Big(- \frac{1}{2\pi i}\int_{\Gamma_\nu} (\tau-A_\nu)^{-1} B_\nu\psi(\tau)d\tau\Big) .$$

Since $(\cdot-A_\nu)^{-1}(I-P_\nu)B_\nu\ \psi(\cdot)$ has an analytic continuation inside Γ_ν (and a zero of order ≥ 2 at ∞ for $\nu = 0$), we see that

$$\frac{1}{2\pi i} \int_{\Gamma_\nu} (\tau-A_\nu)^{-1} (I-P_\nu)B_\nu\ \psi(\tau)d\tau = 0 .$$

Thus

$$\frac{1}{2\pi i}\int_{\Gamma_\nu} \frac{W(\tau)-W(\lambda)}{\tau - \lambda} \psi(\tau)d\tau =$$

$$= C_\nu(\lambda-A_\nu)^{-1} P_\nu\Big(- \frac{1}{2\pi i}\int_{\Gamma_\nu} (\tau-A_\nu)^{-1} P_\nu B_\nu\psi(\tau)d\tau\Big) .$$

The lemma is proved. □

LEMMA 3.4. *Let* T *be the Toeplitz operator with symbol* $W(\lambda)$, *let* R *and* S *denote the right and left indicators of* W, *respectively, and let* U *and* $U^\times$ *be defined as in Theorem 2.1. Then*

$$US = -TU^\times\ ,\ U^\times R = -T^\times U \qquad (3.11)$$

Proof. Letting N_ν and U_ν be as in Lemma 3.3, and defining

$$U_j^x : Z^x \to \mathcal{B}^+ , \quad U_j^x x = -C_j(\lambda - A_j^x)^{-1} P_j^x x , \quad x \in Z^x ,$$

we have

$$U S P_j^x x = \sum_{\nu=0}^{s} U_\nu S_{\nu j} P_j^x x =$$

$$= \sum_{\nu \neq j} U_\nu N_\nu U_j^x P_j^x x + U_j P_j P_j^x x .$$

Put $\psi(\lambda) = (U_j^x P_j^x x)(\lambda) = -C_j(\lambda - A_j^x)^{-1} P_j^x x$, and observe that $\psi(\lambda)$ satisfies the hypothesis of Lemma 3.3.
Hence

$$(U S P_j^x x)(\lambda) =$$

$$= - \sum_{\nu \neq j} \frac{1}{2\pi i} \int_{\Gamma_\nu} \frac{W(\tau) - W(\lambda)}{\tau - \lambda} \psi(\tau) d\tau + (U_j P_j P_j^x x)(\lambda) =$$

$$= - \frac{1}{2\pi i} \int_{\Gamma} \frac{W(\tau) - W(\lambda)}{\tau - \lambda} \psi(\tau) d\tau + (U_j P_j P_j^x x)(\lambda) + \alpha(\lambda) ,$$

where

$$\alpha(\lambda) = \frac{1}{2\pi i} \int_{\Gamma_j} \frac{W(\tau) - W(\lambda)}{\tau - \lambda} \psi(\tau) d\tau .$$

It is not difficult to verify the formula

$$\frac{1}{2\pi i} \int_{\Gamma} \frac{W(\tau) - W(\lambda)}{\tau - \lambda} \varphi(\tau) d\tau = [(P_\Gamma - I) M_W P_\Gamma + P_\Gamma M_W (I - P_\Gamma)] \varphi(\lambda) , \quad \varphi \in \mathcal{B} .$$

As $\psi(\lambda) \in \mathcal{B}^+ = \operatorname{Im} P_\Gamma$, this formula gives

$$\frac{1}{2\pi i} \int_{\Gamma} \frac{W(\tau) - W(\lambda)}{\tau - \lambda} \psi(\tau) d\tau = [(P_\Gamma - I) M_W P_\Gamma] \psi(\lambda) ,$$

and we have

$$(U\,S\,P_j^\times)(\lambda) = (I - P_\Gamma)M_W P_\Gamma \psi(\lambda) + (U_j P_j P_j^\times x)(\lambda) + \alpha(\lambda) \quad .$$

To compute $\alpha(\lambda)$ we distinguish two cases. First assume $\lambda \in \Gamma_j$. Then

$$\alpha(\lambda) = C_j(\lambda-A_j)^{-1}\left(\frac{1}{2\pi i}\int_{\Gamma_j}(\tau-A_j)^{-1}\,B_jC_j(\tau-A_j^\times)^{-1}\,P_j^\times x\,d\tau\right) =$$

$$= C_j(\lambda-A_j)^{-1}\left(\frac{1}{2\pi i}\int_{\Gamma_j}(\tau-A_j)^{-1}d\tau - \frac{1}{2\pi i}\int(\tau-A_j^\times)^{-1}d\tau\right)\cdot P_j^\times\, x =$$

$$= C_j(\lambda-A_j)^{-1}\,(I-P_j)P_j^\times\, x \quad .$$

So for $\lambda \in \Gamma_j$ we have

$$(U_j\,P_j\,P_j^\times\,x)(\lambda) + \alpha(\lambda) =$$

$$= C_j(\lambda-A_j)^{-1}\,P_j^\times\,x$$

$$= W_j(\lambda)C_j(\lambda-A_j^\times)^{-1}\,P_j^\times\,x = -W_j(\lambda)\psi(\lambda).$$

Next, take $\lambda \in \Gamma_k$, $k \neq j$. Then

$$\alpha(\lambda) = \frac{1}{2\pi i}\int_{\Gamma_j}\frac{W(\tau)\psi(\tau)}{\tau-\lambda}\,d\tau \; +$$
$$+\; W(\lambda)\left(-\frac{1}{2\pi i}\int_{\Gamma_j}\frac{\psi(\tau)}{\tau-\lambda}\,d\tau\right) \quad ,$$

and using the definition of $\psi(\lambda)$ and the fact that $\psi(\lambda)$ is analytic outside Ω_j^- having zero at ∞ when $j \neq 0$, we obtain

$$\alpha(\lambda) = \left(-\frac{1}{2\pi i}\int_{\Gamma_j}\frac{C_j(\tau-A_j)^{-1}}{\tau-\lambda}\,d\tau\right)P_j^\times\,x \quad -W(\lambda)\psi(\lambda) =$$

$$= -C_j(\lambda-A_j)^{-1}\,P_j^\times\,x \quad -W(\lambda)\psi(\lambda).$$

Thus for $\lambda \in \Gamma_k$, $k \neq j$, we have

$$(U_j\,P_j\,P_j^\times\,x)(\lambda) + \alpha(\lambda) =$$

$$= - W(\lambda)\psi(\lambda) \quad .$$

We see that

$$USP_j^X x = (I-P_\Gamma)M_W P_\Gamma U_j^X P_j^X x - M_W U_j^X P_j^X x =$$

$$= (I-P_\Gamma)M_W P_\Gamma U_j^X P_j^X x - M_W P_\Gamma U_j^X P_j^X x =$$

$$= - P_\Gamma M_W P_\Gamma U_j^X P_j^X x ,$$

which proves the first identity in (3.11). The second identity is obtained by applying the first identity to $W(\lambda)^{-1}$ in place of $W(\lambda)$. □

The next result shows that the right indicator R and the restriction of the multiplication operator M_W to the subspace $\mathcal{B}^-$ are equivalent after appropriate extension by identity operators.

THEOREM 3.5. *The operators* $R \oplus I_{\mathcal{B}^-}$ *and* $M_{22} \oplus I_{Z^X}$ *are equivalent. More precisely,*

$$E \begin{bmatrix} R & 0 \\ 0 & I_{\mathcal{B}^-} \end{bmatrix} = \begin{bmatrix} M_{22} & 0 \\ 0 & I_{Z^X} \end{bmatrix} F \quad ,$$

where E *and* F *are invertible operators given, together with their inverses, by the following formulas:*

$$E = \begin{bmatrix} M_{21}U^X & M_{22}M_{22}^X \\ I_{Z^X} & N^X \end{bmatrix} : Z^X \oplus \mathcal{B}^- \to \mathcal{B}^- \oplus Z^X \quad ,$$

$$F = \begin{bmatrix} M_{21}^X U & M_{22}^X \\ R & N^X \end{bmatrix} : Z \oplus \mathcal{B}^- \to \mathcal{B}^- \oplus Z^X \quad ,$$

$$E^{-1} = \begin{bmatrix} -N^X & RS \\ I_{\mathcal{B}^-} & -M_{21}U^X \end{bmatrix} : \mathcal{B}^- \oplus Z^X \to Z^X \oplus \mathcal{B}^- \quad ;$$

$$F^{-1} = \begin{bmatrix} -N & S \\ M_{22} & -M_{21}U^{x} \end{bmatrix} : \mathcal{B}^{-} \oplus Z^{x} \to Z \oplus \mathcal{B}^{-} .$$

Proof. We shall apply Theorem 3.1. Put

$$\mathcal{B}_1 = \mathcal{B}_1^{x} = \mathcal{B}^{-} \quad , \quad \mathcal{B}_2 = \mathcal{B}_2^{x} = \mathcal{B}^{+} \quad ,$$

$$Z_1 = \bigoplus_{j=0}^{s} \operatorname{Im} P_j \ (=Z) \quad , \quad Z_2 = \bigoplus_{j=0}^{s} \operatorname{Ker} P_j \quad ,$$

$$Z_1^{x} = \bigoplus_{j=0}^{s} \operatorname{Im} P_j^{x} \ (=Z^{x}) \ , \quad Z_2^{x} = \bigoplus_{j=0}^{s} \operatorname{Ker} P_j^{x} \quad .$$

Note that $Z_1 \oplus Z_2 = Z_1^{x} \oplus Z_2^{x} = \bigoplus_{j=0}^{s} X_j$. For $j = 0,\dots,s$ define

$$\rho_j : \mathcal{B} \to X_j \quad , \quad \rho_j\varphi = \frac{1}{2\pi i} \int_{\Gamma_j} (\tau - A_j)^{-1} B_j\varphi(\tau)d\tau \quad ,$$

$$\rho_j^{x} : \mathcal{B} \to X_j \quad , \quad \rho_j^{x}\varphi = \frac{1}{2\pi i} \int_{\Gamma_j} (\tau - A_j^{x})^{-1} B_j\varphi(\tau)d\tau \quad ,$$

$$\rho, \rho^{x} : \mathcal{B} \to \bigoplus_{j=0}^{s} X_j \ , \ \rho = \operatorname{col}[\rho_j]_{j=0}^{s} \quad , \quad \rho^{x} = \operatorname{col}[\rho_j^{x}]_{j=0}^{s} \quad .$$

Observe that for each j

(3.12) $$(I - \rho_j)\rho_j\varphi_- = 0 \quad , \quad (I - \rho_j^{x})\rho_j^{x}\varphi_- = 0 \ (\varphi_- \in \mathcal{B}^{-}).$$

It follows that ρ and ρ^{x} have the following 2×2 operator matrix representations:

$$\rho = \begin{bmatrix} N & G \\ 0 & H \end{bmatrix} : \mathcal{B}^{-} \oplus \mathcal{B}^{+} \to Z_1 \oplus Z_2 \quad ,$$

$$\rho^{x} = \begin{bmatrix} N^{x} & G^{x} \\ 0 & H^{x} \end{bmatrix} : \mathcal{B}^{-} \oplus \mathcal{B}^{+} \to Z_1^{x} \oplus Z_2^{x} \quad .$$

Also, let $U : Z_1 \to \mathcal{B}^+$ and $U^x : Z_1^x \to \mathcal{B}^+$ be as in Theorem 2.2. Put $D = M_W$ and $D^x = M_{W^{-1}}$, the operators of multiplication by $W(\lambda)$ and $W(\lambda)^{-1}$, respectively. According to Lemma 3.2 the elements in the left lower corner of the 2x2 operator matrix representation for D and D^x (with respect to the direct sum decomposition $\mathcal{B}=\mathcal{B}^- \oplus \mathcal{B}^+$) are of the form required in Theorem 3.1 (to check this use the equality

$$P_\Gamma(W(\lambda)\varphi_-(\lambda)) = \frac{1}{2\pi i}\int_\Gamma \frac{W(\tau)-W(\lambda)}{\tau - \lambda}\varphi_-(\tau)d\tau \quad , \quad \varphi_- \in \mathcal{B}^-) \ .$$

Write

$$I = [K_{ij}]_{i,j=1}^2 : Z_1 \oplus Z_2 \to Z_1^x \oplus Z_2^x \quad ,$$

$$I = [K_{ij}^x]_{i,j=1}^2 \ : \ Z_1^x \oplus Z_2^x \to Z_1 \oplus Z_2 \quad .$$

Note that

$$K_{11} = \operatorname{diag}[P_j^x P_j]_{j=0}^s \ : \ \bigoplus_{j=0}^s \operatorname{Im} P_j \to \bigoplus_{j=0}^s \operatorname{Im} P_j^x \ .$$

It follows that for $k \neq j$ we have

$$P_k^x K_{11} P_j = P_k^x(R_{kj} - \rho_k^x U_j)P_j \quad (= 0) \quad ,$$

where R_{kj} is the (k,j) entry in the right indicator R of $W(\lambda)$, and U_j is defined as in Lemma 3.2.
Further,

$$\frac{1}{2\pi i}\int_{\Gamma_j} (\tau-A_j^x)^{-1} B_j C_j(\tau-A_j)^{-1} d\tau =$$

$$(3.13) = \frac{-1}{2\pi i}\int_{\Gamma_j} (\tau-A_j^x)^{-1} d\tau + \frac{1}{2\pi i}\int_{\Gamma_j} (\tau-A_j)^{-1} d\tau =$$

$$= P_j^x - P_j \quad ,$$

and we have also

$$P_j^x K_{11} P_j = P_{11}^x(R_{jj} - \rho_j^x U_j)P_j \quad .$$

Hence $K_{11} = R - G^{\times} U$. Next we compute

$$(I-P_k^{\times})H^{\times} U P_j = (I-P_k^{\times})\rho_k^{\times} U_j P_j$$

$$= \frac{1}{2\pi i} \int_{\Gamma_k} (\tau-A_k^{\times})^{-1}(I-P_k^{\times}) B_k C_j (\tau-A_j)^{-1}P_j \, d\tau .$$

For $k \neq j$ the functions $(\cdot - A_k^{\times})^{-1}(I-P_k^{\times})B_k$ and $C_j(\cdot-A_j)^{-1} P_j$ are both analytic inside Γ_k (and have a zero at ∞ when $k = 0$). It follows that $(I-P_k^{\times})H^{\times} U P_j = 0$ for $k \neq j$. Next, take $k = j$. In view of (3.13) we have $(I-P_j^{\times})H^{\times} U P_j = -(I-P_j^{\times})P_j$. So, $-H^{\times} U = K_{21}$. Replacing $W(\lambda)$ by $W(\lambda)^{-1}$, we obtain analogously that $K_{11}^{\times} = S - GU^{\times}$, $K_{21}^{\times} = -H U^{\times}$, where S is the left indicator of $W(\lambda)$ (or, what is the same, the right indicator of $W(\lambda)^{-1}$).

We put $K = K^{\times} = I$. Then, clearly, K and $K^{\times}$ have the 2×2 operator matrix representations are required in Theorem 3.1 (with $R^{\times} = S$). Using the equalities

$$(\lambda-A_j)^{-1} B_j W_j(\lambda)^{-1} = (\lambda-A_j^{\times})^{-1}B_j \qquad (j = 0,\dots,s)$$

and the analogous equality involving $W_j(\lambda)$ in place of $W_j(\lambda)^{-1}$, we obtain

$$\rho_j M_W^{-1} = \rho_j^{\times} \quad , \quad \rho_j^{\times} M_W = \rho_j \qquad (j = 0,1,\dots,s).$$

Hence $\rho^{\times} M = K\rho$, and the first identity in condition (i) of Theorem 3.1 is satisfied. The two other identities in (i) of Theorem 3.1 are covered by Lemma 3.4. The conditions (ii) and (iii) in Theorem 3.1 are also satisfied. Thus we can apply Theorem 3.1 to get the result of Theorem 3.5.□

Now we pass to the proofs of the results presented in Section 2.

Proof of Theorem 2.1. Observe that

$$\begin{bmatrix} T & M_{12} \\ M_{21} & M_{22} \end{bmatrix}^{-1} = \begin{bmatrix} T^{\times} & M_{12}^{\times} \\ M_{21}^{\times} & M_{22}^{\times} \end{bmatrix}, \tag{3.14}$$

and therefore T and $M_{22}^{\times}$ are matricially coupled. By Theorem I. 1.1 in [3] we have

$$(3.15)\qquad \begin{bmatrix} T & 0 \\ 0 & I_{\mathcal{B}^-} \end{bmatrix} F_1 = E_1 \begin{bmatrix} M_{22}^{\times} & 0 \\ 0 & I_{\mathcal{B}^+} \end{bmatrix},$$

where E_1 and F_1 are invertible operators given by formulas

$$E_1 = \begin{bmatrix} -M_{12} & TT^{\times} \\ I_{\mathcal{B}^-} & M_{21}^{\times} \end{bmatrix}, \quad F_1 = \begin{bmatrix} M_{12}^{\times} & T^{\times} \\ M_{22}^{\times} & M_{21}^{\times} \end{bmatrix},$$

with the inverses

$$E_1^{-1} = \begin{bmatrix} -M_{21}^{\times} & M_{22}^{\times} M_{22} \\ I_{\mathcal{B}^+} & M_{12} \end{bmatrix}, \quad F_1^{-1} = \begin{bmatrix} M_{21} & M_{22} \\ T & M_{12} \end{bmatrix}.$$

Combining the formula (3.15) with Theorem 3.5 (with $W(\lambda)$ replaced by $W(\lambda)^{-1}$), we obtain Theorem 2.1. □

Proof of Theorem 2.2. First observe that Theorem 3.5 implies

$$(3.16)\qquad \begin{bmatrix} M_{22}^{\times} & M_{21}^{\times} U \\ N^{\times} & R \end{bmatrix}^{-1} = \begin{bmatrix} M_{22} & -M_{21} U^{\times} \\ -N & S \end{bmatrix}.$$

Hence Theorem 2.1 in [3] gives

$$(3.17)\qquad \operatorname{Ker} M_{22}^{\times} = M_{21} U^{\times} \operatorname{Ker} S \ ; \ \operatorname{Ker} S = N^{\times} \operatorname{Ker} M_{22}^{\times} \ ;$$

$$(3.18)\qquad \operatorname{Im} M_{22}^{\times} = N^{-1} \operatorname{Im} S \ ; \quad \operatorname{Im} S = (M_{21}^{\times} U)^{-1} \operatorname{Im} M_{22}^{\times} \ ;$$

$$(3.19)\qquad (M_{22}^{\times})^{I} = M_{22} - M_{21} U^{\times} S^{I} N \ ;$$

$$(3.20)\qquad S^{I} = R - N^{\times} (M_{22}^{\times})^{I} M_{21}^{\times} U.$$

Analogously, the relationship (3.14) gives rise to the equalities

$$(3.21)\qquad \operatorname{Ker} T = M_{12}^{\times} \operatorname{Ker} M_{22}^{\times} \ ; \quad \operatorname{Ker} M_{22}^{\times} = M_{21} \operatorname{Ker} T \ ;$$

(3.22) $\operatorname{Im} T = (M_{21}^{\times})^{-1} \operatorname{Im} M_{22}^{\times}$; $\operatorname{Im} M_{22}^{\times} = M_{12}^{-1} \operatorname{Im} T$;

(3.23) $T^{I} = T^{\times} - M_{12}^{\times}(M_{22}^{\times})^{I} M_{21}^{\times}$;

(3.24) $(M_{22}^{\times})^{I} = M_{22} - M_{21} T^{I} M_{12}$.

Combining (3.17) - (3.18) with (3.21) - (3.22), one proves that

$\operatorname{Ker} T = M_{21}^{\times} M_{21} U^{\times} \operatorname{Ker} S$; $\operatorname{Ker} S = N^{\times} M_{21} \operatorname{Ker} T$;

$\operatorname{Im} T = (N M_{21}^{\times})^{-1} \operatorname{Im} S$; $\operatorname{Im} S = (M_{12} M_{21}^{\times} U)^{-1} \operatorname{Im} T$.

Now

$$M_{12}^{\times} M_{21} U^{\times} = (I - T^{\times} T)U^{\times} = U^{\times} - T^{\times} T U^{\times}$$
$$= U^{\times} + T^{\times} U S ,$$

because of formula (3.11). So for $x \in \operatorname{Ker} S$ we have $M_{12}^{\times} M_{21}U^{\times}x = U^{\times} x$. This yields the desired formula for $\operatorname{Ker} T$.

Next, use that $\operatorname{Im} T = \{ \varphi \in \mathcal{B}^{+} \mid N M_{21}^{\times}\varphi \in \operatorname{Im} S\}$. We have

$$M_{21}^{\times}\varphi = - \frac{1}{2\pi i} \int_{\Gamma} \frac{W(\lambda)^{-1}-W(\tau)^{-1}}{\lambda - \tau} \varphi(\lambda)d\lambda \ , \ \varphi \in \mathcal{B}^{+}$$

(see the proof of Lemma 3.4). We compute

$$N_j M_{21}^{\times}\varphi = - \left(\frac{1}{2\pi i}\right)^2 \int_{\tau\in\Gamma_j} (\tau-A_j)^{-1} P_jB_j\left(\int_{\lambda\in\Gamma} \frac{W(\tau)^{-1}-W(\lambda)^{-1}}{\tau - \lambda} \varphi(\lambda)d\lambda\right)d\tau \ .$$

Put

$$\alpha_{\nu} = - \left(\frac{1}{2\pi i}\right)^2 \int_{\Gamma_j} (\tau-A_j)^{-1} P_jB_j\left(\int_{\Gamma_{\nu}} \frac{W(\tau)^{-1}-W(\lambda)^{-1}}{\tau - \lambda} \varphi(\lambda)d\lambda\right)d\tau \ .$$

For $\tau \in \Gamma_j$, $\lambda \in \Gamma_j$ we have

$$\frac{W(\tau)^{-1}-W(\lambda)^{-1}}{\tau - \lambda} = C_j(\lambda-A_j^{\times})^{-1}(\tau-A_j^{\times})^{-1}B_j \ .$$

So

$$\alpha_j = -\left(\frac{1}{2\pi i}\right)^2 \int_{\Gamma_j} \int_{\Gamma_j} (\tau - A_j)^{-1} P_j B_j C_j (\tau - A_j^\times)^{-1} (\lambda - A_j^\times)^{-1} B_j \varphi(\lambda) d\lambda d\tau.$$

Using the identity $(\tau-A_j)^{-1} B_j C_j (\tau-A_j^\times)^{-1} = (\tau-A_j)^{-1} - (\tau-A_j^\times)^{-1}$,

we find that

$$\alpha_j = \frac{1}{2\pi i} \int_{\Gamma_j} P_j (\lambda - A_j^\times)^{-1} B_j \varphi(\lambda) d\lambda +$$

$$- \frac{1}{2\pi i} \int_{\Gamma_j} P_j P_j^\times (\lambda - A_j^\times)^{-1} B_j \varphi(\lambda) d\lambda$$

$$= \frac{1}{2\pi i} \int_{\Gamma_j} P_j (\lambda - A_j)^{-1} B_j W(\lambda)^{-1} \varphi(\lambda) d\lambda +$$

$$- \frac{1}{2\pi i} \int_{\Gamma_j} P_j P_j^\times (\lambda - A_j^\times)^{-1} B_j \varphi(\lambda) d\lambda .$$

For $\nu \neq j$ we have

$$\alpha_\nu = -\left(\frac{1}{2\pi i}\right)^2 \int_{\tau \in \Gamma_j} \int_{\lambda \in \Gamma_\nu} \frac{P_j (\tau - A_j^\times)^{-1} B_j}{\tau - \lambda} \varphi(\lambda) d\lambda d\tau$$

$$+ \frac{1}{2\pi i} \int_{\lambda \in \Gamma_\nu} P_j (\lambda - A_j)^{-1} B_j W(\lambda)^{-1} \varphi(\lambda) d\lambda .$$

Substituting $(\tau-A_j^\times)^{-1} = P_j^\times(\tau-A_j^\times) + (I-P_j^\times)(\tau-A_j^\times)^{-1}$ one easily computes

$$\left(\frac{1}{2\pi i}\right)^2 \int_{\tau \in \Gamma_j} \int_{\lambda \in \Gamma_\nu} \frac{P_j (\lambda - A_j^\times)^{-1} B_j}{\tau - \lambda} \varphi(\lambda) d\lambda d\tau =$$

$$= \frac{1}{2\pi i} \int_{\lambda \in \Gamma_\nu} P_j P_j^\times (\lambda - A_j^\times)^{-1} B_j \varphi(\lambda) d\lambda .$$

Thus for $j \neq \nu$

$$\alpha_\nu = -\frac{1}{2\pi i} \int_{\Gamma_\nu} P_j P_j^\times (\lambda - A_j^\times)^{-1} B_j \varphi(\lambda) d\lambda$$

$$+ \frac{1}{2\pi i} \int_{\Gamma_\nu} P_j(\lambda - A_j)^{-1} B_j\, W(\lambda)^{-1} \varphi(\lambda) d\lambda \ .$$

It follows that

$$N_j M_{21}^\times \varphi = \frac{1}{2\pi i} \int_\Gamma P_j(\lambda - A_j)^{-1} B_j\, W(\lambda)^{-1} \varphi(\lambda) d\lambda \ +$$

$$- \frac{1}{2\pi i} \int_\Gamma P_j\, P_j^\times(\lambda - A_j^\times)^{-1} B_j\, \varphi(\lambda) d\lambda \ \cdot$$

Since $N\, M_{21}^\times\, \varphi = \mathrm{col}(N_j\, M_{21}^\times \varphi)_{j=0}^s$, we get the desired description of Im T. It remains to establish the formula for the generalized inverse T^I of T. Substituting (3.19) in (3.23) and using the identities

$$M_{12}^\times = -U^\times N^\times \quad , \quad N^\times M_{22} = RN \ ,$$

which follow from (3.16), we see that

$$T^I = T^\times + U^\times R\, N\, M_{21}^\times - U^\times N^\times M_{21}\, U^\times S^I\, N\, M_{21}^\times \ ,$$

and, using again (3.16),

$$N^\times M_{21} U^\times = R\, S - I_{\mathcal{B}^-} \quad .$$

Hence

$$T^I = T^\times + U^\times R\Lambda + U^\times(I - R\, S) S^I \Lambda \quad ,$$

where $\Lambda = N\, M_{21}^\times$. □

In conclusion of this section we remark that using (3.16) - (3.24) one can also express Ker S, Im S and S^I in terms of Ker T, ImT and T^I.

4. The barrier problem

Let $\mathcal{B}$ be an admissible Banach space of Y-valued functions, with the direct sum decomposition $\mathcal{B} = \mathcal{B}^+ \oplus \mathcal{B}^-$, as in Section 2. For a continuous L(Y)-valued function $W(\lambda)$ defined on Γ, consider the Riemann barrier problem:

$$(4.1) \qquad \begin{cases} W(\lambda)\varphi_-(\lambda) = \varphi_+(\lambda) \ , \quad \lambda \in \Gamma \ ; \\ \varphi_- \in \mathcal{B}^- \quad , \quad \varphi_+ \in \mathcal{B}^+ \quad . \end{cases}$$

Using Theorem 3.5, we shall find a general solution to the problem (4.1) in case $W(\lambda)$ is given by its realizations (1.1).

THEOREM 4.1. *Let* $W(\lambda)$ *be given by*

$$(4.2) \qquad W(\lambda) = I + C_j(\lambda - A_j)^{-1} B_j \quad , \quad \lambda \in \Gamma_j \quad , \quad j = 0,\dots,s,$$

where the operators A_j *are such that* $\sigma(A_j)\cap\Gamma_j = \sigma(A_j - B_jC_j)\cap\Gamma_j = \emptyset$ $(j = 0,\dots,s)$. *Let* R *be the right indicator of* $W(\lambda)$ *with respect to its realization* (4.2) *and contour* Γ. *Then the general solution of the Riemann barrier problem* (4.1) *is given by*

$$(4.3)\begin{cases} \varphi_-(\lambda) = C_j(\lambda - A_j^{\times})^{-1} x_j + \\ \qquad + W(\lambda)^{-1}\Big(\sum_{k\neq j} C_k(\lambda - A_k)^{-1} x_k\Big) \quad , \quad \lambda \in \Gamma_j \quad , \ (j = 0,1,\dots,s) \\ \varphi_+(\lambda) = \sum_{k=0}^{s} C_k(\lambda - A_k)^{-1} x_k \quad , \quad \lambda \in \Gamma \quad , \end{cases}$$

where $\mathrm{col}[x_j]_{j=0}^{s}$ *is an arbitrary element in* $\operatorname{Ker} R$. *Moreover, given a solution* $\{\varphi_-, \varphi_+\}$ *of* (4.1), *then* $x_0,\dots,x_s$ *in* (4.3) *are uniquely determined.*

Proof. We shall use the notations introduced in Section 2. Observe that (φ_-, φ_+) is a solution of (4.1) if and only if $\varphi_- \in \operatorname{Ker} M_{22}$. By Theorem 3.5,

$$\operatorname{Ker} M_{22} = \{F\begin{bmatrix} x \\ 0 \end{bmatrix} \mid x \in \operatorname{Ker} R\} ,$$

where

$$F = \begin{bmatrix} M_{21}^{\times}U & M_{22}^{\times} \\ R & N^{\times} \end{bmatrix} : Z \oplus \mathcal{B}^- \to \mathcal{B}^- \oplus Z^{\times} .$$

Take $x \in \operatorname{Ker} R$. Then Lemma 3.4 implies that

$$F\begin{bmatrix} x \\ 0 \end{bmatrix} = M_{21}^{\times} Ux = M_{W^{-1}} Ux - T^{\times} Ux$$

$$= M_{W^{-1}} Ux + U^{\times} Rx = M_{W^{-1}} Ux \quad .$$

It follows that the general solution of (4.1) is given by

$$\varphi_- = M_{W^{-1}} Ux \ , \ \varphi_+ = Ux \ ,$$

where x is an arbitrary element of Ker R. Since F is invertible, it is clear that x is uniquely determined by $\{\varphi_-, \varphi_+\}$. Further, one checks readily that φ_+ and φ_- have the desired form. □

We remark that Theorem 4.1 allows us also to solve the following barrier problem:

$$(4.4) \qquad \begin{cases} W(\lambda)\varphi_+(\lambda) = \varphi_-(\lambda) \ , & \lambda \in \Gamma \ ; \\ \varphi_+ \in \mathcal{B}^+ \ , & \varphi_- \in \mathcal{B}^- \ . \end{cases}$$

Indeed, rewriting the barrier identity in (4.4) as $W^{-1}(\lambda)\varphi_-(\lambda)=\varphi_+(\lambda)$ we obtain a barrier problem of type (4.1) and Theorem 4.1 is applicable. Thus, the general solution of (4.4) is given by the formulas

$$\begin{cases} \varphi_-(\lambda) = C_j(\lambda-A_j)^{-1} x_j + W(\lambda)[\sum_{k\neq j} C_k(\lambda-A_k^{\times})^{-1} x_k] \ , & \lambda \in \Gamma_j \ (j=0,1,\ldots,s) \\ \varphi_+(\lambda) = \sum_{k=0}^{s} C_k(\lambda-A_k^{\times})^{-1} x_k \ , & \lambda \in \Gamma \ , \end{cases}$$

where $\mathrm{col}[x_j]_{j=0}^{s}$ is an arbitrary element in Ker S.

5. Canonical factorization

Let $W(\lambda)$ be a continuous $L(Y)$-valued function defined on Γ and assume that $W(\lambda)$ is invertible for each $\lambda \in \Gamma$. A factorization

$$(5.1) \qquad W(\lambda) = W_-(\lambda)W_+(\lambda) \quad , \lambda \in \Gamma \ ,$$

where $W_{\pm}(\lambda)$ are continuous $L(Y)$-valued functions on Γ, is called *left canonical* (with respect to Γ) if $W_-(\lambda)$ and $W_+(\lambda)$ admit analytic continuations in Ω^- and Ω^+, respectively, such that all operators $W_-(\lambda), \lambda \in \Omega^- \cup \Gamma$, and $W_+(\lambda), \lambda \in \Omega^+ \cup \Gamma$, are invertible. Note that since Ω^- consists of $(s+1)$ disjoint connected components $\Omega_0^-, \Omega_1^-, \ldots, \Omega_s^-$, we can regard W_- as a collection of $(s+1)$ analytic functions $W_-^{(0)}, W_-^{(1)}, \ldots, W_-^{(s)}$ defined in $\Omega_0^-, \Omega_1^-, \ldots, \Omega_s^-$, respectively, and such that the boundary values of $W_-^{(i)}$ on Γ_i coincide with the restriction $W_-|\Gamma_i$. On the other hand, as Ω^+ is connected, the restrictions $W_+|\Gamma_i$ $(i=0,1,\dots,l)$ are the boundary values of the same analytic function W_+ in Ω^+.

It is easy to see that a left canonical factorization, if exists, is unique up to multiplication of $W_+(\lambda)$ by a constant (i.e. independent of λ) invertible operator E on the left and simultaneous multiplication of $W_-(\lambda)$ by E^{-1} on the right.

Interchanging in (2.1) the places of W_- and W_+, we obtain an analogous definition of a *right canonical* factorization $W = W_+ W_-$ of W.

The next theorem gives necessary and sufficient conditions for existence of canonical factorization as well as explicit formulas for the factors $W_\pm$ in case $W(\lambda)$ admits realizations

$$(5.2)\quad W(\lambda) = W_j(\lambda) = I + C_j(\lambda-A_j)^{-1}B_j \quad , \lambda \in \Gamma_j \quad , \quad j = 0,1,..,s.$$

THEOREM 5.1. *Let an* $L(Y)$*-valued function* $W(\lambda)$ *be defined by formulas* (5.2) *and let* $\sigma(A_j) \cap \Gamma_j = \sigma(A_j^\times) \cap \Gamma_j = \emptyset$. *Then* $W(\lambda)$ *admits right (resp. left) canonical factorization with respect to* Γ *if and only if the right (resp. left) indicator* R *(resp.* S*) of* $W(\lambda)$ *with respect to the realizations* (5.2) *is invertible.*

If S *is invertible, then the factors* $W_\pm$ *in* (5.1) *are given by the formulas*

$$(5.3)\qquad W_+(\lambda) = I + C^\times S^{-1}(\lambda-M)^{-1} B, \quad \lambda \in \Omega^+ \quad ;$$

$$(5.4)\quad W_-(\lambda)=I+[C_i C^\times]\left\{\lambda-\begin{bmatrix} A_i & B_i C^\times \\ 0 & M^\times \end{bmatrix}\right\}^{-1}\begin{bmatrix} B_i \\ -S^{-1}B \end{bmatrix} \quad , \lambda\in\Omega_i^- \ ; \ i=0,\dots,s,$$

where M, $M^\times$, $C^\times$, B *are defined by* (1.3)-(1.6).

Analogously, if R *is invertible, then the factors of the right canonical factorization* $W(\lambda) = W_+(\lambda)W_-(\lambda)$ *are given by the formulas*

$$(5.5)\qquad W_+(\lambda) = I - C(\lambda-M)^{-1} R^{-1} B^\times \quad , \quad \lambda \in \Omega^+ \quad ;$$

(5.6) $$W_-(\lambda) = I + [C_i \quad CR^{-1}]\left\{\lambda - \begin{bmatrix} A_i & 0 \\ B^\times C_i & M^\times \end{bmatrix}\right\}^{-1} \begin{bmatrix} B_i \\ B^\times \end{bmatrix}, \quad \lambda \in \Omega_i^-, \quad i=0,\dots,s,$$

where C, $B^\times$ *are defined by* (1.8)-(1.9).

Proof. By the results of Gohberg-Leiterer [8] the Toeplitz operator T with the symbol $W(\lambda)$ is invertible if and only if $W(\lambda)$ admits left canonical factorization. By Theorem 2.1 the invertibility of T is equivalent to the invertibility of the left indicator of $W(\lambda)$, and the part of the Theorem concerning existence of left canonical factorization follows. To prove the existence part for the right canonical factorization, observe that $W(\lambda)$ admits right canonical factorization if and only if $W(\lambda)^{-1}$ admits a left one, and that the right indicator for $W(\lambda)$ coincides with the left indicator for $W(\lambda)^{-1}$.

Now we pass to the verification of formulas (5.3) - (5.6).

Let W_+ and W_- be given by (5.3) and (5.4), respectively. Clearly, W_+ is analytic in Ω^+. Using (1.2), one easily verifies that

(5.7) $$W_+(\lambda)^{-1} = I - C^\times(\lambda - M^\times)^{-1} S^{-1} B,$$

which is also analytic in Ω^+. Multiplying

$$W(\lambda) = I + C_i(\lambda - A_i)^{-1} B_i \quad , \lambda \in \Gamma_i,$$

and $W_+(\lambda)^{-1}$ given by (5.7), it is easily seen that the product coincides with the formula (5.4). It remains to show that the function (5.4) and its inverse admit analytic continuations into Ω_i^-.

The function (5.4) can be written as

$$W_-(\lambda) = I + C_i(\lambda - A_i)^{-1}(I - P_i)B_i + D(\lambda)S^{-1}B \quad , \lambda \in \Gamma_i \quad ,$$

where

$$D(\lambda) = [0\dots0\ C_i(\lambda - A_i)^{-1}\ 0\dots0]S - C_i(\lambda - A_i)^{-1}B_iC^\times(\lambda - M^\times)^{-1} + $$
$$- C^\times(\lambda - M^\times)^{-1} \quad , \quad \lambda \in \Gamma_i$$

is a function whose values are operators $Z^\times \to Y$. In order to prove that $W_-(\lambda)$ admits analytic continuation in Ω_i^- it suffices to show that

$D(\lambda)$ has this property. Substituting the expressions for S, C^{x} and M^{x} we obtain for each $x \in \operatorname{Im} P_k^{x}$ $(k \neq i)$:

$$D(\lambda)x = C_i(\lambda-A_i)^{-1}\left[-\frac{1}{2\pi i}\int_{\Gamma_i}(\mu-A_i)^{-1}B_iC_k(\mu-A_k^{x})^{-1}\,x\,d\mu\right] +$$

$$- [C_i(\lambda-A_i)^{-1}B_i + I]C_k(\lambda-A_k^{x})^{-1}x =$$

$$= C_i\left[-\frac{1}{2\pi i}\int_{\Gamma_i}\frac{(\lambda-A_i)^{-1}-(\mu-A_i)^{-1}}{\mu-\lambda}B_i\,C_k(\mu-A_k^{x})^{-1}\,x\,d\mu\right] +$$

$$- [C_i(\lambda-A_i)^{-1}B_i + I]C_k(\lambda-A_k^{x})^{-1}x.$$

Now $(\mu-A_k^{x})^{-1}x$ is analytic in Ω_i^- (because $k \neq i$), so

$$-\frac{1}{2\pi i}\int_{\Gamma_i}\frac{1}{\mu-\lambda}B_iC_k(\mu-A_k^{x})^{-1}\,x\,d\mu = B_iC_k(\lambda-A_k^{x})^{-1}x \qquad (\lambda \in \Omega_i^-) ,$$

and the expression for $D(\lambda)x$ takes the form

$$C_k(\lambda-A_k^{x})^{-1}x + C_i\left[\frac{1}{2\pi i}\int_{\Gamma_i}\frac{1}{\mu-\lambda}(\mu-A_i)^{-1}B_iC_k(\mu-A_k^{x})^{-1}\,x\,d\mu\right] .$$

Both summands obviously admit analytic continuations into Ω_i^-. If $x \in \operatorname{Im} P_i^{x}$, then

$$D(\lambda)x = C_i[(\lambda-A_i)^{-1}P_i - (\lambda-A_i)^{-1}B_iC_i(\lambda-A_i^{x})^{-1} - (\lambda-A_i^{x})^{-1}]\,x =$$

$$= C_i[(\lambda-A_i)^{-1}P_i - (\lambda-A_i)^{-1}]\,x ,$$

which is again analytic in Ω_i^-.

The function whose values are the inverses of the values of (5.4) can be written as follows:

$$(5.8)\qquad W_-(\lambda)^{-1} = I-[C_i \quad C^{x}]\left\{\lambda-\begin{bmatrix} A_i^{x} & 0 \\ S^{-1}BC_i & M^{x}+S^{-1}BC^{x}\end{bmatrix}\right\}^{-1}\begin{bmatrix} B_i \\ -S^{-1}B\end{bmatrix}, \quad \lambda \in \Gamma_i .$$

Using the equality (which follows from (1.2))

$$(\lambda-(M^{\times}+S^{-1}BC^{\times}))^{-1}S^{-1}=S^{-1}(\lambda-M)^{-1}\quad,$$

one can rewrite the expression for $W_-(\lambda)^{-1}$ in the form

$$W_-(\lambda)^{-1}=I-C_i(I-P_i^{\times})(\lambda-A_i^{\times})^{-1}B+C^{\times}S^{-1}D(\lambda)^{\times}\quad,\quad\lambda\in\Gamma_i\quad,$$

where

$$D(\lambda)^{\times}=S\begin{bmatrix}0\\ \cdot\\ \cdot\\ \cdot\\ 0\\ P_i^{\times}(\lambda-A_i^{\times})^{-1}B_i\\ 0\\ \cdot\\ \cdot\\ \cdot\\ 0\end{bmatrix}-(\lambda-M)^{-1}B[C_i(\lambda-A_i^{\times})^{-1}B_i-I]\quad.$$

Further, one shows (as it was done in the preceding paragraph for $D(\lambda)$) that $P_kD(\lambda)^{\times}$ admits analytic continuation into Ω_i^-, for $k=0,1,\dots,s$. Hence $W_-(\lambda)^{-1}$ is analytic in Ω_i^- as well. So, formulas (5.3) and (5.4) are verified.

To verify (5.5) and (5.6), use the already proved formulas (5.3) and (5.4) for the left canonical factorization $W(\lambda)^{-1}=W_-(\lambda)^{-1}W_+(\lambda)^{-1}$ of the inverse function

$$W(\lambda)^{-1}=I+C_i(\lambda-A_i^{\times})^{-1}(-B_i)\quad,\quad\lambda\in\Gamma_i\quad.$$

Bearing in mind that the right indicator of $W(\lambda)$ coincides with the left indicator of $W(\lambda)^{-1}$, we obtain the inverses of the factors in the right canonical factorization $W=W_+W_-$:

(5.9) $$W_+(\lambda)^{-1}=I+CR^{-1}(\lambda-M^{\times})^{-1}B^{\times}\quad;\quad\lambda\in\Omega^+\quad;$$

(5.10) $$W_-(\lambda)^{-1}=I+[C_i\;\;C]\left\{\lambda-\begin{bmatrix}A_i^{\times} & -B_iC\\ 0 & M\end{bmatrix}\right\}^{-1}\begin{bmatrix}-B_i\\ -R^{-1}B^{\times}\end{bmatrix},\quad\lambda\in\Omega_i^-\quad.$$

Taking inverses of these realizations and using equality (1.7), we obtain the desired formulas (5.5) and (5.6). □

As an illustration of Theorem 5.1 consider the case of a scalar function $W(\lambda)$ (i.e. $Y = \mathbb{C}$). Let ρ_j (resp. ν_j) denote the number of poles (resp. zeros) of $W_j(\lambda)$ in Ω_j^- $(j=0,1,\dots,s)$. If $W(\lambda)$ admits canonical factorization, then, in view of Theorem 5.1, the indicator S is invertible and, in particular, $\dim Z = \dim Z^\times$. This amounts to the equality

$$\sum_{j=1}^{s} \rho_j = \sum_{j=1}^{s} \nu_j \quad . \tag{5.11}$$

Conversely, using the fact that the indicator S satisfies the Lyapunov equation (1.2) and applying Corollary 1.3 from [9], one easily sees that condition (5.11) implies the invertibility of the indicator S and, consequently, the existence of the canonical factorization of $W(\lambda)$.

In conclusion of this section note that the smoothness assumptions imposed on Γ_j in Section 1 are superfluous. This follows from the fact that for the purpose of canonical factorization the given contours Γ_j can be replaced by smooth contours $\tilde{\Gamma}_j$ which are sufficiently close to Γ_j and meet all other hypothesis of Section 1.

6. Unbounded domains.

The results of Sections 2-5 can be extended to the case of unbounded domains Ω^+. In this section we present results on canonical factorization for two types of unbounded domains Ω^+.

Firstly, consider the case that $\Omega^- = [\mathbb{C} \cup \{\infty\}] \setminus \Omega^+$ consists of a finite number of bounded connected components Ω_j^- $(j=1,\dots,s)$, whose boundaries Γ_j are simple closed rectifiable non-intersecting contours positively oriented with respect to Ω^+. The $L(Y)$-valued function $W(\lambda)$ is again defined on $\Gamma = \bigcup_{j=1}^{s} \Gamma_j$ by realizations (1.1), and it is assumed that $\sigma(A_j) \cap \Gamma_j = \sigma(A_j^\times) \cap \Gamma_j = \emptyset$. We define the spaces Z, $Z^\times$, the indicators R, S and the operators M, $M^\times$, C, $C^\times$, B, $B^\times$ precisely as in Section 1 with the only remark that the indices i, j, k which are involved in these definitions take values $1, 2, \dots, s$ (but not the value 0). Then the following analogue of Theorem 5.1 holds.

THEOREM 6.1. *Under the above assumptions and notations the function* $W(\lambda)$ *admits right (resp. left) canonical factorization if and only if the right (resp. left) indicator* R *(resp.* S *) is invertible. If* S *is invertible, then the factors of the left canonical factorization and their inverses are given by formulas* (5.3), (5.4), (5.7), (5.8). *Similarly, if* R *is invertible, then the factors of the right canonical factorization and their inverses are given by* (5.5), (5.6), (5.9), (5.10).

Proof. Choose an auxiliary simple closed rectifiable contour Γ_0 bounding a bounded domain Δ such that $\Gamma_j \subset \Delta$ $(j=1,2,\ldots,s)$. Denote $\tilde{\Omega}^+ = \Delta \setminus [\bigcup_{j=1}^{s} \overline{\Omega_j^-}]$, $\tilde{\Gamma} = \bigcup_{j=0}^{s} \Gamma_j$ (Γ_0 is positively oriented with respect to Ω^+). Introduce a new function $\tilde{W}(\lambda)$ on $\tilde{\Gamma}$ by setting $\tilde{W}(\lambda) = W(\lambda)$, $\lambda \in \Gamma$ and $\tilde{W}(\lambda) = I$, $\lambda \in \Gamma_0$. We claim that $W(\lambda)$ admits right (resp. left) canonical factorization with respect to Γ if and only if $\tilde{W}(\lambda)$ admits right (resp. left) canonical factorization with respect to $\tilde{\Gamma}$. Indeed, assume that $\tilde{W}(\lambda) = \tilde{W}_-(\lambda)W_+(\lambda)$, $\lambda \in \tilde{\Gamma}$ is a left canonical factorization of $\tilde{W}(\lambda)$ and set $W_-(\lambda) = \tilde{W}_-(\lambda)$, $W_+(\lambda) = \tilde{W}_+(\lambda)$ for $\lambda \in \Gamma$. Obviously, $W(\lambda) = W_-(\lambda)W_+(\lambda)$, $\lambda \in \Gamma$ and the function $W_-(\lambda), \lambda \in \Gamma$ meets the requirements for the left factor in factorization (5.1). The factor $W_+(\lambda)$, $\lambda \in \Gamma$ admits an analytic and invertible extension in $\tilde{\Omega}^+$ and since $\tilde{W}_+(\lambda) = [\tilde{W}_0^-(\lambda)]^{-1}$ on Γ_0 the function $W_+(\lambda)$ admits an analytic and invertible extension in the whole Ω^+. Conversely, if $W(\lambda)=W_-(\lambda)W_+(\lambda)$, $\lambda \in \Gamma$ is a left factorization of $W(\lambda)$, we define $\tilde{W}_+(\lambda) = W_+(\lambda)$ for $\lambda \in \tilde{\Gamma}$, $\tilde{W}_-(\lambda) = W_-(\lambda)$ for $\lambda \in \Gamma$ and $\tilde{W}_-(\lambda) = [W_+(\lambda)]^{-1}$ for $\lambda \in \Gamma_0$. Clearly, $\tilde{W}(\lambda) = \tilde{W}_-(\lambda)W_+(\lambda)$, $\lambda \in \tilde{\Gamma}$ and the functions $\tilde{W}_-(\lambda)$, $\tilde{W}_+(\lambda)$ meet all the requirements for the factors in the canonical factorization of $\tilde{W}(\lambda)$.

Using Theorem 5.1 we conclude that $W(\lambda)$ admits left canonical factorization with respect to Γ if and only if the left indicator $\tilde{S} : \tilde{Z}^{\times} \to \tilde{Z}$ of $\tilde{W}(\lambda)$ with respect to $\tilde{\Gamma}$ is invertible, where $\tilde{Z}^{\times} = \bigoplus_{j=0}^{s} \operatorname{Im} P_j^{\times}$, $Z = \bigoplus_{j=0}^{s} \operatorname{Im} P_j$. Now observe that $\operatorname{Im} P_0^{\times} = \operatorname{Im} P_0 = \{0\}$ i.e. $\tilde{Z}^{\times} = Z^{\times} = \bigoplus_{j=1}^{s} \operatorname{Im} P_j^{\times}$, $\tilde{Z} = Z = \bigoplus_{j=1}^{s} \operatorname{Im} P_j$. So, in fact, the operator $\tilde{S}$ coincides with the left indicator $S : Z^{\times} \to Z$ of $W(\lambda)$ with respect to Γ as defined at the very beginning of this Section. Further, the first paragraph of this proof shows that the factorization factors of $W(\lambda)$ are obtained as restrictions of $\tilde{W}(\lambda)$ to the contour Γ. Applying again the observation that $\tilde{Z} = Z$, $\tilde{Z}^x = Z$, we see that formulas (5.3), (5.4), (5.7), (5.8) hold with M, M^x, C^x, B as defined at the beginning of this Section.

The results for the right factorization are obtained similarly. □

Secondly, we are interested in the case that Ω^+ is a strip bounded by the lines $\Gamma_1 = \{\lambda \in \mathbb{C} \mid \operatorname{Im}\lambda = h\}$ and $\Gamma_2 = \{\lambda \in \mathbb{C} \mid \operatorname{Im}\lambda = -h\}$ where h is some fixed positive number. Note that in this case the domains $\Omega_1^- = \{\lambda \in \mathbb{C} \mid \operatorname{Im}\lambda > h\}$ and $\Omega_2^- = \{\lambda \in \mathbb{C} \mid \operatorname{Im}\lambda < -h\}$ are unbounded. Again we assume that the $L(Y)$-valued function $W(\lambda)$ is defined on $\Gamma = \Gamma_1 \cup \Gamma_2$ by

$$W(\lambda) = \begin{cases} W_1(\lambda) = I + C_1(\lambda - A_1)^{-1} B_1 , & \lambda \in \Gamma_1 \\ W_2(\lambda) = I + C_2(\lambda - A_2)^{-1} B_2 , & \lambda \in \Gamma_2 \end{cases} \tag{6.1}$$

and

$$\sigma(A_1) \cap \Gamma_1 = \sigma(A_1^x) \cap \Gamma_1 = \sigma(A_2) \cap \Gamma_2 = \sigma(A_2^x) \cap \Gamma_2 = \emptyset . \tag{6.2}$$

Denote

$$Z = \operatorname{Im} P_1 \oplus \operatorname{Im} P_2 , \qquad Z^x = \operatorname{Im} P_1^x \oplus \operatorname{Im} P_2^x ,$$

where P_j (resp. P_j^x) stands for the spectral projection of A_j (resp. A_j^x) corresponding to the part of $\sigma(A_j)$ (resp. of $\sigma(A_j^x)$) in Ω_j^- $(J = 1,2)$. Introduce the operator matrices $S : Z^x \to Z$ and $R : Z \to Z^x$ as follows:

$$S = \begin{bmatrix} P_1 | \operatorname{Im} P_1^x & \frac{1}{2\pi i} \int_{-\infty}^{\infty} (\lambda + ih - A_1)^{-1} B_1 C_2 (\lambda + ih - A_2^x)^{-1} P_2^x \, d\lambda \\ -\frac{1}{2\pi i} \int_{-\infty}^{\infty} (\lambda - ih - A_2)^{-1} B_2 C_1 (\lambda - ih - A_1^x)^{-1} P_1^x d\lambda & P_2 | \operatorname{Im} P_2^x \end{bmatrix} \tag{6.3}$$

$$R = \begin{bmatrix} P_1^x | \operatorname{Im} P_1 & -\frac{1}{2\pi i} \int_{-\infty}^{\infty} (\lambda + ih - A_1^x)^{-1} B_1 C_2 (\lambda + ih - A_2)^{-1} P_2 \, d\lambda \\ \frac{1}{2\pi i} \int_{-\infty}^{\infty} (\lambda - ih - A_2^x)^{-1} B_2 C_1 (\lambda - ih - A_1)^{-1} P_1 \, d\lambda & P_2^x | \operatorname{Im} P_2 \end{bmatrix} \tag{6.4}$$

The operator S (resp. R) will be referred to as the *left* (resp. *right*) *indicator* of $W(\lambda)$ with respect to $\Gamma = \Gamma_1 \cup \Gamma_2$. We shall see that the existence of the left (resp. right) canonical factorization of $W(\lambda)$ is equivalent to the invertibility of the operator (6.3) (resp. (6.4)). Let us also define explicitly all other operators which will be involved in the formulas for the factorization factors:

$$(6.5)\quad M=\begin{bmatrix} A_1|\operatorname{Im} P_1 & 0 \\ 0 & A_2|\operatorname{Im} P_2\end{bmatrix} : Z\to Z\ ,\quad M^{\times}=\begin{bmatrix} A_1^{\times}|\operatorname{Im} P_1^{\times} & 0 \\ 0 & A_2|\operatorname{IM} P_2^{\times}\end{bmatrix} : Z^{\times}\to Z^{\times}\ ;$$

$$(6.6)\quad C=[C_1|\operatorname{Im} P_1, C_2|\operatorname{Im} P_2] : Z\to Y\ ,\quad C^{\times}=[C_1|\operatorname{Im} P_1^{\times}, C_2|\operatorname{Im} P_2^{\times}] : Z^{\times}\to Y;$$

$$(6.7)\quad B=\begin{bmatrix} P_1B_1 \\ P_2B_2\end{bmatrix} : Y\to Z\quad ,\qquad B^{\times}=-\begin{bmatrix} P_1^{\times}B_1 \\ P_2^{\times}B_2\end{bmatrix} : Y\to Z^{\times}\quad .$$

THEOREM 6.2. *Let the* $L(Y)$*-valued function* $W(\lambda)$ *be defined by* (6.1) *on the boundary* $\Gamma=\Gamma_1\cup\Gamma_2$ *of the strip* Ω^+ *and assume that* (6.2) *holds. Then* $W(\lambda)$ *admits left (resp. right) canonical factorization with respect to* Γ *if and only if the left (resp. right) indicator* S *(resp.* R*) defined by* (6.3) *(resp.* (6.4)) *is invertible.*

If S *(resp.* R *) is invertible, then the factors* $W_{\pm}$ *of the left (resp. right) canonical factorization and their inverses* $W_{\pm}^{-1}$ *are given by formulas* (5.3), (5.4), (5.7), (5.8) *(resp.* (5.5), (5.6), (5.9), (5.10)), *where the operators* M, $M^{\times}$, C, $C^{\times}$ *and* B, $B^{\times}$ *are defined by* (6.5) - (6.7).

Proof. Fix some $\tau>0$ and introduce the auxiliary curves

$$\hat{\Gamma}_1=\{\lambda\in\Gamma_1|\ |\operatorname{Re}\lambda|\leq\tau\}\cup\{\lambda\in\mathbb{C}\ |\ |\lambda-ih|=\tau\quad,\quad \operatorname{Im}\lambda\geq h\}\ ,$$

$$\hat{\Gamma}_2=\{\lambda\in\Gamma_2|\ |\operatorname{Re}\lambda|\leq\tau\}\cup\{\lambda\in\mathbb{C}\ |\ |\lambda+ih|=\tau\quad,\quad \operatorname{Im}\lambda\leq -h\}\ .$$

The orientation on $\hat{\Gamma}_j$ is chosen to be consistent with the orientation on Γ_j $(j=1,2)$, i.e. $\hat{\Gamma}_j$ is negatively oriented with respect to the bounded domain $\hat{\Omega}_j^-$ whose boundary is $\hat{\Gamma}_j$ $(j=1,2)$:

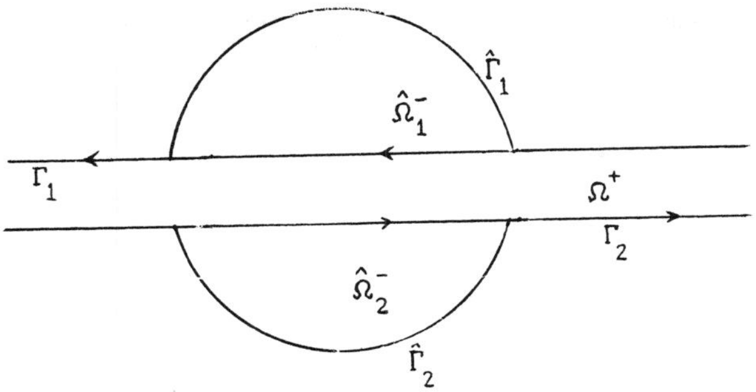

Denote $\hat{\Omega}^+ = [\mathbb{C} \cup \{\infty\}] \setminus [\overline{\hat{\Omega}_1^-} \cup \overline{\hat{\Omega}_2^-}]$ and choose a sufficiently large number τ such that

(6.8) $\sigma(A_j) \cap \Omega_j^- \subset \hat{\Omega}_j^-$, $\sigma(A_j^\times) \cap \Omega_j^- \subset \hat{\Omega}_j^-$ (j=1,2) .

Introduce the function $\hat{W}(\lambda)$ on $\hat{\Gamma} = \hat{\Gamma}_1 \cup \hat{\Gamma}_2$ by setting

$$W(\lambda) = \begin{cases} \hat{W}_1(\lambda) = I + C_1(\lambda - A_1)^{-1}B_1 , & \lambda \in \hat{\Gamma}_1 \\ \hat{W}_2(\lambda) = I + C_2(\lambda - A_2)^{-1}B_2 , & \lambda \in \hat{\Gamma}_2 \end{cases} .$$

We claim that $W(\lambda)$ admits left canonical factorization with respect to Γ if and only if $\hat{W}(\lambda)$ admits such factorization with respect to $\hat{\Gamma}$. Indeed, let $W(\lambda) = W_-(\lambda)W_+(\lambda)$ $(\lambda \in \Gamma)$ be a left canonical factorization of W with respect to Γ (such that $W_+(\infty) = W_-(\infty) = I$). Condition (6.8) implies that the functions $W_j(\lambda)$ and $[W_j(\lambda)]^{-1}$ are analytic in the domain $\Omega_j^- \setminus \overline{\hat{\Omega}_j^-}$ and continuous in its closure. Hence the same property have the functions $[W_j^-(\lambda)]^{-1}W_j(\lambda)$ and $[W_j(\lambda)]^{-1}W_j^-(\lambda)$ (j=1,2). Since $W_+(\lambda) = [W_j^-(\lambda)]^{-1} W_j(\lambda)$ for $\lambda \in \Gamma_j$ (j=1,2), we conclude that $W_+(\lambda)$ admits an analytic and invertible extension in $\hat{\Omega}^+$ which is continuous in $\overline{\hat{\Omega}^+}$ (and takes value I at infinity). Setting $\hat{W}_\pm(\lambda) = W_\pm(\lambda), \lambda \in \hat{\Gamma}$, we obviously obtain the left canonical factorization $\hat{W}(\lambda) = \hat{W}_-(\lambda)W_+(\lambda)(\lambda \in \hat{\Gamma})$ of $\hat{W}$ with respect to $\hat{\Gamma}$. Conversely, let $\hat{W}(\lambda) = \hat{W}_-(\lambda)\hat{W}_+(\lambda)$, $\lambda \in \hat{\Gamma}$ be a left canonical factorization of $\hat{W}$ relative to $\hat{\Gamma}$ (and $\hat{W}_+(\infty) = I$). Set $W_+(\lambda) = \hat{W}_+(\lambda)$ for $\lambda \in \Gamma$. Clearly, $W_+(\lambda)$ meets the requirements for the right factorization factor in (5.1). Further, define $W_-(\lambda) = W_j^-(\lambda) = W_j(\lambda)[W_+(\lambda)]^{-1}$, $\lambda \in \Gamma_j$ (j=1,2).

Then $W(\lambda) = W_-(\lambda)W_+(\lambda)$, $\lambda\in\Gamma$ and using again (6.8), one easily checks that W_- is analytic and invertible in the domain Ω^- and continuous in its closure.

Using Theorem 6.1 we infer that $W(\lambda)$ admits left canonical factorization with respect to Γ if and only if the left indicator $\hat{S}$ of $\hat{W}$ relative to $\hat{\Gamma}$ is invertible, where

$$\hat{S}=\begin{bmatrix} P_1\big|_{\operatorname{Im} P_1^\times} & -\frac{1}{2\pi i}\int\limits_{\hat{\Gamma}_1}(\lambda-A_1)^{-1}B_1C_2(\lambda-A_2^\times)^{-1}P_2^\times d\lambda \\ -\frac{1}{2\pi i}\int\limits_{\hat{\Gamma}_2}(\lambda-A_2)^{-1}B_2C_1(\lambda-A_1^\times)^{-1}P_1^\times d\lambda & P_2\big|_{\operatorname{Im} P_2^\times} \end{bmatrix} : Z^\times \to Z .$$

Note that the spectral projections P_j (resp. $P_j^\times$) $(j=1,2)$ remain unchanged for any contour $\hat{\Gamma}$ satisfying (6.8) and coincide with the spectral projections of the operators A_j (resp. $A_j^\times$) corresponding to $\sigma(A_j)\cap\Omega_j^-$ (resp. $\sigma(A_j^\times)\cap\Omega_j^-$) $(j=1.2)$. Also, one easily sees that the operator $\hat{S}$ is one and the same for any contour $\hat{\Gamma}$ satisfying (6.8) and, in fact, $\hat{S} = S$ as defined by (6.3). Now the formulas for the factorization factors and their inverses follow from Theorem 6.1. The assertions about right factorization are obtained similarly. □

7. <u>The pair equation</u>.

Let Y be a Banach space. Consider the equation of the following type:

$$(7.1)\qquad \begin{cases} \varphi(t) - \int\limits_{-\infty}^{\infty} k_1(t-s)\varphi(s)ds = g(t) , & t < 0 , \\ \varphi(t) - \int\limits_{-\infty}^{\infty} k_2(t-s)\varphi(s)ds = g(t) , & t > 0 . \end{cases}$$

Here $g(t)$ is a given Y-valued function defined on the real line such that

$$\int\limits_{-\infty}^{\infty} e^{h|t|}\,||g(t)||dt < \infty ,$$

where $h \geq 0$ is a fixed number; in short, $g(t)\in e^{-h|t|}L_1(\mathbb{R};Y)$. The unknown function $\varphi(t)$ is also assumed to belong to the Banach space $e^{-h|t|}L_1(\mathbb{R};Y)$. We assume also that the kernels $k_1(t)$ and $k_2(t)$ have the properties that $e^{-ht}k_1(t)$ and $e^{ht}k_2(t)$ are $L(Y)$-valued L_1-functions on the real line, and their Fourier transforms are analytic in a neighborhood of the real line and infinity. This condition was expressed in terms of the functions $k_j(t)$ themselves in [4]. It follows (see Theorem 1.2 in [2] for the case $h = 0$) that $k_j(t)$ $(j=1,2)$ admit exponential representations

$$k_1(t) = \begin{cases} iC_1 e^{-itA_1}(I - P_1)B_1 & , \quad t > 0 , \\ -iC_1 e^{-itA_1} P_1 B_1 & , \quad t < 0 , \end{cases}$$

$$k_2(t) = \begin{cases} iC_2 e^{-itA_2} P_2 B & , \quad t > 0 , \\ -iC_2 {}^{-itA_2} (I - P_2)B & , \quad t < 0 , \end{cases}$$

where $B_j : Y \to X_j$, $A_j : X_j \to X_j$, $C_j : X_j \to Y$ are (bounded linear) operators (with X_1, X_2 denoting some Banach spaces) such that

$$\sigma(A_j) \cap \Gamma_j = \emptyset \quad (j=1,2) \quad , \tag{7.2}$$

and P_j is the Riesz projection of A_j corresponding to the part of the spectrum of A_j in the halfplane Ω_j^- $(j=1,2)$. Here Γ_j, Ω_j^- $(j=1,2)$ are defined as in the part of Section 6 which deals with canonical factorization with respect to the strip Ω^+. In the rest of this Section we use also all other notations introduced in this part of Section 6.

Note that the conditions imposed on the functions $k_j(t)$ $(j=1,2)$ imply that the operator E defined by the left hand side of (7.1):

$$(E\varphi)(t) = \begin{cases} \varphi(t) - \int_{-\infty}^{\infty} k_1(t-s)\varphi(s)ds = g(t) & , \quad t < 0 , \\ \varphi(t) - \int_{-\infty}^{\infty} k_2(t-s)\varphi(s)ds = g(t) & , \quad t > 0 , \end{cases} \tag{7.3}$$

is a bounded linear operator acting $e^{-h|t|}L_1(\mathbb{R},Y) \to e^{-h|t|}L_1(\mathbb{R},Y)$.

Now we start solving equation (7.1). Introduce two functions $\psi_1(t)$ and $\psi_2(t)$ as follows:

$$\psi_1(t) = \begin{cases} 0 , & t < 0 , \\ \varphi(t) - \int_{-\infty}^{\infty} k_1(t-s)\varphi(s)ds , & t > 0 , \end{cases}$$

$$\psi_2(t) = \begin{cases} 0 & t > 0 , \\ \varphi(t) - \int_{-\infty}^{\infty} k_2(t-s)\varphi(s)ds , & t < 0 . \end{cases}$$

Denote also

$$g_1(t) = \begin{cases} g(t) , & t < 0 , \\ 0 , & t > 0 , \end{cases} \tag{7.4}$$

$$g_2(t) = \begin{cases} g(t) , & t > 0 , \\ 0 , & t < 0 . \end{cases} \tag{7.5}$$

With these notations (7.1) can be rewritten in the form

$$\begin{cases} \varphi(t) - \int_{-\infty}^{\infty} k_1(t-s)\varphi(s)ds = g_1(t) + \psi_1(t), & -\infty < t < \infty ; \\ \varphi(t) - \int_{-\infty}^{\infty} k_2(t-s)\varphi(s)ds = g_2(t) + \psi_2(t). & -\infty < t < \infty . \end{cases} \tag{7.6}$$

Multiply the first equation by e^{-ht} and apply the Fourier transform $x(t) \to \int_{-\infty}^{\infty} e^{i\lambda t} x(t)dt$ to both sides:

$$\Phi(\lambda+ih) - K_1(\lambda+ih)\, \Phi(\lambda+ih) = G_1(\lambda+ih) + \Psi_1(\lambda+ih) , \quad -\infty < \lambda < \infty , \tag{7.7}$$

where the capital letters denote the Fourier transform of the function denoted by a corresponding lower case letter. Note that $\Phi(\lambda)$ is analytic in the strip Ω^+, $\Psi_1(\lambda)$ is analytic in the halfplane Ω_1^- and $G_1(\lambda)$ is analytic in $\Omega^+ \cup \overline{\Omega_2^-}$.

Similarly, multypling the second equation in (7.6) by e^{ht} and taking the Fourier transform we obtain

$$\Phi(\lambda-ih) - K_2(\lambda-ih)\,\Phi(\lambda-ih) = G_2(\lambda-ih) + \Psi_2(\lambda-ih) \quad (-\infty < \lambda < \infty)\,. \tag{7.8}$$

Here $\Psi_2(\lambda)$ is analytic in Ω_2^- and $G_2(\lambda)$ is analytic in $\Omega^+ \cup \overline{\Omega_1^-}$.

Introduce the contourwise operator function $W(\lambda)$ defined on $\Gamma = \{\lambda \mid \mathrm{Im}\lambda = \pm h\}$ as

$$W(\lambda) = \begin{cases} I - K_1(\lambda) = I + C_1(\lambda - A_1)^{-1} B_1 & , \quad \lambda \in \Gamma_1\,, \\ I - K_2(\lambda) = I + C_2(\lambda - A_2)^{-1} B_2 & , \quad \lambda \in \Gamma_2\,. \end{cases}$$

Then equations (7.7) and (7.8) can be interpreted as the barrier problem

$$W(\lambda)\,\Phi_+(\lambda) = \Psi_-(\lambda) + G(\lambda) \quad , \quad \lambda \in \Gamma\,, \tag{7.9}$$

where $\Phi_+(\lambda) = \Phi(\lambda) \qquad (\lambda \in \overline{\Omega^+})$,

$$\Psi_-(\lambda) = \begin{cases} \Psi_1(\lambda) & , \quad \lambda \in \Gamma_1\,; \\ \Psi_2(\lambda) & , \quad \lambda \in \Gamma_2\,; \end{cases}$$

$$G(\lambda) = \begin{cases} G_1(\lambda) & , \quad \lambda \in \Gamma_1\,; \\ G_2(\lambda) & , \quad \lambda \in \Gamma_2\,. \end{cases} \tag{7.10}$$

Assume in addition that $\sigma(A_j - B_jC_j) \cap \Gamma_j = \emptyset$ $(j=1,2)$. Then the barrier problem (7.9) has a unique solution $\Phi_+(\lambda)$ for every $g(t) \in e^{-h|t|}L_1(\mathbb{R},Y)$, with $G(\lambda)$ obtained from $g(t)$ by (7.4), (7.5) and

(7.10) if and only if the operator function $W(\lambda)$ admits left canonical factorization with respect to $\Gamma = \Gamma_1 \cup \Gamma_2$ (see [8]). This observation allows us to use Theorem 6.2 in order to express the criterium for unique solvability of (7.9) and hence of (7.1) and produce formulas for the solution in terms of the left indicator S of $W(\lambda)$ with respect to Γ defined by (6.3).

Using the notations (6.5) - (6.7) we define the following functions

$$w_1(t) = \begin{cases} i[C_1, C^{\times}S^{-1}](I-\tilde{P}_1)\exp\left\{-it\begin{bmatrix} A_1^{\times} & 0 \\ BC_1 & M \end{bmatrix}\right\}(I-\tilde{P}_1)\begin{bmatrix} B_1 \\ -B \end{bmatrix}, & t > 0, \\ 0, \quad t < 0 \quad ; \end{cases}$$

$$w_2(t) = \begin{cases} 0, \quad t > 0 \quad , \\ -i[C_2, C^{\times}S^{-1}](I-\tilde{P}_2)\exp\left\{-it\begin{bmatrix} A_2^{\times} & 0 \\ BC_2 & M \end{bmatrix}\right\}(I-\tilde{P}_2)\begin{bmatrix} B_2 \\ -B \end{bmatrix}, & t < 0, \end{cases}$$

where $\tilde{P}_j$ is the spectral projection of $\begin{bmatrix} A_j^{\times} & 0 \\ BC_j & M \end{bmatrix}$ corresponding to the domain Ω_j^- $(j=1,2)$.

Let

$$r(t) = \begin{cases} iC_2 e^{-itA_2^{\times}} P_2^{\times} S^{-1} B, & t > 0, \\ -iC_1 e^{-itA_1^{\times}} P_1^{\times} S^{-1} B, & t < 0, \end{cases}$$

where $P_1^{\times}$ is understood here as the operator $Z^{\times} \to \operatorname{Im} P_1^{\times}$ which is identity on the first component of $Z^{\times} = \operatorname{Im} P_1^{\times} \oplus \operatorname{Im} P_2^{\times}$ and zero on the second. The operator $P_2^{\times}$ is interpreted analogously.

THEOREM 7.1 *The operator* E *defined by* (7.3) *is invertible if and only if the left indicator* S *defined by* (6.3) *is invertible. If this condition holds then for any* $g(t) \in e^{-h|t|}L_1(\mathbb{R};Y)$ *the unique solution of* (7.1) *is given by the formula*

$$\varphi(t) = g(t) + \int_{-\infty}^{\infty} \gamma(t,s)g(s)ds ,$$

where

$$\gamma(t,s) = r(t-s) + w(t-s) + \int_{-\infty}^{\infty} r(t-u)w(u,s)du ,$$

and

$$w(t,s) = \begin{cases} w_1(t,s) , & t < 0 ; \\ w_2(t,s) , & t > 0 . \end{cases}$$

Proof. The first assertion of the theorem is already proved. Let S be invertible. Then the inverses of the factors in the left canonical factorization $W(\lambda) = W_-(\lambda)W_+(\lambda)$ $(\lambda\in\Gamma)$ are given by formulas (5.7) and (5.8). As $W_-(\lambda)^{-1}$ is analytic in $\Omega_1^- \cup \Omega_2^-$, we have

$$[C_j, C^x S^{-1}]\tilde{P}_j \left\{\lambda - \begin{bmatrix} A_j^x & 0 \\ BC_j & M \end{bmatrix}\right\}^{-1} \tilde{P}_j \begin{bmatrix} B_j \\ -B \end{bmatrix} \equiv 0 ,$$

and hence

$$[C_j, C^x S^{-1}]\tilde{P}_j \exp\left\{-it \begin{bmatrix} A_j^x & 0 \\ BC_j & M \end{bmatrix}\right\} \tilde{P}_j \begin{bmatrix} B_j \\ -B \end{bmatrix} \equiv 0 .$$

Now one checks easily that

$$W_-(\lambda)^{-1} = I + \int_{-\infty}^{\infty} e^{i\lambda t} w_j(t)dt , \quad \lambda \in \Gamma_j , \quad j = 1,2.$$

Therefore, for $\lambda \in \Gamma_j$ we obtain

$$W_-(\lambda)^{-1} G_j(\lambda) = (I + \int_{-\infty}^{\infty} e^{i\lambda t} w_j(t)dt) \int_{-\infty}^{\infty} g_j(t)e^{i\lambda t} dt =$$

$$= \int_{-\infty}^{\infty} [g_j(t) + \int_{-\infty}^{\infty} w_j(t - s)g_j(s)ds]e^{i\lambda t} dt .$$

For an $L(Y)$-valued function $A(\lambda)$ which is analytic on $\Gamma_1 \cup \Gamma_2$ and has the form

$$A(\lambda) = \int_{-\infty}^{\infty} a_j(t)e^{i\lambda t} dt \quad , \quad \lambda \in \Gamma_j \quad j = 1,2$$

for some $L(Y)$-valued functions $a_1(t)$ and $a_2(t)$, let $Q_+A(\lambda)$ be the function which admits analytic continuation into Ω^+, has value 0 at infinity, and the difference $A(\lambda) - Q_+A(\lambda)$, $\lambda \in \Gamma_j$ admits analytic continuation into Ω_j^- $(j = 1,2)$. It is easily seen that

$$Q_+A(\lambda) = \int_{-\infty}^{0} a_1(t)e^{i\lambda t} dt + \int_{0}^{\infty} a_2(t)e^{i\lambda t} dt .$$

Applying this formula with $A(\lambda) = W_-(\lambda)^{-1} G(\lambda)$, we have

$$(7.11) \qquad Q_+W_-(\lambda)^{-1}G(\lambda) = \int_{-\infty}^{0} [g_1(t)+\int_{-\infty}^{\infty} w_1(t - s)g_1(s)ds]e^{i\lambda t} dt +$$

$$+ \int_{0}^{\infty} [g_2(t) + \int_{-\infty}^{\infty} w_2(t - s)g_2(s)ds]e^{i\lambda t} dt =$$

$$= \int_{-\infty}^{\infty} [g(t) + \int_{-\infty}^{\infty} w(t,s)g(s)ds]e^{i\lambda t} dt .$$

On the other hand,

$$(7.12) \qquad W_+(\lambda)^{-1} = I + \int_{-\infty}^{\infty} r(t)e^{i\lambda t} dt \quad , \quad \lambda \in \Omega^+ .$$

Now observe that the Fourier transform $\Phi_+(\lambda)$ of the desired function $\varphi(\lambda)$ is given by the formula

$$\Phi_+(\lambda) = W_+(\lambda)^{-1}[Q_+W_-(\lambda)^{-1}G(\lambda)] ,$$

and use (7.11) and (7.12) to derive the required formula for $\varphi(t)$. □

8. Wiener-Hopf equation with two kernels.

Consider the equation

(8.1) $$\varphi(t) - \int_0^\infty k_1(t-s)\varphi(s)ds - \int_{-\infty}^0 k_2(t-s)\varphi(s)ds = g(t) \ , \ -\infty < t < \infty \ ,$$

where $g(t)$ is a given function which belongs to the Banach space $e^{h|t|}L_1(\mathbb{R};Y)$ of all Y-valued functions $f(t)$, $-\infty < t < \infty$ such that

$$\int_{-\infty}^\infty e^{-h|t|}||f(t)||dt < \infty \ .$$

(As before, Y is a fixed Banach space and $h > 0$ is a fixed constant). The solution $\varphi(t)$ is sought also in $e^{h|t|}L_1(\mathbb{R};Y)$.

We assume that the kernels k_1, k_2 have exponential respresentations

$$k_j(t)=\begin{cases} i\, C_j\, e^{-itA_j}(I - Q_j)B_j & , \quad t > 0 \ ; \\ -i\, C_j\, e^{-itA_j}\, Q_jB_j & , \quad t < 0 \ , \end{cases}$$

for $j = 1,2$, where $B_j : Y \to X_j$, $A_j : X_j \to X_j$, $C_j : X_j \to Y$ are linear bounded operators (with some Banach spaces X_1 and X_2) such that

$$\sigma(A_j) \cap \overline{\Omega^+} = \emptyset \ , \ \sigma(A_j - B_jC_j) \cap \Gamma_j = \emptyset \quad (j = 1,2)$$

(Here we continue to use the notations introduced in the second part of Section 6). The projection $Q_j (j=1,2)$ is the spectral projection corresponding to the part of spectra of A_j $(j=1,2)$ lying in the halfplane Ω_1^-. These requirements ensure, in particular, that $k_j \in e^{-h|t|}L_1(\mathbb{R};L(Y))$ $(j=1,2)$ and the operator F defined by the left hand side of (8.1):

(8.2) $$(F\varphi)(t) = \varphi(t) - \int_0^\infty k_1(t - s)\varphi(s)ds - \int_{-\infty}^0 k_2(t-s)\varphi(s)ds,$$

is a linear bounded operator acting in the space $e^{h|t|}L_1(\mathbb{R},Y)$.

Using the procedure from §2 of the Appendix in [6], one can reduce (6.1) to the following barrier problem:

(8.3) $$\begin{cases} [I - K_1(\lambda)]\Phi_-^{(1)}(\lambda) = \Phi_+(\lambda) + G^{(1)}(\lambda) \quad , \quad \lambda \in \Gamma_1 \quad ; \\ [I - K_2(\lambda)]\Phi_-^{(2)}(\lambda)=\Phi_+(\lambda) + G^{(2)}(\lambda) \quad , \quad \lambda \in \Gamma_2 \quad . \end{cases}$$

Here

$$I - K_j(\lambda) = I + C_j(\lambda - A_j)^{-1} B_j \qquad (j=1,2)$$

are analytic operator functions in $\overline{\Omega^+}$,

$$\Phi_-^{(1)}(\lambda) = \int_0^\infty \varphi(t)e^{i\lambda t}\, dt \;;\; \Phi_-^{(2)}(\lambda) = - \int_{-\infty}^0 \varphi(t)e^{i\lambda t} \quad ;$$

$$G^{(1)}(\lambda) = \int_0^\infty g(t)d^{i\lambda t}\, dt \;;\; G^{(2)}(\lambda) = - \int_{-\infty}^0 g(t)e^{i\lambda t}\, dt$$

(so $\Phi_-^{(j)}(\lambda)$ and $G^{(j)}(\lambda)$ are analytic in Ω_j^-, $j = 1,2$); $\Phi_+(\lambda)$ is analytic in Ω^+.

Introduce the contourwise operator function

$$W(\lambda) = I + C_j(\lambda - A_j)^{-1} B_j \quad , \quad \lambda \in \Gamma_j \quad , \quad j = 1,2.$$

As follows from the results in [8], the barrier problem (8.3), and hence the equation (8.1), has a unique solution for every $g(t)\in e^{h|t|}L_1(\mathbb{R};Y)$ if and only if the function $W(\lambda)$ admits the right canonical factorization:

(8.4) $$W(\lambda) = W_+(\lambda)\, W_-(\lambda) \quad , \quad \lambda \in \Gamma_1 \cup \Gamma_2 \quad .$$

Hence we can invoke Theorem 6.2 and obtain that the operator F defined by (8.2) is invertible if and only if the right indicator R given by (6.4) is invertible. Further, assume $W(\lambda)$ admits the factorization (8.4). Then one has

(8.5) $$\Phi_-^{(j)}(\lambda) = W_-(\lambda)^{-1}[Q_-(W_+(\lambda)^{-1}G^{(j)}(\lambda))] \quad , \quad \lambda \in \Omega_j^- \quad , \quad j = 1,2,$$

where for every $L(Y)$-valued function $A(\lambda)$ which is analytic on $\Gamma_1 \cup \Gamma_2$ we define the $L(Y)$-valued function $Q_-A(\lambda)$ by the properties that $Q_-A(\lambda)$ is analytic in $\Omega_1^- \cup \Omega_2^-$ and has value zero at infinity, while

$A(\lambda) - Q_-A(\lambda)$ is analytic in Ω^+. According to Theorem 6.2 the formulas for W_+^{-1}, W_-^{-1} are provided by (5.9) and (5.10), and one can easily check that

$$W_+(\lambda)^{-1} = I + \int_{-\infty}^{\infty} r(t)e^{i\lambda t}\,dt \quad , \quad \lambda \in \Omega^+ \quad ,$$

where

$$r(t) = \begin{cases} iCR^{-1}P_2^\times e^{-itA_2^\times} B_2 & , \quad t > 0 \\ -iCR^{-1}P_1^\times e^{-itA_1^\times} B_1 & , \quad t < 0 \quad . \end{cases}$$

Also,

$$W_-(\lambda)^{-1} = I + \int_{-\infty}^{\infty} w_j(t)e^{i\lambda t}\,dt \quad , \quad \lambda \in \Omega_j^- \; ; \; j = 1,2, \tag{8.6}$$

where

$$w_1(t) = \begin{cases} i[C_1 \;\; C](I - \tilde{P}_1)\exp\left\{-it\begin{bmatrix} A_1^\times & -B_1C \\ 0 & M \end{bmatrix}\right\}(I - \tilde{P}_1)\begin{bmatrix} B_1 \\ R^{-1}B^\times \end{bmatrix} & , \; t > 0 \\ 0 \quad , \quad t < 0 \quad ; \end{cases}$$

$$w_2(t) = \begin{cases} 0 \quad , \quad t > 0 \\ -i[C_2 \;\; C](I - \tilde{P}_2)\exp\left\{-it\begin{bmatrix} A_2^\times & -B_2C \\ 0 & M \end{bmatrix}\right\}(I-\tilde{P}_2)\begin{bmatrix} B_2 \\ R^{-1}B^\times \end{bmatrix} & , \; t < 0, \end{cases}$$

and $\tilde{P}_j$ is the spectral projection of $\begin{bmatrix} A_j^\times & -B_jC \\ 0 & M \end{bmatrix}$ corresponding to the domain Ω_j^- $(j = 1,2)$.

With these definition we have the following result.

THEOREM 8.1. *The operator* F *defined by* (8.2) *is invertible if and only if the right indicator* R *defined by* (6.4) *is invertible. If this condition holds, then for any* $g(t) \in e^{h|t|}L_1(\mathbb{R},Y)$ *the unique solution* $\varphi(t) \in e^{h|t|}L_1(\mathbb{R},Y)$ *of the equation* (8.1) *is given by the formula*

$$(8.7) \qquad \varphi(t) = g(t) + \int_{-\infty}^{\infty} r(t-s)g(s)ds + \int_{-\infty}^{\infty}\gamma(t,s)g(s)ds,$$

where

$$\gamma(t,s) = \begin{cases} w_1^0(t,s) + \int_0^t w_1(t-v)r(v-s)dv, & t > 0 \\ w_2^0(t,s) + \int_t^0 w_2(t-v)r(v-s)dv, & t < 0 \end{cases}$$

and

$$w_1^0(t,s) = \begin{cases} w_1(t-s), & s > 0; \\ 0, & s < 0; \end{cases} \qquad w_2^0(t,s) = \begin{cases} w_2(t-s), & s < 0; \\ 0, & s > 0. \end{cases}$$

Proof. We have

$$W_+^{-1}(\lambda)\, G^{(1)}(\lambda) = \int_{-\infty}^{\infty} [g_1(t) + \int_0^{\infty} r(t-s)g(s)ds]e^{i\lambda t}\, dt, \quad \lambda \in \Gamma_1,$$

$$W_+^{-1}(\lambda)\, G^{(2)}(\lambda) = \int_{-\infty}^{\infty} [-g_2(t) - \int_{-\infty}^{0} r(t-s)g(s)ds]e^{i\lambda t}\, dt, \quad \lambda \in \Gamma_2,$$

where

$$g_1(t) = \begin{cases} g(t), & t > 0; \\ 0, & t < 0; \end{cases} \qquad g_2(t) = \begin{cases} 0, & t > 0; \\ g(t), & t < 0. \end{cases}$$

One checks easily that

$$Q_-(W_+^{-1}\, G^{(1)}) = \int_0^{\infty} [g_1(t) + \int_{-\infty}^{\infty} r(t-s)g(s)ds]e^{i\lambda t}\, dt, \quad \lambda \in \Gamma_1;$$

$$Q_-(W_+^{-1}G^{(2)}) = \int_{-\infty}^{0} [- g_2(t) - \int_{-\infty}^{\infty} r(t - s)g(s)ds]e^{i\lambda t}\, dt \quad , \quad \lambda \in \Gamma_2 \ .$$

Now in view of formulas (8.5) and (8.6) we have to check the equality

$$(8.8) \qquad \int_0^\infty e^{i\lambda t}[g(t) + \int_{-\infty}^{\infty} r(t - s)g(s)ds + \int_{-\infty}^{\infty} \gamma(t,s)g(s)ds]dt =$$

$$= \int_0^\infty e^{i\lambda t}[g_1(t) + \int_{-\infty}^{\infty} r(t - s)g(s)ds]dt +$$

$$+ \int_0^\infty w_1(t)e^{i\lambda t}\, dt\cdot\int_0^\infty [g_1(t) + \int_{-\infty}^{\infty} r(t - s)g(s)ds]e^{i\lambda t}\, dt \ ,$$

as well as the analogous equality for $\Phi_-^{(2)}(\lambda)$. We shall indicate only how to verify (8.8), or, equivalently, the formula

$$\int_0^\infty e^{i\lambda t} \int_{-\infty}^{\infty} [\int_0^t w_1(t - v)r(v - s)dv + w_1^0(t,s)]g(s)ds\, dt =$$

$$= \int_0^\infty w_1(t)e^{i\lambda t}\, dt\cdot\int_0^\infty [g_1(t) + \int_{-\infty}^{\infty} r(t - s)g(s)ds]e^{i\lambda t}\, dt \ .$$

It is enough to check that

$$\int_0^\infty e^{i\lambda t} \int_{-\infty}^{\infty} [\int_0^t w_1(t - v)r(v - s)dv]g(s)ds\, dt =$$

$$= \int_0^\infty w_1(t)e^{i\lambda t}\, dt\cdot\int_0^\infty [\int_{-\infty}^{\infty} r(t - s)g(s)ds]e^{i\lambda t}\, dt$$

and

$$\int_0^\infty e^{i\lambda t} \int_0^t w_1(t - s)g_1(s)ds\, dt = \int_0^\infty w_1(t)e^{i\lambda t}dt\cdot\int_0^\infty g_1(t)e^{i\lambda t}dt \ ;$$

both equalities are easily verifiable. □

9. The discrete case.

Results analogous to those obtained in Sections 7 and 8 hold also for the discrete counterparts of equation (7.1) and (8.1). Namely, for a fixed number $h > 1$, consider the equations

$$(9.1)\qquad \begin{cases} \varphi_j - \sum_{k=-\infty}^{\infty} a_{j-k}\,\varphi_k = g_j \quad , & j \geq 0 \ , \\ \\ \varphi_j - \sum_{k=-\infty}^{\infty} b_{j-k}\,\varphi_k = g_j \quad , & j < 0 \ , \end{cases}$$

where $\{g_j\}_{j=-\infty}^{\infty}$ is a given Y-valued sequence such that

$$\sum_{j=-\infty}^{\infty} h^{|j|}\,||g_j|| < \infty$$

(in short, $\{g_j\}_{j=-\infty}^{\infty} \in h^{-|j|}\ell_1(Y))$, $\{\varphi_j\}_{j=-\infty}^{\infty}$ is a Y-valued sequence from $h^{-|j|}\ell_1(Y)$ to be found, and $\{a_j\}_{j=-\infty}^{\infty}$, $\{b_j\}_{j=-\infty}^{\infty}$ are $L(Y)$-valued sequences such that

$$\sum_{j=-\infty}^{\infty} h^{j}\,||a_j|| < \infty \quad , \quad \sum_{j=-\infty}^{\infty} h^{-j}\,||b_j|| < \infty \ .$$

Assuming that the function $W_1(\lambda) = \sum_{j=-\infty}^{\infty} \lambda^j b_j$ and $W_2(\lambda) = \sum_{j=-\infty}^{\infty} \lambda^j a_j$ are analytic and invertible in a neighbourhood of $\Gamma_1 = \{\lambda \mid |\lambda| = h^{-1}\}$ and $\Gamma_2 = \{\lambda \mid |\lambda| = h\}$, there exist realizations

$$(9.2)\qquad W_i(\lambda) = I + C_i(\lambda - A_i)^{-1}B_i \quad , \quad i = 1,2$$

such that $\sigma(A_i) \cap \Gamma_i = \sigma(A_i - B_iC_i) \cap \Gamma_i = \emptyset$. It turns out that the operator $E : h^{-|j|}\ell_1(Y) \to h^{-|j|}\ell_1(Y)$ defined by $E\{\varphi_j\}_{j=-\infty}^{\infty} = \{g_j\}_{j=-\infty}^{\infty}$, where g_j are given by (9.1) is invertible if and only if the left indicator S of the realizations (9.2) with respect to $\Gamma = \Gamma_1 \cup \Gamma_2$ is invertible. One can write down explicit formulas for the solution of (9.1) in terms of the operators A_i, B_i, C_i and S^{-1}.

The discrete analogue of equation (8.1) is

$$(9.3)\qquad \varphi_j - \sum_{k=0}^{\infty} a_{j-k}\,\varphi_k - \sum_{k=-1}^{-\infty} b_{j-k}\,\varphi_k = g_k \quad , \quad -\infty < j < \infty \ ,$$

where $\{g_j\}_{j=-\infty}^{\infty}$, $\{\varphi_j\}_{j=-\infty}^{\infty} \in h^{|j|}\ell_1(Y)$ and $\{a_j\}_{j=-\infty}^{\infty}$, $\{b_j\}_{j=-\infty}^{\infty} \in h^{-|j|}\ell_1(L(Y))$. The equation (9.3) can be analysed analogously to the analysis of (8.1) in the preceding section, with

$$W_1(\lambda) = \sum_{j=-\infty}^{\infty} \lambda^j b_j \quad \text{and} \quad W_2(\lambda) = \sum_{j=-\infty}^{\infty} \lambda^j a_j \quad .$$

REFERENCES

1. Bart H., Gohberg, I., Kaashoek, M.A.: Minimal factorization of matrix and operator functions. Operator Theory: Advances and Applications, vol. 1, Birkhäuser Verlag, Basel, 1979.

2. Bart H., Gohberg, I., Kaashoek, M.A.: Wiener-Hopf integral equations, Toeplitz matrices and linear systems. In: Toeplitz Centennial. (Ed. I. Gohberg), Operator Theory: Advances and Applications, vol. 4, Birkhaüser Verlag, Basel, 1982; pp. 85-135.

3. Bart, H., Gohberg, I., Kaashoek, M.A.: The coupling method for solving integral equations. In: Topics in Operator Theory, Systems and Networks, the Rehovot Workshop (Ed. H. Dym, I. Gohberg). Operator Theory: Advances and Applications, vol. 12, Birkhaüser Verlag, Basel, 1984, pp.39-73.

4. Bart, H. Kroon, L.S.: An indicator for Wiener-Hopf integral equations with invertible analytic symbol. Integral Equations and Operator Theory, 6/1 (1983), 1-20.

5. Daleckii, Iu. L., Krein,M.G.: Stability of solutions of differential equations in Banach space. Amer. Math. Soc. Transl. 43, American Mathematical Society, Providence R.I., 1974.

6. Gohberg, I.C., Feldman, I.A.: Convolution equations and projection methods of their solution. Amer. Math. Soc. Transl. 41, American Mathematical Society, Providence, R.I., 1974.

7. Gohberg, I., Kaashoek, M.A., Lerer, L., Rodman, L.: Minimal divisors of rational matrix functions with prescribed zero and pole structure. In: Topics in Operator Theory, Systems and Networks, The Rehovot Workshop (Ed. H. Dym, I. Gohberg). Operator theory: Advances and Applications, vol. 12, Birkhaüser Verlag, Basel, 1984, pp. 241-275.

8. Gohberg, I.C., Leiterer, I.: A criterion for factorization of operator functions with respect to a contour. Sov. Math. Doklady 14, No. 2(1973), 425-429.

9. Gohberg, I., Lerer, L., Rodman, L.: Wiener-Hopf factorization of piecewise matrix polynomials. Linear Algebra and Appl. 52/53 (1983), 315-350.

I. Gohberg,
School of Mathematical Science,
Tel-Aviv University,
Tel-Aviv,
Israel

M.A. Kaashoek,
Subfaculteit Wiskunde en Informatica
Vrije Universiteit,
1007 MC Amsterdam,
The Netherlands

L. Lerer,
Department of Mathematics,
Technion-Israel Institute of Technology,
Haifa,
Israel.

L. Rodman,
School of Mathematical Science
Tel-Aviv University,
Tel-Aviv.
Israel.

Operator Theory:
Advances and Applications, Vol. 21

CANONICAL PSEUDO-SPECTRAL FACTORIZATION AND WIENER-HOPF INTEGRAL EQUATIONS

Leen Roozemond [1)]

Wiener-Hopf integral equations with rational matrix symbols that have zeros on the real line are studied. The concept of canonical pseudo-spectral factorization is introduced, and all possible factorizations of this type are described in terms of realizations of the symbol and certain supporting projections. With each canonical pseudo-spectral factorization is related a pseudo-resolvent kernel, which satisfies the resolvent identities and is used to introduce spaces of unique solvability.

0. INTRODUCTION

In this paper we study the invertibility properties of the vector-valued Wiener-Hopf integral equation

$$\varphi(t) - \int_0^\infty k(t-s)\varphi(s)ds = f(t), \qquad t \geq 0, \tag{0.1}$$

assuming the equation is of so-called non-normal type, which means (see [8], [6, § III.12],[7], and the references there) that the symbol has singularities on the real line. We assume additionally that the symbol is rational, and in our analysis we follow the approach of [1], which is based on realization.

First, let us recall the main features of the theory developed in [1, § IV.5] (see also [2]), for the case when the symbol has no singularities on the real line. Take $k \in L_1^{m\times m}(-\infty,\infty)$, and assume that the symbol $W(\lambda) = I - \int_{-\infty}^{\infty} k(t)e^{i\lambda t}dt$ is a rational $m\times m$ matrix function. The symbol can be realized as a transfer function, i.e., it can be written in the form

$$W(\lambda) = I + C(\lambda - A)^{-1}B, \qquad -\infty < \lambda < \infty, \tag{0.2}$$

where A is a square matrix of order n, say, with no real eigenvalues and B and C are matrices of size $n\times m$ and $m\times n$, respectively. In [1] it is assumed that $\det W(\lambda)$ has no real zeros, which is equivalent to the condition that

1) Research supported by the Netherlands Organization for the Advancement of Pure Research (Z.W.O.).

$A^\times := A - BC$ has no eigenvalues on the real line.

It is known (see [5]) that for each $f \in L_p^m[0,\infty)$ the equation (0.1) has a unique solution $\varphi \in L_p^m[0,\infty)$ if and only if $\det W(\lambda)$ has no zeros on the real line and relative to the real line W admits a (right) canonical Wiener-Hopf factorization

$$W(\lambda) = W_-(\lambda)W_+(\lambda). \tag{0.3}$$

The latter means that $W_-(\lambda)$ and $W_-(\lambda)^{-1}$ are holomorphic in the open lower half plane and continuous up to the real line, while $W_+(\lambda)$ and $W_+(\lambda)^{-1}$ are holomorphic in the open upper half plane and also continuous up to the real line. Furthermore we may take $W_-(\infty) = W_+(\infty) = I$.

In terms of the realization (0.2) a canonical Wiener-Hopf factorization exists if and only if on $\mathbb{C}^n$ (with n the order of A) there exists a supporting projection Π (i.e., $\operatorname{Ker}\Pi$ is invariant under A and $\operatorname{Im}\Pi$ is invariant under $A^\times$) such that relative to the decomposition $\mathbb{C}^n = \operatorname{Ker}\Pi \oplus \operatorname{Im}\Pi$ the matrices A and $A^\times$ admit the following partitioning:

$$A = \begin{pmatrix} A_1 & * \\ 0 & A_2 \end{pmatrix}, \quad A = \begin{pmatrix} A_1^\times & 0 \\ * & A_2^\times \end{pmatrix}, \tag{0.4}$$

with the extra property that the eigenvalues of A_1 and $A_1^\times$ are in the open upper half plane and those of A_2 and $A_2^\times$ in the open lower half plane. Furthermore, if such a supporting projection Π exists, then for the factors in (0.3) one may take

$$\begin{aligned} W_-(\lambda) &= I + C(\lambda - A)^{-1}(I - \Pi)B, \\ W_+(\lambda) &= I + C\Pi(\lambda - A)^{-1}B, \end{aligned} \tag{0.5}$$

and for each $f \in L_p^m[0,\infty)$ the unique solution $\varphi \in L_p^m[0,\infty)$ of (0.1) is given by

$$\varphi(t) = f(t) + \int_0^\infty g(t,s)f(s)ds, \qquad t \geq 0, \tag{0.6}$$

where the resolvent kernel g is given by

$$g(t,s) = \begin{cases} iCe^{-itA^\times}\Pi e^{isA^\times}B, & 0 \leq s < t, \\ -iCe^{-itA^\times}(I - \Pi)e^{isA^\times}B, & 0 \leq t < s. \end{cases} \tag{0,7}$$

In this paper we show that with appropriate modifications and the right understanding the theory described above can be carried over to the case when the (determinant of the) symbol $W(\lambda)$ has zeros on the real line. To do this, we first of all replace the notion of canonical Wiener-Hopf factorization by the notion of canonical pseudo-$\mathbb{R}$-spectral factorization. This means a factorization of the symbol $W(\lambda)$ of the form (0.3), where the factors $W_-(\lambda)$, $W_+(\lambda)$ and their inverses $W_-(\lambda)^{-1}$, $W_+(\lambda)^{-1}$ have the same properties as before except now we do not require the inverses $W_-(\lambda)^{-1}$ and $W_+(\lambda)^{-1}$ to be continuous up to the real line. In other words we allow the factors $W_-(\lambda)$ and $W_+(\lambda)$ to have real zeros. In general, in contrast with canonical Wiener-Hopf factorizations, there may be many different non-equivalent canonical pseude-$\mathbb{R}$-spectral factorizations.

In this paper we describe how to get all canonical pseudo-$\mathbb{R}$-spectral factorizations in terms of the realization (0.2). Recall that $\det W(\lambda)$ has zeros on the real line if and only if $A^\times$ has real eigenvalues. To find the canonical pseudo-$\mathbb{R}$-spectral factorizations of $W(\lambda)$ one has to split the spectral subspaces corresponding to the eigenvalues of $A^\times$ on the real line. In fact we prove that a canonical pseudo-$\mathbb{R}$-spectral factorization exists if and only if there exists a supporting projection Π with the same properties as before, except now we have to allow that in (0.4) the entries $A_1^\times$ and $A_2^\times$ also have eigenvalues on the real line. If one has such a supporting projection Π, then the factors W_+ and W_- in the corresponding canonical pseudo-$\mathbb{R}$-spectral factorization are again given by (0.5).

We also show that given a supporting projection Π corresponding to a canonical pseudo-$\mathbb{R}$-spectral factorization, then (0.7) defines a kernel which satisfies the following resolvent identities:

$$(0.8)\qquad \begin{aligned} g(t,s) - \int_0^\infty k(t-u)g(u,s)du &= k(t-s), \qquad s \ge 0,\ t \ge 0,\\ k(t-s) - \int_0^\infty g(t,u)k(u-s)du &= g(t,s), \qquad s \ge 0,\ t \ge 0. \end{aligned}$$

Let K and G be the integral operators with kernel $k(t-s)$ and $g(t,s)$. We use the resolvent identities (0.8) and the specific form of the kernel $g(t,s)$ to introduce spaces of unique solvability. This means that in these spaces equation (0.1) is again uniquely solvable and the solution of (0.1) is given by (0.6). Also in [8], [6] and [7] spaces of unique solvability appear, but because of the use of the realization (0.2), the spaces that we derive in our analysis admit a more detailed description.

A few words about the organization of the paper. The paper consists of six sections. In the first section we introduce the notion of canonical pseudo-Γ-spectral factorization for arbitrary matrix functions and arbitrary Cauchy contours Γ. We introduce pseudo-Γ-spectral subspaces in Section 2. Subspaces of this type will be used later in the contruction of the factorizations. In the third section we give a description of all canonical pseudo-Γ-spectral factorizations in terms of realizations. A special case, non-negative rational matrix functions, is treated in Section 4. In Sections 5 and 6 we study Wiener-Hopf integral equations of non-normal type with rational symbol and we prove the results mentioned in the previous paragraph.

1. CANONICAL PSEUDO-SPECTRAL FACTORIZATIONS

To define canonical pseudo-spectral factorizations we need the notions of minimal factorization and local degree (see [1, Chapter IV]). Let W be a rational $m\times m$ matrix function, and let $\lambda_0 \in \mathbb{C}$. In a deleted neighbourhood of λ_0 we have the following expansion

$$W(\lambda) = \sum_{j=-q}^{\infty} (\lambda-\lambda_0)^j W_j. \tag{1.1}$$

Here q is some positive integer. By the *local degree* of W at λ_0 we mean the number $\delta(W;\lambda_0) = \operatorname{rank}\Omega$, where

$$\Omega = \begin{pmatrix} W_{-q} & \cdots & W_{-1} \\ 0 & \ddots & \vdots \\ \vdots & & \ddots \\ 0 & \cdots\ 0 & W_{-q} \end{pmatrix}.$$

The number $\delta(W;\lambda_0)$ is independent of q, as long as (1.1) holds. We define $\delta(W;\infty) = \delta(\widetilde{W};0)$, where $\widetilde{W}(\lambda) = W(1/\lambda)$. Note that W is analytic in $\mu \in C \cup \{\infty\}$ if and only is $\delta(W;\mu) = 0$. It is well-known (see [1, Chapter IV]) that the local degree has a sublogarithmic property, i.e., whenever W_1 and W_2 are rational $m\times m$ matrix functions, we have $\delta(W_1W_2;\lambda_0) \le \delta(W_1;\lambda_0) + \delta(W_2;\lambda_0)$ for each $\lambda_0 \in \mathbb{C} \cup \{\infty\}$. A factorization $W(\lambda) = W_1(\lambda)W_2(\lambda)$ is called *minimal at* λ_0 if $\delta(W_1W_2;\lambda_0) = \delta(W_1;\lambda_0) + \delta(W_2;\lambda_0)$, and *minimal* if $\delta(W_1W_2;\lambda) = \delta(W_1;\lambda) + \delta(W_2;\lambda)$ for all $\lambda \in \mathbb{C} \cup \{\infty\}$. In other words, a factorization $W(\lambda) = W_1(\lambda)W_2(\lambda)$ is minimal if it is minimal at λ_0 for all $\lambda_0 \in \mathbb{C} \cup \{\infty\}$.

Let W be a rational $m\times m$ matrix function given by the expansion (1.1). We call λ_0 a *zero* of W if in $\mathbb{C}^n$ there exist vectors $x_0,\dots,x_q, x_0 \ne 0$, such that

$$W_{-q}x_i + \dots + W_{-q+i}x_0 = 0 \quad (i = 0,\dots,q).$$

Note that a matrix function may have a pole and a zero at the same point. If $\det W(\lambda)$ does not vanish identically, then λ_0 is a zero of W if and only if

λ_0 is a pole of $W(\lambda)^{-1}$. Minimality of a factorization $W(\lambda) = W_1(\lambda)W_2(\lambda)$ can be understood as the absence of pole-zero cancellations (see [1, Theorem 4.6]).

We shall consider spectral factorizations with respect to a curve Γ. Throughout this paper Γ is a Cauchy contour on the Riemann sphere $\mathbb{C} \cup \{\infty\}$. Thus Γ is the positively oriented boundary of an open set with a finite number of non-intersecting closed rectifiable Jordan curves. We denote the inner (resp. outer) domain of Γ by Ω_Γ^+ (resp. Ω_Γ^-). Associated with Γ is the curve $-\Gamma$. As sets Γ and $-\Gamma$ coincide, but they have opposite orientations. I.e., the inner (resp. outer) domain of $-\Gamma$ is Ω_Γ^- (resp. Ω_Γ^+).

A rational $m \times m$ matrix function W admits a (*right*) *canonical pseudo-Γ-spectral factorization* if W can be represented in the form

$$W(\lambda) = W_-(\lambda)W_+(\lambda) \tag{1.2}$$

where

(a) W_- and W_+ are rational matrix functions, W_- has no poles and no zeros in Ω_Γ^-, W_+ has no poles and no zeros in Ω_Γ^+;

(b) the factorization (1.2) is minimal at each point of Γ.

Since W_+ (resp. W_-) has no poles nor zeros in Ω_Γ^+ (resp. Ω_Γ^-), the factorization (1.2) is minimal at each point of $\Omega_\Gamma^+ \cup \Omega_\Gamma^-$. Hence condition (b) can be replaced by

(b)' the factorization (1.2) is minimal.

Comparing the definitions of canonical pseudo-Γ-spectral factorization and canonical Wiener-Hopf factorization, two major differences appear. First of all, canonical Wiener-Hopf factorization is only defined for rational $m \times m$ matrix functions with no poles and no zeros on Γ. Secondly, the factors in a canonical Wiener-Hopf factorization are required to be continuous up to the boundary Γ.

If a rational $m \times m$ matrix function W has no poles and no zeros on the curve Γ, the notions of canonical pseudo-Γ-spectral factorization and canonical Wiener-Hopf factorization coincide. To see this, suppose W admits a canonical pseudo-Γ-spectral factorization with factors W_- and W_+. If W has no poles and no zeros on Γ, then, because of the minimality condition (b), the factors W_- and W_+ cannot have poles or zeros on Γ. Hence W_-, W_+, W_-^{-1} and W_+^{-1} are continuous up to the boundary Γ, and $W(\lambda) = W_-(\lambda)W_+(\lambda)$ is a canonical Wiener-Hopf factorization with respect to Γ.

Let W be a rational $m \times m$ matrix function. Two canonical pseudo-

Γ-spectral factorizations $W(\lambda) = W_-(\lambda)W_+(\lambda)$ and $W(\lambda) = \widetilde{W}_-(\lambda)\widetilde{W}_+(\lambda)$ are called *equivalent* if there exists an invertible constant $m \times m$ matrix E, such that $W_-(\lambda) = \widetilde{W}_-(\lambda)E$, $W_+(\lambda) = E^{-1}\widetilde{W}_+(\lambda)$. If W admits a canonical Wiener-Hopf factorization, then all canonical Wiener-Hopf factorizations are equivalent (cf. [5]). This is not true for canonical pseudo-Γ-spectral factorizations, as the following examples show.

EXAMPLE 1.1. Let

$$W(\lambda) = \begin{pmatrix} \frac{\lambda}{\lambda+2i} & \frac{3i\lambda}{(\lambda-i)(\lambda+2i)} \\ 0 & \frac{\lambda}{\lambda-i} \end{pmatrix}. \tag{1.3}$$

Then W is a rational 2×2 matrix function, with poles in i and -2i, and a zero in 0. Note that $W(\infty) = \begin{pmatrix} 1 & 0 \\ 0 & 1 \end{pmatrix}$. The matrix function W has many non-equivalent canonical pseudo-$\mathbf{R}$-spectral factorizations. Indeed, put

$$W_-^{(\alpha)}(\lambda) = \begin{pmatrix} \frac{\lambda-i(1+\alpha)}{\lambda-i} & \frac{i(1+\alpha)}{\lambda-i} \\ \frac{-i\alpha}{\lambda-i} & \frac{\lambda+i\alpha}{\lambda-i} \end{pmatrix},$$

$$W_+^{(\alpha)}(\lambda) = \begin{pmatrix} \frac{\lambda+i\alpha}{\lambda+2i} & \frac{i(2-\alpha)}{\lambda+2i} \\ \frac{i\alpha}{\lambda+2i} & \frac{\lambda+i(2-\alpha)}{\lambda+2i} \end{pmatrix}.$$

The function $W_-^{(\alpha)}$ has a pole in i, and a zero in 0. The function $W_+^{(\alpha)}$ has a pole in -2i, and a zero in 0.
A straightforward computation shows that $W(\lambda) = W_-^{(\alpha)}(\lambda)W_+^{(\alpha)}(\lambda)$, and obviously this factorization is minimal since there are no pole-zero cancellations. The factorizations $W(\lambda) = W_-^{(\alpha)}(\lambda)W_+^{(\alpha)}(\lambda)$ and $W(\lambda) = W_-^{(\beta)}(\lambda)W_+^{(\beta)}(\lambda)$ are not equivalent whenever $\alpha \neq \beta$. Indeed, we compute

$$W_+^{(\alpha)}(\lambda)W_+^{(\beta)}(\lambda)^{-1} = W_-^{(\alpha)}(\lambda)^{-1}W_-^{(\beta)}(\lambda) = \begin{pmatrix} \frac{\lambda+i(\alpha-\beta)}{\lambda} & \frac{-i(\alpha-\beta)}{\lambda} \\ \frac{i(\alpha-\beta)}{\lambda} & \frac{\lambda-i(\alpha-\beta)}{\lambda} \end{pmatrix}.$$

Clearly, this is not constant whenever $\alpha \neq \beta$.

EXAMPLE 1.2. (The scalar case) Let W be a rational (scalar) function, with $W(\infty) = 1$. We can write

$$W(\lambda) = \frac{\lambda^{\ell} + b_{\ell-1}\lambda^{\ell-1} + \ldots + b_1\lambda + b_0}{\lambda^{\ell} + a_{\ell-1}\lambda^{\ell} + \ldots + a_1\lambda + a_0}$$

for certain complex numbers $a_0,\ldots,a_{\ell-1},b_0,\ldots,b_{\ell-1}$. We assume that the polynomials $p(\lambda) = \lambda^{\ell} + b_{\ell-1}\lambda^{\ell-1} + \ldots + b_1\lambda + b_0$ and $q(\lambda) = \lambda^{\ell} + a_{\ell-1}\lambda^{\ell-1} + \ldots + a_1\lambda + a_0$ do not have common zeros.

Let Γ be a contour on the Riemann sphere $\mathbb{C} \cup \{\infty\}$. We write $p(\lambda) = (\lambda - \alpha_1)\ldots(\lambda - \alpha_{m_1})(\lambda - \gamma_1)\ldots(\lambda - \gamma_{m_2})(\lambda - \beta_1)\ldots(\alpha - \beta_{m_3})$ and $q(\lambda) = (\lambda - \alpha_1^{x})\ldots(\lambda - \alpha_{n_1}^{x})(\lambda - \gamma_1^{x})\ldots(\lambda - \gamma_{n_2}^{x})(\lambda - \beta_1^{x})\ldots(\lambda - \beta_{n_3}^{x})$. Here $\alpha_1,\ldots,\alpha_{m_1},\alpha_1^{x},\ldots,\alpha_{n_1}^{x}$ are in Ω_{Γ}^{+}, $\beta_1,\ldots,\beta_{m_3},\beta_1^{x},\ldots,\beta_{n_3}^{x}$ are in Ω_{Γ}^{-}, and $\gamma_1,\ldots,\gamma_{m_2},\gamma_1^{x},\ldots,\gamma_{n_2}^{x}$ are on Γ. We have $m_1 + m_2 + m_3 = n_1 + n_2 + n_3 = \ell$.

Suppose $W(\lambda) = W_-(\lambda)W_+(\lambda)$ is a canonical pseudo-Γ-spectral factorization, and $W_-(\infty) = W_+(\infty) = 1$. We write $W_-(\lambda) = p_-(\lambda)q_-(\lambda)^{-1}$, $W_+(\lambda) = p_+(\lambda)q_+(\lambda)^{-1}$, for certain polynomials p_-, q_-, p_+ and q_+. We assume that p_- and q_-, and p_+ and q_+ do not have common zeros. Since the factorization $W(\lambda) = W_-(\lambda)W_+(\lambda)$ is minimal at each $\lambda \in \Gamma$, we have $p(\lambda) = p_-(\lambda)p_+(\lambda)$, $q(\lambda) = q_-(\lambda)q_+(\lambda)$. Furthermore, since $W_-(\infty) = W_+(\infty) = 1$, we have $\deg p_- = \deg q_-$, $\deg p_+ = \deg q_+$. The zeros of p_- are in the set $\{\alpha_1,\ldots,\alpha_{m_1},\gamma_1,\ldots,\gamma_{m_2}\}$, and $\alpha_1,\ldots,\alpha_{m_1}$ are zeros of p_-. The zeros of q_+ are in the set $\{\gamma_1^{x},\ldots,\gamma_{n_2}^{x},\beta_1^{x},\ldots,\beta_{n_3}^{x}\}$, and $\beta_1^{x},\ldots,\beta_{n_3}^{x}$ are zeros of q_+. We also have $\deg p_- + \deg q_+ = \ell = m_1 + m_2 + m_3 = n_1 + n_2 + n_3$. Hence one of the following two cases will occur: (i) $m_1 \le n_1 \le m_1 + m_2$ or (ii) $n_1 \le m_1 \le n_1 + n_2$. Using a combinatorial argument, we get that the total number of canonical pseudo-Γ-spectral factorizations $W(\lambda) = W_-(\lambda)W_+(\lambda)$ such that $W_-(\infty) = W_+(\infty) = 1$ is equal to

(i) $$\sum_{k=0}^{\min(n_2,m_1+m_2-n_1)} \binom{n_2}{k}\binom{m_2}{k+n_1-m_1} \quad \text{if } m_1 + m_2 \ge n_1 \ge m_1,$$

(ii) $$\sum_{k=0}^{\min(m_2,n_1+n_2-m_1)} \binom{m_2}{k}\binom{n_2}{k+m_1-n_1} \quad \text{if } n_1 + n_2 \ge m_1 \ge n_1.$$

2. PSEUDO-Γ-SPECTRAL SUBSPACES

Let $A : X \to X$ be a linear operator acting on a finite dimensional linear space X, and let Γ be a Cauchy contour on the Riemann sphere $\mathbb{C} \cup \{\infty\}$. We call a subspace L of X a *pseudo-Γ-spectral subspace* if L is A-invariant, $A|L$ has no eigenvalues in Ω_{Γ}^{-}, the outer domain of Γ, and L contains all eigenvectors and generalized eigenvectors corresponding to the eigenvalues of A in

Ω_Γ^+, the inner domain of Γ. Denote the spectral projection corresponding to the eigenvalues of A in Ω_Γ^+ (resp. in Ω_Γ^-, on Γ) by P_+ (resp, P_-, P_0). Then L is a pseudo-Γ-spectral subspace if L is A-invariant and $\operatorname{Im} P_+ \subseteq L \subseteq \operatorname{Im} P_+ \oplus \operatorname{Im} P_0$. In other words, L is a pseudo-Γ-spectral subspace if and only if $L = \operatorname{Im} P_+ \oplus K_0$, where K_0 is an A-invariant subspace of $\operatorname{Im} P_0$. It is clear that a pseudo-Γ-spectral subspace always exists, e.g., the spaces $\operatorname{Im} P_+$ and $\operatorname{Ker} P_- = \operatorname{Im} P_+ \oplus \operatorname{Im} P_0$ are pseudo-Γ-spectral subspaces. There exists only one pseudo-Γ-spectral subspace if and only if $\operatorname{Im} P_0 = (0)$, i.e., the operator A has no eigenvalue on Γ. In fact we have

PROPOSITION 2.1. *Let* A *be a* $n \times n$ *matrix, and* Γ *a contour on the Riemann sphere* $\mathbb{C} \cup \{\infty\}$. *One of the following cases will occur:*

(a) *there is exactly one pseudo-Γ-spectral subspace,*

(b) *there are finitely many different pseudo-Γ-spectral subspaces,*

(c) *there is a continuum of pseudo-Γ-spectral subspaces.*

The case (a) *occurs if* A *has no eigenvalues on* Γ. *The case* (b) *occurs if the eigenvalues of* A *on* Γ *do not have more than one Jordan chain. The case* (c) *occurs if* A *has an eigenvalue with geometric multiplicity greater than one.*

PROOF. Suppose A has an eigenvalue $\lambda_0 \in \Gamma$ with geometric multiplicity at least two. Take two linearly independent vectors $x_1, x_2 \in \mathbb{C}^n$ such that $Ax_1 = \lambda_0 x_1$, $Ax_2 = \lambda_0 x_2$. For $\alpha \in \mathbb{C}$ define $L_\alpha = \operatorname{Im} P_+ \oplus \operatorname{span}\{\alpha x_1 + (1-\alpha)x_2\}$. The subspaces L_α, $\alpha \in \mathbb{R}$, are pseudo-Γ-spectral subspaces. Furthermore, $L_\alpha \neq L_\beta$ whenever $\alpha \neq \beta$.

Suppose all eigenvalues of A on Γ have geometric multiplicity one. Denote the eigenvalues on A on Γ by $\lambda_1, \ldots, \lambda_r$, and their algebraic multiplicities by $\alpha_1, \ldots, \alpha_r$. A simple combinatorial argument gives us that there are $\prod_{j=1}^{r} \alpha_j$ different A-invariant subspaces of $\operatorname{Im} P_0$. Hence there are $\prod_{j=1}^{r} \alpha_j$ different pseudo-Γ-spectral subspaces.

Suppose A has no eigenvalues on Γ. Since $\operatorname{Im} P_+ = \operatorname{Ker} P_-$, we have that $\operatorname{Im} P_+$ is the only pseudo-Γ-spectral subspace. □

We shall describe the changes of pseudo-Γ-spectral subspaces under some elementary operations. First, consider the operation of similarity. Suppose $\widetilde{A} = SAS^{-1}$, where $S : X \to \widetilde{X}$ is some invertible linear operator. Then L is a pseudo-Γ-spectral subspace of A is and only if SL is a pseudo-Γ-spectral subspace of $\widetilde{A}$.

Next, assume that $A : X \to X$ is a *dilation* of A. This means that X admits a direct sum decomposition, $X = X_1 \oplus \widetilde{X} \oplus X_2$, such that relative to this decomposition A admits the following partitioning

$$(2.1)\qquad A = \begin{pmatrix} A_{11} & A_{10} & A_{12} \\ 0 & \widetilde{A} & A_{02} \\ 0 & 0 & A_{22} \end{pmatrix}.$$

PROPOSITION 2.2. *Let* A *be the dilation of* $\widetilde{A}$ *given by* (2.1), *and assume that* A_{11} *and* A_{22} *have no eigenvalues on* Γ. *Let* P_+ *and* P_0 *be the spectral projections of* A *corresponding to the eigenvalues inside* Γ *and on* Γ, *respectively, and let* $\widetilde{P}_+$ *and* $\widetilde{P}_0$ *be the analogous projections for* $\widetilde{A}$. *Then the map*

$$(2.2)\qquad \operatorname{Im}\widetilde{P}_+ \oplus \widetilde{K}_0 \mapsto \operatorname{Im}P_+ \oplus P_0\widetilde{K}_0,$$

with $\widetilde{K}_0$ *an* $\widetilde{A}$*-invariant subspace of* $\operatorname{Im}\widetilde{P}_0$, *defines a one to one correspondence between the pseudo-*Γ*-spectral subspaces of* $\widetilde{A}$ *and those of* A.

PROOF. It suffices to show that the map $\widetilde{K}_0 \mapsto P_0\widetilde{K}_0$ defines a one to one correspondence between the $\widetilde{A}$-invariant subspaces of $\operatorname{Im}\widetilde{P}_0$ and the A-invariant subspaces of $\operatorname{Im}P_0$. With respect to the decomposition $X = X_1 \oplus \widetilde{X} \oplus X_2$, P_0 admits the partitioning

$$(2.3)\qquad P_0 = \begin{pmatrix} 0 & V_1 & V_2 \\ 0 & \widetilde{P}_0 & V_3 \\ 0 & 0 & 0 \end{pmatrix}.$$

Suppose $\widetilde{K}_0$ is an $\widetilde{A}$-invariant subspace of $\operatorname{Im}P_0$. We have $AP_0\widetilde{K}_0 = P_0A\widetilde{K}_0$. Using (2.1), we compute that for each $x \in \widetilde{K}_0$

$$AP_0x = P_0Ax = P_0\begin{pmatrix} A_{10}x \\ \widetilde{A}x \\ 0 \end{pmatrix} = \begin{pmatrix} V_1\widetilde{A}x \\ \widetilde{P}_0\widetilde{A}x \\ 0 \end{pmatrix} = P_0\begin{pmatrix} 0 \\ \widetilde{A}x \\ 0 \end{pmatrix}.$$

Thus $AP_0\widetilde{K}_0 = P_0A\widetilde{K}_0 = P_0\widetilde{A}\widetilde{K}_0 \subset P_0\widetilde{K}_0$. Hence $P_0\widetilde{K}_0$ is an A-invariant subspace of $\operatorname{Im}P_0$.

Let K_0 be a subspace of $\operatorname{Im}P_0$. Note that $[x_1,x,x_2]^T \in K_0$ implies that $x_2 = 0$, $x \in \operatorname{Im}\widetilde{P}_0$ and $x_1 = V_1x$. Define $\widetilde{K}_0$ to be the subspace of $\widetilde{X}$ consisting of all vectors x such that $[V_1x,x,0]^T \in K_0$. Obviously $\widetilde{K}_0 \subset \operatorname{Im}\widetilde{P}_0$ and $P_0\widetilde{K}_0 = K_0$. Further, one checks easily that $\widetilde{K}_0$ is $\widetilde{A}$-invariant whenever K_0 is A-invariant. □

3. DESCRIPTION OF ALL CANONICAL PSEUDO-Γ-SPECTRAL FACTORIZATIONS

To describe the canonical pseudo-Γ-spectral factorizations of a rational matrix function $W(\lambda)$ we realize $W(\lambda)$ as the transfer function of a finite dimensional node and next we use the geometric factorization theorem of [1]. But first let us recall the terminology we need from [1].

A *finite dimensional node* is a quintet $\theta = (A,B,C;X,Y)$ of two finite dimensional complex linear spaces X and Y and three linear operators $A : X \to X$, $B : Y \to X$ and $C : X \to Y$. The space X is called the *state space* of the node θ, and the operator A is called the *main operator* of the node θ. In what follows we consider only finite dimensional nodes and therefore the words "finite dimensional" will be omitted.

The *transfer function* of a node $\theta = (A,B,C;X,Y)$ is defined to be the operator function $W(\lambda) = I + C(\lambda - A)^{-1}B$. Here I stands for the identity operator on the (finite dimensional) linear space Y. Note that $W(\lambda)$ is a rational function of which the poles are contained in the set of eigenvalues of A and $W(\infty) = I$. If W is a rational operator function, we call a node $\theta = (A,B,C;X,Y)$ a *realization* of W if $W(\lambda) = I + C(\lambda - A)^{-1}B$. Each rational function W whose values are operators on Y such that $W(\infty) = I$ appears as the tranfer function of a node and thus has a realization.

A node $\theta = (A,B,C;X,Y)$ is called a *minimal realization* of W if θ is a realization and among all realizations of W the state space dimension of is as small as possible. A node $\theta = (A,B,C;X,Y)$ is called a *minimal node* if $\bigcap_{j=0}^{n-1} \operatorname{Ker} CA^j = (0)$, $\bigvee_{j=0}^{n-1} \operatorname{Im} A^jB = X$. Here $n = \dim X$. A node is minimal if and only if it is a minimal realization of its transfer function.

Two nodes $\theta_1 = (A_1,B_1,C_1;X_1,Y)$ and $\theta_2 = (A_2,B_2,C_2;X_2,Y)$ are called *similar* if there exists an invertible linear operator $S : X_2 \to X_1$, called a *node similarity*, such that

$$A_1 = S^{-1}A_2S, \quad B_1 = S^{-1}B_2, \quad C_1 = C_2S.$$

Note that similar nodes have the same transfer function. For minimal nodes there is a converse, namely, two minimal nodes with the same transfer function are similar.

A node $\theta = (A,B,C;X,Y)$ is called a *dilation* of a node $\tilde{\theta} = (\tilde{A},\tilde{B},\tilde{C};\tilde{X},Y)$ if there exists a decomposition $X = X_1 \oplus \tilde{X} \oplus X_2$ such that relative to this decomposition the operators A, B and C have the following partitioning

$$(3.1) \qquad A = \begin{pmatrix} A_1 & A_{10} & A_{12} \\ 0 & \tilde{A} & A_{02} \\ 0 & 0 & A_2 \end{pmatrix}, \quad B = \begin{pmatrix} B_1 \\ \tilde{B} \\ 0 \end{pmatrix}, \quad C = \begin{pmatrix} 0 & \tilde{C} & C_2 \end{pmatrix}.$$

Note that $\tilde{\theta}$ and its dilation θ have the same transfer function. Every realization of a rational operator function is a dilation of a minimal realization, and hence two nodes are realizations of the same function if and only if they are dilations of similar (minimal) nodes.

Let $\theta = (A,B,C;X,Y)$ be a node, and $\lambda_0 \in \mathbb{C}$. The node θ will be called *minimal at the point* λ_0 if

$$\bigcap_{j=0}^{n-1} \operatorname{Ker} CA^jP = \operatorname{Ker} P, \qquad \bigvee_{j=0}^{n-1} \operatorname{Im} PA^jB = \operatorname{Im} P.$$

Here $n = \dim X$, and P denotes the spectral or Riesz projection of A corresponding to the point λ_0. If λ_0 is not an eigenvalue of A, then θ is automatically minimal at λ_0. Further, θ is a minimal node if and only if θ is minimal at each eigenvalue of A (or, equivalently, at each $\lambda_0 \in \mathbb{C}$).

Let $\theta = (A,B,C;X,Y)$ be a node, and let Γ be a Cauchy contour on the Riemann sphere. We call a pair $(L,L^{\times})$ of subspaces of X a *pair of pseudo-*Γ*-spectral subspaces* for θ if L is a pseudo-Γ-spectral subspace for A, and $L^{\times}$ is a pseudo-$(-\Gamma)$-spectral subspace for $A^{\times}$. Here $-\Gamma$ denotes the Cauchy contour which coincides with Γ as a set, but has the opposite orientation. We call a pair (M_1,M_2) of subspaces of a linear space X *matching* if $M_1 \oplus M_2 = X$, i.e., $M_1 \cap M_2 = (0)$ and $M_1 + M_2 = X$.

THEOREM 3.1. *Let W be the transfer function of the node $\theta = (A,B,C;X,Y)$, and assume that θ is minimal at each point of the Cauchy contour Γ. Then there is a one to one correspondence between the right canonical pseudo-Γ-spectral subspaces of W and the matching pairs of pseudo-Γ-spectral subspaces for θ in the following sense:*

(a) *Given a matching pair $(L,L^{\times})$ of pseudo-Γ-spectral subspaces for θ, a canonical pseudo-Γ-spectral factorization $W(\lambda) = W_-(\lambda)W_+(\lambda)$ is obtained by taking*

$$\begin{aligned} W_-(\lambda) &= I + C(\lambda - A)^{-1}(I - \Pi)B, \\ W_+(\lambda) &= I + C\Pi(\lambda - A)^{-1}B, \end{aligned} \tag{3.2}$$

where Π is the projection along L onto $L^{\times}$.

(b) *Given a canonical pseudo-Γ-spectral factorization $W(\lambda) = W_-(\lambda)W_+(\lambda)$, with $W_-(\infty) = W_+(\infty) = I$, there exists a unique matching pair $(L,L^{\times})$ of pseudo-Γ-spectral subspaces for θ such that W_- and W_+ are given by* (3.2).

PROOF. The proof consists of three steps. Firstly, we prove (a). Secondly, we prove (b) for the case where θ is minimal. Thirdly, we prove (b) in the general case, where θ is minimal at each $\lambda \in \Gamma$.

(i) Suppose $(L,L^{\times})$ is a matching pair of pseudo-Γ-spectral subspaces for θ. Denote the projection along L onto $L^{\times}$ by Π. Define W_- and W_+ by (3.2). Using that $\operatorname{Im}\Pi$ is $A^{\times}$-invariant, and $\operatorname{Ker}\Pi$ is A-invariant, we get $A^{\times}\Pi = \Pi A^{\times}\Pi$ and $A(I-\Pi) = (I-\Pi)A(I-\Pi)$. Hence, like in [1], we compute that

$$(3.3)\qquad \begin{aligned} W_-^{-1}(\lambda) &= I - C(I-\Pi)(\lambda - A^{\times})^{-1}B, \\ W_+^{-1}(\lambda) &= I - C(\lambda - A^{\times})^{-1}\Pi B. \end{aligned}$$

Furthermore, we have $W(\lambda) = W_-(\lambda)W_+(\lambda)$. Since $(I-\Pi)A(I-\Pi)$ and $(I-\Pi)A^{\times}(I-\Pi)$ have no eigenvalues in Ω_Γ^-, the function W_- is holomorphic and invertible on Ω_Γ^-. Since $\Pi A\Pi$ and $\Pi A^{\times}\Pi$ have no eigenvalues in Ω_Γ^+, the function W_+ is holomorphic and invertible on Ω_Γ^+. Note that θ is the product of θ_- and θ_+, where

$$\theta_- = ((I-\Pi)A(I-\Pi),(I-\Pi)B,C(I-\Pi);\operatorname{Ker}\Pi, Y),$$

$$\theta_+ = (\Pi A\Pi, \Pi B, C\Pi, \operatorname{Im}\Pi, Y).$$

Further, the transfer functions of θ, θ_- and θ_+ are the functions W, W_- and W_+, respectively. From [1, Theorem 4.2] and the minimality of θ at each $\lambda_0 \in \Gamma$ we conclude that $\delta(W;\lambda_0) = \delta(W_-;\lambda_0) + \delta(W_+;\lambda_0)$ at each $\lambda_0 \in \Gamma$. Hence the factorization $W(\lambda) = W_-(\lambda)W_+(\lambda)$ is minimal at each $\lambda_0 \in \Gamma$.

(ii) Suppose θ is a minimal realization of W, and let $W(\lambda) = W_-(\lambda)W_+(\lambda)$ be a canonical pseudo-Γ-spectral factorization with $W_-(\infty) = W_+(\infty) = I$. Using Theorem 4.8 from [1], we conclude that there exists a unique decomposition $X = L \oplus L^{\times}$ of X into the direct sum of an A-invariant subspace L and an $A^{\times}$-invariant subspace $L^{\times}$ having the following property: If Π denotes the projection along L onto $L^{\times}$ then the functions W_-, W_+ and W_-^{-1}, W_+^{-1} are given by (3.2) and (3.3), respectively. Since W_- is holomorphic and invertible on Ω_Γ^-, the minimality of θ implies that $(I-\Pi)A(I-\Pi)$ and $(I-\Pi)A^{\times}(I-\Pi)$ do not have eigenvalues in Ω_Γ^-. Hence $L \subseteq \operatorname{Im}P_+ \oplus \operatorname{Im}P_0$ and $\operatorname{Im}P_-^{\times} \subseteq L^{\times}$. Similarly, since W_+ is holomorphic and invertible on Ω_Γ^+, we have that $\Pi A\Pi$ and $\Pi A^{\times}\Pi$ do not have eigenvalues in Ω_Γ^+. Hence $\operatorname{Im}P_+ \subseteq L$ and $L^{\times} \subseteq \operatorname{Im}P_-^{\times} \oplus \operatorname{Im}P_0^{\times}$. Thus $(L,L^{\times})$ is a matching pair of pseudo-Γ-spectral subspaces for θ.

(iii) Suppose θ is a realization W which is minimal at each $\lambda \in \Gamma$, and let $W(\lambda) = W_-(\lambda)W_+(\lambda)$ be a canonical pseudo-Γ-spectral factorization with $W_-(\infty) = W_+(\infty) = I$. The node θ is a dilation of a minimal node $\widetilde{\theta} = (\widetilde{A},\widetilde{B},\widetilde{C};\widetilde{X},Y)$. By the previous paragraph we know that there exists a unique projection $\widetilde{\Pi} : \widetilde{X} \to \widetilde{X}$ such that $(\operatorname{Ker}\widetilde{\Pi}, \operatorname{Im}\widetilde{\Pi})$ is a pair of pseudo-Γ-spectral subspaces and

W_- and W_+ are given by

$$W_-(\lambda) = I + \widetilde{C}(\lambda - \widetilde{A})^{-1}(I - \widetilde{\Pi})\widetilde{B},$$

$$W_+(\lambda) = I + \widetilde{C}\widetilde{\Pi}(\lambda - \widetilde{A})^{-1}\widetilde{B}.$$

Let the partitioning of A, B and C be given by (3.1). In particular A and $A^\times$ have the following form:

$$(3.4)\qquad A = \begin{pmatrix} A_1 & A_{10} & A_{12} \\ 0 & \widetilde{A} & A_{02} \\ 0 & 0 & A_2 \end{pmatrix}, \qquad A^\times = \begin{pmatrix} A_1 & A_{10}^\times & A_{12}^\times \\ 0 & \widetilde{A}^\times & A_{02}^\times \\ 0 & 0 & A_2 \end{pmatrix}.$$

Since $\widetilde{\theta}$ is minimal and θ is minimal at each $\lambda \in \Gamma$ the operators A_1 and A_2 do not have eigenvalues on Γ. Using Proposition 2.2, we can associate with $\operatorname{Ker}\widetilde{\Pi}$ a subspace L of X and with $\operatorname{Im}\widetilde{\Pi}$ a subspace $L^\times$ of X such that $(L, L^\times)$ is a pair of pseudo-Γ-spectral subspaces for θ. We shall prove that $L \oplus L^\times = X$ and W_- and W_+ are given by (3.2), where Π is the projection along L onto $L^\times$.

Using (3.4), we have the following partitionings for the spectral projections:

$$P_+ = \begin{pmatrix} P_1 & R_1 & R_2 \\ 0 & \widetilde{P}_+ & R_3 \\ 0 & 0 & P_2 \end{pmatrix}, \qquad P_-^\times = \begin{pmatrix} I - P_1 & -R_1^\times & -R_2^\times \\ 0 & \widetilde{P}_-^\times & -R_3^\times \\ 0 & 0 & I - P_2 \end{pmatrix},$$

$$P_0 = \begin{pmatrix} 0 & V_1 & V_2 \\ 0 & \widetilde{P}_0 & V_3 \\ 0 & 0 & 0 \end{pmatrix}, \qquad P_0^\times = \begin{pmatrix} 0 & V_1^\times & V_2^\times \\ 0 & \widetilde{P}_0^\times & V_3^\times \\ 0 & 0 & 0 \end{pmatrix}.$$

Since $P_+^2 = P_+$ and $(P_-^\times)^2 = P_-^\times$ we have

$$(3.5)\qquad \begin{aligned} P_1R_1 + R_1\widetilde{P}_+ &= R_1, \\ P_1R_2 + R_1R_3 + R_2P_2 &= R_2, \\ \widetilde{P}_+R_3 + R_3P_2 &= R_3, \end{aligned}$$

and

$$(3.6)\qquad \begin{aligned} (I - P_1)R_1^\times + R_1^\times\widetilde{P}_-^\times &= R_1^\times, \\ (I - P_1)R_2^\times - R_1^\times R_3^\times + R_2^\times(I - P_2) &= R_2^\times, \\ \widetilde{P}_-^\times R_3^\times + R_3^\times(I - P_2) &= R_3^\times. \end{aligned}$$

We define operators $S_1,T_1,T_1^\times : \widetilde{X} \to X_1$; $S_2,T_2T_2^\times : X_2 \to X_1$ and $S_3,T_3,T_3^\times : X_2 \to \widetilde{X}$ as follows:

$$S_1 = (I-P_1)R_1\widetilde{P}_+ - P_1R_1^\times\widetilde{P}_-^\times,$$
$$S_2 = (I-P_1)R_2P_2 - P_1R_2^\times(I-P_2),$$
$$S_3 = (I-\widetilde{P}_+)R_3P_2 - (I-\widetilde{P}_-^\times)R_3^\times(I-P_2),$$
$$T_1 = -R_1^\times\widetilde{P}_-^\times\widetilde{P}_+ - R_1,$$
$$T_2 = R_1(I-\widetilde{P}_+)R_3P_2 - R_2,$$
$$T_3 = -R_3,$$
$$T_1^\times = R_1\widetilde{P}_+\widetilde{P}_-^\times + R_1^\times,$$
$$T_2^\times = R_1^\times(I-\widetilde{P}_-^\times)R_3^\times(I-P_2) + R_2^\times,$$
$$T_3^\times = R_3^\times.$$

Define $S,T,T^\times : X_1 \oplus \widetilde{X} \oplus X_2 \to X_1 \oplus \widetilde{X} \oplus X_2$ by

$$S = \begin{pmatrix} I & S_1 & S_2 \\ 0 & I & S_3 \\ 0 & 0 & I \end{pmatrix}, \quad T = \begin{pmatrix} I & T_1 & T_2 \\ 0 & I & T_3 \\ 0 & 0 & I \end{pmatrix}, \quad T^\times = \begin{pmatrix} I & T_1^\times & T_2^\times \\ 0 & I & T_3^\times \\ 0 & 0 & I \end{pmatrix}.$$

The operators X, T and $T^\times$ are invertible, and from (3.5) and (3.6) we get

$$S\begin{pmatrix} P_1 & 0 & 0 \\ 0 & \widetilde{P}_+ & 0 \\ 0 & 0 & P_2 \end{pmatrix} = P_+T, \qquad S\begin{pmatrix} I-P_1 & 0 & 0 \\ 0 & \widetilde{P}_-^\times & 0 \\ 0 & 0 & I-P_2 \end{pmatrix} = P_-^\times T^\times.$$

Hence

$$\operatorname{Im} P_+ = S\begin{pmatrix} \operatorname{Im} P_1 \\ \operatorname{Im} \widetilde{P}_+ \\ \operatorname{Im} P_2 \end{pmatrix}, \qquad \operatorname{Im} P_-^\times = S\begin{pmatrix} \operatorname{Ker} P_1 \\ \operatorname{Im} \widetilde{P}_1^\times \\ \operatorname{Ker} P_2 \end{pmatrix}.$$

We write $\operatorname{Ker}\widetilde{\Pi} = \operatorname{Im}\widetilde{P}_+ \oplus \widetilde{K}_0$ and $\operatorname{Im}\widetilde{\Pi} = \operatorname{Im}\widetilde{P}_-^\times \oplus \widetilde{K}_0^\times$ with $\widetilde{K}_0 \subset \operatorname{Im}\widetilde{P}_0$ and $\widetilde{K}_0^\times \subset \operatorname{Im}\widetilde{P}_0^\times$. Then

$$L = \operatorname{Im} P_+ \oplus P_0\widetilde{K}_0 = S\begin{pmatrix} \operatorname{Im} P_1 \\ \operatorname{Im} \widetilde{P}_+ \\ \operatorname{Im} P_2 \end{pmatrix} \oplus \begin{pmatrix} 0 & V_1 & 0 \\ 0 & I & 0 \\ 0 & 0 & 0 \end{pmatrix}\begin{pmatrix} 0 \\ \widetilde{K}_0 \\ 0 \end{pmatrix},$$

$$L^{\times} = \operatorname{Im} P_-^{\times} \oplus P_0^{\times}\widetilde{K}_0^{\times} = S \begin{pmatrix} \operatorname{Ker} P_1 \\ \operatorname{Im} \widetilde{P}_-^{\times} \\ \operatorname{Ker} P_2 \end{pmatrix} \oplus \begin{pmatrix} 0 & V_1^{\times} & 0 \\ 0 & I & 0 \\ 0 & 0 & 0 \end{pmatrix} \begin{pmatrix} 0 \\ \widetilde{K}_0^{\times} \\ 0 \end{pmatrix}.$$

Hence

$$S^{-1}L = \begin{pmatrix} \operatorname{Im} P_1 \\ \operatorname{Im} \widetilde{P}_+ \\ \operatorname{Im} P_2 \end{pmatrix} \oplus \begin{pmatrix} 0 & V_1 - S_1 & 0 \\ 0 & I & 0 \\ 0 & 0 & 0 \end{pmatrix} \begin{pmatrix} 0 \\ \widetilde{K}_0 \\ 0 \end{pmatrix},$$

(3.7)

$$S^{-1}L^{\times} = \begin{pmatrix} \operatorname{Ker} P_1 \\ \operatorname{Im} \widetilde{P}_-^{\times} \\ \operatorname{Ker} P_2 \end{pmatrix} \oplus \begin{pmatrix} 0 & V_1^{\times} - S_1^{\times} & 0 \\ 0 & I & 0 \\ 0 & 0 & 0 \end{pmatrix} \begin{pmatrix} 0 \\ \widetilde{K}_0^{\times} \\ 0 \end{pmatrix}.$$

From these identities we conclude

$$S^{-1}(L + L^{\times}) = \begin{pmatrix} X_1 \\ \operatorname{Ker} \widetilde{\Pi} + \operatorname{Im} \widetilde{\Pi} \\ X_2 \end{pmatrix} = \begin{pmatrix} X_1 \\ \widetilde{X} \\ X_2 \end{pmatrix} = X,$$

and

$$S^{-1}(L \cap L^{\times}) = \begin{pmatrix} 0 & P_1V_1^{\times} + (I - P_1)V_1 - S_1 & 0 \\ 0 & I & 0 \\ 0 & 0 & 0 \end{pmatrix} \begin{pmatrix} 0 \\ \operatorname{Ker} \widetilde{\Pi} \cap \operatorname{Im} \widetilde{\Pi} \\ 0 \end{pmatrix} = (0).$$

Hence $L \oplus L^{\times} = X$. Denote the projection along L onto $L^{\times}$ by Π. We have, by (3.7),

$$S^{-1}\Pi S = \begin{pmatrix} I - P_1 & * & * \\ 0 & \widetilde{\Pi} & * \\ 0 & 0 & I - P_2 \end{pmatrix},$$

and hence

$$\Pi = \begin{pmatrix} I - P_1 & * & * \\ 0 & \widetilde{\Pi} & * \\ 0 & 0 & I - P_2 \end{pmatrix}. \tag{3.8}$$

From this we get $C\Pi(\lambda - A)^{-1}B = \widetilde{C}\widetilde{\Pi}(\lambda - \widetilde{A})^{-1}\widetilde{B}$ and $C(\lambda - A)^{-1}(I - \Pi)B = \widetilde{C}(\lambda - \widetilde{A})^{-1}(I - \widetilde{\Pi})\widetilde{B}$. Hence the factors W_- and W_+ are given by (3.2).

Finally, let us show the unicity of the pair of pseudo-Γ-spectral subspaces. Assume $(\hat{L},\hat{L}^{\times})$ is a second pair of pseudo-Γ-spectral subspaces, which yields the same factorization $W(\lambda) = W_{-}(\lambda)W_{+}(\lambda)$. Using Proposition 2.2, we associate with $(\hat{L},\hat{L}^{\times})$ a pair $(\hat{L}_0,\hat{L}_0^{\times})$ of pseudo-Γ-spectral subspaces for $\widetilde{\theta}$. By the same reasoning as before we have $\hat{L}_0 \oplus \hat{L}_0^{\times} = \widetilde{X}$, and the pair $(\hat{L}_0,\hat{L}_0^{\times})$ yields the same factorization $W(\lambda) = W_{-}(\lambda)W_{+}(\lambda)$. By the minimality of $\widetilde{\theta}$ we have that $\hat{L}_0 = \operatorname{Ker}\widetilde{\Pi}$ and $\hat{L}_0^{\times} = \operatorname{Im}\widetilde{\Pi}$. Hence $\hat{L} = \operatorname{Ker}\Pi$, $\hat{L}^{\times} = \operatorname{Im}\Pi$. □

EXAMPLE 3.2. (see Example 1.1) Let W be the rational 2×2 matrix function (1.3). We can write $W(\lambda) = I + C(\lambda - A)^{-1}B$, where

$$A = \begin{pmatrix} -2i & 0 \\ 0 & i \end{pmatrix}, \quad B = \begin{pmatrix} -i & i \\ 0 & i \end{pmatrix}, \quad C = \begin{pmatrix} 2 & 1 \\ 0 & 1 \end{pmatrix}.$$

Note that $\theta = (A,B,C;\mathbb{C}^2,\mathbb{C}^2)$ is minimal. Furthermore

$$A^{\times} = \begin{pmatrix} 0 & 0 \\ 0 & 0 \end{pmatrix}.$$

Take $\Gamma = \mathbb{R}$. We then have

$$P_{+} = \begin{pmatrix} 0 & 0 \\ 0 & 1 \end{pmatrix}, \quad P_0 = \begin{pmatrix} 0 & 0 \\ 0 & 0 \end{pmatrix}, \quad P_{-}^{\times} = \begin{pmatrix} 0 & 0 \\ 0 & 0 \end{pmatrix}, \quad P_0^{\times} = \begin{pmatrix} 1 & 0 \\ 0 & 1 \end{pmatrix}.$$

Hence the matching pairs of pseudo-$\mathbb{R}$-spectral subspaces are given by $(\operatorname{span}\left\{\begin{pmatrix}0\\1\end{pmatrix}\right\}, \operatorname{span}\{x\})$, where $x \in \mathbb{C}^2$ and $\begin{pmatrix}0\\1\end{pmatrix}$ are linearly independent. Without loss of generality we may assume $x = x_\alpha = \begin{pmatrix}1\\-\alpha\end{pmatrix}$. We shall compute the corresponding canonical pseudo-$\mathbb{R}$-spectral factorization $W(\lambda) = W_{-}^{(\alpha)}(\lambda)W_{+}^{(\alpha)}(\lambda)$. Denote the projection along $\operatorname{span}\left\{\begin{pmatrix}0\\1\end{pmatrix}\right\}$ onto $\operatorname{span}\{x_\alpha\}$ by Π_α. Then

$$\Pi_\alpha = \begin{pmatrix} 1 & 0 \\ -\alpha & 0 \end{pmatrix},$$

$$W_{-}^{(\alpha)}(\lambda) = I + C(\lambda - A)^{-1}(I - \Pi)B =$$

$$= \begin{pmatrix} 1 & 0 \\ 0 & 1 \end{pmatrix} + \begin{pmatrix} 2 & 1 \\ 0 & 1 \end{pmatrix}\begin{pmatrix} \lambda+2i & 0 \\ 0 & \lambda-i \end{pmatrix}^{-1}\begin{pmatrix} 0 & 0 \\ \alpha & 1 \end{pmatrix}\begin{pmatrix} -i & i \\ 0 & 1 \end{pmatrix} =$$

$$= \begin{pmatrix} \dfrac{\lambda - i(1+\alpha)}{\lambda-i} & \dfrac{i(1+\alpha)}{\lambda-i} \\[2ex] \dfrac{-i\alpha}{\lambda-i} & \dfrac{\lambda+i\alpha}{\lambda-i} \end{pmatrix},$$

and

$$W_+^{(\alpha)}(\lambda) = I + C\Pi(\lambda - A)^{-1}B =$$

$$= \begin{pmatrix}1 & 0\\ 0 & 1\end{pmatrix} + \begin{pmatrix}2 & 1\\ 0 & 1\end{pmatrix}\begin{pmatrix}1 & 0\\ -\alpha & 0\end{pmatrix}\begin{pmatrix}\lambda+2i & 0\\ 0 & \lambda-i\end{pmatrix}^{-1}\begin{pmatrix}-i & i\\ 0 & i\end{pmatrix} =$$

$$= \begin{bmatrix}\dfrac{\lambda+i\alpha}{\lambda+2i} & \dfrac{i(2-\alpha)}{\lambda+2i}\\[2ex] \dfrac{i\alpha}{\lambda+2i} & \dfrac{\lambda+i(2-\alpha)}{\lambda+2i}\end{bmatrix}.$$

Hence the factorizations described in Example 1.1 were all possible canonical pseudo-$\mathbb{R}$-spectral factorizations.

EXAMPLE 3.3. The scalar case (cf. Example 1.2). Let W be a rational (scalar) function with $W(\infty) = 1$. We can write

$$W(\lambda) = \frac{\lambda^{\ell} + b_{\ell-1}\lambda^{\ell-1} + \ldots + b_1\lambda + b_0}{\lambda^{\ell} + a_{\ell-1}\lambda^{\ell-1} + \ldots + a_1\lambda + a_0}$$

for certain complex numbers $a_0,\ldots,a_{\ell-1},b_0,\ldots,b_{\ell-1}$. We assume that the polynomials $p(\lambda) = \lambda^{\ell} + b_{\ell-1}\lambda^{\ell-1} + \ldots + b_1 + b_0$ and $q(\lambda) = \lambda^{\ell} + a_{\ell-1}\lambda^{\ell-1} + \ldots + a_1 + a_0$ do not have common zeros. Put (see [2, Section I.6])

$$A = \begin{bmatrix} 0 & 1 & & \\ \vdots & & \ddots & \\ 0 & & & 1 \\ -a_0 & \cdots & & -a_{\ell-1}\end{bmatrix}, \quad B = \begin{bmatrix}0\\ \vdots\\ 0\\ 1\end{bmatrix}, \quad C = \begin{pmatrix}c_0 \cdots c_{\ell-1}\end{pmatrix}.$$

Here $c_j = b_j - a_j$, $j = 0,1,\ldots,\ell-1$. We compute

$$A^{\times} = A - BC = \begin{bmatrix} 0 & 1 & & \\ \vdots & & \ddots & \\ 0 & & & 1 \\ -b_0 & \cdots & & -b_{\ell-1}\end{bmatrix}.$$

The minimal node $\theta = (A,B,C;\mathbb{C}^{\ell},\mathbb{C})$ is a realization of W. (The minimality follows from the assumption that $p(\lambda)$ and $q(\lambda)$ do not have common zeros.) The matrices A and $A^{\times}$ only have eigenvalues with geometric multiplicity one. The algebraic multiplicity of an eigenvalue μ of A (resp. $A^{\times}$) is equal to the multiplicity of μ as a pole (resp. zero) of W. Suppose μ is an eigenvalue of A (resp. A^x), with algebraic multiplicity m. A basis of the spectral subspace of A (resp. A^x) at μ is given by $\{\xi_j(\mu)\}_{j=0}^{m}$, with

$$\xi_j(\mu)_i = \begin{cases} 0, & i = 1,\dots,j, \\ \binom{i-1}{i-1-j}\mu^{i-1-j}, & i = j+1,\dots,\ell, \end{cases}$$

$$j = 0,1,\dots,m-1.$$

The vectors $\xi_0(\mu),\dots,\xi_{m-1}(\mu)$ are a Jordan-chain of A (resp. A^x) corresponding with the eigenvalue μ.

It is well-known that, whenever $\mu_1,\dots,\mu_t$ are distinct complex numbers and $s_1,\dots,s_t$ are nonnegative integers such that $\Sigma_{j=1}^{t} s_j \le \ell$ (the dimension of the state space), the vectors $\xi_j(\mu_k)$, $j = 0,\dots,s_{k-1}$, $k = 1,\dots,t$ are linearly independent.

A direct consequence of this is the following. If N is A-invariant, and N^x is A^x-invariant, then $\dim N + \dim N^x = \ell$ if and only if $N \oplus N^x = \mathbb{C}^\ell$. Hence there exists a matching pair of pseudo-Γ-spectral subspaces if and only if there eixsts a pair (L,L^x) of pseudo-Γ-spectral subspaces such that $\dim L + \dim L^x = \ell$. We denote by m_1, m_2 and m_3 (resp. n_1, n_2 and n_3) the sum of the algebraic multiplicities in Ω_Γ^+ (the inner domain of Γ), on Γ, and in Ω_Γ^- (the outer domain of Γ). There exists a pair $(L.L^x)$ of pseudo-Γ-spectral subspaces such that $\dim L + \dim L^x = \ell$ if and only if there exist nonnegative integers p,q such that

$$\text{(3.9)} \qquad \begin{cases} p + q = \ell \\ m_1 \le p \le m_1 + m_2 \\ n_3 \le q \le n_2 + n_3. \end{cases}$$

Using $m_1 + m_2 + m_3 = n_1 + n_2 + n_3 = \ell$, we see that p,q satisfying (3.9) exist if and only if one of the following conditions holds: (i) $m_1 + m_2 \ge n_1 \ge m_1$ or (ii) $n_1 + n_2 \ge m_1 \ge n_1$. This is the same result as in Example 1.2.

4. NON-NEGATIVE RATIONAL MATRIX FUNCTIONS

In this section we shall treat the special case of non-negative rational matrix functions. It turns out that a non-negative rational matrix function always admits exactly one canonical pseudo-$\mathbb{R}$-spectral factorization of a special symmetric type. We shall use a number of results from [4], [9],

[10] and [11].

A rational $m \times m$ matrix function W is called *self-adjoint* if $W_*(\lambda) := W(\bar{\lambda})^* = W(\lambda)$. It is called *non-negative* if $0 \leq <W(\lambda)x,x>$ for each $x \in \mathbb{C}^m$, $\lambda \in \mathbb{R}$ (λ not a pole). A non-negative matrix function is automatically self-adjoint.

THEOREM 4.1. *Let W be a non-negative rational $m \times m$ matrix function with $W(\infty) = I$. Then W has exactly one canonical pseudo-$\mathbb{R}$-spectral factorization of the type*

(4.1) $W(\lambda) = N_*(\lambda)N(\lambda)$,

with $N(\infty) = I$. This factorization is obtained in the following way. Choose a minimal realization $\theta = (A,B,C;\mathbb{C}^n,\mathbb{C}^m)$ of W. Let L (resp. L^x) be the span of all eigenvectors and generalized eigenvectors corresponding to the eigenvalues of A (resp. A^x) in the open upper (resp. lower) half plane and of the first half of the Jordan chains corresponding to real eigenvalues of A (resp. A^x). Then (L,L^x) is a matching pair of pseudo-$\mathbb{R}$-spectral subspaces for θ and the factor N in (2.3) is given by $N(\lambda) = I + C\Pi(\lambda-A)^{-1}B$, where Π is the projection along L onto L^x.

PROOF. Let $\theta = (A,B,C;\mathbb{C}^n,\mathbb{C}^m)$ be a minimal realization of W. Define sets σ,σ^x as follows. The set σ (resp. σ^x) consists of all eigenvalues of A (resp. A^x) in the open lower half plane. Hence σ and σ^x do not contain pairs of complex conjugate numbers. By [9, Corollary 4.4], there exists a unique rational $m \times m$ matrix function N such that

(i) $W(\lambda) = N_*(\lambda)N(\lambda)$ is a minimal factorization,

(ii)   the set of non real poles of N is σ,
       the set of non real zeros of N is σ^x,

(iii) $N(\infty) = I$.

In particular, $W(\lambda) = N_*(\lambda)N(\lambda)$ is a canonical pseudo-$\mathbb{R}$-spectral factorization. From [9, Theorem 4.3 and Corollary 4.4] (see also [11, Theorem 1]) it is clear that the factor N is this factorization is obtained in the way described in the second part of the theorem. Finally, if we have a canonical pseudo-$\mathbb{R}$-spectral factorization (4.1), then this factorization has the properties (i), (ii) and (iii) mentioned above, and hence there is only one such factorization.   □

EXAMPLE 4.2. We consider

(4.2) $W(\lambda) = \dfrac{\lambda^2}{\lambda^2+1}$.

In this case W is a non-negative rational (scalar) function. A minimal realization $\theta = (A,B,C;\mathbb{C}^2,\mathbb{C})$ of W is given by

$$(4.3) \qquad A = \begin{pmatrix} i & 0 \\ 0 & -i \end{pmatrix}, \quad B = \begin{pmatrix} 1 \\ 1 \end{pmatrix}, \quad C = \begin{pmatrix} \tfrac{1}{2}i & -\tfrac{1}{2}i \end{pmatrix}.$$

Note that

$$(4.4) \qquad A^{\times} = \begin{bmatrix} \tfrac{1}{2}i & \tfrac{1}{2}i \\ -\tfrac{1}{2}i & -\tfrac{1}{2}i \end{bmatrix}.$$

We see that A has no real eigenvalues, and $A^{\times}$ has precisely one eigenvalue, namely at 0, which has geometric multiplicity one. A corresponding Jordan chain is

$$x_0 = \begin{pmatrix} 1 \\ -1 \end{pmatrix}, \quad x_1 = \begin{pmatrix} -i \\ -i \end{pmatrix}.$$

According to Theorem 4.1, we have a matching pair $(L,L^{\times})$ of pseudo-$\mathbb{R}$-spectral subspaces, given by $L = \operatorname{span}\left\{\begin{pmatrix} 1 \\ 0 \end{pmatrix}\right\}$, $L^{\times} = \operatorname{span}\left\{\begin{pmatrix} 1 \\ -1 \end{pmatrix}\right\}$. The corresponding canonical pseudo-$\mathbb{R}$-spectral factorization is

$$W(\lambda) = \frac{\lambda}{\lambda - i} \cdot \frac{\lambda}{\lambda + i}.$$

Note that this is the only canonical pseudo-$\mathbb{R}$-spectral factorization of $W(\lambda)$ (of which the factors have the value 1 at infinity).

5. WIENER-HOPF INTEGRAL EQUATIONS OF NON-NORMAL TYPE

In this section we apply the theory of canonical pseudo-Γ-spectral factorization to Wiener-Hopf integral equations of non-normal type. Consider the Wiener-Hopf integral equation

$$(5.1) \qquad \varphi(t) - \int_0^\infty k(t-s)\varphi(s)ds = f(t), \quad t \geq 0,$$

where $k \in L_1^{m\times m}(-\infty,\infty)$. We shall assume that the *symbol* of (5.1),

$$(5.2) \qquad W(\lambda) = I - \int_{-\infty}^{\infty} k(t)e^{i\lambda t}dt, \quad \lambda \in \mathbb{R},$$

is a rational $m \times m$ matrix function.

The rationality of the symbol allows us to see W as the transfer function of a node $\theta = (A,B,C;\mathbb{C}^n,\mathbb{C}^m)$. That is, $W(\lambda) = I + C(\lambda - A)^{-1}B$. Throughout this section, we assume that A has no real eigenvalues. (Note that this is equivalent to the requirement that θ is minimal at each $\lambda \in \mathbb{R}$.) Denote by P_+ the spectral projection of A corresponding to the eigenvalues of

A in the upper half plane. Using $W(\lambda) = I + C(\lambda - A)^{-1}B$, and taking the inverse Fourier transform of (5.2), we get

$$(5.3)\qquad k(t) = \begin{cases} iCe^{-itA}(I - P_+)B, & t > 0, \\ -iCe^{-itA}P_+B, & t < 0. \end{cases}$$

Assume that the symbol W of the equation (5.1) is realized by the node $\theta = (A,B,C,\mathbb{C}^n,\mathbb{C}^m)$, where A has no real eigenvalues. Suppose that W has a canonical pseudo-$\mathbb{R}$-spectral factorization $W(\lambda) = W_-(\lambda)W_+(\lambda)$. Then there is a corresponding matching pair $(L,L^\times)$ of pseudo-$\mathbb{R}$-spectral subspaces for the node θ. Let Π be the projection along $L = \operatorname{Im} P_+$ onto $L^\times$, and define

$$(5.4)\qquad g(t,s) = \begin{cases} iCe^{-itA^\times}\Pi e^{isA^\times}B, & 0 \le s < t, \\ -iCe^{-itA^\times}(I - \Pi)e^{isA^\times}B, & 0 \le t < s. \end{cases}$$

From the proof of Theorem 3.1 it follows (cf. formulas (3.1), (3.4) and (3.8)) that the definition of g does not depend on the particular realization of W (as long as we assume that the main operator A does not have real eigenvalues). We call g the *pseudo-resolvent kernel* corresponding to the canonical pseudo-$\mathbb{R}$-spectral factorization $W(\lambda) = W_-(\lambda)W_+(\lambda)$. In case $A^\times$ has no real eigenvalues, formula (5.4) gives the resolvent kernel as given in [2].

THEOREM 5.1. *Let* k *be the kernel of* (5.1), *and let* g *be the pseudo-resolvent kernel corresponding to a canonical pseudo-*$\mathbb{R}$*-spectral factorization of the symbol. The following resolvent identities hold*

$$(5.5)\qquad g(t,s) - \int_0^\infty k(t-u)g(u,s)du = k(t-s), \qquad s \ge 0,\ t \ge 0$$

$$(5.6)\qquad k(t-s) - \int_0^\infty g(t,u)k(u-s)du = g(t,s), \qquad s \ge 0,\ t \ge 0.$$

Here the integrals are considered as improper integrals, converging for each $s \ge 0$, $t \ge 0$.

PROOF. We shall prove (5.5) in case $0 \le s < t$. The other case and (5.6) are left to the reader.

Assume $0 \le s < t$, and take $T > t$. We have $\int_0^T k(t-u)g(u,s) = I_1 + I_2 + I_3$, with

$$I_1 = \int_0^s iCe^{-i(t-u)A}(I - P_+)B\cdot -iCe^{-iuA^\times}(I - \Pi)e^{isA^\times}Bdu,$$

$$I_2 = \int_s^t iCe^{-i(t-u)A}(I - P_+)B\cdot iCe^{-iuA^\times}\Pi e^{isA^\times}Bdu,$$

$$I_3 = \int_t^T -iCe^{-i(t-u)A}P_+B\cdot iCe^{-iuA^\times}\Pi e^{iSA^\times}Bdu.$$

We rewrite

$$I_1 = -iCe^{-itA} \int_0^s (I - P_+)\frac{d}{du}\left[e^{iuA}e^{-iuA^x}\right](I - \Pi)e^{isA^x}Bdu$$

$$= -iCe^{-itA}(I - P_+)e^{isA}e^{-isA^x}(I - \Pi)e^{isA^x}B +$$

$$+iCe^{-itA}(I - P_+)(I - \Pi)e^{isA^x}B$$

$$= -iCe^{-itA}(I - P_+)e^{isA}e^{-isA^x}(I - \Pi)e^{isA^x}B,$$

using that $(I - P_+)(I - \Pi) = 0$. Similarity, we get

$$I_2 = iCe^{-itA}(I - P_+)\left[e^{itA}e^{-itA^x} - e^{isA}e^{-isA^x}\right]\Pi e^{isA^x}B,$$

$$I_3 = iCe^{-itA}P_+\left[e^{itA}e^{-itA^x} - e^{iTA}e^{-iTA^x}\right]\Pi e^{isA^x}B.$$

Hence

$$I_1 + I_2 + I_3 = -k(t - s) + g(s,t) - iCe^{-itA}P_+e^{iTA}e^{-iTA^x}\Pi e^{isA}B.$$

We have $\lim_{T\to\infty} P_+e^{iTA}e^{-iTA^x}\Pi = 0$, since P_+e^{iTA} decreases exponentially, and $e^{-iTA^x}\Pi$ grows polynomially for $T \to \infty$. Thus (5.5) has been proven in case $0 \le s < t$. □

By K and G we denote the integral operators with kernels $k(t - s)$ and $g(t,s)$, given by (5.3) and (5.4), respectively. In case $W(\lambda)$ is invertible for each $\lambda \in \mathbb{R}$, the operator $I - K : L_p^m[0,\infty) \to L_p^m[0,\infty)$ is invertible with inverse $I + G$. Although the resolvent identities hold, this inversion result is not true in case $\det W(\lambda)$ has zeros on the real line. In fact, it may happen that $(I - K)\varphi_0 = 0$ for some function φ_0, while $[(I + G)(I - K)]\varphi_0 = \varphi_0$.

EXAMPLE 5.2. (see Example 4.2) Consider the kernel $k(t) = \frac{1}{2}e^{-|t|}$, $t \in \mathbb{R}$. The symbol is given by $W(\lambda) = \lambda^2/(\lambda^2+1)$. We take the following minimal realization of W: $\theta = (A,B,C;\mathbb{C}^2,\mathbb{C})$ with A, B and C given by (4.3). In that case A^x is given by (4.4). There is only one matching pair of pseudo-$\mathbb{R}$-spect spectral subspaces, namely span $\left\{\begin{pmatrix}1\\0\end{pmatrix}\right\}$, span $\left\{\begin{pmatrix}1\\-1\end{pmatrix}\right\}$. If we denote the projection along span $\left\{\begin{pmatrix}1\\0\end{pmatrix}\right\}$ onto span $\left\{\begin{pmatrix}1\\-1\end{pmatrix}\right\}$ by Π, we have

$$(5.7)\qquad \Pi = \begin{pmatrix}0 & -1\\0 & 1\end{pmatrix}.$$

Furthermore, we have

$$e^{itA^x} = \exp\left\{it\begin{bmatrix}1 & -i\\-1 & -i\end{bmatrix}\begin{bmatrix}0 & 1\\0 & 0\end{bmatrix}\begin{bmatrix}1 & -i\\-1 & -i\end{bmatrix}^{-1}\right\} =$$

$$= \begin{pmatrix} 1 & -i \\ -1 & -i \end{pmatrix} \exp \begin{pmatrix} 0 & it \\ 0 & 0 \end{pmatrix} \begin{pmatrix} \frac{1}{2} & -\frac{1}{2} \\ \frac{1}{2}i & \frac{1}{2}i \end{pmatrix}$$

$$= \begin{pmatrix} 1-\frac{1}{2}t & -\frac{1}{2}t \\ \frac{1}{2}t & 1+\frac{1}{2}t \end{pmatrix} .$$

The pseudo-resolvent kernel is given by

$$g(t,s) = \begin{cases} iCe^{-itA^{\times}}\Pi e^{isA^{\times}}B = 1+s, & 0 \le s < t, \\ -iCe^{-itA^{\times}}(I+\Pi)e^{isA^{\times}}B = 1+t, & 0 \le t < s. \end{cases}$$

Since the symbol has zeros on the real line, it is clear that $I - K : L_p[0,\infty) \to L_p[0,\infty)$ cannot be invertible with inverse $I+G$. In fact, the integral operator G is not defined for each $f \in L_p[0,\infty)$ and for $\varphi_0(t) = 1+t$, $t \ge 0$, the function $(I-K)\varphi_0 = 0$, while from the resolvent identities one expects $[(I+G)(I-K)]\varphi_0 = \varphi_0$.

6. PAIRS OF FUNCTION SPACES OF UNIQUE SOLVABILITY

To deal with the phenomenon described at the end of the previous section we shall define function spaces of unique solvability.

As before, we assume the symbol of (5.1) to be rational. Let K be the integral operator with kernel k(t-s). Assume g to be the pseudo-resolvent kernel corresponding to a canonical pseudo-$\mathbb{R}$-spectral factorization of the symbol. Let G be the integral operator with kernel g(t,s). A pair of function spaces $(L,L^{\times})$, where $L \subset L^m_{1,loc}[0,\infty)$ and $L^{\times} \subset L^m_{1,loc}[0,\infty)$, is called a *pair of function spaces of unique solvability* (for (5.1) corresponding to the pseudo-resolvent kernel g) if $I - K : L \to L^{\times}$ and $I+G : L^{\times} \to L$ are defined and invertible, and $(I-K)^{-1} = I+G$. More precisely, this means the following:
(α) for $\varphi \in L$ and $f \in L^{\times}$ the integrals $\int_0^\infty k(t-s)\ (s)ds$ and $\int_0^\infty g(t,s)f(s)ds$ exist as improper integrals for each $t \ge 0$,
(β) for each $\varphi \in L$ and each $f \in L^{\times}$ we have $(I-K)\varphi \in L^{\times}$ and $(I+G)f \in L$,
(γ) for each $\varphi \in L$ and each $f \in L^{\times}$ we have $(I+G)(I-K)\varphi = \varphi$ and $(I-K)(I+G)f = f$. We shall construct (maximal) pairs $(L,L^{\times})$ in terms of a realization of the symbol of (5.1).

THEOREM 6.1. *Let* $\theta = (A,B,C;\mathbb{C}^n,\mathbb{C}^m)$ *be a realization of the symbol of* (5.1) *such that* A *has no real eigenvalues, and let* $(L,L^{\times})$ *be a matching pair of pseudo-*$\mathbb{R}$*-spectral subspaces for* θ. *Denote by* P_+ *the spectral pro-*

jection of A *corresponding to the eigenvalues of* A *in the upper half plane, and let* Π *be the projection along* L *onto* $L^{\times}$. *Define function spaces* $L_\theta, L_\theta^{\times} \subset L^m_{1.loc}[0,\infty)$ *by*

(b) $\varphi \in L_\theta$ *if and only if*

(i) $\lim_{t\to\infty} \int_0^t P_+ e^{isA} B\varphi(s)ds$ *exists*,

(ii) $\lim_{t\to\infty} (I+\Pi)e^{itA^{\times}}e^{-itA}\left[\int_0^t (I-P_+)e^{isA}B\varphi(s)ds - \int_t^\infty P_+e^{isA}B\varphi(s)ds\right] = 0,$

(b) $f \in L_\theta^{\times}$ *if and only if*

(i) $\lim_{t\to\infty} \int_0^t (I-\Pi)^{isA^{\times}}Bf(s)ds$ *exists*,

(ii) $\lim_{t\to\infty} P_+e^{itA}e^{-itA^{\times}}\left[\int_0^t \Pi e^{isA^{\times}}Bf(s)ds - \int_t^\infty (I-\Pi)e^{isA^{\times}}Bf(s)ds\right] = 0.$

Then $(L_\theta, L_\theta^{\times})$ *is a pair of function spaces of unique solvability. If* θ *is a minimal realization, then* $(L_\theta, L_\theta^{\times})$ *is maximal with respect to inclusion.*

The maximality in the last part of the theorem has to be understood in the following way. For each pair $(L, L^{\times})$ of function spaces of unique solvability we have $L \subset L_\theta$ and $L^{\times} \subset L_\theta^{\times}$ whenever θ is minimal. The proof of Theorem 6.1 will be based on three lemmas for k and equivalent versions for g. In what follows it is assumed that $\theta = (A,B,C;\mathbb{C}^n,\mathbb{C}^m)$ is a realization of the symbol of (5.1) such that A has no real eigenvalues and Π is as in Theorem 6.1.

LEMMA 6.2. *Let* k *be given by* (5.3) *and let* $\varphi \in L^m_{1,loc}[0,\infty)$. *The improper integral* $\int_0^\infty k(t-s)\varphi(s)ds$ *exists for each* $t \geq 0$ *if* $\lim_{t\to\infty} \int_0^t P_+e^{isA}B\varphi(s)ds$ *exists. One may replace* "if" *by* "if and only if" *whenever* θ *is minimal.*

PROOF. Take $T \geq t \geq 0$. Then

$$\int_0^T k(t-s)\varphi(s)ds = \int_0^t iCe^{-i(t-s)}B\varphi(s)ds - \int_0^T iCe^{-i(t-s)}P_+B\varphi(s)ds.$$

Hence $\lim_{T\to\infty} \int_0^T k(t-s)\varphi(s)ds$ exists if and only if $\lim_{T\to\infty} Ce^{-itA}\int_0^T P_+e^{isA}B\varphi(s)ds$ exists. This is the case whenever $\lim_{T\to\infty} \int_0^T P_+e^{isA}B\varphi(s)ds$ exists.

Suppose θ is minimal. Define $C^{n-1}[0,\infty)$ to be the (n-1)-times continuously differentiable $\mathbb{C}^m$-valued functions, whose derivatives, up to the order (n-1), are bounded. Define a norm $|||\cdot|||$ on $C^{n-1}[0,\infty)$ by $|||f||| = \Sigma_{j=0}^{n-1} \|f^{(j)}\|_\infty$. With this norm, $C^{n-1}[0,\infty)$ is a normed linear space. Define $\Lambda : \operatorname{Im} P_+ \to C^{n-1}[0,\infty)$ by $\Lambda x = Ce^{-itA}x$, $t \geq 0$. Clearly, Λ is continuous.

Since θ is minimal, we have

$$\Omega = \begin{pmatrix} A \\ CA \\ \vdots \\ \vdots \\ CA^{n-1} \end{pmatrix} : \mathbb{C}^n \to \mathbb{C}^{mn}$$

is left invertible. Denote the left inverse by Ω^+. Define $\Lambda^+ : C^{n-1}[0,\infty) \to \operatorname{Im} P_+$ by

$$\Lambda^+ f = P_+\Omega^+ \begin{pmatrix} f(0) \\ if'(0) \\ \vdots \\ \vdots \\ i^{n-1}f^{(n-1)}(0) \end{pmatrix} .$$

We have Λ^+ is continuous, and $\Lambda^+\Lambda x = x$ for each $x \in \operatorname{Im} P_+$. Hence, from $\lim_{T\to\infty} Ce^{-itA} \int_0^T P_+e^{isA}B\varphi(s)ds$ exists we conclude $\lim_{T\to\infty} \int_0^T P_+e^{isA}B\varphi(s)ds$ exists. □

For g we have the following analogue of Lemma 6.2 (the proof is the same and is omitted).

LEMMA 6.2'. *Let* g *be given by* (5.4) *and let* $f \in L^m_{1,loc}[0,\infty)$. *The improper integral* $\int_0^\infty g(t,s)f(d)ds$ *exists for each* $t \geq 0$ *if* $\lim_{t\to\infty} \int_0^t (I-\Pi)e^{isA^\times}Bf(s)ds$ *exists. One may replace* "if" *by* "if and only if" *whenever* θ *is minimal.*

LEMMA 6.3. *Let* k *and* g *be given by* (5.3) *and* (5.4), *respectively, and let* $\varphi \in L^m_{1,loc}[0,\infty)$. *Assume that* $\lim_{t\to\infty} \int_0^t P_+e^{isA}B\varphi(s)ds$ *exists. Then for* $f = (I-K)\varphi$ *the improper integral* $\int_0^\infty g(t,s)f(s)ds$ *exists for each* $t \geq 0$ *if*

$$\lim_{t\to\infty} (I-\Pi)e^{itA^\times}e^{-itA}\left[\int_0^t (I-P_+)e^{isA}B\varphi(s)ds - \int_t^\infty P_+e^{isA}B\varphi(s)ds\right]$$

exists. One may replace "if" *by* "if and only if" *whenever* θ *is minimal.*

PROOF. We compute for $t \geq 0$ the integral $\int_0^t e^{isA^\times}Bf(s)ds = I_1 + I_2 + I_3$, where

$$I_1 = \int_0^t e^{isA}B\varphi(s)ds,$$

$$I_2 = \int_0^t e^{isA^\times}B\cdot -iCe^{-isA} \int_0^s e^{irA}B\varphi(r)drds,$$

$$I_3 = \int_0^t e^{isA^\times}B\cdot iCe^{-isA}x ds.$$

Here $x = \int_0^\infty P_+ e^{isA} B\varphi(s)ds$. We rewrite

$$I_2 = \int_0^t \frac{d}{ds}\left[e^{isA^\times} e^{-isA}\right] \int_0^s e^{irA} B\varphi(r)drds =$$
$$= e^{itA^\times} e^{-itA} \int_0^t e^{isA} B\varphi(s)ds - \int_0^t e^{isA^\times} B\varphi(s)ds.$$

Similarly, we get

$$I_3 = -e^{itA^\times} e^{-itA} x + x.$$

Hence

$$\int_0^t e^{isA^\times} Bf(s)ds = e^{itA^\times} e^{-itA}\left[\int_0^t e^{isA} B\ (s)ds - x\right] + x. \tag{6.1}$$

We conclude that $\lim_{t\to\infty} \int_0^t (I-\Pi)e^{isA^\times} Bf(s)ds$ exists if and only if $\lim_{t\to\infty} (I-\Pi)e^{itA^\times} e^{-itA}\left[\int_0^t e^{isA} B\ (s)ds - x\right]$ exists. We rewrite this to

$$\lim_{t\to\infty} (I-\Pi)e^{itA^\times} e^{-itA}\left[\int_0^t (I-P_+)e^{isA} B\varphi(s)ds - \int_t^\infty P_+ e^{isA} B\varphi(s)ds\right]$$

exists. Using Lemma 6.2', we prove the lemma. □

The analogue of Lemma 6.3 is

LEMMA 6.3'. *Let* k *and* g *be given by* (5.3) *and* (5.4), *respectively, and let* $f \in L^m_{1,loc}[0,\infty)$. *Assume that* $\lim_{t\to\infty} \int_0^t (I-\Pi)e^{isA^\times} Bf(s)ds$ *exists. Then for* $\varphi = (I+G)f$ *the improper integral* $\int_0^\infty k(t-s)\varphi(s)ds$ *exists for each* $t \geq 0$ *if*

$$\lim_{t\to\infty} P_+ e^{itA} e^{-itA^\times}\left[\int_0^t \Pi e^{isA^\times} Bf(s)ds - \int_t^\infty (I-\Pi)e^{isA^\times} Bf(s)ds\right]$$

exists. One may replace "if" *by* "if and only if" *whenever* θ *is minimal.*

LEMMA 6.4. *Let* k *and* g *be given by* (5.3) *and* (5.4), *respectively, and let* $\varphi \in L^m_{1,loc}[0,\infty)$. *Assume that* $x = \lim_{t\to\infty} \int_0^t P_+ e^{isA} B\varphi(s)ds$ *exists. Assume that for* $f = (I-K)\varphi$ *the limit* $y = \lim_{t\to\infty} \int_0^t (I-\Pi)e^{isA^\times} Bf(s)ds$ *exists. Then* $(I+G)(I-K)\varphi = \varphi$ *if* $y = x$, *or equivalently,*

$$\lim_{t\to\infty} (I-\Pi)e^{itA^\times} e^{-itA}\left[\int_0^t (I-P_+)e^{isA} B\varphi(s)ds - \int_t^\infty P_+ e^{isA} B\varphi(s)ds\right] = 0.$$

One may replace "if" *by* "if and only if" *whenever* θ *is minimal.*

PROOF. Using (6.1), we write

$$(Gf)(t) = iCe^{-itA^\times}\left[\int_0^t e^{isA^\times} Bf(s)ds - y\right] =$$
$$= iCe^{-itA}\left[\int_0^t e^{isA} B\varphi(s)ds - x\right] + iCe^{-itA^\times}(x-y) =$$

$$= (K\varphi)(t) + iCe^{-itA^{x}}(x-y), \qquad t \geq 0.$$

Since $(I+G)(I-K)\varphi = \varphi + GF - K\varphi$, and

$$(6.2) \qquad (Gf)(t) - (K\varphi)(t) = iCe^{-itA^{x}}(x-y), \qquad t \geq 0,$$

we have $(I+G)(I-K)\varphi = \varphi$ if and only if $iCe^{-itA^{x}}(x-y) = 0$. The latter equality holds when $x = y$. When θ is minimal, $iCe^{-itA^{x}}(x-y) = 0$ for each $t \geq 0$ if and only if $x = y$. We use (6.1) to rewrite

$$y = \lim_{t\to\infty} \int_0^t (I-\Pi)e^{isA^{x}}Bf(s)ds =$$

$$= \lim_{t\to\infty} (I-\Pi)e^{itA^{x}}e^{-itA}\left[\int_0^t (I-P_+)e^{isA}B\varphi(s)ds - \int_t^\infty P_+e^{isA}B\varphi(s)ds\right] + x.$$

From this we get the one but last statement in the theorem.

LEMMA 6.4'. *Let* f *and* g *be given by* (5.3) *and* (5.4), *respectively, and let* $f \in L^m_{1,loc}[0,\infty)$. *Assume that* $y = \lim_{t\to\infty} \int_0^t (I-\Pi)e^{isA^{x}}Bf(s)ds$ *exists. Assume that for* $\varphi = (I+G)f$ *the limit* $x = \lim_{t\to\infty} \int_0^t P_+e^{isA}B\varphi(s)ds$ *exists. Then* $(I-K)(I+G)f = f$ *if* $x = y$, *or equivalently,*

$$\lim_{t\to\infty} P_+e^{itA}e^{-itA^{x}}\left[\int_0^t \Pi e^{isA^{x}}Bf(s)ds - \int_t^\infty (I-\Pi)e^{isA^{x}}Bf(s)ds\right] = 0.$$

One may replace "if" *by* "if and only if" *whenever* θ *is minimal.*

PROOF of Theorem 6.1. Suppose $\varphi \in L_\theta$. By Lemma 6.2 we have that $K\varphi$ is defined. From Lemma 6.3 and 6.4 we get $(I-K)\varphi \in L^{x}_\theta$. The identity $(I+G)(I-K)\varphi = \varphi$ follows from Lemma 6.4. Similarly we prove that for each $f \in L^{x}_\theta$ we have Gf is defined, $(I+G)f \in L_\theta$ and $(I-K)(I+G)f = f$.

Next assume θ to be minimal and let (L, L^{x}) be an arbitrary pair of function spaces of unique solvability. Take $\varphi \in L$. By Lemma 6.2 we have that $\lim_{t\to\infty} \int_0^\infty P_+e^{isA}B\varphi(s)ds$ exists. By Lemmas 6.3 and 6.4 we get

$$\lim_{t\to\infty} (I-\Pi)e^{itA^{x}}e^{-itA}\left[\int_0^t (I-P_+)e^{isA}B\varphi(s)ds - \int_t^\infty P_+e^{isA}B\varphi(s)ds\right] = 0.$$

Hence $\varphi \in L_\theta$. Similarly we prove $L^{x} \subset L^{x}_\theta$. □

Let θ and Π be as in Theorem 6.1. Without going into details we mention a few concrete function spaces contained in L_θ and L^{x}_θ. Let α_0 be the length of the longest Jordan chain corresponding to a real eigenvalue of $(I-\Pi)A^{x}(I-\Pi)$ (note that $(I-\Pi)e^{itA^{x}}$ is bounded by a polynomial of degree $\alpha_0 - 1$ for $t \geq 0$). Then

$$\frac{1}{(t+1)^\alpha} L_p^m[0,\infty) \subset L_\theta \text{ for } \alpha > \alpha_0 - 1, \quad 1 \le p \le \infty,$$

$$\frac{1}{(t+1)^\beta} L_p^m[0,\infty) \subset L_\theta^x \text{ for } \beta > \alpha_0 - 1 + \frac{1}{p}, \quad 1 \le p \le \infty.$$

In particular,

$$e^{-ht} L_p^m[0,\infty) \subset L_\theta \quad \text{for } h > 0,\ 1 \le p \le \infty,$$

$$e^{-ht} L_p^m[0,\infty) \subset L_\theta^x \quad \text{for } h > 0,\ 1 \le p \le \infty.$$

EXAMPLE 6.6. (see Examples 4.2 and 5.2) Consider the kernel $k(t) = \frac{1}{2}e^{-|t|}$, $t \in \mathbb{R}$. Its symbol is $W(\lambda) = \lambda^2/(\lambda^2+1)$. From the Examples 4.2 and 5.2 we know that there is a unique pseudo-resolvent kernel. In this case the maximal pair (L, L^x) of function spaces of unique solvability is given by

(a) $\varphi \in L$ if and only if $\varphi \in L_{1,loc}[0,\infty)$ and

(i) $\lim_{t\to\infty} \int_0^t e^{-s}\varphi(s)ds$ exists,

(ii) $\lim_{t\to\infty} \left[\int_0^t e^{-t+s}\varphi(s)ds - \int_t^\infty e^{t-s}\varphi(s)ds\right] = 0,$

(b) $f \in L^x$ if and only if $f \in L_{1,loc}[0,\infty)$ and

(i) $\lim_{t\to\infty} \int_0^t f(s)ds$ exists.

To see this, we use the minimal realization $\theta = (A,B,C;\mathbb{C}^2,\mathbb{C})$ given by (4.3). In that case A^x is given by (4.4). We have a unique matching pair (L, L^x) of pseudo-$\mathbb{R}$-spectral subspaces. The projection Π along L onto L^x is given by (5.7). We compute

$$(I-\Pi)e^{isA^x} = \begin{pmatrix} 1 & 1 \\ 0 & 0 \end{pmatrix} \begin{pmatrix} 1-\frac{1}{2}s & -\frac{1}{2}s \\ \frac{1}{2}s & 1+\frac{1}{2}s \end{pmatrix} = \begin{pmatrix} 1 & 1 \\ 0 & 0 \end{pmatrix},$$

$$P_+e^{isA} = \begin{pmatrix} 1 & 0 \\ 0 & 0 \end{pmatrix} \begin{pmatrix} e^{-s} & 0 \\ 0 & e^{s} \end{pmatrix} = \begin{pmatrix} e^{-s} & 0 \\ 0 & 0 \end{pmatrix}.$$

For each $\varphi \in L_{1,loc}[0,\infty)$ we have

$$\int_0^t P_+e^{isA}B\varphi(s)ds = \begin{pmatrix} \int_0^t e^{-s}\varphi(s)ds \\ 0 \end{pmatrix}.$$

If $\lim_{t\to\infty} \int_0^t P_+e^{isA}B\varphi(s)ds$ exists, we have

$$(I-\Pi)e^{itA^x}e^{-itA}\left[\int_0^t (I-P_+)e^{isA}B\varphi(s) - \int_t^\infty P_+e^{isA}B\varphi(s)ds\right] =$$

$$= \begin{pmatrix} \int_0^t e^{-t+s}\varphi(s)ds - \int_t^\infty e^{t-s}\varphi(s)ds \\ 0 \end{pmatrix}.$$

Hence $\varphi \in L_\theta$ if and only if $\varphi \in L_{1,loc}[0,\infty)$ and

(i) $\lim_{t\to\infty} \int_0^t e^{-s}\varphi(s)ds$ exists,

(ii) $\lim_{t\to\infty} \left[\int_0^t e^{-t+s}\varphi(s)ds - \int_t^\infty e^{t-s}\varphi(s)ds\right] = 0.$

Note that $L_p[0,\infty) \subset L_\theta$ whenever $1 \le p \le \infty$.

To describe L_θ^x, we compute for $f \in L_{1,loc}[0,\infty)$

$$\int_0^t (I-\Pi)e^{isA^x}Bf(s)ds = \begin{pmatrix} 2\int_0^t f(s)ds \\ 0 \end{pmatrix} .$$

Suppose that $\lim_{t\to\infty} \int_0^t (I-\Pi)e^{isA^x}Bf(s)ds$ exists. We have

$$e^{-itA^x}\left[\int_0^t \Pi e^{isA^x}Bf(s)ds - \int_t^\infty (I-\Pi)e^{isA^x}Bf(s)ds\right] =$$

$$= \begin{pmatrix} -\int_0^t (1+s)f(s)ds - (2+t)\int_t^\infty f(s)ds \\ \int_0^t (1+s)f(s)ds + t\int_t^\infty f(s)ds \end{pmatrix} .$$

Hence $f \in L_\theta^x$ if and only if $f \in L_{1,loc}[0,\infty)$ and

(i) $\lim_{t\to\infty} \int_0^t f(s)ds$ exists,

(ii) $\lim_{t\to\infty} e^{-t}\left[\int_0^t (1+s)f(s)ds + (2+t)\int_t^\infty f(s)ds\right] = 0.$

It turns out that the second condition is superfluous. For if $\lim_{t\to\infty} \int_0^t f(s)ds$ exists, then $\lim_{t\to\infty} e^{-t}(2+t)\int_t^\infty f(s)ds = 0$ and $\lim_{t\to\infty} e^{-t}\int_0^t f(s)ds = 0$. Furthermore we have

$$\int_0^t sf(s)ds = t\int_0^t f(s)ds - \int_0^t\int_0^s f(r)drds.$$

The function $F(t) = \int_0^t f(s)ds$, $t \ge 0$, is absolutely continuous and bounded. Hence $\lim_{t\to\infty} \int_0^t sf(s)ds = 0$, and $f \in L_\theta^x$ if and only if $f \in L_{1,loc}[0,\infty)$ and

(i) $\lim_{t\to\infty} \int_0^t f(s)ds$ exists.

Note that $L_1[0,\infty) \subset L_\theta^x$, but $L_p[0,\infty) \not\subset L_\theta^x$ whenever $1 < p \le \infty$. We do have $e^{-ht}L_p[0,\infty) \subset L_\theta^x$ whenever $h > 0$, $1 \le p \le \infty$. We conclude that the only solution in $L_p[0,\infty)$ of $(I-K)\varphi = 0$ is $\varphi = 0$. The equation $(I-K)\varphi = f$ is not always solvable in L (and hence in $L_p[0,\infty)$) for $f \in L_p[0,\infty)$, $1 < p \le \infty$.

ACKNOWLEDGEMENT

Sincere thanks are due to M.A. Kaashoek and I. Gohberg for useful suggestions, many discussions on the subject of this paper, and advice concerning preparation of the final version of this paper.

REFERENCES

1. Bart, H, Gohberg, I., Kaashoek, M.A.: Minimal factorization of matrix and operator functions. Operator Theory: Advances and Applications. Vol. 1, Birkhäuser Verlag,Basel, 1979.

2. Bart, H., Gohberg, I. Kaashoek, M.A.: 'Wiener-Hopf integral equations, Toeplitz matrices and linear systems.' In: Toeplitz Centennial (Ed. I. Gohberg), Operator Theory: Advances and Applications, Vol. 4, Birkhäuser Verlag, Basel, 1982; 85-135.

3. Bart, H., Gohberg, I., Kaashoek, M.A.: 'Wiener-Hopf factorization of analytic operator functions and realization', Wiskundig Seminarium der Vrije Universiteit, Rapport nr. 231, Amsterdam, 1983.

4. Gohberg, I., Lancaster, P., Rodman, L.: Matrix polynomials, Academic Press, New York N.Y., 1982.

5. Gohberg, I.C., Krein, M.G.: 'Systems of integral equations on a half line with kernels depending on the difference of arguments', Uspehi Mat. Nauk 13 (1958) no. 2 (80), 3-72 (Russian) = Amer. Math. Soc. Transl. (2) 14 (1960), 217-287.

6. Gohberg, I., Krupnik, N.: Einführung in die Theorie der eindimensionalen singulären Integraloperatoren, Birkhäuser Verlag, Basel, 1979.

7. Michlin, S.G., Prössdorf, X.: Singuläre Integraloperatoren. Akademie-Verlag, Berlin, 1980.

8. Prössdorf, S.: Einige Klassen singulärer Gleichungen. Akademie-Verlag, Berlin, 1974.

9. Ran, A.C.M.: Minimal factorization of selfadjoint rational matrix functions. Integral Equations and Operator Theory 5 (1982), 850-869.

10. Ran, A.C.M.: Semidefinite invariant subspaces; stability and applications. Ph.D. Thesis Vrije Universiteit, Amsterdam, 1984.

11. Rodman, L.: Maximal invariant neutral subspaces and an application to the algebraic Riccati equation. Manuscripta Math. 43 (1983) 1-12.

L. Roozemond,
Subfaculteit Wiskunde en Informatica,
Vrije Universiteit,
De Boelelaan 1081,
1081 HV Amsterdam
The Netherlands.

Operator Theory:
Advances and Applications, Vol. 21

MINIMAL FACTORIZATION OF INTEGRAL OPERATORS AND CASCADE DECOMPOSITIONS OF SYSTEMS

I. Gohberg and M.A. Kaashoek

A minimal factorization theory is developed for integral operators of the second kind with a semi-separable kernel. Explicit formulas for the factors are given. The results are natural generalizations of the minimal factorization theorems for rational matrix functions. LU- and UL-factorization appear as special cases. In the proofs connections with cascade decompositions of systems with well-posed boundary conditions play an essential role.

0. INTRODUCTION

This paper deals with integral operators T which act on the space $L_2([a,b],\mathbb{C}^m)$ of all square integrable $\mathbb{C}^m$-valued functions on [a,b] and which are of the form

$$(0.1)\qquad (T\varphi)(t) = \varphi(t) + \int_a^b k(t,s)\varphi(s)ds, \quad a \le t \le b,$$

with k a semi-separable $m \times m$ matrix kernel. The latter means that k admits a representation of the following type:

$$(0.2)\qquad k(t,s) = \begin{cases} F_1(t)G_1(s), & a \le s < t \le b, \\ -F_2(t)G_2(s), & a \le t < s \le b. \end{cases}$$

Here for $\nu = 1,2$ the functions $F_\nu(\cdot)$ and $G_\nu(\cdot)$ are matrix functions of sizes $m \times n_\nu$ and $n_\nu \times m$, respectively, which are square integrable on $a \le t \le b$. The numbers n_1 and n_2 may vary with T and the representation (0.2) of its kernel.

The main topic of the paper concerns factorizations $T = T_1T_2$, where T_1 and T_2 are integral operators of the same type as T, and the factorization $T = T_1T_2$ has certain additional minimality properties. Roughly speaking, minimal factorization allows to factor an integral operator into a product of simpler ones and it excludes trivial factorizations as $T = T_1(T_1^{-1}T)$ for example. The concepts of minimal factorization for integral operators introduced and studied in this paper is a natural generalization of the concept of minimal factorization for rational matrix functions.

To explain this in more detail consider a factorization

$$(0.3) \qquad W_0(\lambda) = W_1(\lambda)W_2(\lambda),$$

where $W_0(\lambda)$, $W_1(\lambda)$ and $W_2(\lambda)$ are rational $m \times m$ matrix functions. For the factorization in (0.3) minimality means that there is no "pole-zero cancellation" between the factors (see, e.g., [2], §4.3). Let us assume that the functions $W_0(\cdot)$, $W_1(\cdot)$ and $W_2(\cdot)$ are analytic at ∞ with the value at ∞ equal to the $m \times m$ identity matrix I_m. Then ([2], §2.1) we may represent these functions in the form:

$$W_\nu(\lambda) = I_m + C_\nu(\lambda - A_\nu)B_\nu, \qquad \nu = 0,1,2,$$

where A_ν is a square matrix of order ℓ_ν, say, and B_ν and C_ν are matrices of sizes $\ell_\nu \times m$ and $m \times \ell_\nu$, respectively. The equality in (0.3) is now equivalent to the factorization $T_0 = T_1T_2$, where for $\nu = 0,1,2$ the operator T_ν is the integral operator on $L_2([a,b],\mathbb{C}^m)$ defined by

$$(T_\nu\varphi)(t) = \varphi(t) + \int_a^t C_\nu e^{(t-s)A_\nu}B_\nu\varphi(s)ds, \qquad a \le t \le b.$$

In this way theorems about minimal factorization of rational matrix functions (see [2,3]) can be seen as theorems about minimal factorization of Volterra integral operators of the second kind with semi-separable kernels of a special type.

In this paper we extend this minimal factorization theory to the class of all integral operators of the second kind with semi-separable kernels. The extension is done in two different ways, corresponding to two different notions of minimal realization (see [8,9,10,16]), and leading to two different types of minimal factorization, one of which will be called SB-minimal factorization (for reasons which will be clear later) and the other just minimal factorization. For both concepts we describe all minimal factorizations and we give explicit formulas for the factors. We also analyze the advantages and disadvantages of the two types of minimal factorization.

Special attention is paid to LU- and UL-factorization. Such factorizations appear here as examples of SB-minimal factorization. Our theorems for LU- and UL-factorization are related to those of [11], §IV.8 and [18], and they generalize the results in [13] (see also [1]) which concern the positive definite case. We note that the class of kernels considered in the present paper differs from the class of kernels treated in [11, 18]; for example, in our case we allow the kernels to have discontinuities on the diagonal.

The paper is based on the fact ([5], §I.4) that an integral operator with a semi-separable kernel can be viewed as the input/output operator of a time varying linear system with well-posed boundary conditions. For example, the integral operator (0.1) with kernel (0.2) is the input/output operator of the system

$$\begin{cases} \dfrac{d}{dt}\begin{bmatrix} x_1(t) \\ x_2(t) \end{bmatrix} = \begin{bmatrix} G_1(t) \\ G_2(t) \end{bmatrix} u(t), & a \le t \le b, \\ y(t) = F_1(t)x_1(t) + F_2(t)x_2(t) + u(t), & a \le t \le b, \\ x_1(a) = 0, \quad x_2(b) = 0. \end{cases}$$

Using this connection we reduce the problem of minimal factorization of integral operators to a problem of cascade decomposition of time varying systems. The system theory language makes more transparent the problem and also shows better the analogy with the rational matrix case.

The paper consists of three chapters. In the first chapter we state (without proofs) the main minimal factorization theorems. This is done for different classes of integral operators. In Chapter I we also give a number of illustrative examples. The proofs of the theorems are given in the third chapter; they are based on a decomposition theory for time varying systems with well-posed boundary conditions which we develop in the second chapter.

A few words about notation. All linear spaces appearing below are vector spaces over $\mathbb{C}$. The identity operator on a linear space is always denoted by I. The symbol I_m is used for the $m \times m$ identity matrix. An $m \times n$ matrix A will be identified with the linear operator from $\mathbb{C}^n$ into $\mathbb{C}^m$ given by the canonical action of W with respect to the standard bases in $\mathbb{C}^n$ and $\mathbb{C}^m$. The symbol χ_E stands for the characteristic function of a set E; thus $\chi_E(t) = 1$ for $t \in E$ and $\chi_E(t) = 0$ for $t \notin E$.

I. MAIN RESULTS

In this chapter we state (without proofs) the main results of this paper. In the first four sections the main theorems and examples for minimal factorization are given. The SB-minimal factorization theorems appear in the next three sections. The last section is devoted to LU- and UL-factorization.

I.1 Minimal representation and degree

Let Y be a finite dimensional inner product space. By $L_2([a,b],Y)$ we denote the Hilbert space of all square integrable Y-valued functions on $a \le t \le b$. An operator T on $L_2([a,b],Y)$ will be called a *(SK)-integral operator* if T is an integral operator of the second kind with a semi-separable kernel, i.e., T has the form

$$(1.1)\qquad (T\varphi)(t) = \varphi(t) + \int_a^b k(t,s)\varphi(s)ds, \qquad a \le t \le b,$$

and its kernel k admits the following representation:

$$(1.2)\qquad k(t,s) = \begin{cases} F_1(t)G_1(s), & a \le s < t \le b, \\ -F_2(t)G_2(s), & a \le t < s \le b. \end{cases}$$

Here $F_\nu(t) : X_\nu \to Y$ and $G_\nu(t) : Y \to X_\nu$ are linear operators, the space X_ν is a finite dimensional inner product space, and as functions F_ν and G_ν are square integrable on $a \le t \le b$ ($\nu = 1,2$). The spaces X_1 and X_2 may vary with T and the representation (1.2) of its kernel. In what follows we keep the space Y fixed.

The kernel k of a (SK)-integral operator T may also be represented in the form

$$(1.3)\qquad k(t,s) = \begin{cases} C(t)(I-P)B(s), & a \le s < t \le b, \\ -C(t)PB(s), & a \le t < s \le b, \end{cases}$$

where $B(t) : Y \to X$, $C(t) : X \to Y$ and $P : X \to X$ are linear operators, the space X is a finite dimensional inner product space and the functions $B(\cdot)$ and $C(\cdot)$ are square integrable on $a \le t \le b$. Indeed, starting from (1.2) we may write k as in (1.3) by taking $X = X_1 \oplus X_2$ and

$$(1.4)\qquad B(t) = \begin{pmatrix} G_1(t) \\ G_2(t) \end{pmatrix}, \quad C(t) = \begin{pmatrix} F_1(t) & F_2(t) \end{pmatrix}, \quad P = \begin{pmatrix} 0 & 0 \\ 0 & I \end{pmatrix}.$$

If (1.3) holds for the kernel k of T, then we call the triple $\Delta = (B(\cdot),C(\cdot);P)$ a *representation* of T. The space X will be called the *internal space* of the representation Δ, and we shall refer to P as the *internal operator*. Note that in (1.4) the operator P is a projection, but this is not a condition and we allow the internal operator to be an arbitrary operator.

Let T be an (SK)-integral operator. A representation Δ of T will be called a *minimal representation* of T if among all representations of T the dimension of the internal space of Δ is as small as possible. We define the *degree* of T (notation: $\delta(T)$) to be the dimension of the internal space of a minimal representation of T. The degree satisfies a sublogarithmic property.

1.1 LEMMA. *Let* T_1 *and* T_2 *be (SK)-integral operators on* $L_2([a,b],Y)$. *Then the product* T_1T_2 *is a (SK)-integral operator and*

$$(1.5)\qquad \delta(T_1T_2) \le \delta(T_1) + \delta(T_2).$$

I.2 Minimal factorization (1)

Let T be a (SK)-integral operator on $L_2([a,b],Y)$. We call $T = T_1T_2$ a *minimal factorization* (of T) if T_1 and T_2 are (SK)-integral operators on $L_2([a,b],Y)$ and the degree of T is the sum of the degrees of T_1 and T_2, i.e.,

$$(2.1)\qquad \delta(T) = \delta(T_1) + \delta(T_2).$$

To construct minimal factorizations of T we need the notion of a decomposing projection.

Let $\Delta = (B(\cdot),C(\cdot);P)$ be a representation of T. Let X be the internal space of Δ. By definition the *fundamental operator* of Δ is the unique absolutely continuous solution $\Omega(t): X \to X$, $a \le t \le b$, of the initial value problem:

$$(2.2)\qquad \dot{\Omega}(t) = -B(t)C(t)\Omega(t),\quad a \le t \le b,\quad \Omega(a) = I.$$

A projection Π of the internal space X of Δ will be called a *decomposing projection* for Δ if the following conditions are satisfied:

(i) $\Pi P\Pi = \Pi P$,

(ii) $\det(I - \Pi + \Omega(t)\Pi) \neq 0,\quad a \le t \le b,$

(iii) $(I-\Pi)P\Pi = P(I-\Pi)(I-\Pi+\Omega(b)\Pi)^{-1}\Pi(I-P).$

The function $\Omega(\cdot)$ in (ii) and (iii) is the fundamental operator of Δ. Condition (i) means that $\operatorname{Ker}\Pi$ is invariant under P. Let us write P and $\Omega(t)$ as 2×2 block matrices relative to the decomposition $X = \operatorname{Ker}\Pi \oplus \operatorname{Im}\Pi$:

$$P = \begin{pmatrix} P_{11} & P_{12} \\ P_{21} & P_{22} \end{pmatrix},\qquad \Omega(t) = \begin{pmatrix} \Omega_{11}(t) & \Omega_{12}(t) \\ \Omega_{21}(t) & \Omega_{22}(t) \end{pmatrix}.$$

Then the conditions (i) - (iii) are equivalent to

(i)' $P_{21} = 0$,

(ii)' $\det \Omega_{22}(t) \neq 0,\quad a \le t \le b,$

(iii)' $P_{12} = -P_{11}\Omega_{12}(b)\Omega_{22}(b)^{-1}(I - P_{22})$.

2.1 THEOREM. *Let* T *be a (SK)-integral operator on* $L_2([a,b],Y)$, *and let* $\Delta = (B(\cdot),C(\cdot);P)$ *be a minimal representation of* T. *If* Π *is a decomposing projection for* Δ, *then a minimal factorization* $T = T_1T_2$ *is obtained by taking* T_1 *and* T_2 *to be the (SK)-integral operators on* $L_2([a,b],Y)$ *of which the kernels are given by*

$$(2.3)\qquad k_1(t,s) = \begin{cases} C(t)(I-P)(I-\Pi(s))B(s), & a \le s < t \le b, \\ -C(t)P(I-\Pi(s))B(s), & a \le t < s \le b, \end{cases}$$

$$(2.4)\qquad k_2(t,s) = \begin{cases} C(t)\Pi(t)(I-P)B(s), & a \le s < t \le b, \\ -C(t)\Pi(t)PB(s), & a \le t < s \le b, \end{cases}$$

respectively. Here

$$(2.5)\qquad \Pi(t) = \Omega(t)\Pi(I-\Pi+\Omega(t)\Pi)^{-1}\Pi, \qquad a \le t \le b,$$

with $\Omega(t)$ *being the fundamental operator of* Δ.

Furthermore, if Δ *runs over all possible minimal representations of* T, *then all minimal factorizations of* T *may be obtained in this way, that is, given a minimal factorization* $T = T_1T_2$, *there exists a minimal representation* $\Delta = (B(\cdot),C(\cdot);P)$ *of* T *and a decomposing projection* Π *for* Δ *such that the kernels of* T_1 *and* T_2 *are given by* (2.3) *and* (2.4), *respectively.*

Let $\Delta = (B(\cdot),C(\cdot);P)$ be a minimal representation of the (SK)-integral operator T. We say that Δ *generates the minimal factorization* $T = T_1T_2$ if there exists a decomposing projection Π for Δ such that the kernels of T_1 and T_2 are given by (2.3) and (2.4), respectively. In general, there is no single minimal representation of T which generates all minimal factorizations of T. In other words, to get all minimal factorizations of T it is, in general, necessary to consider various minimal representations of T. This statement is substantiated by the next example.

On $L_2[0,1]$ we consider the (SK)-integral operators:

$$(T_0\varphi)(t) = \varphi(t) + \int_0^1 \varphi(s)ds,$$

$$(T_1\varphi)(t) = \varphi(t) + \int_0^t \frac{1}{1+s}\varphi(s)ds,$$

$$(T_2\varphi)(t) = \varphi(t) + \int_t^1 \frac{1}{1+t}\varphi(s)ds,$$

$$(T_3\varphi)(t) = \varphi(t) - \int_t^1 \frac{1}{1+s}\varphi(s)ds,$$

$$(T_4\varphi)(t) = \varphi(t) + \int_0^t \frac{2}{1+t}\varphi(s)ds + \int_t^1 \frac{3}{1+t}\varphi(s)ds.$$

It is straightforward to check (see also the analysis here below) that

$$(2.6) \qquad T_0 = T_1T_2, \qquad T_0 = T_3T_4.$$

Let us prove that the two factorizations in (2.6) are minimal.

First of all, note that

$$\Delta = \left(\begin{pmatrix} 1 \\ -1 \end{pmatrix}, (1 \ \ 1) ; \begin{pmatrix} 0 & 0 \\ 0 & 1 \end{pmatrix} \right),$$

$$\Delta^* = \left(\begin{pmatrix} 1 \\ -1 \end{pmatrix}, (1 \ \ 1) ; \begin{pmatrix} 1 & -1 \\ 0 & 3 \end{pmatrix} \right),$$

are representations of T_0. The internal space of both Δ and Δ^* has dimension two and it is not difficult to check (see [9], § 3) that T_0 has no representation with an internal space of a smaller dimension. Hence both Δ and Δ^* are minimal representations of T_0.

Next, consider on $\mathbb{C}^2$ the projection

$$(2.7) \qquad \Pi = \begin{pmatrix} 0 & 0 \\ 0 & 1 \end{pmatrix}.$$

We claim that Π is a decomposing projection for Δ and Δ^*. Obviously, condition (i)' is satisfied for Δ and Δ^*. Furthermore,

$$(2.8) \qquad \Omega(t) = \begin{pmatrix} 1-t & -t \\ t & 1+t \end{pmatrix}, \qquad 0 \le t \le 1,$$

is the fundamental operator for both Δ and Δ^*. Now, it is simple to see that conditions (ii)' and (iii)' are fulfilled for both Δ and Δ^*. Hence Π is decomposing for Δ and Δ^*.

Note that for Π given by (2.7) and $\Omega(t)$ by (2.8) the function $\Pi(t)$ in (2.5) is given by

$$(2.9) \qquad \Pi(t) = \begin{pmatrix} 0 & -t(1+t)^{-1} \\ 0 & 1 \end{pmatrix}, \qquad 0 \le t \le 1.$$

Using this one sees that Theorem 2.1 applied to Δ and Π yields the first factorization in (2.6), and the second factorization in (2.6) is obtained by applying Theorem 2.1 to Δ^* and Π. Thus both factorizations in (2.6) are minimal.

The fact that two different minimal representation of T_0 were used to construct the minimal factorizations in (2.6) is no coincidence. In § III.1 we shall prove that the operator T_0 in the present example has no single minimal representation which generates both factorizations in (2.6).

Let T be an arbitrary (SK)-integral operator. The question whether there exists a single minimal representation of T which generates all minimal factorizations of T is related to the question of uniqueness of a minimal representation. Two representations $\Delta_1 = (B_1(\cdot),C_1(\cdot);P_1)$ and $\Delta_2 = (B_2(\cdot),C_2(\cdot);P_2)$ of T are called *similar* if there exists an invertible operator $S : X_1 \to X_2$, acting between the internal spaces of Δ_1 and Δ_2, such that

$$(2.10)\qquad B_2(\cdot) = SB_1(\cdot),\quad C_2(\cdot) = C_1(\cdot)S^{-1},\quad P_2 = SP_1S^{-1}.$$

It may happen that T has non-similar minimal representations. In fact, the minimal representations Δ and Δ^* of the operator T_0 in the preceding example are not similar. We say that T has *up to similarity a unique minimal representation* if any two minimal representations of T are similar. Similar minimal representations of T generate the same set of minimal factorizations of T. Thus, if T has up to similarity a unique minimal representation, then Theorem 2.1 implies that each minimal representation of T generates all minimal factorizations of T. It is not known whether the converse of the latter statement holds true.

I.3 Minimal factorization of Volterra integral operators (1)

This section concerns Volterra integral operators on $L_2([a,b],Y)$:

$$(3.1)\qquad (T\varphi)(t) = \varphi(t) + \int_a^t k(t,s)\varphi(s)ds,\qquad a \le t \le b.$$

Throughout the section we assume that the kernel k admits a representation of the form:

$$(3.2)\qquad k(t,s) = F(t)G(s),\qquad a \le s \le t \le b,$$

where $F(t) : X \to Y$ and $G(t) : Y \to X$ are linear operators acting between finite dimensional inner product spaces and as functions $F(\cdot)$ and $G(\cdot)$ are analytic on $[a,b]$. The space X may vary with T and the representation of its kernel. Note that the kernel k of T is semi-separable, and hence T is a (SK)-integral operator. In §III.5 we shall see that the degree of T is given by

$$(3.3)\qquad \delta(T) = \operatorname{rank}(F \otimes G).$$

Here F and G are the analytic functions in the representation (3.2) of the kernel k of T and $F \otimes G$ stands for the finite rank integral operator on $L_2([a,b],Y)$ with kernel $F(t)G(s)$, that is,

$$(3.4)\qquad (F \otimes G)\varphi(t) = F(t)\int_a^b G(s)\varphi(s)ds, \qquad a \le t \le b.$$

Operators T of the type described in the previous paragraph will be called *(SK)-integral operators in the class (AC)* (the letters AC stand for analytic causal). Let T be an operator in this class. Later (in §III.5) we shall prove that a minimal representation Δ of T always has the form $(B(t),C(t);0)$ with $B(\cdot)$ and $C(\cdot)$ analytic on $[a,b]$. In particular, the internal operator P of Δ is the zero operator. Note that $P = 0$ implies that the conditions (i) and (iii) in the definition of a decomposing projection are automatically fulfilled. Hence a projection Π is a decomposing projection for a minimal representation Δ of T if and only if

$$(3.5)\qquad \det(I - \Pi + \Omega(t)\Pi) \ne 0, \qquad a \le t \le b,$$

where $\Omega(t)$ is the fundamental operator of Δ.

3.1 THEOREM. *Let* T *be a (SK)-integral operator on* $L_2([a,b],Y)$ *in the class (AC), and let* $\Delta = (B(\cdot),C(\cdot);0)$ *be a minimal representation of* T.

(i) *If* Π *is a decomposing projection for* Δ, *then a minimal factorization* $T = T_1T_2$ *is obtained by taking*

$$(3.6)\qquad (T_1\varphi)(t) = \varphi(t) + \int_a^t C(t)(I - \Pi(s))B(s)\varphi(s)ds, \qquad a \le t \le b,$$

$$(3.7)\qquad (T_2\varphi)(t) = \varphi(t) + \int_a^t C(t)\Pi(t)B(s)\varphi(s)ds, \qquad a \le t \le b.$$

Here

$$(3.8)\qquad \Pi(t) = \Omega(t)\Pi(I - \Pi + \Omega(t)\Pi)^{-1}\Pi, \qquad a \le t \le b,$$

with $\Omega(t)$ *being the fundamental operator of* Δ.

(ii) *If* $T = T_1T_2$ *is a minimal factorization, then the factors are in the class (AC) and there exists a unique decomposing projection* Π *for* Δ *such that* T_1 *and* T_2 *are given by* (3.6) *and* (3.7), *respectively.*

Note that according to part (ii) of the previous theorem all minimal factorizations of a (SK)-integral operator T in the class (AC) may be generated from a single minimal representation of T. The latter statement does not remain true if one drops the analyticity condition on the functions $F(\cdot)$ and $G(\cdot)$ in (3.2). To see this consider on $L_2[0,1]$ the integral operator

$$(3.9)\qquad (T\varphi)(t) = \varphi(t) + \int_0^t (1 + \chi_{[0,\frac{1}{2})}(t)\chi_{[0,\frac{1}{4})}(s))\varphi(s)ds, \qquad 0 \le t \le 1.$$

Here χ_E stands for the characteristic function of the set E. We shall construct two minimal factorizations

$$(3.10)\qquad T = T_1^{(\nu)}T_2^{(\nu)}, \qquad \nu = 1,2,$$

and we shall prove (in §III.5) that there does not exist a single minimal representation of T which generates the two minimal factorizations in (3.10). Introduce the following functions on $0 \le t \le 1$:

$$g_1 = \chi_{[0,\frac{1}{4})}, \qquad g_2 = \chi_{[0,\frac{1}{4})\cup[\frac{1}{2},1]},$$

$$r_1(t) = \begin{cases} \dfrac{e^{2t}-1}{e^{2t}+1}, & 0 \le t \le \frac{1}{4}, \\ (c-1)e^{-t+\frac{1}{4}}+1, & \frac{1}{4} \le t \le \frac{1}{2}, \\ de^{-t+\frac{1}{2}}, & \frac{1}{2} \le t \le 1, \end{cases}$$

$$r_2(t) = \begin{cases} r_1(t) & 0 \le t \le \frac{1}{2}, \\ \dfrac{de^{-t+\frac{1}{2}}}{1+d-de^{-t+\frac{1}{2}}}, & \frac{1}{2} \le t \le 1. \end{cases}$$

Here c and d are chosen in such a way that $r_1(\cdot)$ and $r_2(\cdot)$ are continuous on $[0,1]$. For $\nu = 1,2$ let $T_1^{(\nu)}$ and $T_2^{(\nu)}$ be the integral operators on $L_2[0,1]$ defined by

$$(T_1^{(\nu)}\varphi)(t) = \varphi(t) + \int_0^t (1+r_\nu(s)g_\nu(s))\varphi(s)ds,$$

$$(T_2^{(\nu)}\varphi)(t) = \varphi(t) + \int_0^t (\chi_{[0,\frac{1}{2})}(t) - r_\nu(t))g_\nu(s)\varphi(s)ds.$$

To prove that (3.10) holds, consider the following Riccati equation:

$$(3.11) \qquad \begin{cases} \dot{r}(t) = \chi_{[0,\frac{1}{2})}(t) + r(t)g(t)\chi_{[0,\frac{1}{2})}(t) - r(t) - r(t)^2 g(t), & 0 \le t \le 1, \\ r(0) = 0. \end{cases}$$

One checks that for $g = g_1$ the function $r_1(\cdot)$ is a solution of (3.11) and for $g = g_2$ the function $r_2(\cdot)$ is a solution of (3.11). Using this a direct computation shows that (3.10) holds.

Next, let us prove that the two factorizations in (3.10) are minimal. For $\nu = 1,2$ put

$$(3.12) \qquad \Delta_\nu = \left(\begin{bmatrix} 1 \\ g_\nu(\cdot) \end{bmatrix}, \left(1 \quad \chi_{[0,\frac{1}{2})}(\cdot)\right); \begin{bmatrix} 0 & 0 \\ 0 & 0 \end{bmatrix} \right).$$

Since $g_1(t) = g_2(t) = \chi_{[0,\frac{1}{4})}(t)$ for $0 \le t < \frac{1}{2}$, one sees that Δ_1 and Δ_2 are representations of T. It is readily checked that T does not have a representation with a one dimensional internal space. Thus $\delta(T) = 2$. From the definitions of $T_1^{(\nu)}$ and $T_2^{(\nu)}$ it is clear that $\delta(T_1^{(\nu)}) = \delta(T_2^{(\nu)}) = 1$. So

$$\delta(T) = \delta(T_1^{(\nu)}) + \delta(T_2^{(\nu)}), \qquad \nu = 1,2,$$

and hence the factorizations in (3.10) are minimal. In § III.5 we shall prove that the operator T in (3.9) has no single minimal representation with decomposing projections yielding the minimal factorization in (3.10).

It is interesting to note the differences between the above example and the example given at the end of the previous section. In the example in the previous section the kernels are analytic functions, but the internal operators are allowed to be complicated ; in the example given above the kernels are not analytic, but now the internal operators are all required to be equal to the zero operator. In both cases there is not a single minimal representation which generates all minimal factorizations.

Let T be a (SK)-integral operator in the class (AC). We shall see in § III.5 that up to similarity T has a unique minimal representation. From this fact it follows that each minimal representation Δ of T generates all minimal factorizations of T (cf., the last paragraph of § I.2). Theorem 3.1 (ii) adds to this that there is a one-one correspondence between all minimal factorizations of T and all decomposing projections of a given minimal representation of T.

I.4 Stationary causal operators and transfer functions

This section concerns Volterra integral operators on $L_2([a,b],Y)$ which can be represented in the form

$$(4.1)\qquad (T\varphi)(t) = \varphi(t) + \int_a^t Ce^{(t-s)A}B\varphi(s)ds, \qquad a \le t \le b.$$

Here $A : X \to X$, $B : Y \to X$ and $C : X \to Y$ are linear operators and X is a finite dimensional inner product space. The space X may vary with T and the representation (4.1). Note that the kernel of the integral operator T in (4.1) is semi-separable, and hence T is a (SK)-integral operator. The degree of T is given by

$$(4.2)\qquad \delta(T) = \operatorname{rank}\begin{pmatrix} CB & \dots & CA^{n-1}B \\ \vdots & & \vdots \\ CA^{n-1}B & \dots & CA^{2n-1}B \end{pmatrix},$$

where n is equal to the dimension of the space X.

An operator T of the type described in the previous paragraph will be called a *(SK)-integral operator in the class (STC)* (the letters STC stand for stationary causal). Obviously, the class (STC) is contained in the class (AC). We shall prove (in §III.6) that each minimal representation Δ of a (SK)-integral operator in the class (STC) is of the form

$$(4.3)\qquad \Delta_0 = (e^{-(t-a)A}B, Ce^{(t-a)A}; 0).$$

In (4.3) one may replace a by any number α; in particular, a may be replaced by 0. A projection Π is decomposing for the representation (4.3) if and only if

$$(4.4)\qquad \det(I - \Pi + e^{-(t-a)A}e^{(t-a)(A-BC)}\Pi) \ne 0, \qquad a \le t \le b.$$

To describe all minimal factorizations with factors in the class (STC) we need the notion of a supporting projection (cf. [2], §1.1). Let $\Delta = (e^{-(t-\alpha)A}B, Ce^{(t-\alpha)A}; 0)$ be a representation of T. A projection Π of the internal space X of Δ is called a *supporting projection* for Δ if

$$(4.5)\qquad A\,\operatorname{Ker}\Pi \subset \operatorname{Ker}\Pi, \qquad (A - BC)\,\operatorname{Im}\Pi \subset \operatorname{Im}\Pi.$$

The definition does not depend on α. If (4.5) holds, then (4.4) is satisfied, and hence a supporting projection for Δ_0 in (4.3) is a decomposing projection. The converse of the latter statement does not hold (see the example after Theorem 4.1).

4.1 THEOREM. *Let* T *be a (SK)-integral operator on* $L_2([a,b],Y)$ *in the class (STC), and let* $\Delta = (e^{-tA}B, Ce^{tA}; 0)$ *be a minimal representation of* T.

(i) *If* Π *is a supporting projection for* Δ, *then a minimal factorization* $T = T_1T_2$ *is obtained by taking*

$$(4.6)\qquad (T_1\varphi)(t) = \varphi(t) + \int_a^t Ce^{(t-s)A}(I-\Pi)B\varphi(s)ds, \quad a \le t \le b,$$

$$(4.7)\qquad (T_2\varphi)(t) = \varphi(t) + \int_a^t C\Pi e^{(t-s)A}B\varphi(s)ds, \qquad a \le t \le b.$$

The factors T_1 *and* T_2 *are (SK)-integral operators in the class (STC).*

(ii) *If* $T = T_1T_2$ *is a minimal factorization of* T *with factors* T_1 *and* T_2 *in the class (STC), then there exists a unique supporting projection* Π *for* Δ *such that* T_1 *and* T_2 *are given by* (4.6) *and* (4.7), *respectively.*

Note that the definition of a supporting projection does not involve the interval $[a,b]$. It follows that under the conditions of Theorem 4.1 (i) the factorization $T = T_1T_2$ in Theorem 4.1 (i) also holds when $[a,b]$ is replaced by any other interval.

An integral operator T in the class (STC) may have minimal factorizations $T = T_1T_2$ for which the factors are not in the class (STC). To see this, define T_0 on $L_2[0,1]$ by

$$(T_0\varphi)(t) = \varphi(t) + \int_0^t (2+t-s)e^{t-s}\varphi(s)ds, \qquad 0 \le t \le 1.$$

The triple

$$\Delta_0 = \left(\begin{pmatrix} e^{-t} \\ (1-t)e^{-t} \end{pmatrix}, \begin{pmatrix} (1+t)e^{t} & e^{t} \end{pmatrix} ; \begin{pmatrix} 0 & 0 \\ 0 & 0 \end{pmatrix} \right)$$

is a minimal representation of T_0. Note that Δ_0 is of the form (4.3) with

$$A = \begin{pmatrix} 1 & 0 \\ 1 & 1 \end{pmatrix}, \qquad B = \begin{pmatrix} 1 \\ 1 \end{pmatrix}, \qquad C = \begin{pmatrix} 1 & 1 \end{pmatrix}, \qquad a = 0.$$

Consider on $\mathbb{C}^2$ (the internal space of Δ_0) the projection

$$(4.8)\qquad \Pi = \begin{pmatrix} 0 & 0 \\ 0 & 1 \end{pmatrix}.$$

Using (4.4) with $a = 0$ one computes that Π is a decomposing projection for Δ_0. Indeed

$$\Omega(t) := e^{-tA}e^{t(A-BC)} = \begin{pmatrix} e^{-t} & -te^{-t} \\ -te^{-t} & (1+t^2)e^{-t} \end{pmatrix},$$

and hence

$$\det(I - \Pi + \Omega(t)\Pi) = \det \begin{pmatrix} 1 & -te^{-t} \\ 0 & (1+t^2)e^{-t} \end{pmatrix} \neq 0$$

for $0 \leq t \leq 1$. So Π is a decomposing projection for Δ_0, and we may apply Theorem 3.1 (i) to Δ_0 and Π. Put

$$(T_1\varphi)(t) = \varphi(t) + \int_0^t (1+t)\left(\frac{s+1}{s^2+1}\right)e^{t-s}\varphi(s)ds, \qquad 0 \leq t \leq 1,$$

$$(T_2\varphi)(t) = \varphi(t) + \int_0^t \left(\frac{1-t}{1+t^2}\right)(1-s)e^{t-s}\varphi(s)ds, \qquad 0 \leq t \leq 1.$$

It follows that $T_0 = T_1T_2$ is a minimal factorization of T_0. But the factors T_1 and T_2 are not in the class (STC). Note that the projection Π in (4.8) is a decomposing projection for Δ_0 but not a supporting projection.

Another example of the type mentioned above is provided by the following integral operator:

$$\left(T\begin{bmatrix}\varphi_1\\ \varphi_2\end{bmatrix}\right)(t) = \begin{bmatrix}\varphi_1(t)\\ \varphi_2(t)\end{bmatrix} + \int_0^t \begin{bmatrix}1 & 0\\ t-s & 1\end{bmatrix}\begin{bmatrix}\varphi_1(s)\\ \varphi_2(s)\end{bmatrix}ds, \qquad 0 \leq t \leq 1.$$

The operator T acts on $L_2([0,1],\mathbb{C}^2)$, and obviously T is a (SK)-integral operator in the class (STC). One can show that this operator has no non-trivial minimal factorization with factors in the class (STC). On the other hand, if

$$\left(T_1\begin{bmatrix}\varphi_1\\ \varphi_2\end{bmatrix}\right)(t) = \begin{bmatrix}\varphi_1(t)\\ \varphi_2(t)\end{bmatrix} + \int_0^t \begin{bmatrix}1 & 0\\ t & 0\end{bmatrix}\begin{bmatrix}\varphi_1(s)\\ \varphi_2(s)\end{bmatrix}ds,$$

$$\left(T_2\begin{bmatrix}\varphi_1\\ \varphi_2\end{bmatrix}\right)(t) = \begin{bmatrix}\varphi_1(t)\\ \varphi_2(t)\end{bmatrix} + \int_0^t \begin{bmatrix}0 & 0\\ -s & 1\end{bmatrix}\begin{bmatrix}\varphi_1(s)\\ \varphi_2(s)\end{bmatrix}ds,$$

on $0 \leq t \leq 1$, then $T = T_1T_2$ and this factorization is a non-trivial minimal factorization of T.

We conclude this section with a remark in which we compare Theorem 4.1 to the main minimal factorization theorem in [2,3]. In fact, we shall see that Theorem 4.1 is just the integral operator version of Theorem 4.8 in [2]. In what follows $Y = \mathbb{C}^m$. Let (TF) be the class of all rational $m \times m$ matrix functions W such that W is analytic at ∞ and $W(\infty)$ is equal to the $m \times m$ identity matrix I_m. A function W in the class (TF) may be viewed as the transfer function (see [2], §1.1) of a causal time invariant linear system. This means that W may be written in the form

$$(4.9) \qquad W(\lambda) = I_m + C(\lambda - A)^{-1}B,$$

where $A : X \to X$, $B : \mathbb{C}^m \to X$ and $\mathbb{C} : X \to \mathbb{C}^m$ are linear operators and X is a finite dimensional inner product space. The right hand side of (4.9) is said to be a *realization* of W. The space X, which may vary with the realization and with W, is called the *state space* of the realization.

Take W in the class (TF), and assume that W is given by (4.9). Define T_W to be the (SK)-integral operator on $L_2([a,b],\mathbb{C}^m)$ with kernel

$$k(t,s) = \begin{cases} Ce^{(t-s)A}B, & a \le s \le t \le b, \\ 0 \quad\quad\quad , & a \le t \le s \le b. \end{cases}$$

Thus T_W is given by the right hand side of (4.1). The definition of T_W does not not depend on the particular choice of the realization of W. To see this, note that

$$Ce^{(t-s)A}B = \sum_{n=0}^{\infty} \frac{1}{n!}(t-s)^n W_{n+1},$$

where W_n is the n-th coefficient in the Taylor expansion of W at ∞. The map

$$(4.10) \qquad W \to T_W$$

is a bijection of the class (TF) onto the class (STC) and this map preserves the multiplication in (TF) and (STC), that is,

$$(4.11) \qquad T_{W_1W_2} = T_{W_1} T_{W_2}$$

for W_1 and W_2 in (TF).

For W in the class (TF) the *McMillan degree* $\delta(W)$ of W is the smallest possible state space dimension of a realization of W. A realization of W is called a *minimal realization* of W if its state space dimension is equal to $\delta(W)$. If $W = W_1W_2$ with W_1 and W_2 in (TF), then

$$(4.12) \qquad \delta(W) \le \delta(W_1) + \delta(W_2),$$

and $W = W_1W_2$ is called a *minimal factorization* if equality holds in (4.12). Let W be given by (4.9). Then $\delta(W)$ is also equal to the right hand side of (4.2) (cf. [2], §4.2). This implies that

$$(4.13) \qquad \delta(W) = \delta(T_W),$$

and hence $I_m + C(\lambda - A)^{-1}B$ is a minimal realization of W if and only if $\Delta = (e^{-tA}B, Ce^{tA}; 0)$ is a minimal representation of T_W. Furthermore, the right hand side of (4.11) is a minimal factorization in the class (STC) if and only if W_1W_2 is a minimal factorization in the class (TF). Hence for functions in the class (TF) Theorem 4.1 translates into the following theorem.

4.2 THEOREM. *Let* $W(\lambda) = I_m + C(\lambda - A)^{-1}B$ *be a minimal realization of the rational* $m \times m$ *matrix function* W, *and let* X *be the state space of the realization.*

(i) *Let* Π *be a projection of* X *such that*

$$(4.14) \qquad A \operatorname{Ker}\Pi \subset \operatorname{Ker}\Pi, \qquad (A - BC)\operatorname{Im}\Pi \subset \operatorname{Im}\Pi.$$

Then a minimal factorization $W = W_1W_2$ *of* W *is obtained by taking*

$$(4.15) \qquad W_1(\lambda) = I_m + C(\lambda - A)^{-1}(I - \Pi)B,$$

$$(4.16) \qquad W_2(\lambda) = I_m + C\Pi(\lambda - A)^{-1}B.$$

(ii) *Conversely, if* $W = W_1W_2$ *is a minimal factorization of* W, *then there exists a unique projection* Π *of* X *for which* (4.14) *holds such that the factors* W_1 *and* W_2 *are given by* (4.15) *and* (4.16), *respectively.*

Theorem 4.2 is the minimal factorization theorem for rational matrix functions as it appears in [2], §4.2 and [3], §2.3.

I.5 SB-minimal factorization (1)

Let T be a (SK)-integral operator on $L_2([a,b],Y)$. From §I.1 we know that T has representations $(B(\cdot),C(\cdot);P)$ for which P is a projection. Such a representation we shall call a *SB-representation*. (The letters SB refer to separable boundary conditions; see [8], §7). Thus a representation Δ of T is a SB-representation if and only if the internal operator of Δ is a projection.

A representation Δ of T will be called a *SB-minimal representation* of T if Δ is a SB-representation of T and among all SB-representations of T the dimension of the internal space of Δ is as small as possible. We define the *SB-degree* of T (notation: $\varepsilon(T)$) to be the dimension of the internal space of a SB-minimal representation of T. The SB-degree of T is also given by

$$\varepsilon(T) = \min\{\operatorname{rank}(F_1 \otimes G_1) + \operatorname{rank}(F_2 \otimes G_2)\}. \tag{5.1}$$

Here $F \otimes G$ stands for the finite rank integral operator with kernel $F(t)G(s)$ (see formula (3.4)) and in (5.1) the minimum is taken over all possible representations of the kernel k of T in the form (1.2).

Obviously, $\varepsilon(T) \geq \delta(T)$, but it may happen that $\varepsilon(T) > \delta(T)$. In fact, if T is the (SK)-integral operator on $L_2[0,1]$ given by

$$(T\varphi)(t) = 2\int_0^t \varphi(s)ds + \int_t^1 \varphi(s)ds, \qquad 0 \leq t \leq 1,$$

then $\varepsilon(T) = 2$ and $\delta(T) = 1$ (cf. the example at the end of §7 in [8]). If T belongs to the class (AC), then (see §I.3) the internal operator of a minimal representation of T is equal to the zero operator (which is a projection), and thus in this case a minimal representation is automatically SB-minimal. It follows that *for a (SK)-integral operator in the class (AC) the SB-degree* $\varepsilon(T)$ *is equal to the degree* $\delta(T)$.

Like the degree the SB-degree has a sublogarithmic behaviour:

$$\varepsilon(T_1T_2) \leq \varepsilon(T_1) + \varepsilon(T_2). \tag{5.2}$$

Given a (SK)-integral operator T on $L_2([a,b],Y)$ we call $T = T_1T_2$ a *SB-minimal factorization* of T if T_1 and T_2 are (SK)-integral operators on $L_2([a,b],Y)$ and the SB-degree of T is the sum of the SB-degrees of T_1 and T_2, i.e.,

$$\varepsilon(T) = \varepsilon(T_1) + \varepsilon(T_2). \tag{5.3}$$

In the same way as for minimal factorizations one may obtain SB-minimal factorizations of T from decomposing projections of SB-minimal representations of T. In fact, *Theorem* 2.1 *remains true if everywhere in the theorem the word minimal is replaced by SB-minimal* (see §III.3).

In general, the SB-minimal factorizations of T cannot be generated from a single SB-minimal representation, but one has to employ various SB-minimal representations of T. To see this one can use the (SK)-integral operator T on $L_2[0,1]$ defined by (3.9) and the two factorization in (3.10). First of all, since the representation Δ_ν in (3.12) is a SB-representation, we have $\varepsilon(T) = \delta(T) = 2$. From the definitions of $T_1^{(\nu)}$ and $T_2^{(\nu)}$ one sees that $\varepsilon(T_1^{(\nu)}) = \varepsilon(T_2^{(\nu)}) = 1$. So

$$\varepsilon(T) = \varepsilon(T_1^{(\nu)}) + \varepsilon(T_2^{(\nu)}), \qquad \nu = 1,2,$$

and it follows that the two factorizations in (3.10) are also SB-minimal factorizations. From $\varepsilon(T) = \delta(T)$ we know that a SB-minimal representation of T is automatically a minimal representation, and hence there does not exist a single SB-minimal representation which generates the two SB-minimal factorizations in (3.10).

I.6 SB-minimal factorization in the class (USB)

Let T be a (SK)-integral operator on $L_2([a,b],Y)$. We say that T belongs to the *class (USB)* if up to similarity T has a unique SB-minimal representation. In the terminology of [8] this is equivalent to the requirement that the kernel k of T is both lower unique and upper unique. If the kernel k of T admits a representation (1.2) in which the functions $F_1(\cdot)$, $G_1(\cdot)$, $F_2(\cdot)$ and $G_2(\cdot)$ are analytic on [a,b], then T is in the class (USB) (cf. [8], §5). The (SK)-integral operator T defined by (3.9) is not in the class (USB); in fact, the representations Δ_1 and Δ_2 of T defined by (3.12) are SB-minimal and not similar.

6.1 THEOREM. *Let* T *be a (SK)-integral operator on* $L_2([a,b],Y)$ *in the class (USB), and let* $\Delta = (B(\cdot),C(\cdot);P)$ *be a SB-minimal representation of* T.

(i) *If* Π *is a decomposing projection for* Δ, *then a SB-minimal factorization* $T = T_1T_2$ *is obtained by taking* T_1 *and* T_2 *to be the (SK)-integral operators on* $L_2([a,b],Y)$ *of which the kernels are given by*

$$(6.1)\qquad k_1(t,s) = \begin{cases} C(t)(I-P)(I-\Pi(s))B(s), & a \le s < t \le b, \\ -C(t)P(I-\Pi(s))B(s), & a \le t < s \le b, \end{cases}$$

$$(6.2)\qquad k_2(t,s) = \begin{cases} C(t)\Pi(t)(I-P)B(s), & a \le s < t \le b, \\ -C(t)\Pi(t)PB(s), & a \le t < s \le b, \end{cases}$$

respectively. Here

$$(6.3)\qquad \Pi(t) = \Omega(t)\Pi(I-\Pi+\Omega(t)\Pi)^{-1}\Pi, \qquad a \le t \le b,$$

with $\Omega(t)$ *being the fundamental operator of* Δ.

(ii) *If* $T = T_1T_2$ *is a SB-minimal factorization, then the factors* T_1

and T_2 *are in the class (USB) and there exists a unique decomposing projection* Π *for* Δ *such that the kernels of* T_1 *and* T_2 *are given by* (6.1) *and* (6.2), *respectively.*

For minimal factorization the analogue of the uniqueness statement in Theorem 6.1 (ii) is not known. To make the problem more precise, let T be a (SK)-integral operator and assume that up to similarity T has a unique minimal representation, Δ say. Let $T = T_1T_2$ be a minimal factorization. Then we know (see the last paragraph of Section I.2) that there exists a decomposing projection Π for Δ which yields the factorization $T = T_1T_2$. The question whether Π is unique is open.

I.7 Analytic semi-separable kernels

A (SK)-integral operator T on $L_2([a,b],Y)$ is said to be in the *class* (*A*) (the letter A stands for analytic) if the kernel k of T admits a representation (1.2) in which the functions $F_1(\cdot)$, $G_1(\cdot)$, $F_2(\cdot)$ and $G_2(\cdot)$ are analytic on [a,b]. In that case the kernel k of T is lower unique and upper unique (see [8], §5), and hence the class (A) is contained in the class (USB). The SB-degree $\varepsilon(T)$ of an operator T in the class (A) is equal to

$$(7.1)\qquad \varepsilon(T) = \operatorname{rank}(F_1 \otimes G_1) + \operatorname{rank}(F_2 \otimes G_2),$$

where $F_1(\cdot)$, $G_1(\cdot)$, $F_2(\cdot)$ and $G_2(\cdot)$ are the analytic functions in the representation (1.2) of its kernel (and $F \otimes G$ is defined by (3.4)).

Let T be a (SK)-integral operator in the class (A), and let $\Delta = (B(\cdot),C(\cdot);P)$ be a SB-minimal representation of T. It is easy to prove that in this case the functions $B(\cdot)$ and $C(\cdot)$ are analytic on [a,b]. Thus $B(t)C(t)$ depends analytically on $t \in [a,b]$, and hence (see [4], §VI.1) the solution of the equation (2.2) is analytic on [a,b]. In other words the fundamental operator $\Omega(t)$ of a SB-minimal representation of T is analytic on [a,b]. From these remarks it is clear that *Theorem* 6.1 *remains valid if in the theorem the class (USB) is replaced by the class (A).*

I.8 LU- and UL-factorizations (1)

Consider on $L_2([a,b],Y)$ the integral operator

$$(8.1)\qquad (T\varphi)(t) = \varphi(t) + \int_a^b k(t,s)\varphi(s)ds, \qquad a \le t \le b, \quad \text{a.e.},$$

and assume that the kernel k is square integrable on $[a,b]\times[a,b]$. The integral operator T is said to admit a *LU-factorization* (see [11], §IV.7) if T can be factored as $T = T_-T_+$, where

$$(T_-\varphi)(t) = \varphi(t) + \int_a^t k_-(t,s)\varphi(s)ds, \qquad a \le t \le b, \quad \text{a.e.},$$

$$(T_+\varphi)(t) = \varphi(t) + \int_t^b k_+(t,s)\varphi(s)ds, \qquad a \le t \le b, \quad \text{a.e.},$$

and the kernels k_- and k_+ are square integrable on $a \le s < t \le b$ and $a \le t < s \le b$, respectively. A factorization of the form $T = T_+T_-$ is called a *UL-factorization.*

8.1 THEOREM. *Let* T *be a* (*SK*)*-integral operator on* $L_2([a,b],Y)$ *with kernel* k, *and let* $\Delta = (B(\cdot),C(\cdot);P)$ *be a SB-representation of* T. *The fundamental operator of* Δ *is denoted by* $\Omega(t)$. *The following statements are equivalent*:

(1) T *admits a LU-factorization*;

(2) *the internal operator* P *of* Δ *is a decomposing projection for* Δ;

(3) *the following Riccati differential equation has a solution* $R(t) : \operatorname{Im} P \to \operatorname{Ker} P$ *on* $a \le t \le b$:

$$(8.2)\qquad \begin{cases} \dot R(t) = (I - P + R(t)P)(-B(t)C(t))(R(t)P-P), & a \le t \le b, \\ R(a) = 0; \end{cases}$$

(4) $\det(I - P + \Omega(t)P) \ne 0$ *for* $a \le t \le b$;

(5) *the operator* $T_\alpha : L_2([a,\alpha],Y) \to L_2([a,\alpha],Y)$, *defined by*

$$(8.3)\qquad (T_\alpha\varphi)(t) = \varphi(t) + \int_a^\alpha k(t,s)\varphi(s)ds, \qquad a \le t \le \alpha,$$

is invertible for each $a < \alpha \le b$.

Furthermore, in that case $T = (I + K_-)(I + K_+)$ *with*

$$(8.4)\qquad (K_-\varphi)(t) = \int_a^t C(t)(I - P + R(s)P)B(s)\varphi(s)ds,$$

$$(8.5)\qquad (K_+\varphi)(t) = \int_t^b C(t)(R(t)P - P)B(s)\varphi(s)ds,$$

where $R(t) : \operatorname{Im} P \to \operatorname{Ker} P$, $a \le t \le b$, *is the solution of the Riccati equation* (8.2), *which is equal to*

$$(8.6)\qquad R(t) = (I - P)\{I - P + \Omega(t)P\}^{-1}P.$$

8.2 THEOREM. *Let* T *be a (SK)-integral operator on* $L_2([a,b],Y)$ *with kernel* k, *and let* $\Delta = (B(\cdot),C(\cdot);P)$ *be a SB-representation of* T. *The fundamental operator of* Δ *is denoted by* $\Omega(t)$. *The following statements are equivalent:*

(1) T *admits a UL-factorization;*

(2) *the following Riccati differential equation has a solution* $R(t) : \operatorname{Ker} P \to \operatorname{Im} P$ *on* $a \le t \le b$:

$$(8.7)\qquad \begin{cases} \dot{R}(t) = (P + R(t)(I-P))(-B(t)C(t))(R(t)(I-P) - (I-P)), & a \le t \le b, \\ R(b) = 0; \end{cases}$$

(3) $\det[(I-P)\Omega(t) + P\Omega(b)] \ne 0$ *for* $a \le t \le b$;

(4) *the operator* $T^\beta : L_2([\beta,b],Y) \to L_2([\beta,b],Y)$, *defined by*

$$(8.8)\qquad (T^\beta\varphi)(t) = \varphi(t) + \int_\beta^b k(t,s)\varphi(s)ds, \quad \beta \le t \le b,$$

is invertible for each $a \le \beta < b$;

Furthermore, in that case $T = (I + H_+)(I + H_-)$ *with*

$$(8.9)\qquad (H_-\varphi)(t) = \int_a^t C(t)(I - P - R(t)(I-P))B(s)\varphi(s)ds,$$

$$(8.10)\qquad (H_+\varphi)(t) = -\int_t^b C(t)(P + R(s)(I-P))B(s)\varphi(s)ds,$$

where $R(t) : \operatorname{Ker} P \to \operatorname{Im} P$, $a \le t \le b$, *is the solution of the Riccati equation* (8.7), *which is equal to*

$$(8.11)\qquad R(t) = -P\{I - P + P\Omega(b)\Omega(t)^{-1}\}^{-1}(I-P).$$

For the positive definite case the main statements of Theorems 8.1 and 8.2 were proved in [13]. The equivalence of the statements (1) and (5) in Theorem 8.1 and the equivalence of the statements (1) and (4) in Theorem 8.2 are well-known (see [11], § IV.7). For the equivalence of the statements (1), (4) and (5) in Theorem 8.1 and for the equivalence of the statements (1), (3) and (4) in Theorem 8.2 it is not necessary that the internal operator P of the representation Δ is a projection, and hence these equivalences hold for an arbitrary representation of T.

From the formulas (8.4) and (8.5) it follows that the factors in a LU-factorization of a (SK)-integral operator are again (SK)-integral operators. Furthermore, the equivalence of (1) and (2) in Theorem 8.1 implies that a

LU-factorization of a (SK)-integral operator (assuming it exists) may be obtained from any SB-minimal representation. It follows that a LU-factorization is a SB-minimal factorization. A similar result holds for UL-factorization (see §III.8)

8.3 COROLLARY. *A LU- or UL-factorization of a (SK)-integral operator is a SB-minimal factorization.*

In general, a LU- (or UL-) factorization is not a minimal factorization. For example, consider the (SK)-integral operator T_0 on $L_2[0,1]$ defined by

$$(8.12)\qquad (T_0\varphi)(t) = \varphi(t) + 2\int_0^t \varphi(s)ds + \int_t^1 \varphi(s)ds, \qquad 0 \le t \le 1.$$

The operator T_0 admits a LU-factorization. In fact, $T_0 = T_-T_+$ with

$$(T_-\varphi)(t) = \varphi(t) + 2\int_0^t (2-e^{-s})^{-1}\varphi(s)ds, \qquad 0 \le t \le 1,$$

$$(T_+\varphi)(t) = \varphi(t) + (2e^t-1)^{-1}\int_t^1 \varphi(s)ds, \qquad 0 \le t \le 1.$$

Next, note that T_0 is represented by (1,1;-1). Hence the degree $\delta(T) = 1$. But then it follows that a minimal representation of T_0 has no non-trivial decomposing projection. Thus the LU-factorization $T_0 = T_-T_+$ is not generated by a minimal representation of T_0. Hence this factorization is not minimal.

II. CASCADE DECOMPOSITION OF SYSTEMS

In this chapter we develop a cascade decomposition theory for time varying linear systems with well-posed boundary conditions. The first section has a preliminary character. In Sections 2 and 3 we define cascade decompositions and decomposing projections. The main results are stated in Section 4 and and are proved in Sections 5 - 7. The last section is devoted to cascade decompositions of inverse systems.

II.1 Preliminaries about systems with boundary conditions

In this chapter we consider time varying linear systems with boundary conditions of the following form:

$$(1.1)\qquad \begin{cases} \dot{x}(t) = A(t)x(t) + B(t)u(t), & a \le t \le b,\\ y(t) = C(t)x(t) + u(t), & a \le t \le b,\\ N_1x(a) + N_2x(b) = 0. \end{cases}$$

Here $A(t) : X \to X$, $B(t) : Y \to X$ and $C(t) : X \to Y$ are linear operators acting between finite dimensional inner product spaces. As a function of t the *main coefficient* $A(t)$ is assumed to be integrable on $a \le t \le b$. The *input coefficient* $B(t)$ and the *output coefficient* $C(t)$ are square integrable on $a \le t \le b$. Throughout this chapter the *input/output space* Y is kept fixed; the *state space* X may differ per system. Note that the *external coefficient* (i.e., the coefficient of $u(t)$ in the second equation of (1.1)) is assumed to be equal to the identity operator on the space Y. If the main coefficient, input coefficient and output coefficient do not depend on t, then (1.1) is called *time invariant*. The boundary conditions of (1.1) are given in terms of two linear operators N_1 and N_2 acting on X. The system is called *causal* if $N_1 = I$ (where I stands for the identity operator on X) and $N_2 = 0$; if $N_1 = 0$ and $N_2 = I$, then (1.1) is said to be *anticausal*. In what follows the system (1.1) will be denoted by θ and for the sake of brevity we write

$$\theta = (A(t),B(t),C(t);N_1,N_2)_a^b. \tag{1.2}$$

The indices will be omitted when it is clear on which time interval the system is considered.

Let θ be as in (1.2). The system

$$\theta^\times = (A(t)-B(t)C(t),B(t),-C(t);N_1,N_2)_a^b$$

is called the *inverse system* associated with θ (see [5], §II.2). Note that one obtains $\theta^\times$ by interchanging in (1.1) the roles of the input and output. The main coefficient of $\theta^\times$ will be denoted by $A^\times(t)$; thus

$$A^\times(t) = A(t)-B(t)C(t), \qquad a \le t \le b. \tag{1.3}$$

The system θ is called a *boundary value system* if the boundary conditions of θ are *well-posed*. The latter means that the homogeneous boundary value problem

$$\dot{x}(t) = A(t)x(t), \quad a \le t \le b, \quad N_1x(a)+N_2x(b) = 0,$$

has the trivial solution only. Well-posedness of the boundary conditions is equivalent to the requirement that $\det(N_1+N_2U(b)) \ne 0$. Here $U(t) : X \to X$, $a \le t \le b$, is the *fundamental operator* of the system θ, i.e., the unique absolutely continuous solution of

$$\dot{U}(t) = A(t)U(t), \qquad a \le t \le b, \qquad U(a) = I.$$

The causal and anticausal systems are the classical examples of boundary value systems.

A boundary value system θ has a well-defined *input/output map* T_θ, namely (see [14,15]; [5], §I.2) the integral operator

$$y(t) = (T_\theta u)(t) = u(t) + \int_a^b k_\theta(t,s)u(s)ds, \qquad a \le t \le b,$$

of which the kernel k_θ is given by

$$(1.4) \qquad k_\theta(t,s) = \begin{cases} C(t)U(t)(I-P_\theta)U(s)^{-1}B(s), & a \le s < t \le b, \\ -C(t)U(t)P_\theta U(s)^{-1}B(s), & a \le t < s \le b, \end{cases}$$

where $P_\theta = (N_1 + N_2U(b))^{-1}N_2U(b)$. It follows that T_θ is a (SK)-integral operator on $L_2([a,b],Y)$. The operator P_θ is called the *canonical boundary value operator* of θ.

Let θ have well-posed boundary conditions. It does not follow that the inverse system $\theta^\times$ also has well-posed boundary conditions. In fact, the latter happens if and only if T_θ is invertible, and in that case (see [5], Theorem II.2.1)

$$(1.5) \qquad T_{\theta^\times} = (T_\theta)^{-1}.$$

Two time varying systems θ_1 and θ_2,

$$(1.6) \qquad \theta_\nu = (A_\nu(t), B_\nu(t), C_\nu(t); N_1^{(\nu)}, N_2^{(\nu)})_a^b, \qquad \nu = 1,2,$$

with state spaces X_1 and X_2, respectively, are called *similar* (notation: $\theta_1 \simeq \theta_2$) if there exist an invertible operator $E : X_1 \to X_2$ and an absolutely continuous function $S(t) : X_1 \to X_2$, $a \le t \le b$, of which the values are invertible operators such that

$$(1.7) \qquad A_2(t) = S(t)A_1(t)S(t)^{-1} + \dot{S}(t)S(t)^{-1},$$

$$(1.8) \qquad B_2(t) = S(t)B_1(t),$$

$$(1.9) \qquad C_2(t) = C_1(t)S(t)^{-1},$$

$$(1.10) \qquad N_1^{(2)} = EN_1^{(1)}S(a)^{-1}, \qquad N_2^{(2)} = EN_2^{(1)}S(b)^{-1},$$

almost everywhere on $a \le t \le b$. This notion of similarity appears in a natural way when in (1.1) the state $x(t)$ is replaced by $z(t) = S(t)x(t)$ (see [5], §I.5). We shall refer to $S(t)$, $a \le t \le b$, as a *similarity transformation* between θ_1 and θ_2. Formula (1.7) implies that the fundamental operators $U_1(t)$ and $U_2(t)$ of θ_1 and θ_2, respectively, are related in the following way:

$$U_2(t) = S(t)U_1(t)S(a)^{-1}, \qquad a \le t \le b.$$

Well-posedness of the boundary conditions is preserved under a similarity transformation and similar systems with well-posed boundary conditions have similar canonical boundary value operators. In fact, if P_1 and P_2 are the canonical boundary value operators of θ_1 and θ_2, respectively, then the above formulas imply that $P_2 = S(a)P_1S(a)^{-1}$. It follows that similar systems with well-posed boundary conditions have the same input/output map.

The system $\theta_0 = (A_0(t),B_0(t),C_0(t);N_1^{(0)},N_2^{(0)})_a^b$ is called a *reduction* of the system $\theta = (A(t),B(t),C(t);N_1,N_2)_a^b$ if (see [6,7]) the state space X of θ admits a decomposition, $X = X_1 \oplus X_0 \oplus X_2$, such that relative to this decomposition the coefficients of θ and its boundary value operators are partitioned in the following way:

$$(1.11) \qquad A(t) = \begin{pmatrix} * & * & * \\ 0 & A_0(t) & * \\ 0 & 0 & * \end{pmatrix}, \qquad B(t) = \begin{pmatrix} * \\ B_0(t) \\ 0 \end{pmatrix}$$

$$(1.12) \qquad C(t) = \begin{pmatrix} 0 & C_0(t) & * \end{pmatrix};$$

$$(1.13) \qquad N_\nu = E\begin{pmatrix} * & * & * \\ 0 & N_\nu^{(0)} & * \\ 0 & 0 & * \end{pmatrix}, \qquad \nu = 1,2.$$

Here (1.11) and (1.12) hold a.e. on $a \le t \le b$. The operator E appearing in (1.13) is some invertible operator on X. The symbols $*$ denote unspecified entries. If, in addition, $\dim X > \dim X_0$, then θ_0 is called a *proper* reduction of θ. The boundary value system θ is said to be *irreducible* if none of the systems similar to θ admits a proper reduction. We say that the system θ is a (*proper*) *dilation* of θ_0 if θ_0 is a (proper) reduction of θ. If the dilation θ of θ_0 has well-posed boundary conditions, then the same is true for θ_0 and the systems θ and θ_0 have the same input/output map.

II.2 Cascade decompositions

In this section we define the notion of a cascade decomposition of two boundary value systems. First we recall the definition of a cascade connection. For $\nu = 1,2$ let $\theta_\nu = (A_\nu(t),B_\nu(t),C_\nu(t);N_1^{(\nu)},N_2^{(\nu)})_a^b$ be a time varying linear system with boundary conditions. Introduce

$$(2.1)\qquad A(t) = \begin{pmatrix} A_1(t) & B_1(t)C_2(t) \\ 0 & A_2(t) \end{pmatrix}, \qquad B(t) = \begin{pmatrix} B_1(t) \\ B_2(t) \end{pmatrix},$$

$$(2.2)\qquad C(t) = \begin{pmatrix} C_1(t) & C_2(t) \end{pmatrix},$$

$$(2.3)\qquad N_j = \begin{pmatrix} N_j^{(1)} & 0 \\ 0 & N_j^{(2)} \end{pmatrix}, \qquad j = 1,2.$$

By definition ([5], §II.1) the *cascade connection* of θ_1 and θ_2 is the system

$$\theta_1\theta_2 = (A(t),B(t),C(t);N_1,N_2)_a^b,$$

where $A(\cdot),B(\cdot),C(\cdot),N_1,N_2$ are given by (2.1) - (2.3). The system $\theta_1\theta_2$ appears in a natural way if one connects the output of θ_2 to the input of θ_1.

Let θ be a boundary value system. We say that θ admits a *cascade decomposition* if θ is similar to a cascade connection $\theta_1\theta_2$ (notation $\theta \simeq \theta_1\theta_2$). Assume $\theta \simeq \theta_1\theta_2$. Since similarity preserves the well-posedness of the boundary conditions, we can apply [5], Theorem II.1.1 to show that the compounding systems θ_1 and θ_2 have well-posed boundary conditions. It follows that

$$(2.4)\qquad T_\theta = T_{\theta_1}T_{\theta_2}.$$

Thus cascade decompositions yield factorizations of the input/output map.

II.3 Decomposing projections

To describe how cascade decompositions may be constructed we introduce the concept of a decomposing projection for a system. Let $\theta = (A(t), B(t),C(t);N_1,N_2)_a^b$ be a boundary value system. As before, $U(t)$ denotes the fundamental operator of θ and P is its canonical boundary value operator. By $U^\times(t)$ we denote the fundamental operator of the inverse system $\theta^\times$. Thus

$$(3.1)\qquad \dot{U}^\times(t) = A^\times(t)U^\times(t), \qquad a \le t \le b, \qquad U^\times(a) = I.$$

A projection Π of the state space X of θ is called a *decomposing projection* for θ if

(i) $\Pi P \Pi = \Pi P$,

(ii) $\det(I - \Pi + U(t)^{-1}U^{\times}(t)\Pi) \neq 0, \quad a \le t \le b$,

(iii) $(I - \Pi)P\Pi = P(I - \Pi)\{I - \Pi + U(b)^{-1}U^{\times}(b)\Pi\}^{-1}\Pi(I - P)$.

In §III.2 we shall see that the notion of a decomposing projection for a system is a natural extension of the corresponding concept for representations as defined in §I.2. The following lemma provides an alternative definition of a decomposing projection.

3.1. LEMMA. *Let* $\theta = (A(t),B(t),C(t);N_1,N_2)_a^b$ *be a boundary value system. Let* $U(t)$ *be the fundamental operator of* θ *and let* P *be its canonical boundary value operator. A projection* Π *of the state space* X *of* θ *is a de composing projection for* θ *if and only if* $\operatorname{Ker}\Pi$ *is invariant under* P *and the following Riccati differential equation is solvable on* $a \le t \le b$:

$$(3.2)\qquad \begin{cases} \dot R(t) = (I - \Pi + R(t)\Pi)(-U(t)^{-1}B(t)C(t)U(t))(R(t)\Pi - \Pi), & a \le t \le b, \\ R(a) = 0 , \end{cases}$$

and its solution $R(t) : \operatorname{Im}\Pi \to \operatorname{Ker}\Pi$, $a \le t \le b$, *satisfies the additional boundary condition*

$$(3.3)\qquad P(I - \Pi)R(b)\Pi(I - P) = (I - \Pi)P\Pi.$$

Furthermore, in that case the unique solution of (3.2) *is given by*

$$(3.4)\qquad R(t) = (I - \Pi)\{I - \Pi + U(t)^{-1}U^{\times}(t)\Pi\}^{-1}\Pi, \quad a \le t \le b,$$

where $U^{\times}(t)$ *denotes the fundamental operator of the inverse system* $\theta^{\times}$.

PROOF. Let Π be a projection of the state X of θ. Obviously, condition (i) in the definition of a decomposing projection is equivalent to the requirement that $\operatorname{Ker}\Pi$ is invariant under P. Next consider the RDE (3.2). From

$$(3.5)\qquad \frac{d}{dt}U(t)^{-1}U^{\times}(t) = (-U(t)^{-1}B(t)C(t)U(t))U(t)^{-1}U^{\times}(t)$$

it follows that $\Omega(t) = U(t)^{-1}U^{\times}(t)$, $a \le t \le b$, is the fundamental operator of the RDE (3.2). But then we can apply the main result of [17] to show that the RDE (3.2) is solvable if and only if

$$(3.6)\qquad \Pi U(t)^{-1}U^{\times}(t)\Pi : \operatorname{Im}\Pi \to \operatorname{Im}\Pi$$

is invertible for each $a \le t \le b$, and in that case its unique solution on $a \le t \le b$ is given by

$$R(t) = -(I-\Pi)U(t)^{-1}U^{\times}(t)\Pi(\Pi U(t)^{-1}U^{\times}(t)\Pi)^{-1}\Pi, \quad a \le t \le b. \tag{3.7}$$

The invertibility for each $a \le t \le b$ of the operator (3.6) is equivalent to condition (ii). Furthermore, (3.4) is just a reformulation of (3.7). Since $R(\cdot)$ is given by (3.4) it is clear that condition (iii) in the definition of a decomposing projection is the same as the extra boundary condition in (3.3). □

II.4 Main decomposition theorems

Let $\theta = (A(t),B(t),C(t);N_1,N_2)_a^b$ be a boundary value system. The fundamental operator of θ is denoted by $U(t)$ and P is its canonical boundary value operator. Let Π be a decomposing projection for θ. Write $U(t)^{-1}B(t)$, $C(t)U(t)$ and P as block matrices relative to the decomposition $X = \operatorname{Ker}\Pi \oplus \operatorname{Im}\Pi$:

$$U(t)^{-1}B(t) = \begin{pmatrix} B_1(t) \\ B_2(t) \end{pmatrix}, \qquad C(t)U(t) = \begin{pmatrix} C_1(t) & C_2(t) \end{pmatrix}, \tag{4.1}$$

$$P = \begin{pmatrix} P_{11} & P_{12} \\ P_{21} & P_{22} \end{pmatrix}. \tag{4.2}$$

Introduce the following systems:

$$\ell_\Pi(\theta) := (0,B_1(t)+R(t)B_2(t),C_1(t);I-P_{11},P_{11})_a^b,$$

$$r_\Pi(\theta) := (0,B_2(t),-C_1(t)R(t)+C_2(t);I-P_{22},P_{22})_a^b.$$

Here $R(t)$ is the solution of the RDE (3.2) (or, equivalently, $R(t)$ is given by (3.4)) and the other terms come frome (4.1) and (4.2). We call $\ell_\Pi(\theta)$ the *left factor* and $r_\Pi(\theta)$ the *right factor* associated with Π and θ. This terminology is justified by the next theorem.

4.1. THEOREM. *If Π is a decomposing projection for θ, then*

$$\theta \simeq \ell_\Pi(\theta)r_\Pi(\theta), \tag{4.3}$$

and the similarity $S(t)$ between θ and $\ell_\Pi(\theta)r_\Pi(\theta)$ is given by

$$S(t) = \begin{pmatrix} I_{\operatorname{Ker}\Pi} & R(t) \\ 0 & I_{\operatorname{Im}\Pi} \end{pmatrix} U(t)^{-1} : \operatorname{Ker}\Pi \oplus \operatorname{Im}\Pi \to \operatorname{Ker}\Pi \oplus \operatorname{Im}\Pi, \quad a \le t \le b,$$

where $R(\cdot)$ *is the solution of the RDE* (3.2) *or, equivalently,* $R(\cdot)$ *is given by* (3.4).

The next theorem shows that up to similarity all cascade decompositions of θ may be obtained from decomposing projections.

4.2. THEOREM. *Assume* $\theta \simeq \theta_1\theta_2$ *is a cascade decomposition of the boundary value system* θ. *Then there exists a decomposing projection* Π *for* θ *such that*

$$(4.4) \qquad \ell_\Pi(\theta) \simeq \theta_1, \quad r_\Pi(\theta) \simeq \theta_2.$$

Formula (4.4) leads to the following definition. Two cascade decompositions $\theta \simeq \theta_1\theta_2$ and $\theta \simeq \theta_1'\theta_2'$ of θ are said to be *equivalent* if $\theta_1 \simeq \theta_1'$ and $\theta_2 \simeq \theta_2'$.

4.3. THEOREM. *Let* θ *be an irreducible system. Then the map*

$$(4.5) \qquad \Pi \to \ell_\Pi(\theta)r_\Pi(\theta)$$

defines a one-one correspondence between all decomposing projections Π *for* θ *and all non-equivalent cascade decompositions of* θ. *Furthermore, the factors* $\ell_\Pi(\theta)$ *and* $r_\Pi(\theta)$ *are irreducible.*

The proofs of Theorems 4.1 - 4.3 will be given in the next three sections.

The systems $\ell_\Pi(\theta)$ and $r_\Pi(\theta)$ are boundary value systems and the kernels of their input/output maps are, respectively, given by

$$(4.6) \qquad k_1(t,s) = \begin{cases} C(t)U(t)(I-P)(I-\Pi(s))U(s)^{-1}B(s), & a \le s < t \le b, \\ -C(t)U(t)P(I-\Pi(s))U(s)^{-1}B(s), & a \le t < s \le b, \end{cases}$$

$$(4.7) \qquad k_2(t,s) = \begin{cases} C(t)U(t)\Pi(t)(I-P)U(s)^{-1}B(s), & a \le s < t \le b, \\ -C(t)U(t)\Pi(t)PU(s)^{-1}B(s), & a \le t < s \le b. \end{cases}$$

Here for $a \le t \le b$ the operator $\Pi(t) : X \to X$ is given by

$$\Pi(t) = \Pi - R(t)\Pi = U(t)^{-1}U^\times(t)\Pi\{I - \Pi + U(t)^{-1}U^\times(t)\}^{-1}\Pi.$$

Since $\Pi R(t) = 0$, the operator $\Pi(t)$ is a projection for each $t \in [a,b]$. Note that Theorem 4.1 implies that

$$(4.8) \qquad T_\theta = T_{\ell_\Pi(\theta)} \cdot T_{r_\Pi(\theta)}.$$

II.5 Proof of Theorem II.4.1

In this section we derive Theorem 4.1 as a corollary of a more general cascade decomposition theorem. Let $\theta = (A(t),B(t),C(t);N_1,N_2)_a^b$ be a given boundary value system. As usual $U(t)$ and $U^\times(t)$ denote the fundamental operators of θ and $\theta^\times$. The operator P is the canonical boundary value operator of θ and $A^\times(t) = A(t) - B(t)C(t)$.

Let $X = X_1 \oplus X_2$ be a direct sum decomposition of the state space X of θ. Relative to this decomposition we write:

$$(5.1) \qquad A(t) = \begin{bmatrix} A_{11}(t) & A_{12}(t) \\ A_{21}(t) & A_{22}(t) \end{bmatrix}, \qquad A^\times(t) = \begin{bmatrix} A_{11}^\times(t) & A_{12}^\times(t) \\ A_{21}^\times(t) & A_{22}^\times(t) \end{bmatrix},$$

$$(5.2) \qquad U(t) = \begin{bmatrix} U_{11}(t) & U_{12}(t) \\ U_{21}(t) & U_{22}(t) \end{bmatrix}, \qquad U^\times(t) = \begin{bmatrix} U_{11}^\times(t) & U_{12}^\times(t) \\ U_{21}^\times(t) & U_{22}^\times(t) \end{bmatrix},$$

$$(5.3) \qquad B(t) = \begin{bmatrix} B_1(t) \\ B_2(t) \end{bmatrix}, \qquad C(t) = \begin{pmatrix} C_1(t) & C_2(t) \end{pmatrix},$$

$$(5.4) \qquad P = \begin{bmatrix} P_{11} & P_{12} \\ P_{21} & P_{22} \end{bmatrix}.$$

Consider the following Riccati differential equations:

$$(5.5) \qquad \begin{cases} \dot{R}_{21}(t) = -A_{21}(t) - R_{21}(t)A_{11}(t) + A_{22}(t)R_{21}(t) + R_{21}(t)A_{12}(t)R_{21}(t), \\ \hfill a \le t \le b, \\ R_{21}(a) = 0 \end{cases}$$

$$(5.6) \qquad \begin{cases} \dot{R}_{12}(t) = -A_{12}^\times(t) - R_{12}(t)A_{22}^\times(t) + A_{11}^\times(t)R_{12}(t) + R_{12}(t)A_{21}^\times(t)R_{12}(t), \\ \hfill a \le t \le b, \\ R_{12}(a) = 0. \end{cases}$$

Note that $U(t)$ and $U^\times(t)$ are the fundamental operators of the RDE's (5.5) and (5.6), respectively. It follows (cf. [17]) that (5.5) is solvable on $a \le t \le b$ if and only if $\det U_{11}(t) \ne 0$, $a \le t \le b$, and in that case the unique solution of (5.5) is given by

$$(5.7) \qquad R_{21}(t) = -U_{21}(t)U_{11}(t)^{-1}, \qquad a \le t \le b.$$

Similarly, equation (5.6) is solvable on $a \le t \le b$ if and only if

$\det U_{22}^{\times}(t) \neq 0$, $a \le t \le b$, and in that case the unique solution of (5.6) is equal to

$$(5.8)\qquad R_{12}(t) = -U_{12}^{\times}(t)U_{22}^{\times}(t)^{-1}, \qquad a \le t \le b.$$

5.1 THEOREM. *Assume that the Riccati equations* (5.5) *and* (5.6) *are solvable on* $a \le t \le b$ *and that their solutions have the following extra property:*

$$(5.9)\qquad \det(I - R_{12}(t)R_{21}(t)) \neq 0, \qquad a \le t \le b.$$

Furthermore assume that

$$(5.10)\qquad P_{21} = 0, \qquad P_{12} = P_{11}Q(I - P_{22}),$$

where

$$(5.11)\qquad Q = U_{11}(b)^{-1}(I - R_{12}(b)R_{21}(b))^{-1}(U_{12}(b) + R_{12}(b)U_{22}(b)).$$

Then θ *admits a cascade decomposition,* $\theta \simeq \theta_1\theta_2$, *with similarity*

$$(5.12)\qquad S(t) = \begin{bmatrix} I & R_{12}(t) \\ R_{21}(t) & I \end{bmatrix} : X_1 \oplus X_2 \to X_1 \oplus X_2, \qquad a \le t \le b,$$

and the compounding systems are given by

$$\theta_1 = (\tilde{A}_1(t), \tilde{B}_1(t), \tilde{C}_1(t); N_1^{(1)}, N_2^{(1)})_a^b,$$

$$\theta_2 = (\tilde{A}_2(t), \tilde{B}_2(t), \tilde{C}_2(t); N_1^{(2)}, N_2^{(2)})_a^b,$$

where

$$\tilde{A}_1(t) = (A_{11}(t) + R_{12}(t)A_{21}(t) - A_{12}(t)R_{21}(t) + \\ - R_{12}(t)A_{22}(t)R_{21}(t) - \dot{R}_{12}(t)R_{21}(t))\Delta(t)^{-1},$$

$$\tilde{B}_1(t) = B_1(t) + R_{12}(t)B_2(t),$$

$$\tilde{C}_1(t) = (C_1(t) - C_2(t)R_{21}(t))\Delta(t)^{-1},$$

$$N_1^{(1)} = I - P_{11}, \qquad N_2^{(1)} = P_{11}U_{11}(b)^{-1}\Delta(b)^{-1},$$

$$\tilde{A}_2(t) = (A_{22}(t) + R_{21}(t)A_{12}(t) - A_{21}(t)R_{12}(t) + \\ - R_{21}(t)A_{11}(t)R_{12}(t) - \dot{R}_{21}(t)R_{12}(t))\nabla(t)^{-1},$$

$$\tilde{B}_2(t) = R_{21}(t)B_1(t) + B_2(t),$$

$$\tilde{C}_2(t) = (-C_1(t)R_{12}(t) + C_2(t))\nabla(t)^{-1},$$

$$N_1^{(2)} = I - P_{22}, \qquad N_2^{(2)} = P_{22}(U_{22}(b) - U_{21}(b)U_{11}(b)^{-1}U_{12}(b))^{-1}.$$

Here $\Delta(t) = I - R_{12}(t)R_{21}(t)$ *and* $\nabla(t) = I - R_{21}(t)R_{12}(t)$.

PROOF. The solvability of the Riccati equation (5.5) implies that $\det U_{11}(b) \neq 0$. Thus $U_{11}(b)^{-1}$ exists. Next, observe that the invertibility of $U_{11}(b)$ and $U(b)$ yields the invertibility of the Schur complement $U_{22}(b) - U_{21}(b)U_{11}(b)^{-1}U_{12}(b)$ (cf. [2], Remark 1.2). From (5.9) we may conclude that both $\Delta(t)$ and $\nabla(t)$ are invertible for each $a \leq t \leq b$. It follows that the various operators and operator functions in Theorem 5.1 are well-defined.

Since $\Delta(t)$ and $\nabla(t)$ are invertible, the same is true for $S(t)$. In fact,

$$S(t)^{-1} = \begin{pmatrix} \Delta(t)^{-1} & -R_{12}(t)\nabla(t)^{-1} \\ -R_{21}(t)\Delta(t)^{-1} & \nabla(t)^{-1} \end{pmatrix}, \qquad a \leq t \leq b.$$

Thus $S(t)$ establishes a similarity transformation in the state space $X = X_1 \oplus X_2$. Let $\tilde{\theta}$ be the system which one obtains when the similarity transformation $S(t)$, $a \leq t \leq b$, is applied to θ. So

$$\tilde{\theta} = (\tilde{A}(t),\tilde{B}(t),\tilde{C}(t);I - P,P\tilde{U}(b)^{-1})_a^b,$$

where

$$\tilde{A}(t) = S(t)A(t)S(t)^{-1} + \dot{S}(t)S(t)^{-1},$$

$$\tilde{B}(t) = S(t)B(t), \qquad \tilde{C}(t) = C(t)S(t)^{-1}, \qquad \tilde{U}(t) = S(t)U(t).$$

From $S(a) = I$, it follows that $\tilde{U}(t)$ is the fundamental operator of $\tilde{\theta}$, and hence $\tilde{\theta} \simeq \theta$ indeed.

The fact that $R_{21}(\cdot)$ is the solution of the Riccati equation (5.5) on $a \leq t \leq b$ implies that relative to the decomposition $X = X_1 \oplus X_2$ the operator $\tilde{A}(t)$ admits the following partitioning:

$$\tilde{A}(t) = \begin{pmatrix} \tilde{A}_1(t) & * \\ 0 & \tilde{A}_2(t) \end{pmatrix}, \qquad a \le t \le b.$$

Here $\tilde{A}_1(t)$ and $\tilde{A}_2(t)$ are the main coefficients of θ_1 and θ_2, respectively. Next, observe that

$$\tilde{A}^\times(t) = S(t)A^\times(t)S(t)^{-1} + \dot{S}(t)S(t)^{-1}.$$

Since $R_{12}(\cdot)$ is the solution of the Riccati equation (5.6) on $a \le t \le b$, it follows that $\tilde{A}^\times(t)$ admits the following partitioning:

$$\tilde{A}^\times(t) = \begin{pmatrix} * & 0 \\ * & * \end{pmatrix}, \qquad a \le t \le b.$$

By a direct computation one shows that

$$\tilde{B}(t) = \begin{pmatrix} \tilde{B}_1(t) \\ \tilde{B}_2(t) \end{pmatrix}, \qquad \tilde{C}(t) = \begin{pmatrix} \tilde{C}_1(t) & \tilde{C}_2(t) \end{pmatrix},$$

where for $\nu = 1,2$ the operator functions $\tilde{B}_\nu(\cdot)$ and $\tilde{C}_\nu(\cdot)$ are the input coefficient and output coefficient of θ_ν, respectively. We conclude that $\tilde{\theta}$ and the cascade connection $\theta_1\theta_2$ have the same coefficients.

Next we compare the boundary value operators of $\tilde{\theta}$ and $\theta_1\theta_2$. Put

$$E = \begin{pmatrix} I & P_{11}Q \\ 0 & I \end{pmatrix} : X_1 \oplus X_2 \to X_1 \oplus X_2,$$

where Q is given by (5.11). We shall prove that

$$(5.13) \qquad E(I-P) = \begin{pmatrix} N_1^{(1)} & 0 \\ 0 & N_1^{(2)} \end{pmatrix}, \qquad EP\tilde{U}(b)^{-1} = \begin{pmatrix} N_2^{(1)} & 0 \\ 0 & N_2^{(2)} \end{pmatrix}.$$

The first equality in (5.13) is a direct consequence of (5.10) and of the definitions of $N_1^{(1)}$ and $N_1^{(2)}$. To prove the second equality in (5.13) we first compute $\tilde{U}(b)^{-1}$. Since $\tilde{U}(b) = S(b)U(b)$, we have

$$\tilde{U}(b) = \begin{pmatrix} \Delta(b)U_{11}(b) & \Delta(b)U_{11}(b)Q \\ 0 & U_{22}(b) - U_{21}(b)U_{11}(b)^{-1}U_{12}(b) \end{pmatrix}.$$

Here we used (5.7) and the fact that Q is given by (5.11). It follows that

$$EP\tilde{U}(b)^{-1} = \begin{pmatrix} P_{11} & P_{11}Q \\ 0 & P_{22} \end{pmatrix} \tilde{U}(b)^{-1} = \begin{pmatrix} T_{11} & T_{12} \\ 0 & T_{22} \end{pmatrix},$$

where

$$T_{11} = P_{11}U_{11}(b)^{-1}\Delta(b)^{-1} = N_2^{(1)},$$

$$T_{12} = -P_{11}Q(U_{22}(b) - U_{21}(b)U_{11}(b)^{-1}U_{12}(b))^{-1} + $$
$$+ P_{11}Q(U_{22}(b) - U_{21}(b)U_{11}(b)^{-1}U_{12}(b))^{-1} = 0,$$

$$T_{22} = P_{22}(U_{22}(b) - U_{21}(b)U_{11}(b)^{-1}U_{12}(b))^{-1} = N_2^{(2)}.$$

This proves the second equality in (5.13). From (5.13) and the fact that $\tilde{\theta}$ and $\theta_1\theta_2$ have the same coefficients it is clear that $\tilde{\theta} \simeq \theta_1\theta_2$, and thus $\theta \simeq \theta_1\theta_2$. □

Theorem 5.1 simplifies considerably if the main coefficient $A(\cdot)$ of θ is zero. In fact, if $A(\cdot) = 0$, then $U(t) = I$, $a \le t \le b$, and the Riccati equation (5.5) reduces to a trivial equation which is solvable on $a \le t \le b$ and has $R_{21}(t) = 0$, $a \le t \le b$, as its unique solution. It follows that in this case condition (5.9) is automatically fulfilled and condition (5.10) simplifies to

$$(5.14) \qquad P_{21} = 0, \qquad P_{12} = P_{11}R_{12}(b)(I - P_{22}).$$

Furthermore, the compounding systems θ_1 and θ_2 in the cascade decomposition $\theta \simeq \theta_1\theta_2$ are now given by

$$\theta_1 = (0, B_1(t) + R_{12}(t)B_2(t), C_1(t); I - P_{11}, P_{11})_a^b,$$

$$\theta_2 = (0, B_2(t), -C_1(t)R_{12}(t) + C_2(t); I - P_{22}, P_{22})_a^b.$$

<u>PROOF of Theorem 4.1</u>. Let $\theta = (A(t), B(t), C(t); N_1, N_2)_a^b$ be a boundary value system, and let Π be a decomposing projection for θ. Introduce the system

$$(5.15) \qquad \theta_\square = (0, U(t)^{-1}B(t), C(t)U(t); I - P, P)_a^b.$$

Here $U(t)$ is the fundamental operator of θ and P is its canonical boundary value operator. Note that the main coefficient of $\theta_\square$ is identically zero and the main coefficient of the inverse system $\theta_\square^\times$ is equal to $-U(t)B(t)C(t)U(t)$. Put $X = X_1 \oplus X_2$, where $X_1 = \operatorname{Ker}\Pi$ and $X_2 = \operatorname{Im}\Pi$. The fact that Π is a decomposing projection for θ is precisely equal to the statement that the conditions of

Theorem 5.1 are fulfilled for the system $\theta_{\square}$ (cf. the paragraph preceding this proof and Lemma 3.1). It follows that $\theta_{\square} \simeq \theta_1\theta_2$, and one checks that $\theta_1 = \ell_{\Pi}(\theta)$ and $\theta_2 = r_{\Pi}(\theta)$. Since $\theta \simeq \theta_{\square}$ one gets the cascade decomposition (4.3). Furthermore, the fact that the similarity between θ and $\theta_{\square}$ is given by $U(t)^{-1}$ implies that the similarity in (4.3) has the desired form. □

II.6 Proof of Theorem II.4.2

We first prove two auxiliary results concerning decomposing projections. The first shows that decomposing projections are preserved under similarity.

6.1. PROPOSITION. *Let* $S(t) : X_1 \to X_2$, $a \le t \le b$, *be a similarity transformation between the boundary value systems* θ_1 *and* θ_2, *and let* Π_1 *be a decomposing projection for* θ_1. *Then* $\Pi_2 := S(a)\Pi_1 S(a)^{-1}$ *is a decomposing projection for* θ_2 *and*

(6.1) $$\ell_{\Pi_1}(\theta_1) \simeq \ell_{\Pi_2}(\theta_2), \qquad r_{\Pi_1}(\theta_1) \simeq r_{\Pi_2}(\theta_2).$$

PROOF. Let θ_1 and θ_2 be as in (1.6). Since $S(t) : X_1 \to X_2$, $a \le t \le b$, is a similarity transformation between θ_1 and θ_2 we know that the formulas (1.7) - (1.9) hold true. Further, the fundamental operators $U_1(t)$ and $U_2(t)$ and the canonical boundary value operators P_1 and P_2 of θ_1 and θ_2, respectively, are related in the following way:

(6.2) $$U_2(t) = S(t)U_1(t)S(a)^{-1}, \qquad P_2 = S(a)P_1S(a)^{-1}.$$

Now, assume that Π_1 is a decomposing projection for θ_1, and define $\Pi_2 := S(a)\Pi_1(a)^{-1}$. Obviously, Π_2 is a projection. The operator $S(a) : X_1 \to X_2$ has the following block matrix representation:

$$S(a) = \begin{pmatrix} E & 0 \\ 0 & F \end{pmatrix} : \operatorname{Ker}\Pi_1 \oplus \operatorname{Im}\Pi_1 \to \operatorname{Ker}\Pi_2 \oplus \operatorname{Im}\Pi_2.$$

Here $E : \operatorname{Ker}\Pi_1 \to \operatorname{Ker}\Pi_2$ and $F : \operatorname{Im}\Pi_1 \to \operatorname{Im}\Pi_2$ are invertible operators. For $\nu = 1,2$ write P_ν as a 2×2 operator matrix relative to the decomposition $X_\nu = \operatorname{Ker}\Pi_\nu \oplus \operatorname{Im}\Pi_\nu$:

$$P = \begin{pmatrix} P_{11}^{(\nu)} & P_{12}^{(\nu)} \\ P_{21}^{(\nu)} & P_{22}^{(\nu)} \end{pmatrix}.$$

From the second identity in (6.2) it follows that

$$(6.3)\qquad P^{(2)}_{11} = E P^{(1)}_{11} E^{-1}, \qquad P^{(2)}_{12} = E P^{(1)}_{12} F^{-1}$$

$$(6.4)\qquad P^{(2)}_{21} = F P^{(1)}_{21} E^{-1}, \qquad P^{(2)}_{22} = F P^{(1)}_{22} F^{-1}.$$

Since Π_1 is a decomposing projection for θ_1 we know that $P^{(1)}_{21} = 0$. Hence $P^{(2)}_{21} = 0$.

From formulas (1.7) - (1.9) and the first identity in (6.2) it follows that

$$(6.5)\qquad U_2(t)^{-1}B_2(t) = S(a)U_1(t)^{-1}B_1(t),$$

$$(6.6)\qquad C_2(t)U_2(t) = C_1(t)U_1(t)S(a)^{-1}.$$

Next, consider the RDE:

$$(6.7)\qquad \begin{cases} \dot{R}_\nu(t) = (I - \Pi_\nu + R_\nu(t)\Pi_\nu)(-U_\nu(t)^{-1}B_\nu(t)C_\nu(t)U_\nu(t))(R_\nu(t)\Pi_\nu - \Pi_\nu), \\ \hfill a \le t \le b, \\ R\ (a) = 0, \qquad P^{(\nu)}_{11}R_\nu(b)(I - P^{(\nu)}_{22}) = P^{(\nu)}_{12}. \end{cases}$$

Here $\nu = 1,2$. We know that for $\nu = 1$ the RDE (6.7) has a solution $R_1(t) : \operatorname{Im}\Pi_1 \to \operatorname{Ker}\Pi_1$. Put

$$R_2(t) = ER_1(t)F^{-1} : \operatorname{Im}\Pi_2 \to \operatorname{Ker}\Pi_2, \qquad a \le t \le b.$$

From (6.3) - (6.6) it follows that $R_2(\cdot)$ is the solution of the RDE (6.7) for the case $\nu = 2$. We have now proved that Π_2 is a decomposing projection for θ_2. It remains to prove (6.1). Using (6.5), (6.6) and the first identity in (6.3) it is straightforward to check that $E : \operatorname{Ker}\Pi_1 \to \operatorname{Ker}\Pi_2$ provides a similarity transformation which transforms $\ell_{\Pi_1}(\theta_1)$ into $\ell_{\Pi_2}(\theta_2)$. In the same way, using (6.5), (6.6) and the second identity in (6.4) one checks that $F : \operatorname{Im}\Pi_1 \to \operatorname{Im}\Pi_2$ provides a similarity between $r_{\Pi_1}(\theta_1)$ and $r_{\Pi_2}(\theta_2)$. □

Next we show that the factorization projections introduced in [5], §III.1 are decomposing projections. Let $\theta = (A(t),B(t),C(t);N_1,N_2)_a^b$ be a boundary value system. Put $A^\times(t) = A(t) - B(t)C(t)$, $a \le t \le b$. Following [5] we call a projection Π of the state space X of θ a *factorization projection* for θ if Π commutes with the boundary value operators N_1 and N_2 and

$$(6.8)\qquad A(t)\operatorname{Ker}\Pi \subset \operatorname{Ker}\Pi, \qquad A^\times(t)\operatorname{Im}\Pi \subset \operatorname{Im}\Pi$$

for almost all $t \in [a,b]$. Let Π be such a projection. Write $A(t)$, $B(t)$, $C(t)$, N_1 and N_2 as block matrices relative to the decomposition $X = \operatorname{Ker}\Pi \oplus \operatorname{Im}\Pi$:

$$(6.9) \qquad A(t) = \begin{pmatrix} A_{11}(t) & A_{12}(t) \\ 0 & A_{22}(t) \end{pmatrix}, \qquad B(t) = \begin{pmatrix} B_1(t) \\ B_2(t) \end{pmatrix},$$

$$(6.10) \qquad C(t) = \begin{pmatrix} C_1(t) & C_2(t) \end{pmatrix},$$

$$(6.11) \qquad N_j = \begin{pmatrix} N_j^{(1)} & 0 \\ 0 & N_j^{(2)} \end{pmatrix}, \qquad j = 1,2.$$

Note that the triangular form of $A(t)$ follows from the first inclusion in (6.8). The diagonal forms (6.11) for N_1 and N_2 are a consequence of the fact that Π commutes with N_1 and N_2. Using the partitionings (6.9) - (6.11), we introduce the following systems:

$$pr_{I-\Pi}(\theta) = (A_{11}(t), B_1(t), C_1(t); N_1^{(1)}, N_2^{(1)})_a^b,$$

$$pr_{\Pi}(\theta) = (A_{22}(t), B_2(t), C_2(t); N_1^{(2)}, N_2^{(2)})_a^b.$$

The symbol pr stands for projection. From [5], Theorem III.1.1 it follows that

$$(6.12) \qquad \theta = pr_{I-\Pi}(\theta) pr_{\Pi}(\theta).$$

6.2 PROPOSITION. *Let Π be a factorization projection for θ. Then Π is a decomposing projection for θ and*

$$(6.13) \qquad \ell_{\Pi}(\theta) \simeq pr_{I-\Pi}(\theta), \qquad r_{\Pi}(\theta) \simeq pr_{\Pi}(\theta).$$

PROOF. Note that the second inclusion in (6.8) implies that $A_{12}(t) = B_1(t)C_2(t)$ a.e. on $a \le t \le b$. Let $U(t)$ be the fundamental operator of θ. From the first identity in (6.9) and the remark just made it follows that

$$U(t) = \begin{pmatrix} U_{11}(t) & V(t) \\ 0 & U_{22}(t) \end{pmatrix}, \qquad a \le t \le b,$$

where $U_{11}(t)$ and $U_{22}(t)$ are the fundamental operators of $pr_{I-\Pi}(\theta)$ and $pr_{\Pi}(\theta)$, resepectively, and $V(t) = U_{11}(t)H(t)$ with

$$H(t) = \int_a^t U_{11}(s)^{-1} B_1(s) C_2(s) U_{22}(s) ds.$$

By a direct computation one shows that $H(\cdot)$ is the solution of the RDE:

$$(6.14)\qquad \begin{cases} \dot{H}(t) = (I - \Pi + H(t)\Pi)(-U(t)^{-1}B(t)C(t)U(t))(H(t)\Pi - \Pi), & a \le t \le b, \\ H(a) = 0. \end{cases}$$

From the proof of Theorem II.1.1 in [5] we may conclude that the canonical boundary value operator P of θ is given by

$$(6.15)\qquad P = \begin{pmatrix} P_{11} & P_{12} \\ 0 & P_{22} \end{pmatrix},$$

with

$$(6.16)\qquad P_{12} = P_{11}H(b)(I - P_{22}).$$

Thus P has the desired triangular form and the solution of (6.14) satisfies the additional boundary condition (6.16). Hence Π is a decomposing projection for θ. Note that

$$(I - \Pi + H(t)\Pi)U(t)^{-1}B(t) = U_{11}(t)^{-1}B_1(t),$$

$$C(t)U(t)(\Pi - H(t)\Pi) = C_2(t)U_{22}(t).$$

It follows that the left and right factor corresponding to Π and θ are given by

$$\ell_\Pi(\theta) = (0, U_{11}(t)^{-1}B_1(t), C_1(t)U_{11}(t); I - P_{11}, P_{11})_a^b,$$

$$r_\Pi(\theta) = (0, U_{22}(t)^{-1}B_2(t), C_2(t)U_{22}(t); I - P_{22}, P_{22})_a^b.$$

Thus $\ell_\Pi(\theta)$ is similar to $pr_{I-\Pi}(\theta)$ with similarity $U_{11}(t)$ and $r_\Pi(\theta)$ is similar to $pr_{I-\Pi}(\theta)$ with similarity $U_{22}(t)$.

PROOF of Theorem 4.2. Assume $\theta \simeq \theta_1\theta_2$. Write $\theta_0 = \theta_1\theta_2$. From Theorem III.1.1 in [5] we know that θ_0 has a factorization projection Π_0 such that

$$(6.17)\qquad pr_{I-\Pi_0}(\theta_0) = \theta_1, \qquad pr_{\Pi_0}(\theta_0) = \theta_2.$$

From Proposition 6.2 we know that Π_0 is a decomposing projection for θ_0. Since $\theta \simeq \theta_0$ we can apply Proposition 6.1 to show that θ has a decomposing projection Π such that

$$\ell_\Pi(\theta) \simeq \ell_{\Pi_0}(\theta_0), \qquad r_\Pi(\theta) \simeq r_{\Pi_0}(\theta_0).$$

Now use (6.13) for θ_0 and Π_0 and (6.17). This yields (4.4). □

II.7 Proof of Theorem II.4.3

Let θ be an irreducible system. According to Theorem 4.1 the right hand side of (4.5) is a cascade decomposition of θ for each decomposing projection Π for θ. From Theorem 4.2 it follows that each cascade decomposition of θ is equivalent to a cascade decomposition of the form $\ell_\Pi(\theta)r_\Pi(\theta)$. Thus the map (4.5) is well-defined and onto.

To prove that the map (4.5) is one-one we use the irreducibility of θ. Let Π_1 and Π_2 be two decomposing projections for θ, and let $R_1(t)$: $\operatorname{Im}\Pi_1 \to \operatorname{Ker}\Pi_1$ and $R_2(t) : \operatorname{Im}\Pi_2 \to \operatorname{Ker}\Pi_2$ be the solutions of the corresponding RDE's. Assume

$$(7.1)\qquad \ell_{\Pi_1}(\theta) \simeq \ell_{\Pi_2}(\theta), \qquad r_{\Pi_1}(\theta) \simeq r_{\Pi_2}(\theta).$$

We have to show that $\Pi_1 = \Pi_2$. Since the left and right factors of θ have zero main coefficients, the similarities in (7.1) do not depend on time. Let $S_1 : \operatorname{Ker}\Pi_1 \to \operatorname{Ker}\Pi_2$ give the first similarity in (7.1) and $S_2 : \operatorname{Im}\Pi_1 \to \operatorname{Im}\Pi_2$ the second. Introduce

$$S = \begin{pmatrix} S_1 & 0 \\ 0 & S_2 \end{pmatrix} : \operatorname{Ker}\Pi_1 \oplus \operatorname{Im}\Pi_1 \to \operatorname{Ker}\Pi_2 \oplus \operatorname{Im}\Pi_2.$$

Obviously, S is a similarity between $\ell_{\Pi_1}(\theta)r_{\Pi_1}(\theta)$ and $\ell_{\Pi_2}(\theta)r_{\Pi_2}(\theta)$ (cf. [5], Proposition II.1.3). But then

$$S(t) = U(t)\begin{pmatrix} I_{\operatorname{Ker}\Pi_2} & -R_2(t) \\ 0 & I_{\operatorname{Im}\Pi_2} \end{pmatrix} S \begin{pmatrix} I_{\operatorname{Ker}\Pi_1} & R_1(t) \\ 0 & I_{\operatorname{Im}\Pi_1} \end{pmatrix} U(t)^{-1}$$

is a similarity of θ with itself. Since θ is irreducible, a self-similarity of θ is trivial, i.e., $S(t) = I$ for $a \le t \le b$ (see [7], Theorem 2.2; also [6], Proposition II.3.2 and II.3.3). In particular, $S = S(a) = I$, which implies that $\Pi_1 = \Pi_2$. Thus the map (4.5) is one-one.

Let θ be irreducible, and let Π be a decomposing projection for θ. It remains to prove that $\ell_\Pi(\theta)$ and $r_\Pi(\theta)$ are also irreducible. We know (see [7], Theorem 2.4) that $\ell_\Pi(\theta)$ is similar to a system θ_1 which is a dilation of an irreducible system θ_{10}. Similarly, $r_\Pi(\theta)$ is similar to a system θ_2 which is a dilation of an irreducible system θ_{20}. Clearly, $\ell_\Pi(\theta)r_\Pi(\theta) \simeq \theta_1\theta_2$ (cf. [5],

Proposition II.1.3). Next, apply Lemma 7.1 below. It follows that $\theta_1\theta_2$ is a dilation of $\theta_{10}\theta_{20}$, and hence θ is similar to a dilation of $\theta_{10}\theta_{20}$. But θ is irreducible, and so the latter dilation cannot be proper. It follows that θ_1 is not a proper dilation of θ_{10} and θ_2 is not a proper dilation of θ_{20}. Thus θ_1 and θ_2 are irreducible. Since irreducibility is preserved under similarity, we conclude that $\ell_\Pi(\theta)$ and $r_\Pi(\theta)$ are irreducible. □

7.1 LEMMA. *Let* $\nu = 1,2$ *let the system* θ_ν *be a dilation of* $\theta_{\nu 0}$. *Then the cascade connection* $\theta_1\theta_2$ *is a dilation of* $\theta_{10}\theta_{20}$.

PROOF. For $\nu = 1,2$ let

$$\theta_\nu = (A_\nu(t),B_\nu(t),C_\nu(t);N_1^{(\nu)},N_2^{(\nu)})_a^b,$$

$$\theta_{\nu 0} = (A_{\nu 0}(t),B_{\nu 0}(t),C_{\nu 0}(t);N_{10}^{(\nu)},N_{20}^{(\nu)})_a^b .$$

Since θ_ν is a dilation of $\theta_{\nu 0}$, the state space X_ν of θ_ν admits a decomposition $X_\nu = X_{\nu 1} \oplus X_{\nu 0} \oplus X_{\nu 2}$ such that relative to this decomposition

$$A_\nu(t) = \begin{pmatrix} * & * & * \\ 0 & A_{\nu 0}(t) & * \\ 0 & 0 & * \end{pmatrix}, \qquad B_\nu(t) = \begin{pmatrix} * \\ B_{\nu 0}(t) \\ 0 \end{pmatrix},$$

$$C_\nu(t) = \begin{pmatrix} 0 & C_{\nu 0}(t) & * \end{pmatrix},$$

$$N_j^\nu = E^{(\nu)} \begin{pmatrix} * & * & * \\ 0 & N_{j0}^{(\nu)} & * \\ 0 & 0 & * \end{pmatrix}, \qquad j = 1,2.$$

Here $E^{(\nu)}$ is an invertible operator on the space X_ν. Put $\hat{X} = X_1 \oplus X_2$, $\hat{X}_1 = X_{11} \oplus X_{21}$, $\hat{X}_0 = X_{10} \oplus X_{20}$ and $\hat{X}_2 = X_{12} \oplus X_{22}$. Then $\hat{X}$ is the state space of $\theta_1\theta_2$ and $\hat{X}_0$ is the state space of $\theta_{10}\theta_{20}$. Furthermore

$$\hat{X} = \hat{X}_1 \oplus \hat{X}_0 \oplus \hat{X}_2. \tag{7.2}$$

Put

$$\theta_1\theta_2 = (A(t),B(t),C(t);N_1,N_2)_a^b ,$$

$$\theta_{10}\theta_{20} = (A_0(t),B_0(t),C_0(t);N_{10},N_{20})_a^b .$$

Recall that $A(t)$, $B(t)$, $C(t)$, N_1 and N_2 are given by (2.1) - (2.3). Consider the partitioning of $A(t)$, $B(t)$, $C(t)$, N_1 and N_2 relative to the decomposition

$$\hat{X} = X_{11} \oplus X_{21} \oplus X_{10} \oplus X_{20} \oplus X_{12} \oplus X_{22}.$$

One finds that

$$A(t) = \begin{pmatrix} * & 0 & * & * & * & * \\ 0 & * & 0 & * & 0 & * \\ 0 & 0 & A_{10}(t) & B_{10}(t)C_{20}(t) & * & * \\ 0 & 0 & 0 & A_{20}(t) & 0 & * \\ 0 & 0 & 0 & 0 & * & 0 \\ 0 & 0 & 0 & 0 & 0 & * \end{pmatrix},$$

$$B(t) = \begin{pmatrix} * \\ * \\ B_{10}(t) \\ B_{20}(t) \\ 0 \\ 0 \end{pmatrix}, \qquad C(t) = \begin{pmatrix} 0 & 0 & C_{10}(t) & C_{20}(t) & * & * \end{pmatrix}.$$

$$N_j = E \begin{pmatrix} * & 0 & * & 0 & * & 0 \\ 0 & * & 0 & * & 0 & * \\ 0 & 0 & N_{j0}^{(1)} & 0 & * & 0 \\ 0 & 0 & 0 & N_{j0}^{(2)} & 0 & * \\ 0 & 0 & 0 & 0 & * & 0 \\ 0 & 0 & 0 & 0 & 0 & * \end{pmatrix}, \qquad j = 1,2,$$

where

$$E = \begin{pmatrix} E^{(1)} & 0 \\ 0 & E^{(2)} \end{pmatrix}.$$

It follows that relative the decomposition (7.2)

$$A(t) = \begin{pmatrix} * & * & * \\ 0 & A_0(t) & * \\ 0 & 0 & * \end{pmatrix}, \qquad B(t) = \begin{pmatrix} * \\ B_0(t) \\ 0 \end{pmatrix},$$

$$C(t) = \begin{pmatrix} 0 & C_0(t) & * \end{pmatrix},$$

$$N_j = E \begin{pmatrix} * & * & * \\ 0 & N_{j0} & * \\ 0 & 0 & * \end{pmatrix}.$$

This shows that $\theta_1\theta_2$ is a dilation of $\theta_{10}\theta_{20}$. □

II.8 Decomposing projections for inverse systems

The aim of this section is to prove the following theorem.

8.1 THEOREM. *Let the system θ and its inverse system $\theta^\times$ have well-posed boundary conditions. Then Π is a decomposing projection for θ if and only if $I-\Pi$ is a decomposing projection for $\theta^\times$, and in that case*

$$(8.1) \qquad (\ell_\Pi(\theta))^\times \simeq r_{I-\Pi}(\theta^\times), \qquad (r_\Pi(\theta))^\times \simeq \ell_{I-\Pi}(\theta^\times).$$

PROOF. Let $\theta = (A(t),B(t),C(t);N_1,N_2)_a^b$. Put $\Omega(t) = U(t)^{-1}U^\times(t)$, $a \le t \le b$, where $U(t)$ and $U^\times(t)$ are the fundamental operators of θ and $\theta^\times$, respectively. Furthermore, let P and $P^\times$ be the canonical boundary value operators of θ and $\theta^\times$, respectively. It is known (see [5], §§ II.3 and II.4) that

$$(8.2) \qquad P^\times = (I-P+P\Omega(b))^{-1}P\Omega(b).$$

Now assume that Π is a decomposing projection for θ. Write P, $P^\times$ and $\Omega(t)$ as 2×2 operator matrices relative to the decomposition $X = \operatorname{Ker}\Pi \oplus \operatorname{Im}\Pi$:

$$P = \begin{pmatrix} P_{11} & P_{12} \\ P_{21} & P_{22} \end{pmatrix}, \qquad P^\times = \begin{pmatrix} P_{11}^\times & P_{12}^\times \\ P_{21}^\times & P_{22}^\times \end{pmatrix},$$

$$\Omega(t) = \begin{pmatrix} \Omega_{11}(t) & \Omega_{12}(t) \\ \Omega_{21}(t) & \Omega_{22}(t) \end{pmatrix}, \qquad a \le t \le b.$$

The fact that Π is a decomposing projection for θ is equivalent to the following statements:

(8.3) $$P_{12} = 0,$$

(8.4) $$\det \Omega_{22}(t) \neq 0, \qquad a \le t \le b,$$

(8.5) $$P_{12} = -P_{11}\Omega_{12}(b)\Omega_{22}(b)^{-1}(I - P_{22}).$$

For $a \le t \le b$ set

(8.6) $$R(t) = -\Omega_{12}(t)\Omega_{22}(t)^{-1},$$

(8.7) $$\Delta(t) = \Omega_{11}(t) - \Omega_{12}(t)\Omega_{22}(t)^{-1}\Omega_{12}(t).$$

Using (8.3) and (8.5) one finds that

(8.8) $$\begin{pmatrix} I & P_{11}R(b) \\ 0 & I \end{pmatrix}(I - P + P\Omega(b)) = \begin{pmatrix} I - P_{11} + P_{11}\Delta(b) & 0 \\ P_{22}\Omega_{21}(b) & I - P_{22} + P_{22}\Omega_{22}(b) \end{pmatrix}.$$

It follows that $I - P_{11} + P_{11}\Delta(b)$ and $I - P_{22} + P_{22}\Omega_{22}(b)$ are invertible. Denote the right hand side of (8.8) by Σ. Then we see from (8.2), (8.3) and (8.5) that

$$P^{\times} = \Sigma^{-1} \cdot \begin{pmatrix} P_{11}\Delta(b) & 0 \\ P_{22}\Omega_{21}(b) & P_{22}\Omega_{22}(b) \end{pmatrix},$$

and hence

$$P^{\times}_{11} = (I - P_{11} + P_{11}\Delta(b))^{-1}P_{11}\Delta(b), \qquad P^{\times}_{12} = 0,$$

$$P^{\times}_{21} = (I - P_{22} + P_{22}\Omega_{22}(b))^{-1}P_{22}\Omega_{21}(b)(I - P_{11} + P_{11}\Delta(b))^{-1}(I - P_{11}),$$

$$P^{\times}_{22} = (I - P_{22} + P_{22}\Omega_{22}(b))^{-1}P_{22}\Omega_{22}(b).$$

We prove now that $I - \Pi$ is a decomposing projection for $\theta^{\times}$. From $P^{\times}_{12} = 0$, it is clear that $(I - \Pi)P^{\times}(I - \Pi) = (I - \Pi)P^{\times}$, and hence for $I - \Pi$ the condition (i) in the definition of a decomposing projection is satisfied. Next, observe that

$$\det(\Pi + U^{\times}(t)^{-1}U(t)(I - \Pi)) = \det(U^{\times}(t)^{-1}U(t))\det(I - \Pi + U(t)^{-1}U^{\times}(t)\Pi).$$

Since Π is a decomposing projection for θ, the last determinant in this identity is $\neq 0$ for each $a \le t \le b$. But then

$$\det(\Pi + U^{\times}(t)^{-1}U(t)(I-\Pi)) \neq 0, \qquad a \le t \le b,$$

which proves the second condition for a decomposing projection. To check the third condition, note that

$$P^{\times}\Pi\{\Pi + U^{\times}(b)^{-1}U(b)(I-\Pi)\}^{-1}(I-\Pi)(I-P^{\times})$$

$$= P^{\times}\Pi\{I-\Pi+\Omega(b)\Pi\}^{-1}\Omega(b)(I-\Pi)(I-P^{\times})$$

$$= \begin{pmatrix} 0 & 0 \\ 0 & P_{22}^{\times} \end{pmatrix} \begin{pmatrix} \Delta(b) & 0 \\ \Omega_{22}(b)^{-1}\Omega_{21}(b) & I \end{pmatrix} \begin{pmatrix} I-P_{11}^{\times} & 0 \\ 0 & 0 \end{pmatrix}$$

$$= \begin{pmatrix} 0 & 0 \\ P_{22}^{\times}\Omega_{22}(b)^{-1}\Omega_{21}(b)(I-P_{11}^{\times}) & 0 \end{pmatrix}.$$

Now use the formulas for $P_{11}^{\times}$, $P_{21}^{\times}$ and $P_{22}^{\times}$ derived above. One sees that

$$P_{21}^{\times} = P_{22}^{\times}\Omega_{22}(b)^{-1}\Omega_{21}(b)(I-P_{11}^{\times}),$$

and hence for $I-\Pi$ condition (iii) also holds true.

Write $U(t)^{-1}B(t)$, $C(t)U(t)$, $U^{\times}(t)^{-1}B(t)$, $C(t)U^{\times}(t)$ as block matrices relative to the decomposition $X = \operatorname{Ker}\Pi \oplus \operatorname{Im}\Pi$:

$$U(t)^{-1}B(t) = \begin{pmatrix} B_1(t) \\ B_2(t) \end{pmatrix}, \qquad U^{\times}(t)^{-1}B(t) = \begin{pmatrix} B_1^{\times}(t) \\ B_2^{\times}(t) \end{pmatrix},$$

$$C(t)U(t) = \begin{pmatrix} C_1(t) & C_2(t) \end{pmatrix}, \qquad C(t)U^{\times}(t) = \begin{pmatrix} C_1^{\times}(t) & C_2^{\times}(t) \end{pmatrix}.$$

Then

$$\ell_{\Pi}(\theta) = (0, B_1(t) + R(t)B_2(t), C_1(t); I-P_{11}, P_{11})_a^b,$$

$$r_{\Pi}(\theta) = (0, B_2(t), -C_1(t)R(t) + C_2(t); I-P_{22}, P_{22})_a^b,$$

$$\ell_{I-\Pi}(\theta^{\times}) = (0, B_2^{\times}(t) + R^{\times}(t)B_1^{\times}(t), -C_2^{\times}(t); I-P_{22}^{\times}, P_{22}^{\times})_a^b,$$

$$r_{I-\Pi}(\theta^{\times}) = (0, B_1^{\times}(t), C_2^{\times}(t)R^{\times}(t) - C_1^{\times}(t); I-P_{11}^{\times}, P_{11}^{\times})_a^b.$$

Here $R(t)$ is given by (8.6) and

$$R^{\times}(t) = \Pi\{\Pi + U^{\times}(t)^{-1}U(t)(I-\Pi)\}^{-1}(I-\Pi) : \operatorname{Ker}\Pi \to \operatorname{Im}\Pi.$$

A simple computation shows that $R^\times(t) = \Omega_{22}(t)^{-1}\Omega_{21}(t)$ for $a \le t \le b$, and hence

$$\begin{pmatrix} I & R(t) \\ 0 & I \end{pmatrix} \Omega(t) \begin{pmatrix} I & 0 \\ -R^\times(t) & I \end{pmatrix} = \begin{pmatrix} \Delta(t) & 0 \\ 0 & \Omega_{22}(t) \end{pmatrix}$$

for $a \le t \le b$. The latter identity can be rewritten in the form:

$$\begin{pmatrix} I & R(t) \\ 0 & I \end{pmatrix} U(t)^{-1} = \begin{pmatrix} \Delta(t) & 0 \\ 0 & \Omega_{22}(t) \end{pmatrix} \begin{pmatrix} I & 0 \\ R^\times(t) & I \end{pmatrix} U^\times(t)^{-1} . \tag{8.9}$$

It follows that for $a \le t \le b$

$$B_1(t) + R(t)B_2(t) = \Delta(t)B_1^\times(t),$$

$$B_2(t) = \Omega_{22}(t)(B_2^\times(t) + R^\times(t)B_1^\times(t)),$$

$$C_1(t) = (C_1^\times(t) - C_2^\times(t)R^\times(t))\Delta(t)^{-1},$$

$$-C_1(t)R(t) + C_2(t) = C_2^\times(t)\Omega_{22}(t)^{-1}.$$

Next we use (cf. (3.5)) that

$$\dot\Omega(t) = (-U(t)^{-1}B(t)C(t)U(t))\Omega(t), \qquad a \le t \le b.$$

From this identity one may conclude that

$$\dot\Delta(t) = -(B_1(t) + R(t)B_2(t))C_1(t)\Delta(t), \qquad a \le t \le b,$$

$$\dot\Omega_{22}(t) = -B_2(t)(C_2(t) - C_1(t)R(t))\Omega_{22}(t), \qquad a \le t \le b.$$

Also, note that $\Delta(a) = I$ and $\Omega_{22}(a) = I$. It is now clear that

$$(\ell_\Pi(\theta))^\times = (\dot\Delta(t)\Delta(t)^{-1}, B_1(t) + R(t)B_2(t), -C_1(t); I - P_{11}, P_{11})_a^b ,$$

$$(r_\Pi(\theta))^\times = (\dot\Omega_{22}(t)\Omega_{22}(t)^{-1}, B_2(t), C_1(t)R(t) - C_2(t); I - P_{22}, P_{22})_a^b .$$

With $E_1 = (I - P_{11} + P_{11}\Delta(b))^{-1}$ and $E_2 = (I - P_{22} + P_{22}\Omega_{22}(b))^{-1}$ we have

$$E_1(I - P_{11}) = I - P_{11}^\times, \qquad E_1P_{11}\Delta(b) = P_{11}^\times,$$

$$E_2(I - P_{22}) = I - P_{22}^\times, \qquad E_2P_{22}\Omega_{22}(b) = P_{22}^\times.$$

It follows that $\Delta(t)^{-1}$, $a \le t \le b$, provides a similarity transformation between $(\ell_{\Pi}(\theta))^{\times}$ and $r_{I-\Pi}(\theta^{\times})$ and $\Omega_{22}(t)^{-1}$, $a \le t \le b$, provides a similarity transformation between $(r_{\Pi}(\theta))^{\times}$ and $\ell_{I-\Pi}(\theta^{\times})$. This proves formula (8.1) and the "only if part" of the theorem. To prove the "if part" one applies the foregoing to $I-\Pi$ and $\theta^{\times}$. □

III. PROOFS OF THE MAIN THEOREMS

In this chapter we prove the results stated in Chapter I. The proofs are based on the cascade decomposition theory developed in the previous chapter. The last section is devoted to causal/anticausal decompositions of systems.

III.1 A factorization lemma

1.1. LEMMA. *Let* Π *be a decomposing projection for the boundary value system* $\theta = (A(t),B(t),C(t);N_1,N_2)_a^b$, *and let* T, T_1 *and* T_2 *be the input/output maps of* θ, $\ell_{\Pi}(\theta)$ *and* $r_{\Pi}(\theta)$, *respectively. Then* $T = T_1T_2$ *and the kernels of* T_1 *and* T_2 *are given by*

$$(1.1)\qquad k_1(t,s) = \begin{cases} C(t)U(t)(I-P)(I-\Pi(s))U(s)^{-1}B(s), & a \le s < t \le b,\\ -C(t)U(t)P(I-\Pi(s))U(s)^{-1}B(s), & a \le t < s \le b,\end{cases}$$

$$(1.2)\qquad k_2(t,s) = \begin{cases} C(t)U(t)\Pi(t)(I-P)U(s)^{-1}B(s), & a \le s < t \le b,\\ -C(t)U(t)\Pi(t)PU(s)^{-1}B(s), & a \le t < s \le b,\end{cases}$$

respectively. Here

$$(1.3)\qquad \Pi(t) = U(t)^{-1}U^{\times}(t)\Pi(I-\Pi+U(t)^{-1}U^{\times}(t)\Pi)^{-1}\Pi, \qquad a \le t \le b.$$

Furthermore, $U(t)$ *and* $U^{\times}(t)$ *denote the fundamental operators of* θ *and the inverse system* $\theta^{\times}$, *respectively, and* P *is the canonical boundary value operator of* θ.

PROOF. Since Π is a decomposing projection, the operator $\Pi(t)$ is well-defined. Note that

$$\begin{aligned}\Pi(t) &= \Pi\Pi(t) + (I-\Pi)\Pi(t)\\ &= \Pi + (I-\Pi)U(t)^{-1}U^{\times}(t)\Pi(I-\Pi+U(t)^{-1}U^{\times}(t)\Pi)^{-1}\Pi\\ &= \Pi - (I-\Pi)(I-\Pi+U(t)^{-1}U^{\times}(t)\Pi)^{-1}\Pi = \Pi - R(t)\Pi,\end{aligned}$$

with $R(t) = (I-\Pi)(I-\Pi+U(t)^{-1}U^{\times}(t)\Pi)^{-1}\Pi$. It follows that

$$(I-\Pi(s))U(s)^{-1}B(s) = (I-\Pi+R(s)\Pi)U(s)^{-1}B(s),$$

$$C(t)U(t)\Pi(t) = C(t)U(t)(\Pi-R(t)\Pi).$$

Note that $R(\cdot)$ is precisely the function $R(\cdot)$ appearing in the definitions of $\ell_{\Pi}(\theta)$ and $r_{\Pi}(\theta)$. The fact that Π is a decomposing projection for θ also implies that

$$(I-\Pi)P(I-\Pi) = P(I-\Pi), \qquad \Pi P\Pi = \Pi P.$$

Using these observations and formula (II.1.4) it is now easily checked that the kernels of T_1 and T_2 are given by (1.1) and (1.2), respectively. From formula (II.4.3) it follows that $T = T_1T_2$. □

III.2 Minimal factorization (2)

In this section the results presented in Section I.1 and I.2 are proved.

Let T be a (SK)-integral operator on $L_2([a,b],Y)$. A boundary value system θ is called a *realization* of T if the input/output map of θ is equal to T. A realization θ of T is said to be a *minimal realization* of T if among all realizations of T the state space dimension of θ is as small as possible. Of course, minimal realizations are irreducible systems; the converse is not true (see [10]). It may happen that a (SK)-integral operator has various non-similar minimal realizations (see [10], § 4).

Let us describe the connections between (minimal) realizations and (minimal) representations (as defined in § I.1). Given a representation $\Delta = (B(\cdot),C(\cdot);P)$ of T we define $\theta(\Delta)$ to be the system

(2.1) $$\theta(\Delta) = (0,B(t),C(t);I-P,P)_a^b .$$

Obviously, $\theta(\Delta)$ is a realization of T, the state space of $\theta(\Delta)$ is the internal space of Δ and the canonical boundary value operator of $\theta(\Delta)$ is equal to the internal operator of Δ. Conversely, given a realization $\theta = (A(t),B(t),C(t);N_1,N_2)_a^b$ of T we set

(2.2) $$\Delta(\theta) = (U(\cdot)^{-1}B(\cdot),C(\cdot)U(\cdot);P),$$

where $U(t)$ is the fundamental operator of θ and P is the canonical boundary value operator of θ. The triple $\Delta(\theta)$ is a representation of T. Note that

(2.3) $\theta(\Delta(\theta)) \simeq \theta, \quad \Delta(\theta(\Delta)) = \Delta.$

In the first part of (2.3) we may replace $\simeq$ by = whenever the main coefficient of θ is equal to zero (almost everywhere on [a,b]).

From (2.3) it follows that θ is a minimal realization of T if and only if $\Delta(\theta)$ is a minimal representation of T, and also Δ is a minimal representation of T if and only if $\theta(\Delta)$ is a minimal realization of T. In particular, we see that *the degree* $\delta(T)$ *of* T *is equal to the dimension of the state space of a minimal realization of* T.

PROOF of Lemma I.1.1. For $\nu = 1,2$ let θ_ν be a minimal realization of T_ν. Let X_ν denote the state space of θ_ν. Thus $\delta(T_\nu) = X_\nu$ ($\nu = 1,2$). The input/output map of $\theta_1\theta_2$ is equal to T_1T_2 ([5], Theorem II.1.1). Thus T_1T_2 is a (SK)-integral operator. Since $X_1 \oplus X_2$ is the state space of $\theta_1\theta_2$, we have

$$\delta(T_1T_2) \le \dim X_1 + \dim X_2 = \delta(T_1) + \delta(T_2),$$

which proves the lemma. □

We shall use the following lemma in the proof of Theorem I.2.1.

2.1. LEMMA. *Let* Π *be a decomposing projection of the boundary value system* θ, *and let* T, T_1 *and* T_2 *be the input/output maps of* θ, $\ell_\Pi(\theta)$ *and* $r_\Pi(\theta)$, *respectively. Then* θ *is a minimal realization of* T *if and only if* $\ell_\Pi(\theta)$ *and* $r_\Pi(\theta)$ *are minimal realizations of* T_1 *and* T_2, *respectively, and* $T = T_1T_2$ *is a minimal factorization.*

PROOF. Assume θ is a minimal realization of T. Formula (II.4.3) implies $T = T_1T_2$. Since $\operatorname{Ker}\Pi$ is the state space of $\ell_\Pi(\theta)$, we have $\delta(T_1) \le \dim \operatorname{Ker}\Pi$. Similarly, $\delta(T_2) \le \dim \operatorname{Im}\Pi$. Thus (cf. (I.1.5))

$$\delta(T) \le \delta(T_1) + \delta(T_2) \le \dim \operatorname{Ker}\Pi + \dim \operatorname{Im}\Pi = \delta(T).$$

Here the last equality is a consequence of the fact that θ is a minimal realization of T. It follows that (I.2.1) holds, $\delta(T_1) = \dim \operatorname{Ker}\Pi$ and $\delta(T_2) = \dim \operatorname{Im}\Pi$. Thus $T = T_1T_2$ is a minimal factorization and $\ell_\Pi(\theta)$ and $r_\Pi(\theta)$ are minimal realizations of T_1 and T_2, respectively.

To prove the converse, assume that $\ell_\Pi(\theta)$ and $r_\Pi(\theta)$ are minimal realizations of T_1 and T_2, respectively, and let $T = T_1T_2$ be a minimal factorization. Then

$$\delta(T) = \delta(T_1) + \delta(T_2) = \dim \operatorname{Ker} \Pi + \dim \operatorname{Im} \Pi.$$

So $\delta(T)$ is equal to the state space dimension of θ, which implies that θ is a minimal realization of T. □

PROOF of Theorem I.2.1. Let $\Delta = (B(\cdot),C(\cdot);P)$ be a minimal representation of T, and let $\Omega(t)$ be the fundamental operator of Δ. Assume Π is a decomposing projection for Δ. Consider the system $\theta(\Delta)$ defined by (2.1). The fundamental operator $U(t)$ of $\theta(\Delta)$ is identically equal to I and the fundamental operator $U^{\times}(t)$ of the inverse system $\theta(\Delta)^{\times}$ is precisely $\Omega(t)$. It follows that Π is a decomposing projection for $\theta(\Delta)$. Let T_1 and T_2 be the input/output maps corresponding to $\ell_{\Pi}(\theta(\Delta))$ and $r_{\Pi}(\theta(\Delta))$, respectively. Since $\theta(\Delta)$ is a minimal realization of T, Lemma 2.1 implies that $T = T_1T_2$ is a minimal factorization. From Lemma 1.1 it follows that the kernels of T_1 and T_2 are given by the formulas (I.2.3) and (I.2.4), respectively. This proves the first part of the theorem.

To prove the second part, let $T = T_1T_2$ be a minimal factorization. We want to construct a minimal representation of T that generates the factorization $T = T_1T_2$. For $\nu = 1,2$ let θ_ν be a minimal realization of T_ν. Let X_ν be the state space of θ_ν $(\nu = 1,2)$. Put $\theta = \theta_1\theta_2$. Obviously, the projection

$$\Pi = \begin{pmatrix} 0 & 0 \\ 0 & I \end{pmatrix} : X_1 \oplus X_2 \to X_1 \oplus X_2$$

is a factorization projection (see §II.6) for $\theta = \theta_1\theta_2$ and $pr_{I-\Pi}(\theta) = \theta_1$, $pr_{\Pi}(\theta) = \theta_2$. Next, apply Proposition II.6.2. We conclude that Π is a decomposing projection for θ and

$$(2.4) \qquad \ell_{\Pi}(\theta) \simeq \theta_1, \qquad r_{\Pi}(\theta) \simeq \theta_2.$$

It follows that $\ell_{\Pi}(\theta)$ is a minimal realization of T_1 and $r_{\Pi}(\theta)$ is a minimal realization of T_2. So, by Lemma 2.1, the system θ is a minimal realization of T. Now, put $\Delta_0 = \Delta(\theta)$ (see formula (2.2)). Then Δ_0 is a minimal representation of T. The fundamental operator of Δ_0 is $U(t)^{-1}U^{\times}(t)$, where $U(t)$ and $U^{\times}(t)$ are the fundamental operators of θ and $\theta^{\times}$, respectively (cf. formula (II.3.5)). It follows that Π is a decomposing projection for Δ_0. Using Lemma 1.1 one sees that $T = T_1T_2$ is the corresponding factorization. So Δ_0 is a minimal representation of T which generates the factorization $T = T_1T_2$. □

Let T_0, T_1, T_2, T_3 and T_4 be the integral operators on $L_2[0,1]$ appearing in formula (I.2.6). We shall now prove that T_0 does not have a minimal representation that generates the minimal factorizations

$$(2.5)\qquad T_0 = T_1T_2, \qquad\qquad T_0 = T_3T_4.$$

To do this consider the following two systems:

$$\theta = \left(\begin{bmatrix} 0 & 0 \\ 0 & 0 \end{bmatrix}, \begin{bmatrix} 1 \\ -1 \end{bmatrix}, (\,1 \;\; 1\,)\,;\; \begin{bmatrix} 1 & 0 \\ 0 & 0 \end{bmatrix}, \begin{bmatrix} 0 & 0 \\ 0 & 1 \end{bmatrix}\right)_0^1,$$

$$\theta^* = \left(\begin{bmatrix} 0 & 0 \\ 0 & 0 \end{bmatrix}, \begin{bmatrix} 1 \\ -1 \end{bmatrix}, (\,1 \;\; 1\,)\,;\; \begin{bmatrix} 0 & 1 \\ 0 & -2 \end{bmatrix}, \begin{bmatrix} 1 & -1 \\ 0 & 3 \end{bmatrix}\right)_0^1.$$

The canonical boundary value operators of θ and θ^* are given by

$$P_\theta = \begin{bmatrix} 0 & 0 \\ 0 & 1 \end{bmatrix}, \qquad P_{\theta^*} = \begin{bmatrix} 1 & -1 \\ 0 & 3 \end{bmatrix}.$$

Note that P_θ and P_{θ^*} are not similar. In fact, P_θ is a projection and P_{θ^*} is not. It follows that θ and θ^* are not similar.

From what we proved in § I.2 it is clear that θ and θ^* are minimal realizations of T_θ. Furthermore, the projection

$$\Pi = \begin{bmatrix} 0 & 0 \\ 0 & 1 \end{bmatrix},$$

which acts on $\mathbb{C}^2$, is a decomposing projection for both θ and θ^*. The left and right factors of θ and θ^* corresponding to Π are given by:

$$\ell_\Pi(\theta) = (0,(1+t)^{-1},1;1,0)_0^1,$$

$$r_\Pi(\theta) = (0,-1,(1+t)^{-1};0,1)_0^1,$$

$$\ell_\Pi(\theta^*) = (0,(1+t)^{-1},1;0,1)_0^1,$$

$$r_\Pi(\theta^*) = (0,-1,(1+t)^{-1};-2,3)_0^1.$$

Note that the input/output maps of $\ell_\Pi(\theta)$, $r_\Pi(\theta)$, $\ell_\Pi(\theta^*)$ and $r_\Pi(\theta^*)$ are the operators T_1, T_2, T_3 and T_4, respectively.

For $j = 1,2,3$ and 4 the operator T_j has the property that all its minimal realizations are similar. We shall give the proof of this statement for the operator T_4 (for the other operators one can use analogous arguments). Let θ_0 be a minimal realization of T_4. We want to show that $\theta_0 \simeq r_\Pi(\theta^*)$. Without loss of generality we may assume that $\theta_0 = (0,b(t),c(t);1-p,p)_0^1$. Since $T_{\theta_0} = T_4$ we have

$$(2.6) \qquad c(t)(1-p)b(s) = 2(1+t)^{-1}, \quad 0 \le s < t \le 1,$$

$$(2.7) \qquad -c(t)pb(s) = 3(1+t)^{-1}, \quad 0 \le t < s \le 1.$$

We can apply Theorems 3.2 and 3.3 in [8] to show that the kernel

$$h(t,s) = \begin{cases} 2(1+t)^{-1}, & 0 \le s < t \le 1, \\ -3(1+t)^{-1}, & 0 \le t < s \le 1, \end{cases}$$

is both lower unique and upper unique. It follows that

$$c(t)(1-p)b(s) = 2(1+t)^{-1}, \qquad c(t)pb(s) = -3(1+t)^{-1}$$

almost everywhere on the full square $[0,1]\times[0,1]$. In particular, $1-p \neq 0$ and

$$-3(1+t)^{-1} = c(t)pb(s) = \frac{p}{1-p}\, c(t)(1-p)b(s) = \frac{2p}{1-p}\,(1+t)^{-1}.$$

Hence $p = 3$ and $c(t)b(s) = -(1+t)^{-1}$ on $[0,1]\times[0,1]$. The latter identity implies (see [8], §1) that there exists a non-zero constant γ such that $b(t) = \gamma$ and $c(t) = -\gamma^{-1}(1+t)^{-1}$ a.e. on $0 \le t \le 1$. Thus $\theta_0 \simeq r_\Pi(\theta^*)$.

Now, let Δ be a minimal representation of T and assume that Π_1 and Π_2 are decomposing projections for Δ which yield the minimal factorizations in (2.5). Put $\Sigma = \theta(\Delta)$ (cf. (2.1)). Then (cf. the first paragraph of the proof of Theorem I.2.1) the system Σ is a minimal realization of T, the projections Π_1 and Π_2 are decomposing projections for Σ and

$$T_{\ell_{\Pi_1}(\Sigma)} = T_1, \quad T_{r_{\Pi_1}(\Sigma)} = T_2, \quad T_{\ell_{\Pi_2}(\Sigma)} = T_3, \quad T_{r_{\Pi_2}(\Sigma)} = T_4.$$

Since Σ is a minimal realization of T, we can use Lemma 2.1 to show that $\ell_{\Pi_1}(\Sigma)$ and $r_{\Pi_1}(\Sigma)$ are minimal realizations of T_1 and T_2, respectively. But then the result of the preceding paragraph implies that $\ell_{\Pi_1}(\Sigma) \simeq \ell_\Pi(\theta)$ and $r_{\Pi_1}(\Sigma) \simeq r_\Pi(\theta)$. In an analogous way one proves that $\ell_{\Pi_2}(\Sigma) \simeq \ell_\Pi(\theta^*)$ and $r_{\Pi_2}(\Sigma) \simeq r_\Pi(\theta^*)$. Next, apply Theorem II.4.1 and [5], Proposition II.1.3:

$$\theta \simeq \ell_\Pi(\theta)r_\Pi(\theta) \simeq \ell_{\Pi_1}(\Sigma)r_{\Pi_1}(\Sigma) \simeq \Sigma;$$

$$\theta^* \simeq \ell_\Pi(\theta^*)r_\Pi(\theta^*) \simeq \ell_{\Pi_2}(\Sigma)r_{\Pi_2}(\Sigma) \simeq \Sigma.$$

So $\theta \simeq \theta^*$, which is a contradiction. Hence T does not have a single minimal representation that yields both minimal factorizations in (2.5).

III.3 SB-minimal factorization (2)

In this section we prove the following theorem, which is the analogue of Theorem I.2.1 for SB-minimal factorization (cf. § I.5).

3.1. THEOREM. *Let* T *be a (SK)-integral operator on* $L_2([a,b],Y)$, *and let* $\Delta = (B(\cdot),C(\cdot),P)$ *be a SB-minimal representation of* T. *If* Π *is a decomposing projection for* Δ, *then a SB-minimal factorization* $T = T_1T_2$ *is obtained by taking* T_1 *and* T_2 *to be the (SK)-integral operators on* $L_2([a,b],Y)$ *of which the kernels are given by*

$$(3.1)\qquad k_1(t,s) = \begin{cases} C(t)(I-P)(I-\Pi(s))B(s), & a \le s < t \le b,\\ -C(t)P(I-\Pi(s))B(s), & a \le t < s \le b,\end{cases}$$

$$(3.2)\qquad k_2(t,s) = \begin{cases} C(t)\Pi(t)(I-P)B(s), & a \le s < t \le b,\\ -C(t)\Pi(t)PB(s), & a \le t < s \le b,\end{cases}$$

respectively. Here

$$(3.3)\qquad \Pi(t) = \Omega(t)\Pi(I-\Pi+\Omega(t)\Pi)^{-1}\Pi, \qquad a \le t \le b,$$

with $\Omega(t)$ *being the fundamental operator of* Δ.

Furthermore, if Δ *runs over all possible SB-minimal representations of* T, *then all SB-minimal factorizations of* T *are obtained in this way.*

We prove Theorem 3.1 by reducing the theorem to statements about systems with separable boundary conditions. Recall (see [8], § 7) that a boundary value system θ is said to have *separable boundary conditions* if and only if its canonical boundary value operator P_θ is a projection. In that case θ is called a *SB-system*. We shall need the following lemma.

3.2. LEMMA. *Let* $\theta \simeq \theta_1\theta_2$ *be a cascade decomposition. Then* θ *is a SB-system if and only if the compounding systems* θ_1 *and* θ_2 *are SB-systems.*

PROOF. Since the class of SB-systems is preserved under similarity we may assume that $\theta = \theta_1\theta_2$. But then we can use the proof of Theorem II.1.1 in [5] to show that the canonical boundary value operator P of θ has the following form ([5], formula (II.1.5)):

$$P = \begin{pmatrix} P_1 & P_1Q(I - P_2) \\ 0 & P_2 \end{pmatrix} . \tag{3.4}$$

Here P_1 and P_2 are the canonical boundary value operators of θ_1 and θ_2, respectively, and Q is an operator which we shall not specify. From (3.4) it is clear that P is a projection if and only if P_1 and P_2 are projections, which proves the lemma. □

Let T be a (SK)-integral operator on $L_2([a,b],Y)$. A system θ is called a *SB-realization* of T if θ is a SB-system and T is equal to the input/output map of θ. We say that θ is a *SB-minimal realization* if among all SB-realizations of T the state space dimension of θ is as small as possible. Since the class of SB-systems is closed under similarity and reduction, a SB-minimal realization is automatically irreducible. (Note that SB-minimality as defined in [8], §7 does not effect the external coefficient, and hence a SB-realization of a (SK)-integral operator is SB-minimal in the sense of [8] if and only if it is SB-minimal in the sense defined here.)

If Δ is a SB-representation of T (see § I.5), then the systemd $\theta(\Delta)$ (defined by (2.1)) is a SB-realization of T. Conversely, if θ is a SB-realization of T, then $\Delta(\theta)$ (see (2.2)) is a SB-representation of T. It follows that Δ is a SB-minimal representation of T if and only if $\theta(\Delta)$ is a SB-minimal realization of T, and, similarly, θ is a SB-minimal realization of T if and only if $\Delta(\theta)$ is a SB-minimal representation of T. In particular, *the SB-degree* $\varepsilon(T)$ *of* T *is equal to the dimension of the state space of a SB-minimal realization of* T. The next lemma concerns formula (I.5.2).

3.3. LEMMA. *If* $T = T_1T_2$ *is the product of two (SK)-integral operators on* $L_2([a,b],Y)$, *then*

$$\varepsilon(T) \leq \varepsilon(T_1) + \varepsilon(T_2). \tag{3.5}$$

PROOF. For $\nu = 1,2$ let θ_ν be a SB-minimal realization of T_ν. Let X_ν be the state space of θ_ν ($\nu = 1,2$). Then $X_1 \oplus X_2$ is the state space of $\theta_1\theta_2$ and $\varepsilon(T_\nu) = \dim X_\nu$ ($\nu = 1,2$). From Lemm 3.2 and [5], Theorem II.1.1 it follows that $\theta_1\theta_2$ is a SB-realization of T. Thus

$$\varepsilon(T) \le \dim X_1 + \dim X_2 = \varepsilon(T_1) + \varepsilon(T_2). \qquad \square$$

The next lemma is the analogue of Lemma 2.1 for SB-systems. Recall that $T = T_1T_2$ is a SB-minimal factorization if and only if

$$(3.6) \qquad \varepsilon(T) = \varepsilon(T_1) + \varepsilon(T_2).$$

3.4. LEMMA. *Let Π be a decomposing projection for the SB-system θ, and let T, T_1 and T_2 be the input/output maps of θ, $\ell_\Pi(\theta)$ and $r_\Pi(\theta)$, respectively. Then θ is a SB-minimal realization of T if and only if $\ell_\Pi(\theta)$ and $r_\Pi(\theta)$ are SB-minimal realizations of T_1 and T_2, respectively, and $T = T_1T_2$ is a SB-minimal factorization.*

PROOF. Assume θ is a SB-minimal realization of T. Formula (II.4.3) implies that $T = T_1T_2$. From Lemma 3.2 we may conclude that $\ell_\Pi(\theta)$ and $r_\Pi(\theta)$ are SB-systems. Thus $\varepsilon(T_1) \le \dim \operatorname{Ker} \Pi$ and $\varepsilon(T_2) \le \dim \operatorname{Im} \Pi$. According to (3.5) this implies that

$$\varepsilon(T) \le \varepsilon(T_1) + \varepsilon(T_2) \le \dim X,$$

where X is the state space of θ. Since θ is a SB-realization of T, we have $\varepsilon(T) = \dim X$. Hence (3.6) holds, $\varepsilon(T_1) = \dim \operatorname{Ker} \Pi$ and $\varepsilon(T_2) = \dim \operatorname{Im} \Pi$, which proves the "only if part" of the lemma.

To prove the "if part", assume that $\ell_\Pi(\theta)$ and $r_\Pi(\theta)$ are SB-minimal realizations of T_1 and T_2, respectively, and let $T = T_1T_2$ be a SB-minimal factorization. Then

$$\varepsilon(T) = \varepsilon(T_1) + \varepsilon(T_2) = \dim \operatorname{Ker} \Pi + \dim \operatorname{Im} \Pi.$$

So $\varepsilon(T)$ is equal to the state space dimension of θ. Hence θ is a SB-minimal realization of T. $\square$

PROOF of Theorem 3.1. Let $\Delta = (B(\cdot), C(\cdot); P)$ be a SB-minimal representation of T, and let Π be a decomposing projection for Δ. Then the system $\theta(\Delta)$ (defined by (2.1)) is a SB-minimal realization of T and Π is a decomposing projection for $\theta(\Delta)$. Let T_1 and T_2 be the input/output maps corresponding to $\ell_\Pi(\theta(\Delta))$ and $r_\Pi(\theta(\Delta))$, respectively. Lemma 3.4 implies that $T = T_1T_2$ is a SB-minimal realization. From Lemma 1.1 it follows that the kernels of T_1 and T_2 are given by the formulas (3.1) and (3.2), respectively. This proves the first part of the theorem.

To prove the second part, let $T = T_1T_2$ be a SB-minimal factorization. For $\nu = 1,2$ let θ_ν be a SB-minimal realization of T_ν. Put $\theta = \theta_1\theta_2$. Then θ is a SB-system (by Lemma 3.2). Now proceed as in the proof of Theorem I.2.1 (see §III.2). So there exists a decomposing projection Π for θ such that (2.4) holds true. It follows that $\ell_\Pi(\theta)$ and $r_\Pi(\theta)$ are SB-minimal realizations of T_1 and T_2, respectively. Thus θ is a SB-minimal realization of T (by Lemma 3.4). Now, put $\Delta_0 = \Delta(\theta)$ (see formula (2.2)). Then Δ_0 is a SB-minimal representation of T, the projection Π is a decomposing projection for Δ_0 and using Lemma 1.1 one sees that $T = T_1T_2$ is the corresponding factorization. Thus Δ_0 is a SB-minimal representation of T which generates the factorization $T = T_1T_2$. □

III.4 Proof of Theorem I.6.1

To prove Theorem I.6.1 we shall need Lemma 4.1 below. Let k be a semi-separable kernel on $[a,b] \times [a,b]$. Thus k is the kernel of an integral operator on $L_2([a,b],Y)$ and k admits a representation of the form (I.1.2). For $a < \gamma < b$ we let k_γ and k^γ denote the following restrictions:

$$k_\gamma(t,s) = k(t,s), \quad \gamma \le t \le b, \quad a \le s \le \gamma,$$

$$k^\gamma(t,s) = k(t,s), \quad a \le t \le \gamma, \quad \gamma \le s \le b.$$

Since k is semi-separable, the kernels k_γ and k^γ are finite rank kernels. Given a finite rank kernel h we denote by rank(h) the rank of the corresponding integral operator (see [8], §1).

4.1. LEMMA. *Let* k_1 *and* k_2 *be semi-separable kernels on* $[a,b] \times [a,b]$. *Put*

$$k(t,s) := k_1(t,s) + k_2(t,s) + \int_a^b k_1(t,\alpha)k_2(\alpha,s)d\alpha. \tag{4.1}$$

Then k *is again a semi-separable kernel and for* $a < \gamma < b$

$$\operatorname{rank}(k_\gamma) \le \operatorname{rank}((k_1)_\gamma) + \operatorname{rank}((k_2)_\gamma), \tag{4.2}$$

$$\operatorname{rank}(k^\gamma) \le \operatorname{rank}(k_1^\gamma) + \operatorname{rank}(k_2^\gamma). \tag{4.3}$$

PROOF. The semi-separability of k follows from [5], Theorem II.1.1. We prove (4.2). For $\nu = 1,2$ let

$$k_\nu(t,s) = \begin{cases} F_1^{(\nu)}(t)G_1^{(\nu)}(s), & a \le s < t \le b, \\ F_2^{(\nu)}(t)G_2^{(\nu)}(s), & a \le t < s \le b. \end{cases}$$

Here

$$F_1^{(\nu)}(t) : X_1^{(\nu)} \to Y, \qquad G_1^{(\nu)}(t) : Y \to X_1^{(\nu)},$$

$$F_2^{(\nu)}(t) : X_2^{(\nu)} \to Y, \qquad G_2^{(\nu)}(t) : Y \to X_2^{(\nu)}$$

are linear operators acting between finite dimensional linear spaces, which as functions of t are square integrable on $a \le t \le b$. Take $a \le s < \gamma < t \le b$. Then

$$\begin{aligned} k(t,s) = {} & F_1^{(1)}(t)G_1^{(1)}(s) + F_1^{(2)}(t)G_1^{(2)}(s) + \\ & + F_1^{(1)}(t)\left(\int_a^s G_1^{(1)}(\alpha)F_2^{(2)}(\alpha)d\alpha\right)G_2^{(2)}(s) \\ & + F_1^{(1)}(t)\left(\int_s^t G_1^{(1)}(\alpha)F_1^{(2)}(\alpha)d\alpha\right)G_1^{(2)}(s) \\ & + F_2^{(1)}(t)\left(\int_t^b G_2{}^{(1)}(\alpha)F_1^{(2)}(\alpha)d\alpha\right)G_1^{(2)}(s). \end{aligned}$$

Put

$$\begin{aligned} H(s) := {} & G_1^{(1)}(s) + \left(\int_a^s G_1^{(1)}(\alpha)F_2^{(2)}(\alpha)d\alpha\right)G_2^{(2)}(s) + \\ & + \left(\int_s^\gamma G_1^{(1)}(\alpha)F_1^{(2)}(\alpha)d\alpha\right)G_1^{(2)}(s), \qquad a \le s \le \gamma, \end{aligned}$$

$$\begin{aligned} K(t) := {} & F_1^{(2)}t) + F_1^{(1)}(t)\left(\int_\gamma^t G_1^{(1)}(\alpha)F_1^{(2)}(\alpha)d\alpha\right) + \\ & + F_2^{(1)}(t)\left(\int_t^b G_2^{(1)}(\alpha)F_1^{(2)}(\alpha)d\alpha\right), \qquad \gamma \le t \le b. \end{aligned}$$

Then

$$k_\gamma(t,s) = F_1^{(1)}(t)H(s) + K(t)G_1^{(2)}(s), \qquad a \le s < \gamma < t \le b. \tag{4.4}$$

Introduce the following auxiliary operators

$$\Gamma_\nu : L_2([a,\gamma],Y) \to X_1^{(\nu)}, \qquad \Gamma_\nu\varphi = \int_a^\gamma G_1^{(\nu)}(s)\varphi(s)ds;$$

$$\Lambda_\nu : X_1^{(\nu)} \to L_2([\gamma,b],Y), \qquad (\Lambda_\nu x)(t) = F_1^{(\nu)}(t)x;$$

$$\Gamma : L_2([a,\gamma],Y) \to X_1^{(1)}, \qquad \Gamma\varphi = \int_a^\gamma H(s)\varphi(s)ds;$$

$$\Lambda : X_1^{(2)} \to L_2([\gamma,b],Y), \qquad (\Lambda x)(t) = K(t)x.$$

We have (cf. (4.4)):

$$(4.5) \qquad \operatorname{rank}(k_\gamma) = \operatorname{rank}(\Lambda_1\Gamma + \Lambda\Gamma_2),$$

$$(4.6) \qquad \operatorname{rank}((k_\nu)_\gamma) = \operatorname{rank}\Lambda_\nu\Gamma_\nu, \quad \nu = 1,2.$$

We shall prove that

$$(4.7) \qquad \operatorname{Im}\Gamma \subset \operatorname{Im}\Gamma_1, \qquad \operatorname{Ker}\Lambda_2 \subset \operatorname{Ker}\Lambda.$$

Let Π be a projection of $X_1^{(1)}$ along $\operatorname{Im}\Gamma_1$. So $\Pi\Gamma_1 = 0$. This implies that $\Pi G_1^{(1)}(s) = 0$ a.e. on $a \le s \le \gamma$. It follows that $\Pi H(s) = 0$ a.e. on $a \le s \le \gamma$, and hence $\Pi\Gamma = 0$. We see that $\operatorname{Im}\Gamma \subset \operatorname{Ker}\Pi = \operatorname{Im}\Gamma_1$. Next, take $x \in \operatorname{Ker}\Lambda_2$. So $F_1^{(2)}(t)x = 0$ a.e. on $\gamma \le t \le b$. It follows that $K(t)x = 0$ a.e. on $\gamma \le t \le b$. Thus $\Lambda x = 0$, and hence the second inclusion in (4.7) holds true.

Now, use (4.7) in (4.5). We get

$$\operatorname{rank}(k_\gamma) \le \operatorname{rank}(\Lambda_1\Gamma) + \operatorname{rank}(\Lambda\Gamma_2) \le \operatorname{rank}(\Lambda_1\Gamma_1) + \operatorname{rank}(\Lambda_2\Gamma_2)$$

$$= \operatorname{rank}((k_1)_\gamma) + \operatorname{rank}((k_2)_\gamma).$$

This proves (4.2). Formula (4.3) is proved in an analogous way. □

Let k be the semi-separable kernel of the (SK)-integral operator T. In what follows $\ell(k)$ and $u(k)$ denote the lower and upper order of k as defined in [8], §4. If $\Delta = (B(\cdot),C(\cdot);P)$ is a SB-minimal representation of T, then (cf. the proof of Theorem 7.1 in [8])

$$(4.8) \qquad \ell(k) = \dim\operatorname{Ker}P, \quad u(k) = \dim\operatorname{Im}P.$$

It follows that the SB-degree $\varepsilon(T)$ of T is given by

$$(4.9) \qquad \varepsilon(T) = \ell(k) + u(k),$$

which is just another form of formula (I.5.1).

PROOF of Theorem I.6.1. Part (i) of the theorem is contained in Theorem 3.1. So we have to prove part (ii). Let T be in the class (USB), and let $T = T_1T_2$ be a SB-minimal factorization. First we prove that T_1 and T_2 are in the class (USB). Let k, k_1 and k_2 be the kernels of T, T_1 and T_2, respectively. Then k, k_1 and k_2 are related as in (4.1). We have

$$(4.10) \qquad \ell(k) = \ell(k_1) + \ell(k_2), \quad u(k) = u(k_1) + u(k_2).$$

Indeed, from (4.8) and (3.4) it follows that $\ell(k) \le \ell(k_1) + \ell(k_2)$ and $u(k) \le u(k_1) + u(k_2)$. Since $T = T_1T_2$ is a SB-minimal factorization we know that $\varepsilon(T) = \varepsilon(T_1) + \varepsilon(T_2)$. Now use (4.9) and the two identities in (4.10) are clear. The fact that T belongs to the class (USB) is equivalent to the statement that k is lower unique and upper unique (see [8]), and hence it follows (see [12]) that

$$\ell(k) = \operatorname{rank}(k_\gamma), \quad u(k) = \operatorname{rank}(k^\gamma), \quad a < \gamma < b.$$

Fix $a < \gamma < b$, and apply Lemma 4.1. We get

$$\ell(k) = \operatorname{rank}(k_\gamma) \le \operatorname{rank}((k_1)_\gamma) + \operatorname{rank}((k_2)_\gamma)$$

$$\le \ell(k_1) + \ell(k_2) = \ell(k);$$

$$u(k) = \operatorname{rank}(k^\gamma) \le \operatorname{rank}(k_1^\gamma) + \operatorname{rank}(k_2^\gamma) \le u(k_1) + u(k_2) = u(k).$$

It follows that for $\nu = 1,2$

$$\operatorname{rank}((k_\nu)_\gamma) = \ell(k_\nu), \quad a < \gamma < b,$$

$$\operatorname{rank}(k_\nu^\gamma) = u(k_\nu), \quad a < \gamma < b.$$

But then we can apply [8], Theorems 3.2 and 3.3 to show that k_ν is lower unique and upper unique. So T_1 and T_2 are in the class (USB).

Next, let Δ be a SB-minimal representation of T. We have to show that there exists a unique decomposing projection Π for Δ that yields the factorization $T = T_1T_2$. Put $\theta_0 = \theta(\Delta)$ (see (2.1)). Then θ_0 is a SB-minimal realization of T. It suffices to show that θ_0 has a unique decomposing projection Π such that T_1 is the input/output map of $\ell_\Pi(\theta_0)$ and T_2 is the input/output map of $r_\Pi(\theta_0)$. We already know (see the proof of the second part of Theorem 3.1) that there exists a SB-minimal realization θ of T and a decomposing projection τ for θ such that T_1 is the input/output map of $\ell_\tau(\theta)$ and T_2 is the input/output of $r_\tau(\theta)$. Since T is the class (USB), the SB-minimal representations Δ and $\Delta(\theta)$ of T are similar. But then

$$\theta_0 = \theta(\Delta) \simeq \theta(\Delta(\theta)) \simeq \theta.$$

Thus $\theta_0 \simeq \theta$, and we can apply Proposition II.6.1. So there exists a decomposing projection Π for θ_0 such that

$$\ell_\Pi(\theta) \simeq \ell_\tau(\theta), \qquad r_\Pi(\theta_0) \simeq r_\tau(\theta).$$

It follows that T_1 and T_2 are the input/output maps of $\ell_\Pi(\theta_0)$ and $r_\Pi(\theta_0)$, respectively. It remains to prove the uniqueness of Π.

Let ρ be a second decomposing projection for θ_0 that yields the factorization $T = T_1T_2$. Thus T_1 is the input/output map of $\ell_\rho(\theta_0)$ and T_2 is the input/output map of $r_\rho(\theta_0)$. From Lemma 3.4 we know that $\ell_\Pi(\theta_0)$ and $\ell_\rho(\theta_0)$ are SB-minimal realizations of T_1 and $r_\Pi(\theta_0)$ and $r_\rho(\theta_0)$ are SB-minimal realizations of T_2. Now use that T_1 and T_2 are in the class (USB). So

$$(4.11) \qquad \ell_\Pi(\theta_0) \simeq \ell_\rho(\theta_0), \qquad r_\Pi(\theta_0) \simeq r_\rho(\theta_0).$$

In other words the cascade decompositions $\theta_0 \simeq \ell_\Pi(\theta_0)r_\Pi(\theta_0)$ and $\theta_0 \simeq \ell_\rho(\theta_0)r_\rho(\theta_0)$ are equivalent. Now recall that the class of SB-systems is closed under similarity and reduction. So the SB-minimal system θ_0 is irreducible. But then we can apply Theorem II.4.3 to show that $\rho = \Pi$. □

III.5 Minimal factorization of Volterra integral operators (2)

In this section we prove the results announced in §I.3. We begin with a lemma.

5.1. LEMMA. *Let* T *be a (SK)-integral operator on* $L_2([a,b],Y)$ *in the class (AC), and let* $\Delta = (B(\cdot),C(\cdot),P)$ *be a minimal representation of* T. *Then* $B(\cdot)$ *and* $C(\cdot)$ *are analytic on* $[a,b]$ *and* $P = 0$. *In particular,* $\varepsilon(T) = \delta(T)$.

PROOF. Let k be the kernel of T. Since Δ is a representation of T, we have

$$(5.1) \qquad C(t)(I-P)B(s) = k(t,s), \qquad a \le s < t \le b,$$

$$(5.2) \qquad C(t)PB(s) = 0, \qquad a \le t < s \le b.$$

Let X be the internal space of Δ, and put $X_1 = \operatorname{Im}(I-P)$. Introduce

$$B_1(t) = (I-P)B(t) : Y \to X_1, \qquad C_1(t) = C(t)|_{X_1} : X_1 \to Y.$$

Then $\Delta_1 = (B_1(\cdot),C_1(\cdot);0)$ is a SB-representation of T, and hence $\varepsilon(T) \le \dim X_1$. Since Δ is a minimal representation of T, we have $\delta(T) = \dim X$. So

$$\delta(T) = \dim X \ge \dim X_1 \ge \varepsilon(T) \ge \delta(T).$$

We conclude that $\varepsilon(T) = \delta(T)$ and $X_1 = X$. In particular, $I - P$ is invertible and $\tilde{\Delta} = ((I-P)B(\cdot),C(\cdot);0)$ is a SB-minimal representation of T.

According to our hypotheses the kernel k admits a representation of the form:

$$k(t,s) = F(t)G(s), \quad a \le s \le t \le b,$$

with $F(\cdot)$ and $G(\cdot)$ analytic on [a,b]. Without loss of generality (cf. [8], §1) we may assume that the operators

$$(5.3) \qquad \int_a^b F(t)^* F(t)dt, \qquad \int_a^b G(t)G(t)^* dt$$

are invertible. It follows that $\Delta_0 = (G(\cdot),F(\cdot);0)$ is a SB-minimal realization of T. So $\tilde{\Delta}$ and Δ_0 are two SB-minimal representations of T. But $T \in (AC) \subset (USB)$, and thus Δ_0 and $\tilde{\Delta}$ are similar. So there exists an invertible operator S such that $(I-P)B(t) = SG(t)$ and $C(t) = F(t)S^{-1}$ a.e. on $a \le t \le b$. According to (5.2) this implies that

$$(5.4) \qquad F(t)[S^{-1}P(I-P)^{-1}S]G(s) = 0$$

for $a \le t < s \le b$. But $F(\cdot)$ and $G(\cdot)$ are analytic on [a,b]. Hence (5.4) holds for all (t,s) in $[a,b] \times [a,b]$. Now use that the operators (5.3) are invertible. We obtain $S^{-1}P(I-P)^{-1}S = 0$. Thus $P = 0$. Furthermore, $B(\cdot) = SG(\cdot)$ and $C(\cdot) = F(\cdot)S^{-1}$ are analytic on [a,b]. □

Note that in Lemma 5.1 and its proof we used the convention that a function in $L_2[a,b]$ is analytic on [a,b] if one of the elements in its equivalence class is analytic.

In Lemma 5.1 the analyticity condition on the kernel k of T plays an essential role. Without this analyticity condition it may happen that a Volterra (SK)-integral operator T has a minimal representation with a non-zero internal operator; in fact, this may happen even if the kernel k of T is lower unique. For example, take

$$k_0(t,s) = \chi_{[\frac{2}{3},1]}(t)\chi_{[0,\frac{1}{3}]}(s), \qquad 0 \le s \le t \le 1.$$

Then k_0 is lower unique (because of [8], Theorem 3.2) and

$$\Delta_0 = (2\chi_{[0,\frac{1}{3}]},\chi_{[\frac{2}{3},1]};\tfrac{1}{2})$$

is a minimal representation of the corresponding integral operator T_0. Note that the internal operator of Δ_0 is non-zero. In terms of realizations this example shows that a Volterra (SK)-integral operator with a lower unique kernel may have a non-causal minimal realization.

PROOF of Theorem I.3.1. Part (i) of the theorem is a straightforward application of the first part of Theorem I.2.1 (where one only has to take $P = 0$). Because of Lemma 5.1 the functions $B(\cdot)$ and $C(\cdot)$ in the minimal representation Δ are analytic on $[a,b]$. It follows that the fundamental operator $\Omega(\cdot)$ of Δ is analytic on $[a,b]$. Hence the integral operators T_1 and T_2 given by (I.3.6) and (I.3.7) are in the class (AC).

To prove part (ii), let T be a (SK)-integral operator in the class (AC), and let $T = T_1T_2$ be a minimal factorization of T. First we prove that T_1 and T_2 are in the class (AC). From the second part of Theorem I.2.1 we know that there exist a minimal representation $\Delta_0 = (B_0(\cdot),C_0(\cdot);P_0)$ of T and a decomposing projection Π_0 for Δ_0 which yields the factorization $T = T_1T_2$. According to Lemma 5.1 the internal operator $P_0 = 0$, and hence we can apply part (i) to Δ_0 and Π_0. So the remark made at the end of the previous paragraph implies that T_1 and T_2 are in the class (AC).

Since $T = T_1T_2$ is a minimal factorization with factors in the class (AC), we have (cf. Lemma 5.1)

$$(5.5) \qquad \varepsilon(T) = \delta(T) = \delta(T_1) + \delta(T_2) = \varepsilon(T_1) + \varepsilon(T_2).$$

Hence $T = T_1T_2$ is also a SB-minimal factorization. Next, observe that $T \in (AC) \subset (USB)$. So we can apply part (ii) of Theorem I.6.1 (which is proved in the previous section) to finish the proof. □

5.2. COROLLARY. *Let* T *be in the class* (*AC*). *Then* $T = T_1T_2$ *is a minimal factorization if and only if* $T = T_1T_2$ *is a SB-minimal factorization.*

PROOF. The proof of the "only if part" is given by (5.5). To prove the "if part", let $T = T_1T_2$ be a SB-minimal factorization. Then there exist (see the second part of Theorem III.3.1) a SB-minimal representation Δ of T and a decomposing projection Π for Δ which yields the factorization $T = T_1T_2$. By Lemma 5.1 the representation Δ is a minimal representation of T. So

Theorem I.2.1 (i) implies that $T = T_1T_2$ is a minimal factorization. □

Let T be the integral operator defined by formula (I.3.9), and for $\nu = 1,2$ put

$$(T_1^{(\nu)}\varphi)(t) = \varphi(t) + \int_0^t (1 + r_\nu(s)g_\nu(s))\varphi(s)ds,$$

$$(T_2^{(\nu)}\varphi)(t) = \varphi(t) + \int_0^t (\chi_{[0,\frac{1}{2})}(t) - r_\nu(t))g_\nu(s)\varphi(s)ds,$$

where the functions $r_\nu(\cdot)$ and $g_\nu(\cdot)$ are as in §I.3. We already know that

$$(5.6) \qquad T = T_1^{(1)}T_2^{(1)}, \qquad T = T_1^{(2)}T_2^{(2)},$$

and these two factorizations are minimal factorizations. We shall now prove that T does not have a minimal representation that generates the minimal factorizations in (5.6).

For $\nu = 1,2$ let θ_ν be the following causal system

$$\theta_\nu = \left(\begin{bmatrix} 0 & 0 \\ 0 & 0 \end{bmatrix}, \begin{bmatrix} 1 \\ g_\nu(t) \end{bmatrix}, \begin{bmatrix} 1 & \chi_{[0,\frac{1}{2})}(t) \end{bmatrix}; \begin{bmatrix} 1 & 0 \\ 0 & 1 \end{bmatrix}, \begin{bmatrix} 0 & 0 \\ 0 & 0 \end{bmatrix} \right)_0^1.$$

Thus $\theta_\nu = \theta(\Delta_\nu)$, where Δ_ν is given by (I.3.12). It follows that θ_1 and θ_2 are minimal realizations of the operator T (given by (I.3.9)). We know that for $g = g_1$ the function $r_1(\cdot)$ is a solution of the RDE (I.3.11) and for $g = g_2$ the function $r_2(\cdot)$ is a solution of (I.3.11). This implies (cf. Lemma II.3.1) that the projection

$$\Pi = \begin{bmatrix} 0 & 0 \\ 0 & 1 \end{bmatrix} : \mathbb{C}^2 \to \mathbb{C}^2$$

is a decomposing projection for both θ_1 and θ_2. The corresponding left and right factors are given by:

$$\ell_\Pi(\theta_1) = (0, 1 + r_1(t)g_1(t), 1; 1, 0)_0^1,$$

$$r_\Pi(\theta_1) = (0, g_1(t), -r_1(t) + \chi_{[0,\frac{1}{2})}(t); 1, 0)_0^1,$$

$$\ell_\Pi(\theta_2) = (0, 1 + r_2(t)g_2(t), 1; 1, 0)_0^1,$$

$$r_\Pi(\theta_2) = (0, g_2(t), -r_2(t) + \chi_{[0,\frac{1}{2})}(t); 1, 0)_0^1.$$

For $\nu = 1,2$ the operator $T_1^{(\nu)}$ is the input/output map of $\ell_\Pi(\theta_\nu)$ and $T_2^{(\nu)}$ is the input/output map of $r_{\Pi}(\theta_\nu)$.

For $\nu = 1,2$ and $j = 1,2$ the integral operator $T_j^{(\nu)}$ has no non-similar minimal realizations. We shall prove this for $T_1^{(1)}$. Let θ_0 be an arbitrary minimal realization of $T_1^{(1)}$. We want to show that $\theta_0 \simeq \ell_{\Pi}(\theta_1)$. Without loss of generality (apply a similarity if necessary) we may assume that $\theta_0 = (0, b(t),c(t);1-p,p)_0^1$. Since $T_{\theta_0} = T_1^{(1)}$ we have

$$(5.7) \qquad c(t)(1-p)b(s) = 1+r_1(s)g_1(s), \qquad 0 \le s \le t \le 1, \quad \text{a.e.},$$

$$(5.8) \qquad c(t)pb(s) = 0, \qquad 0 \le t < s \le 1, \quad \text{a.e.}.$$

Now use that the kernel $k_1(t,s) = 1+r_1(s)g_1(s)$, $0 \le t \le 1$, $0 \le s \le 1$, is lower unique. So there exists a non-zero constant γ such that

$$(5.9) \qquad 1 = c(t)(1-p)\gamma, \qquad 1+r_1(t)g_1(t) = \frac{1}{\gamma}b(t), \qquad 0 \le t \le 1, \quad \text{a.e.}.$$

Assume $p \neq 0$. Then (5.9) and (5.8) imply that

$$1+r_1(s)g_1(s) = \left(\frac{1-p}{p}\right) c(t)pb(s) = 0, \qquad 0 \le t < s \le 1, \quad \text{a.e.},$$

which is impossible. So $p = 0$. But then (5.9) shows that $\theta_0 \simeq \ell_{\Pi}(\theta_1)$.

Now, let θ be a minimal realization of T, and assume that there exist decomposing projections Π_1 and Π_2 for θ such that for $\nu = 1,2$ the operator $T_1^{(\nu)}$ is the input/output map of $\ell_{\Pi_\nu}(\theta)$ and $T_2^{(\nu)}$ is the input/output map of $r_{\Pi_\nu}(\theta)$. Since θ is a minimal realization of T, we can apply Lemma 2.1 to show that $\ell_{\Pi_\nu}(\theta)$ and $r_{\Pi_\nu}(\theta)$ are minimal realizations of $T_1^{(\nu)}$ and $T_2^{(\nu)}$, respectively ($\nu = 1,2$). But then the result of the preceding paragraph shows that

$$\ell_{\Pi}(\theta_1) \simeq \ell_{\Pi_1}(\theta), \qquad r_{\Pi}(\theta_1) \simeq r_{\Pi_1}(\theta),$$

$$\ell_{\Pi}(\theta_2) \simeq \ell_{\Pi_2}(\theta), \qquad r_{\Pi}(\theta_2) \simeq r_{\Pi_2}(\theta),$$

and thus $\theta_\nu \simeq \ell_{\Pi}(\theta_\nu)r_{\Pi}(\theta_\nu) \simeq \ell_{\Pi_\nu}(\theta)r_{\Pi_\nu}(\theta) \simeq \theta$ for $\nu = 1,2$. In particular, $\theta_1 \simeq \theta_2$. Contradiction. Thus there is no single minimal realization θ of T which has decomposing projections that yield the minimal factorizations in (5.6), and hence T does not have a minimal representation that generates the minimal factorizations in (5.6).

III.6 Proof of Theorem I.4.1

In this section we shall prove Theorem I.4.1. We begin with two lemmas.

6.1. LEMMA. *A minimal representation* Δ *of a (SK)-integral operator on* $L_2([a,b],Y)$ *in the class (STC) is of the form*

(6.1) $$\Delta = (e^{-(t-a)A}B, Ce^{(t-a)A}; 0).$$

PROOF. Let T belong the class (STC). Since the class (STC) is contained in the class (AC), the degree of T is equal to the SB-degree of T (Lemma 5.1). From formula (I.4.1) it is clear that T has a representation of the form (6.1). Moreover, without loss of generality we may assume that

(6.2) $$\bigcap_{j=1}^{n} \operatorname{Ker} CA^{j-1} = (0), \qquad X = \operatorname{Im}\begin{pmatrix} B & AB & \cdots & A^{n-1}B \end{pmatrix} .$$

Here n is the dimension of the internal space X of Δ. The identities in (6.2) imply that the representation (6.1) is a SB-minimal representation. So $\varepsilon(T) = \dim X$. But $\delta(T) = \varepsilon(T)$. It follows that T has a minimal representation of the form (6.1).

Next, assume that Δ in (6.1) and $\Delta_0 = (B(\cdot), C(\cdot); P)$ are minimal representations of T. Since (STC) $\subset$ (AC), Lemma 5.1 implies that P = 0 and both Δ and Δ_0 are SB-minimal representations of T. But (STC) $\subset$ (AC) $\subset$ (USB). It follows that Δ and Δ_0 are similar, and so there exists an invertible operator S such that

$$B(t) = Se^{-(t-a)A}B, \qquad C(t) = Ce^{(t-a)A}S .$$

Put $A_0 = SAS^{-1}$, $B_0 = SB$ and $C_0 = CS^{-1}$. Then $\Delta_0 = (e^{-(t-a)A_0}B_0, C_0e^{(t-a)A_0}; 0)$ and the lemma is proved. □

6.2. LEMMA. *Let* T *be a (SK)-integral operator on* $L_2([a,b],Y)$ *in the class (STC), and let* $\Delta = (e^{-(t-a)A}B, Ce^{(t-a)A}; 0)$ *be a minimal representation of* T. *Then* Π *is a supporting projection for* Δ *if and only if* Π *is a decomposing projection for* Δ *and the corresponding factorization* $T = T_1T_2$ *has factors* T_1 *and* T_2 *in the class (STC).*

PROOF. Consider the time invariant causal system $\theta = (A,B,C;I,0)_a^b$. Note that $\Delta = \Delta(\theta)$ (cf. formula (2.2)). Let Π be a supporting projection for Δ. Then Π is a factorization projection for θ, and we can apply Proposition

II.6.2 to show that Π is a decomposing projection for θ. Formula (II.6.13) implies that the input/output maps of $\ell_\Pi(\theta)$ and $r_\Pi(\theta)$ are input/output maps of time invariant causal systems and hence belong to the class (STC). It follows that the factorization $T = T_1T_2$ generated by Π and Δ has its factors in in the class (STC). This proves the "only if part" of the lemma.

To prove the "if part", let Π be a decomposing projection for Δ and assume that the corresponding factorization $T = T_1T_2$ has its factors in the class (STC). Then Π is a decomposing projection for the system θ. Furthermore T_1 and T_2 are the input/output maps of $\ell_\Pi(\theta)$ and $r_\Pi(\theta)$, respectively. For $\nu = 1,2$ let $\theta_\nu = (e^{-(t-a)A_\nu}B_\nu, Ce^{(t-a)A_\nu};0)$ be a minimal representation of T_ν and consider $\theta_\nu = (A_\nu,B_\nu,C_\nu;I,0)_a^b$. Let X_ν be the state space of θ_ν $(\nu = 1,2)$. Put $\theta_0 = \theta_1\theta_2$. Note that θ_0 is a time invariant causal system. From Lemma 2.1 it follows that θ_0 is a minimal realization of T. Furthermore, the projection

$$\rho = \begin{pmatrix} 0 & 0 \\ 0 & 1 \end{pmatrix} : X_1 \oplus X_2 \to X_1 \oplus X_2$$

is a factorization projection for θ_0, $pr_{I-\rho}(\theta_0) = \theta_1$ and $pr_\rho(\theta_0) = \theta_2$. Now θ_0 and θ are both time invariant causal minimal realizations of T. It follows that θ_0 and θ are similar and a similarity between θ_0 and θ is time independent (cf. [7], §5). Put $\Pi_0 = S\rho S^{-1}$, where S is a similarity between θ_0 and θ. Then Π_0 is a factorization projection for θ, and hence Π_0 is a supporting projection for Δ. According to Proposition II.6.1 and II.6.2 the projection Π_0 is also a decomposing projection for θ and

$$\ell_{\Pi_0}(\theta) \simeq \ell_\rho(\theta_0) \simeq pr_{I-\rho}(\theta_0) = \theta_1,$$

$$r_{\Pi_0}(\theta) \simeq r_\rho(\theta_0) \simeq pr_\rho(\theta_0) = \theta_2.$$

So Π_0 is a decomposing projection for Δ and $T = T_1T_2$ is the corresponding factorization. By Theorem I.3.1 (ii) there is only one decomposing projection for Δ which generates the factorization $T = T_1T_2$. Thus $\Pi = \Pi_0$ and Π is a supporting projection for Δ. □

PROOF of Theorem I.4.1. To prove part (i), let Π be a supporting projection of the minimal representation $\Delta = (e^{-tA}B, Ce^{tA};0)$. Put $\theta = (A,B,C;I,0)_a^b$. Then θ is a minimal realization of T and Π is a factorization projection for θ. It is straightforward to check that the input/output maps of $pr_{I-\Pi}(\theta)$ and $pr_\Pi(\theta)$ are the integral operators T_1 and T_2 given by (I.4.6) and (I.4.7), respectively. Thus $T = T_1T_2$. Since $\ell_{\Pi}(\theta) \simeq pr_{I-\Pi}(\theta)$ and $r_{\Pi}(\theta) \simeq pr_{\Pi}(\theta)$ (Proposition I.6.2), we can apply Lemma 2.1 to show that the factorization $T = T_1T_2$ is minimal. Obviously T_1 and T_2 are in the class (STC).

Replacing Δ by the representation (6.1) one sees that part (ii) of the theorem is an immediate corollary of Theorem I.3.1 (ii) and Lemma 6.2. □

III.7 A remark about minimal factorization and inversion

Let T be a (SK)-integral operator on $L_2([a,b],Y)$. If T is invertible, then

$$\delta(T) = \delta(T^{-1}), \qquad \varepsilon(T) = \varepsilon(T^{-1}). \tag{7.1}$$

To prove the first equality in (7.1), let θ be a minimal realization of T. Thus $\delta(T) = \dim X$, where X is the state space of θ. The fact that T is invertible implies that the inverse system $\theta^{\times}$ has well-posed boundary conditions and T^{-1} is the input/output map of $\theta^{\times}$ (see [5], Theorem II.2.1). Since θ and $\theta^{\times}$ have the same state space, it follows that $\delta(T^{-1}) \le \delta(T)$. By applying the latter inequality to T^{-1} instead of T, one obtains $\delta(T) \le \delta(T^{-1})$. This proves the first identity in (7.1); the second is proved in a similar way (one takes θ to be a SB-minimal realization of T).

7.1. COROLLARY. *Let $T = T_1T_2$ be a minimal (resp. SB-minimal) factorization. If T_1 and T_2 are invertible, then $T^{-1} = T_2^{-1}T_1^{-1}$ is again a minimal (resp. SB-minimal) factorization.*

PROOF. Assume $T = T_1T_2$ is a minimal factorization. Thus $\delta(T) = \delta(T_1) + \delta(T_2)$. Now apply (7.1) to T and the factors T_1 and T_2. We get $\delta(T^{-1}) = \delta(T_1^{-1}) + \delta(T_2^{-1})$. Hence $T^{-1} = T_2^{-1}T_1^{-1}$ is a minimal factorization. The SB-minimal case is proved in the same way. □

III.8 LU- and UL-factorizations (2)

In this section we prove the results about LU- and UL-factorization stated in Section I.8.

PROOF of Theorem I.8.1. The equivalence of the statements (1) and (5) is well-known (see [11], Section IV.7). Assume that (5) holds. Put $\theta = \theta(\Delta)$ (see formuls (2.1)). Then T is the input/output map of the system θ and the kernel k of T is as in (I.1.3). It follows that T_{α} is the input/output map of the system

$$\theta_\alpha = (0,B(t),C(t);I-P,P)_a^\alpha .$$

Note that $\Omega(t)$ is the fundamental operator of the inverse system $\theta_\alpha^\times$ associated with θ_α. Since T_α is invertible, we can apply the results of § II.4 in [5] to show that $\det(I-P+P\Omega(\alpha)) \neq 0$. Equivalently, $\det(I-P+\Omega(\alpha)P) \neq 0$, and we see that (5) implies (4).

Since Δ is a SB-representation, the operator P is a projection. Thus (I.8.2) is a well-defined Riccati differential equation. The fundamental operator of (I.8.2) is also equal to $\Omega(t)$. It follows that the equation (I.8.2) is solvable on $a \le t \le b$ if and only if (see [17]) the operator

(8.1) $\quad P\Omega(t)P : \operatorname{Im} P \to \operatorname{Im} P$

is invertible for each $a \le t \le b$. Furthermore, in that case the solution of (8.1) is given by

(8.2) $\quad R(t) = -(I-P)\Omega(t)P(P\Omega(t)P)^{-1} : \operatorname{Im} P \to \operatorname{Ker} P.$

Obviously, the operator (8.1) is invertible if and only if $\det(I-P+\Omega(t)P) \neq 0$, and hence the statements (3) and (4) are equivalent. Moreover we see that the solution of (I.8.2) (assuming it exists) may also be written in the form (I.8.6).

From the definition of a decomposing projection it is clear that the projection P is a decomposing projection for Δ is and only if $\det(I-P+\Omega(t)P) \neq 0$ for $a \le t \le b$. Hence the statements (2) and (4) are also equivalent.

It remains to prove that (2) implies (1). Let P be a decomposing projection for Δ. Put $\theta = \theta(\Delta)$ (see formula (2.1)). Then P is a decomposing projection for θ and the corresponding left and right factor are given by

$$\ell_P(\theta) = (0,(I-P+R(t)P)B(t),C(t)|_{\operatorname{Ker} P};I_{\operatorname{Ker} P},0)_a^b,$$

$$r_P(\theta) = (0,PB(t),C(t)(P-R(t)P)|_{\operatorname{Im} P};0,I_{\operatorname{Im} P},0)_a^b.$$

Here $R(t) : \operatorname{Im} P \to \operatorname{Ker} P$ is the solution of the Riccati differential equation (I.8.2), which exists because (2) is equivalent to (4) and (4) is equivalent to (3). Next, observe that the input/output map of $\ell_P(\theta)$ is the operator $I+K_-$ with K_- given by (I.8.4). Similarly, the input/output map of $r_P(\theta)$ is equal to $I+K_+$ with K_+ given by (I.8.5). Since P is a decomposing projection, Theorem II.4.1 implies

(8.3) $\theta \simeq \ell_P(\theta) r_P(\theta)$,

and hence $T = (I + K_-)(I + K_+)$ which is the desired LU-factorization. □

PROOF of Theorem 8.2. The equivalence of the statements (1) and (4) is well-known (see [11], Chapter IV). Take $a \leq \beta < b$. Put $\theta = \theta(\Delta)$ (see formula (2.1)). We know that T is the input/output map of the system θ. So the kernel k of T is as in (I.1.3). It follows that T^β is the input/output map of the system

$$\theta^\beta = (0, B(t), C(t); I - P, P)_\beta^b .$$

The function $Z(t) := \Omega(t)\Omega(\beta)^{-1}$, $\beta \leq t \leq b$, is the fundamental operator of the inverse system $(\theta^\beta)^\times$ associated with θ^β. Since T^β is invertible, the results of §II.4 in [5] imply that

$$\det(I - P + P\Omega(b)\Omega(\beta)^{-1}) \neq 0.$$

But then $\det((I - P)\Omega(\beta) + P\Omega(b)) \neq 0$. Since $a \leq \beta < b$ is arbitrary, we see that (4) implies (3).

Since P is a projection, the Riccati differential equation (I.8.7) is well-defined. Assume (3) holds. Then $I - P + P\Omega(b)\Omega(t)^{-1}$ is invertible. For $a \leq t \leq b$ define R(t) by (I.8.11). A direct computation shows that $R(\cdot)$ is the solution of the equation (I.8.7).

Next, assume that the equation (I.8.7) is solvable, and let R(t), $a \leq t \leq b$, be its solution. Put $Q = -R(a)$, and define $\Pi = (I - P) + Q(I - P)$. Since $Q : \operatorname{Ker} P \to \operatorname{Im} P$, the operator Π is a projection and $\operatorname{Ker}\Pi = \operatorname{Im} P$. We shall prove that Π is a decomposing projection for $\theta := \theta(\Delta)$. To do this we use Lemma II.3.1. Obviously, $\operatorname{Ker}\Pi$ is invariant under P. Put

(8.4) $G(t) = (R(t) + Q)(I - P)\Pi : \operatorname{Im}\Pi \to \operatorname{Ker}\Pi, \quad a \leq t \leq b.$

Since $(I - P)Q = 0$ it is straightforward to check that

(8.5) $I - \Pi + G(t)\Pi = P + R(t)(I - P)$,

(8.6) $G(t)\Pi - \Pi = R(t)(I - P) - (I - P)$,

(8.7) $G(a) = 0, \quad G(b) = Q(I - P)\Pi.$

It follows that G is the solution of the following Riccati differential equation:

$$(8.8)\quad \begin{cases} \dot{G}(t) = (I - \Pi + G(t)\Pi)(-B(t)C(t))(G(t)\Pi - \Pi), & a \le t \le b, \\ G(a) = 0. \end{cases}$$

Next, observe that

$$P(I - \Pi)G(b)\Pi(I - P) = Q(I - P)\Pi = (I - \Pi)P\Pi.$$

So Lemma II.3.1 implies that Π is a decomposing projection for θ. The left and right factor corresponding to Π are given by

$$\ell_{\Pi}(\theta) = (0,(I - \Pi + G(t)\Pi)B(t),C(t)\big|_{\operatorname{Ker}\Pi};0,I_{\operatorname{Ker}\Pi})_a^b,$$

$$r_{\Pi}(\theta) = (0,\Pi B(t),C(t)(\Pi - G(t)\Pi)\big|_{\operatorname{Im}\Pi};I_{\operatorname{Im}\Pi},0)_a^b.$$

Let T_1 and T_2 be the input/output maps of $\ell_{\Pi}(\theta)$ and $r_{\Pi}(\theta)$, respectively. A simple computation, using (8.5) and (8.6), shows that $T_1 = I + H_+$ with H_+ given by (I.8.10) and $T_2 = I + H_-$ with H_- given by (I.8.9). It follows that $T = (I + H_+)(I + H_-)$, which proves (1). □

<u>PROOF of Corollary 8.3</u>. Let $T = T_-T_+$ be a LU-factorization. Let Δ be a SB-minimal representation of T with external operator P. From Theorem I.8.1 we know that P is a decomposing projection for Δ and (because of the unicity of the LU-factorization) $T = T_-T_+$ is the corresponding factorization. Since Δ is SB-minimal, it follows (cf. Theorem 3.1) that $T = T_-T_+$ is a SB-minimal factorization.

Next, let $T = T_+T_-$ be a UL-factorization. Then $T^{-1} = T_-^{-1}T_+^{-1}$ and this factorization is a LU-factorization. So, by the result of the preceding paragraph, the factorization $T^{-1} = T_-^{-1}T_+^{-1}$ is SB-minimal. Now apply Corollary 7.1 to T^{-1}. We conclude that $T = T_+T_-$ is also a SB-minimal factorization. □

III.9 Causal/anticausal decompositions

A boundary value system θ is said to admit a *causal/anticausal decomposition* if $\theta \simeq \theta_-\theta_+$ with θ_- a causal system and θ_+ an anti-causal system. Note that θ_- and θ_+ are SB-systems; in fact, the canonical boundary value operator of θ_- is the zero operator and for θ_+ this operator is the identity operator. Thus, if θ admits a causal/anticausal decomposition, then θ must be a SB-system (cf. Lemma 3.2). The following theorem is the system theoretical version of Theorem I.8.1.

9.1. THEOREM. *Let* $\theta = (A(t),B(t),C(t);N_1,N_2)_a^b$ *be a SB-system, and let* $U(t)$ *and* $U^\times(t)$ *denote the fundamental operators of* θ *and the inverse system* $\theta^\times$. *The following statements are equivalent:*

(1) θ *admits a causal/anticausal decomposition;*

(2) *the canonical boundary value operator* P *of* θ *is a decomposing projection for* θ;

(3) *the following Riccati differential equation has a solution* $R(t) : \operatorname{Im} P \to \operatorname{Ker} P$ *on* $a \le t \le b$:

$$(9.1) \qquad \begin{cases} \dot{R}(t) = (I - P + R(t)P)(-U(t)^{-1}B(t)C(t)U(t))(R(t)P - P), & a \le t \le b, \\ R(a) = 0; \end{cases}$$

(4) $\det (I - P + U(t)^{-1}U^\times(t)P) \neq 0$ *for* $a \le t \le b$.

Furthermore, in that case $\theta \simeq \theta_-\theta_+$ *with*

$$(9.2) \qquad \theta_- = (0,(I - P + R(t)P)U^{-1}(t)B(t),C(t)U(t)\big|_{\operatorname{Ker} P};I_{\operatorname{Ker} P},0)_a^b,$$

$$(9.3) \qquad \theta_+ = (0,PU(t)^{-1}B(t),C(t)U(t)(P - R(t)P)\big|_{\operatorname{Im} P};0,I_{\operatorname{Im} P})_a^b,$$

where $R(t) : \operatorname{Im} P \to \operatorname{Ker} P$, $a \le t \le b$, *is the solution of* (9.1), *which is equal to*

$$R(t) = (I - P)\{I - P + U(t)^{-1}U^\times(t)P\}^{-1}P.$$

PROOF. Let T be the input/output map of θ. Put $\Delta = \Delta(\theta)$ (see formula (2.2)). Then Δ is a SB-representation of T and the internal operator of Δ is equal to the canonical boundary value operator P of θ. Note that the fundamental operator $\Omega(t)$ of Δ is given by

$$\Omega(t) = U(t)^{-1}U^\times(t), \qquad a \le t \le b.$$

It follows that P is a decomposing projection for θ if and only if P is a decomposing projection for Δ. The equivalence of the statements (2), (3) and (4) is now clear from Theorem I.8.1. It remains to prove the equivalence of (1) and (2).

Assume P is a decomposing projection for θ. From the definitions of the left and right factor we see that $\ell_P(\theta) = \theta_-$ and $r_P(\theta) = \theta_+$, where θ_- and θ_+ are given by (9.2) and (9.3). Formula (II.4.3) implies that $\theta \simeq \theta_-\theta_+$. Obviously, θ_- is a causal system and θ_+ is an anticausal system. So (2) implies (1).

Next, assume that $\theta \simeq \theta_-\theta_+$ is a causal/anticausal decomposition of θ. Let T_- and T_+ be the input/output maps of θ_- and θ_+, respectively. Then $T = T_-T_+$ is a LU-factorization. So we can apply Theorem I.8.1 to show that P is a decomposing projection for Δ. Hence (2) holds. □

A boundary value system θ is said to admit an *anticausal/causal decomposition* if $\theta \simeq \theta_+\theta_-$ with θ_+ an anticausal system and θ_- a causal system. In order that such a decomposition exists it is necessary that θ is an SB-system (Lemma 3.2).

9.2. THEOREM. *Let* $\theta = (A(t),B(t),C(t);N_1,N_2)_a^b$ *be a SB-system, and let* $U(t)$ *and* $U^\times(t)$ *be the fundamental operators of* θ *and the inverse system* $\theta^\times$. *Let* P *denote the canonical boundary value operator of* θ. *The following statements are equivalent:*

(1) θ *admits an anticausal/causal decomposition;*

(2) *the following Riccati differential equation has a solution* $R(t) : \operatorname{Ker} P \to \operatorname{Im} P$ *on* $a \le t \le b$:

$$(9.4)\qquad \begin{cases} \dot R(t) = (P + R(t)(I-P))(-U(t)^{-1}B(t)C(t)U(t))(R(t)(I-P) - (I-P)), \\ \qquad\qquad\qquad\qquad\qquad\qquad\qquad\qquad\qquad a \le t \le b, \\ R(b) = 0; \end{cases}$$

(3) $\det [(I-P)U(t)^{-1}U^\times(t) + PU(b)^{-1}U^\times(b)] \ne 0$ *for* $a \le t \le b$.

Furthermore, in that case $\theta \simeq \theta_+\theta_-$ *with*

$$(9.5)\qquad \theta_- = (0,(I-P)U(t)^{-1}B(t),C(t)U(t)(I-P-R(t)(I-P))\big|_{\operatorname{Ker} P};I_{\operatorname{Ker} P},0)_a^b,$$

$$(9.6)\qquad \theta_+ = (0,(P+R(t)(I-P))U(t)^{-1}B(t),C(t)U(t)\big|_{\operatorname{Im} P};0,I_{\operatorname{Im} P})_a^b,$$

where $R(t) : \operatorname{Ker} P \to \operatorname{Im} P$, $a \le t \le b$, *is the solution of* (9.4), *which is equal to*

$$R(t) = -P\{I - P + PU(b)^{-1}U^\times(b)U^\times(t)^{-1}U(t)\}^{-1}(I-P).$$

PROOF. As in the proof of Theorem 9.1 put $\Delta = \Delta(\theta)$, and apply Theorem I.8.2 to Δ. This yields the equivalence of the statements (2) and (3).

Assume (2) holds. Define $\Pi = (I-P) + Q(I-P)$, where $Q = -R(a)$. From the proof of Theorem I.8.2 we know that Π is a decomposing projection for θ and

$$\ell_\Pi(\theta) = (0,(I-\Pi+G(t)\Pi)U(t)^{-1}B(t),C(t)U(t)\big|_{\operatorname{Ker}\Pi};0,I_{\operatorname{Ker}\Pi})_a^b,$$

$$r_\Pi(\theta) = (0,\Pi U(t)^{-1}B(t),C(t)U(t)(\Pi - G(t)\Pi)\big|_{\operatorname{Im}\Pi};I_{\operatorname{Im}\Pi},0)_a^b,$$

where $G(t) = (R(t)+Q)(I-P)\Pi : \operatorname{Im}\Pi \to \operatorname{Ker}\Pi$. Since $\operatorname{Ker}\Pi = \operatorname{Im}P$, it is clear from (8.5) that $\ell_\Pi(\theta) = \theta_+$ with θ_+ given by (9.6). We shall prove that $r_\Pi(\theta)$ is similar to θ_-. Define $S : \operatorname{Ker}P \to \operatorname{Im}\Pi$ by $Sx = \Pi x$, $x \in \operatorname{Ker}P$. Note that

$$(I-P)Sx = (I-P)(I-P+Q(I-P))x = (I-P)x = x$$

for each $x \in \operatorname{Ker}P$. Thus S is an invertible operator and $S^{-1}\Pi = I-P$. It follows that

$$S^{-1}\Pi U(t)^{-1}B(t) = (I-P)U(t)^{-1}B(t),$$

$$C(t)U(t)(\Pi - G(t)\Pi)S = C(t)U(t)(I-P-R(t)(I-P)).$$

To prove the last equality one uses (8.6) and $\Pi S = \Pi$. We see that the operator S establishes a similarity between $r_\Pi(\theta)$ and θ_-. We conclude that

$$\theta \simeq \ell_\Pi(\theta)r_\Pi(\theta) \simeq \theta_+\theta_-,$$

and (1) is proved.

Next, assume (1) holds. Then T has a UL-factorization, and we can apply Theorem I.8.2 to $\Delta = \Delta(\theta)$ to show that (2) holds. □

In general, causal/anticausal decompositions (or anticausal/causal decompositions) of time invariant systems cannot be made within the class of time invariant systems. One has to allow that the factors vary in time. To see this consider the following example. Put

$$\theta = \left(\begin{pmatrix} 0 & 0 \\ 0 & 0 \end{pmatrix}, \begin{pmatrix} 1 \\ -1 \end{pmatrix}, \begin{pmatrix} 2 & 1 \end{pmatrix}; \begin{pmatrix} 1 & 0 \\ 0 & 0 \end{pmatrix}, \begin{pmatrix} 0 & 0 \\ 0 & 1 \end{pmatrix} \right)_0^1. \tag{9.7}$$

Note that θ is a SB-system. One computes that

$$P = \begin{pmatrix} 0 & 0 \\ 0 & 1 \end{pmatrix}, \quad U^\times(t) = \begin{pmatrix} -1+2e^{-t} & -1+e^{-t} \\ 2-2e^{-t} & 2-e^{-t} \end{pmatrix},$$

$$I-P+U(t)^{-1}U^\times(t)P = \begin{pmatrix} 1 & -1+e^{-t} \\ 0 & 2-e^{-t} \end{pmatrix}.$$

Now apply Theorem 9.1. It follows that $\theta \simeq \theta_-\theta_+$ with

$$\theta_- = (0,(2-e^{-t})^{-1},2;1,0)_0^1,$$
$$\theta_+ = (0,-1,(2e^t-1)^{-1};0,1)_0^1.$$

Thus θ admits a cascade decomposition in which the factors are time varying. It is easy to check that θ does not have a cascade decomposition with time invariant factors.

REFERENCES

1. Anderson, B.D.O. and Kailath, T: Some integral equations with non-symmetric separable kernels, *SIAM J. Appl. Math.* 20 (4) (1971), 659-669.

2. Bart, H., Gohberg, I. and Kaashoek, M.A.: *Minimal Factorization of Matrix and Operator Functions.* Operator Theory: Advances and Applications, Vol. 1, Birkhäuser Verlag, Basel etc., 1979.

3. Bart, H., Gohberg, I., Kaashoek, M.A. and Van Dooren, P.: Factorizations of transfer functions, *SIAM J. Control Opt.* 18 (6) (1980), 675-696.

4. Daleckii, Ju. L. and Krein, M.G.: *Stability of solutions of differential equations in Banach space*, Transl. Math. Monographs, Vol. 43, Amer. Math. Soc., Providence R.I., 1974.

5. Gohberg, I. and Kaashoek, M.A.: Time varying linear systems with boundary conditions and integral operators, I. The transfer operator and its properties, *Integral Equations and Operator Theory* 7 (1984), 325-391.

6. Gohberg, I. and Kaashoek, M.A.: Time varying linear systems with boundary conditions and integral operators, II. Similarity and reduction, Report Nr. 261, Department of Mathematics and Computer Science, Vrije Universiteit, Amsterdam, 1984.

7. Gohberg, I. and Kaashoek, M.A.: Similarity and reduction for time varying linear systems with well-posed boundary conditions, *SIAM J. Control Opt.*, to appear.

8. Gohberg, I. and Kaashoek, M.A.: Minimal representations of semi-separable kernels and systems with separable boundary conditions, *J. Math. Anal. Appl.*, to appear.

9. Gohberg, I. and Kaashoek, M.A.: On minimality and stable minimality of time varying linear systems with well-posed boundary conditions, *Int. J. Control.* to appear.

10. Gohberg, I. and Kaashoek, M.A.: Various minimalities for systems with boundary conditions and integral operators, *Proceedings MTNS*, 1985, to appear.

11. Gohberg, I. and Krein, M.G.: *Theory and applications of Volterra operators in Hilbert space,* Transl. Math. Monographs, Vol. 24, Amer. Math. Soc., Providence R.I., 1970.

12. Kaashoek, M.A. and Woerdeman, H.J.: Unique minimal rank extensions of triangular operators, to appear.

13. Kailath, T.: Fredholm resolvents, Wiener-Hopf equations, and Riccati differential equations, *IEEE Trans. Information Theory*, vol. IT-15 (6) (1969), 665-672.

14 Krener, A.J.: Acausal linear systems, *Proceedings* 18-*th IEEE CDC*, Ft. Lauderdale, 1979.

15. Krener, A.J.: Boundary value linear systems, *Astérisque* 75/76 (1980), 149-165.

16. Krener, A.J.: Acausal realization theory, Part I; Linear deterministic systems, submitted for publication, 1986.

17. Levin, J.J.: On the matrix Riccati equation, *Proc. Amer. Math. Soc.* 10 (1959),519-524.

18. Schumitzky, A.: On the equivalence between matrix Riccati equations and Fredholm resolvents, *J. Computer and Systems Sciences* 2 (1) (1968), 76-87.

I. Gohberg
Dept. of Mathematical Sciences
The Raymond and Beverly Sackler
Faculty of Exact Sciences
Tel-Aviv University
Ramat-Aviv
Isreal

M.A. Kaashoek
Subfaculteit Wiskunde en Informatica
Vrije Universiteit
Postbus 7161
1007 MC Amsterdam
The Netherlands

PART II

NON-CANONICAL WIENER-HOPF FACTORIZATION

EDITORIAL INTRODUCTION

To explain the background of this part of the book consider

$$W(\lambda) = I_m - \int_{-\infty}^{\infty} e^{i\lambda t}k(t)dt, \qquad -\infty < \lambda < \infty,$$

where k is an $m\times m$ matrix-valued function of which the entries are in $L_1(-\infty,\infty)$ and I_m stands for the $m\times m$ identity matrix. A (*right Wiener-Hopf factorization*) of W relative to the real line is a factorization

$$(1) \qquad W(\lambda) = W_-(\lambda)\begin{pmatrix} \left(\frac{\lambda-i}{\lambda+i}\right)^{\kappa_1} & & \\ & \ddots & \\ & & \left(\frac{\lambda-i}{\lambda+i}\right)^{\kappa_m}\end{pmatrix} W_+(\lambda), \qquad -\infty < \lambda < \infty,$$

such that $\kappa_1,\dots,\kappa_m$ are integers, the factors W_- and W_+ are of the form

$$W_-(\lambda) = I_m - \int_{-\infty}^{0} e^{i\lambda t}k_1(t)dt, \qquad \operatorname{Im}\lambda \le 0,$$

$$W_+(\lambda) = I_m - \int_{0}^{\infty} e^{i\lambda t}k_2(t)dt, \qquad \operatorname{Im}\lambda \ge 0,$$

where k_1 and k_2 are $m\times m$ matrix functions with entries in $L_1(-\infty,0]$ and $L_1[0,\infty)$, respectively, and

$$\det W_-(\lambda) \ne 0 \quad (\operatorname{Im}\lambda \le 0), \qquad \det W_+(\lambda) \ne 0 \quad (\operatorname{Im}\lambda \ge 0).$$

The integers $\kappa_1 \ge \dots \ge \kappa_m$ are uniquely determined by W and called the (*right*) *factorization indices* of the matrix function. The existence of the factorization (1) is proved in the Gohberg-Krein paper (Amer. Math. Soc. Transl. (2) 14 (1960), 217-287) on factorization of matrix-valued functions. Its construction is done in two steps. The first involves the case when

$$\int_{-\infty}^{\infty} \|k(t)\| dt < 1.$$

In this case the factorization is just a canonical one (i.e., the indices are all zero) and the factors are obtained by an iterative procedure. The second step concerns matrix-valued functions that are analytic in the open upper half plane and continuous up to the boundary (infinity included). For matrix functions of the latter type the factors are obtained by a repeated applica tion of a special algorithm based on elementary matrix transformations and the factorization indices appear at the end of this procedure as multiplicities of zeros of column functions. In general, the two steps are combined by an appromixation of the original function W with a rational matrix-valued function. This construction of the factorization does not yield explicit formulas for the factors and it leads to the values of the indices in an implicit way only.

As in Part I let us represent the function W in the form

$$(2) \qquad W(\lambda) = I_m + C(\lambda I_n - A)^{-1}B, \qquad -\infty < \lambda < \infty,$$

where $A : \mathbb{C}^n \to \mathbb{C}^n$, $B : \mathbb{C}^m \to \mathbb{C}^n$ and $C : \mathbb{C}^n \to \mathbb{C}^m$ are linear operators and A has no eigenvalue on the real line. The main problem dealt with in this part of the book consists of constructing explicit formulas for the factors W_- and W_+ and the factorization indices $\kappa_1, \ldots, \kappa_m$ in terms of the three operators A, B and C. This problem is also considered for the case when the spaces $\mathbb{C}^n$ and $\mathbb{C}^m$ in the realization (2) are replaced by arbitrary Banach spaces and W is analytic on the real line and at infinity.

The first paper in this part "*Explicit Wiener-Hopf factorization and realization*", by H. Bart, I. Gohberg and M.A. Kaashoek, gives the solution of the problem mentioned above. As in the case of canonical factorization an important role is played by the spectral subspaces

$$(3) \qquad M = \operatorname{Im}\left(\frac{1}{2\pi i}\int_{\Gamma} (\lambda - A)^{-1} d\lambda\right),$$

$$(4) \qquad M^{\times} = \operatorname{Ker}\left(\frac{1}{2\pi i}\int_{\Gamma} (\lambda - A^{\times})^{-1} d\lambda\right),$$

where $A^{\times} = A - BC$ and Γ is a suitable contour in the open upper half plane around the parts of the spectra of A and $A^{\times}$ in the open upper half plane. The analysis of the spaces $M \cap M^{\times}$ and $M + M^{\times}$ leads to explicit formulas for the factorization indices and to the construction of incoming and outgoing bases.

The latter are used in a sophisticated way to obtain final formulas for the factors W_- and W_+. Since in general the factorization (1) is not minimal, one cannot use invariant subspace methods directly, but first one has to dilate the original realization in an appropriate way. The results obtained in this paper are also valid for the case when W is analytic on the real line and at infinity.

The second paper "*Invariants for Wiener-Hopf equivalence of analytic operator functions*", by H. Bart, I. Gohberg and M.A. Kaashoek, concerns operator functions that are analytic on the real line and at infinity. The main result gives a necessary condition for Wiener-Hopf equivalence of two such operator functions. On the basis of this the authors show that the necessary and sufficient conditions for the existence of a Wiener-Hopf factorization are

$$\dim(M \cap M^\times) < \infty, \qquad \operatorname{codim}(M + M^\times) < \infty,$$

where M and $M^\times$ are given by (3) and (4), respectively. In this way they also show that the minimality condition in the canonical factorization theorem (Theorem 1 in the Editorial Introduction of Part I) can be omitted.

The third paper "*Multiplication by diagonals and reduction to canonical factorization*", by H. Bart, I. Gohberg and M.A. Kaashoek, developes further the calculus for matrix functions in realized form. The operation of multiplication by diagonal rational matrix functions is analyzed in geometrical terms. As a corollary one obtains a method to reduce rational matrix functions to functions that admit canonical factorization.

The last paper "*Symmetric Wiener-Hopf factorization of self-adjoint rational matrix functions and realization*", by M.A. Kaashoek and A.C.M. Ran, solves the main problem of the present Part II for the case when the original matrix function W has selfadjoint values on the real line and the factorization is also required to be of symmetric type. The method developed in the first paper of this part is improved and adapted for symmetric factorization problems. It turns out that this already allows one to obtain final formulas for symmetric Wiener-Hopf factorization.

Operator Theory:
Advances and Applications, Vol. 21

EXPLICIT WIENER-HOPF FACTORIZATION AND REALIZATION

H. Bart, I. Gohberg, M.A. Kaashoek

Explicit formulas for Wiener-Hopf factorization of rational matrix and analytic operator functions relative to a closed contour are constructed. The formulas are given in terms of a realization of the functions. Also formulas for the factorization indices are given.

0. INTRODUCTION

Singular integral equations, Wiener-Hopf integral equations, equations with Toeplitz matrices and other types of equations can be solved when a Wiener-Hopf factorization of the symbol of the equation is known (see, e.g., [GK1,GF,GKr,K]), and in general the solutions of the equations can be obtained as explicit as the factors in the factorization are known. Recall that a <u>Wiener-Hopf factorization</u> relative to a simple closed contour Γ of a continuous operator function W on Γ is a representation of W in the form:

$$(0.1)\qquad W(\lambda) = W_-(\lambda)\{\Pi_0 + \sum_{j=1}^{r} \left(\frac{\lambda-\varepsilon_1}{\lambda-\varepsilon_2}\right)^{\kappa_j}\Pi_j\}W_+(\lambda), \quad \lambda \in \Gamma.$$

Here $\Pi_1,\ldots,\Pi_r$ are mutually disjoint one dimensional projections and $\Sigma_{j=0}^{r}\Pi_j$ is the identity operator. The point ε_1 lies in the inner domain of Γ and ε_2 is in the outer domain of Γ. The operator functions W_- and W_+ are analytic on the inner and outer domain of Γ, respectively, both W_- and W_+ are continuous up to the boundary and their values are invertible operators. The numbers $\kappa_1,\ldots,\kappa_r$ are non-zero integers, which are called <u>factorization indices</u>.

If W is a rational matrix function with no poles and zeros on Γ, then W admits a Wiener-Hopf factorization relative to Γ, and there exists an algorithm which allows one to find the factors in a finite number of steps (see, e.g., [CG], Section I.2). Only in the scalar case explicit formulas for the factors are available. In general, an arbitrary continuous operator or matrix function does not admit a Wiener-Hopf factorization, but necessary and sufficient conditions for its existence are known (see [CG,GL]).

The main goal of the present paper is to produce explicit formulas for Wiener-Hopf factorization, including formulas for the indices, for the case when on Γ the operator function W can be written in the form:

$$W(\lambda) = I + C(\lambda - A)^{-1}B. \tag{0.2}$$

Here $A: X \to X$, $B: Y \to X$ and $C: X \to Y$ are (bounded linear) operators acting between Banach spaces X and Y and the symbol I stands for the identity operator on Y. This covers the case of all rational matrix functions (that are analytic and invertible at infinity) and of all operator functions that are analytic in a neighbourhood of Γ. For canonical Wiener-Hopf factorization, i.e., all factorization indices are equal to zero, this goal was already achieved in [BGK1], Section 4.4. In [BGK1] we obtained the canonical Wiener-Hopf factorization explicity on the basis of a geometrical principle of factorization, which gives the factors in terms of invariant subspaces of the operators A and $A-BC$.

The present paper is a continuation of [BGK1]. The same geometrical principle of factorization is used, but the construction of the factors is much more involved. It turns out that one cannot apply directly the factorization principle to the operators A, B and C appearing in (0,2), but starting with the representation (0,2) one has to prepare in a special way a new triple of operators $\tilde{A},\tilde{B},\tilde{C}$ for which the representation (0.2) is also valid and which can be used to obtain the factors in terms of invariant subspaces of $\tilde{A}$ and $\tilde{A}-\tilde{B}\tilde{C}$.

Precisely what kind of special triples one has to look for ('centralized singularities') is explained in Section 3 of Chapter I. The first two sections of that chapter contain preliminary material. In Chapter II we introduce the so-called incoming characteristics. These have to do with the positive factorization indices. Also Wiener-Hopf factorizations are constructed explicitly for the case when all indices are non-negative. The dual concepts (outgoing characteristics) and Wiener-Hopf factorizations with non-positive indices are treated in Chapter III. Finally, Chapter IV contains the main result dealing with Wiener-Hopf factorizations without any a priori restriction on the sign of the factorization indices.

An earlier version of this paper appeared in the report [BGK2]. The main results were also announced in [BGK3]. We added Sections II.3 and III.3 which concern the cases of indices of one sign. The material presented in these sections makes our main factorization theorem more transparent and many years ago it was the starting point of this study. In the paper [KR], which also appears in this volume, the construction of the factorization of the present paper is developed further and applied to the selfadjoint case.

I. PRELIMINARIES

I.1. Preliminaries about transfer functions

In this paper we shall often think about analytic operator functions as transfer functions. In the present section we collect together some of the basic terminology related to this notion. A *node* (or a *system*) is a quintet $\Theta = (A,B,C;X,Y)$, where $A: X \to X$, $B: Y \to X$ and $C: X \to Y$ are (bounded linear) operators acting between complex Banach spaces X,Y. The space X is called the *state space*; the space Y is called the *input/output space*. The node Θ is called *finite dimensional* if both X and Y are finite dimensional spaces. The operator A is referred to as the *state space* or *main operator*. The spectrum of the operator A is denoted by $\sigma(A)$. The operator function

$$(1.1) \qquad W(\lambda) = I + C(\lambda-A)^{-1}B$$

is called the *transfer function* of the node $\Theta = (A,B,C;X,Y)$

and is denoted by $W_\Theta(\lambda)$. Here $\lambda-A$ stands for $\lambda I-A$ as usual and the transfer function has to be considered as an operator function which is analytic outside $\sigma(A)$ and at infinity. Part of our terminology is taken from systems theory where the transfer function describes the input/output behaviour of the linear dynamical system

$$\dot{x}(t) = Ax(t) + Bx(t), \qquad y(t) = Cx(t) + x(t).$$

The idea of a node also appears in the theory of characteristic operator functions as explained, for instance, in [Br] (see also [BGK1], Section 1.4).

If λ is not in the spectrum of A and $W(\lambda)$ is given by (1.1), then $W(\lambda)$ is invertible if and only if λ is not in the spectrum of the operator $A-BC$, and in that case

$$W(\lambda)^{-1} = I - C[\lambda-(A-BC)]^{-1}B.$$

The operator $A-BC$ is called the <u>associate</u> (<u>main</u>) <u>operator</u> of the node $\Theta = (A,B,C;X,Y)$ and is denoted by $A^\times$. Note that $A^\times$ depends not only on A but also on the other operators appearing in the node Θ.

Two nodes $\Theta_i = (A_i,B_i,C_i;X_i,Y)$, $i = 1,2$, are said to be <u>similar</u>, if there exists an invertible operator $S: X_1 \to X_2$, called <u>node</u> (or <u>system</u>) <u>similarity</u> between Θ_1 and Θ_2, such that

$$A_1 = S^{-1}A_2S, \qquad B_1 = S^{-1}B_2, \qquad C_1 = C_2S.$$

Similar nodes have the same transfer function, but the transfer function does not determine the node up to similarity.

The node $\Theta = (A,B,C;X,Y)$ is said to be a <u>dilation</u> of the node $\Theta_0 = (A_0,B_0,C_0;X_0,Y)$ if the state space X of Θ admits a decomposition $X = X_1 \oplus X_0 \oplus X_2$, where X_0 is the state space of Θ_0 and X_1,X_2 are closed subspaces of X, such that with respect to this decomposition the operators A, B and C have the following matrix representations:

$$A = \begin{pmatrix} * & * & * \\ 0 & A_0 & * \\ 0 & 0 & * \end{pmatrix}, \quad B = \begin{pmatrix} * \\ B_0 \\ 0 \end{pmatrix}, \quad C = [0 \quad C_0 \quad *].$$

One checks easily that this implies that Θ and Θ_0 have the same transfer function (on a neighbourhood of ∞).

Given a node $\Theta = (A,B,C;X,Y)$ we define

$$(1.2) \qquad \operatorname{Ker}(C|A) = \bigcap_{j=0}^{\infty} \operatorname{Ker} CA^j, \quad \operatorname{Im}(A|B) = \bigvee_{j=0}^{\infty} \operatorname{Im} A^j B.$$

Here $\bigvee_{j=0}^{\infty} N_j$ stands for the linear hull of the spaces $N_1, N_2, \ldots$. We say that the node Θ is <u>minimal</u> if $\operatorname{Ker}(C|A) = (0)$ and $\operatorname{Im}(A|B)$ is dense in X. This notion is of particular interest if Θ is a finite dimensional. The following two important results hold true: (1) Two minimal finite dimensional nodes with the same transfer function are similar (this result is known as the state space isomorphism theorem); (2) any finite dimensional node is a dilation of a minimal (finite dimensional) node (which by the state space isomorphism theorem is unique up to similarity). The proofs of these results can be found, e.g., in [BGK1], Section 3.2).

For the definition of the product of two nodes and the geometrical principle of factorization for transfer functions we refer to Section 1.1 of [BGK1].

The transfer function of a finite dimensional node may be seen as a rational matrix function which is analytic and has the value I at infinity. Conversely, any such function is the transfer function of a finite dimensional node. This is the so-called realization theorem (see [KFA]). Similarly, any operator function W which is analytic in an open neighbourhood of ∞ on the Riemann sphere and has the value I at infinity may be written in the form (1.1) and in that case we call the right hand side of (1.1) a <u>realization</u> of W. This term will also be used for the corresponding node $\Theta = (A,B,C;X,Y)$. In Chapter 2 of [BGK1] one finds a further discussion of these and related realization results.

I.2. Preliminaries about Wiener-Hopf factorization

Throughout this paper Γ is a contour on the Riemann sphere $\mathbb{C} \cup \{\infty\}$. By assumption Γ forms the positively oriented boundary of an open set with a finite number of components in $\mathbb{C} \cup \{\infty\}$ and Γ consists of a finite number of non-intersecting closed rectifiable Jordan curves. The inner domain of Γ we denote by Ω_Γ^+, and Ω_Γ^- stands for the open set on the Riemann sphere consisting of all points outside Γ. We fix two points $\varepsilon_1, \varepsilon_2$ in the complex plane such that $\varepsilon_1 \in \Omega_\Gamma^+$ and $\varepsilon_2 \in \Omega_\Gamma^-$.

Consider a continuous operator-valued function $W: \Gamma \to L(Y)$, where $L(Y)$ denotes the space of all bounded linear operators on the Banach space Y. A (right) Wiener-Hopf factorization of W with respect to Γ (and the points $\varepsilon_1, \varepsilon_2$) is a representation of W in the form

$$(2.1) \qquad W(\lambda) = W_-(\lambda)\{\Pi_0 + \sum_{j=1}^{r} \left(\frac{\lambda-\varepsilon_1}{\lambda-\varepsilon_2}\right)^{\kappa_j}\Pi_j\}W_+(\lambda)$$

for each $\lambda \in \Gamma$, where the factors have the following properties. By definition, $\Pi_1, \ldots, \Pi_r$ are mutually disjoint one dimensional projections of Y and $\Sigma_{j=0}^{r}\Pi_j$ is the identity operator on Y. The operator functions W_- and W_+ are holomorphic on the open sets Ω_Γ^- and Ω_Γ^+, respectively, both W_- and W_+ are continuous up to the boundary Γ and their values are invertible operators on Y. The numbers $\kappa_1, \ldots, \kappa_r$ are non-zero integers, called the (right) factorization indices, which are assumed to be arranged in increasing order $\kappa_1 \leqq \kappa_2 \leqq \cdots \leqq \kappa_r$.

In general, a Wiener-Hopf factorization (assuming it exists) is not unique, but the factorization indices are determined uniquely by W. If in (2.1) the projection Π_0 is the identity operator on Y, i.e., if

$$(2.2) \qquad W(\lambda) = W_-(\lambda)W_+(\lambda), \qquad \lambda \in \Gamma,$$

then the factorization is called <u>canonical</u> with respect to Γ. By interchanging in (2.1) and (2.2) the factors $W_-(\lambda)$ and $W_+(\lambda)$ one obtains a <u>left Wiener-Hopf factorization</u> and a <u>canonical left Wiener-Hopf factorization</u>, respectively.

If $W(\lambda)$ is a rational matrix function with no poles and zeros on Γ, then W admits a Wiener-Hopf factorization with respect to Γ and there exists an algorithm to construct the factors (see [GF,CG]). In general, a Wiener-Hopf factorization does not exist, but necessary and sufficient conditions for its existence are known. For example, if Y is a Hilbert space and Γ is closed and bounded, then W admits a right Wiener-Hopf factorization with respect to Γ if and only if the corresponding Toeplitz operator (acting on the space of all Y-valued L_2-functions on Γ with an analytic continuation to Ω_Γ^+) is Fredholm (see [GL]).

Now assume that W is the transfer function of the node $\Theta = (A,B,C;X,Y)$, i.e.,

$$(2.3) \qquad W(\lambda) = I + C(\lambda-A)^{-1}B, \qquad \lambda \in \Gamma.$$

To ensure that W is continuous on Γ we require that Γ does not intersect the spectrum $\sigma(A)$ of A, which implies that W is analytic on a neighbourhood of Γ. Conversely, if W is analytic on a neighbourhood of Γ (and is normalized to I at ∞ whenever $\infty \in \Gamma$), then one can construct explicitly operators A, B, and C such that (2.3) holds (see [BGK1], Section 2.3). An $m \times m$ rational matrix function W, which is analytic and has the value I at ∞, can be represented in the form (2.3) with A, B and C acting on finite dimensional spaces and the node $\Theta = (A,B,C;X,Y)$ being minimal. (In the latter case the continuity of W on Γ is equivalent to the statement that A has no eigenvalues on Γ.) By assumption $W(\lambda)$ has to be invertible for each $\lambda \in \Gamma$. Since $\sigma(A) \cap \Gamma = \emptyset$, this implies that the spectrum of the associate operator $A^\times = A-BC$ does not meet Γ (cf. Corollary 2.7 in [BGK1]). So we shall assume that W is the transfer function of a node $\Theta = (A,B,C;X,Y)$ <u>without</u>

<u>spectrum on the contour</u> Γ, which means that A and $A^\times$ have no spectrum on Γ.

Our aim is to find a Wiener-Hopf factorization of the function (2.3) in terms of the operators A, B and C. For canonical factorization we solved this problem in [BGK1]. The following theorem holds.

THEOREM 2.1. *Let W be the transfer function of a node $\Theta = (A,B,C;X,Y)$ without spectrum on Γ. Let M be the spectral subspace of A corresponding to the part of $\sigma(A)$ in Ω_Γ^+, and let $M^\times$ be the spectral subspace of $A^\times$ corresponding to the part of $\sigma(A^\times)$ in Ω_Γ^-. Then W admits a canonical right Wiener-Hopf factorization with respect to Γ if and only if*

$$(2.4) \qquad X = M \oplus M^\times.$$

If (2.4) is satisfied and Π is the projection of X along M onto $M^\times$, then a canonical Wiener-Hopf factorization of W with respect to Γ is given by $W(\lambda) = W_-(\lambda)W_+(\lambda)$, where

$$W_-(\lambda) = I + C(\lambda-A)^{-1}(I-\Pi)B,$$

$$W_+(\lambda) = I + C\Pi(\lambda-A)^{-1}B,$$

$$W_-(\lambda)^{-1} = I - C(I-\Pi)(\lambda-A^\times)^{-1}B,$$

$$W_+(\lambda)^{-1} = I - C(\lambda-A^\times)^{-1}\Pi B.$$

Note that the space M introduced in the above theorem is the largest A-invariant subspace such that the spectrum of $A|M$ is in Ω_Γ^+. Similarly, $M^\times$ is the largest $A^\times$-invariant subspace such that $\sigma(A^\times|M^\times)$ is in Ω_Γ^-. If Ω_Γ^+ is a bounded open set in $\mathbb{C}$, then

$$M = \operatorname{Im}\Big(\frac{1}{2\pi i}\int_\Gamma (\lambda-A)^{-1}d\lambda\Big),$$

$$M^\times = \operatorname{Ker}\Big(\frac{1}{2\pi i}\int_\Gamma (\lambda-A^\times)^{-1}d\lambda\Big).$$

We call M, $M^\times$ *the pair of spectral subspaces associated with*

Θ and Γ.

The sufficiency of the condition (2.4) and the formulas for W_- and W_+ are covered by Theorem 1.5 in [BGK1]. Theorem 4.9 in [BGK1] yields the necessity of condition (2.4) for the case when Θ is a minimal finite dimensional node. (The fact that in [BGK1] the set Ω_Γ^+ is assumed to be bounded is not important and the proofs of Theorems 1.5 and 4.9 in [BGK1] go through for the more general curves considered here.) The necessity of the condition (2.4) for general nodes is proved in [BGK4].

Let us now consider the case when $M \cap M^\times$ is non-trivial and/or $M + M^\times$ has a non-trivial complement in X. In that case W does not admit a canonical factorization. We have the following theorem.

THEOREM 2.2. *Let W be the transfer function of a node Θ without spectrum on Γ, and let M, $M^\times$ be the pair of spectral subspaces associated with Θ and Γ. Then W admits a right Wiener-Hopf factorization with respect to Γ if and only if*

(2.5) $$\dim(M \cap M^\times) < \infty, \quad \dim(X/M+M^\times) < \infty.$$

The sufficiency of the condition (2.5) in Theorem 2.2 is a corollary of the main factorization theorem proved in this chapter (cf. Theorem 3.1 in Ch.IV). The necessity of (2.5) is proved in [BGK2]. Note that (2.5) is automatically fulfilled if X is finite dimensional. In particular, (2.5) holds if Θ is a minimal node of a rational matrix function.

I.3. Reduction of factorization to nodes with centralized singularities

In this section $\Theta = (A,B,C;X,Y)$ is a node without spectrum on Γ. Let M, $M^\times$ be the pair of spectral subspaces associated with Θ and Γ. Under the condition that

(3.1) $$\dim(M \cap M^\times) < \infty, \quad \dim(X/M+M^\times) < \infty,$$

we want to give an explicit construction of the factors in a Wiener-Hopf factorization of $W(\lambda) = I + C(\lambda-A)^{-1}B$ in terms of A, B and C. To indicate how this may be done, note that condition (2.4) in Theorem 2.1 is equivalent to the statement that the state space X admits a decomposition into two closed subspaces, $X = X_1 \oplus X_2$, such that relative to this decomposition the operators A and $A^\times$ can be written in the form

$$A = \begin{pmatrix} A_1 & * \\ 0 & A_2 \end{pmatrix}, \quad A^\times = \begin{pmatrix} A_1^\times & 0 \\ * & A_2^\times \end{pmatrix},$$

where A_1 and $A_1^\times$ have their spectra in Ω_Γ^+ and the spectra of A_2 and $A_2^\times$ are in Ω_Γ^-. Furthermore, in this language the factors in the canonical factorization $W(\lambda) = W_-(\lambda)W_+(\lambda)$ are given by

$$W_-(\lambda) = I + C_1(\lambda-A_1)^{-1}B_1,$$

$$W_+(\lambda) = I + C_2(\lambda-A_2)^{-1}B_2,$$

where

$$B = \begin{pmatrix} B_1 \\ B_2 \end{pmatrix}, \quad C = [C_1 \quad C_2]$$

are the operator matrices of B and C relative to the decomposition $X = X_1 \oplus X_2$.

For the non-canonical case we shall see that in order to find a Wiener-Hopf factorization (relative to Γ and the points $\varepsilon_1, \varepsilon_2$) of the function $W(\lambda) = I + C(\lambda-A)^{-1}B$, one has to look for a decomposition of the space X into four closed subspaces, $X = X_1 \oplus X_2 \oplus X_3 \oplus X_4$, with X_2 and X_3 finite dimensional, such that with respect to this decomposition the operators A, $A^\times$, B and C can be written in the form

$$(3.2)\quad A = \begin{pmatrix} A_1 & * & * & * \\ 0 & A_2 & 0 & * \\ 0 & 0 & A_3 & * \\ 0 & 0 & 0 & A_4 \end{pmatrix}, \quad A^\times = \begin{pmatrix} A_1^\times & 0 & 0 & 0 \\ * & A_2^\times & 0 & 0 \\ * & 0 & A_3^\times & 0 \\ * & * & * & A_4^\times \end{pmatrix},$$

(3.3) $B = \begin{pmatrix} B_1 \\ B_2 \\ B_3 \\ B_4 \end{pmatrix}, \quad C = [C_1 \quad C_2 \quad C_3 \quad C_4],$

and the following properties hold:

(i) the spectra of A_1, $A_1^\times$ are in Ω_Γ^+;

(ii) the spectra of A_4, $A_4^\times$ are in Ω_Γ^-;

(iii) the operators $A_2-\varepsilon_1$ and $A_2^\times-\varepsilon_2$ have the same nilpotent Jordan normal form in bases

$\{d_{jk}\}_{k=1,j=1}^{\alpha_j\ \ t}$ and $\{e_{jk}\}_{k=1,j=1}^{\alpha_j\ \ t}$, respectively, i.e.,

(3.4a) $(A_2-\varepsilon_1)d_{jk} = d_{j,k+1}, \quad k = 1,\dots,\alpha_j,$

(3.4b) $(A_2^\times-\varepsilon_2)e_{jk} = e_{j,k+1}, \quad k = 1,\dots,\alpha_j,$

where $d_{j,\alpha_j+1} = e_{j,\alpha_j+1} = 0$ and $j = 1,\dots,t$, and

the two bases are related by

(3.5a) $e_{jk} = \sum_{\upsilon=0}^{k-1} \binom{k-1}{\upsilon}(\varepsilon_1-\varepsilon_2)^{\upsilon} d_{j,k-\upsilon},$

(3.5b) $d_{jk} = \sum_{\upsilon=0}^{k-1} \binom{k-1}{\upsilon}(\varepsilon_2-\varepsilon_1)^{\upsilon} e_{j,k-\upsilon},$

where $k = 1,\dots,\alpha_j$, $j = 1,\dots,t$;

(iv) rank B_2 = rank C_2 = t with t as in (iii);

(v) the operators $A_3^\times-\varepsilon_1$ and $A_3-\varepsilon_2$ have the same nilpotent Jordan form in bases $\{f_{jk}\}_{k=1,j=1}^{\omega_j \ s}$ and $\{g_{jk}\}_{k=1,j=1}^{\omega_j \ s}$, respectively, i.e.,

(3.6a) $(A_3^{\times}-\varepsilon_1)f_{jk} = f_{j,k+1}, \ k = 1,\dots,\omega_j,$

(3.6b) $(A_3-\varepsilon_2)g_{jk} = g_{j,k+1}, \ k = 1,\dots,\omega_j,$

where $f_{j,\omega_j+1} = g_{j,\omega_j+1} = 0$ and $j = 1,\dots,s$, and the two bases are related by

(3.7a) $$g_{jk} = \sum_{\upsilon=0}^{k-1} \binom{\upsilon+\omega_j-k}{\upsilon}(\varepsilon_2-\varepsilon_1)^{\upsilon} f_{j,k-\upsilon},$$

(3.7b) $$f_{jk} = \sum_{\upsilon=0}^{k-1} \binom{\upsilon+\omega_j-k}{\upsilon}(\varepsilon_1-\varepsilon_2)^{\upsilon} g_{j,k-\upsilon},$$

where $k = 1,\dots,\omega_j$ and $j = 1,\dots,s$;

(vi) rank B_3 = rankC_3 = s with s as in (v).

We shall suppose that the numbers α_j and ω_j appearing in (iii) and (v) are ordered in the following way:

$$\alpha_1 \geqq \alpha_2 \geqq \cdots \geqq \alpha_t, \quad \omega_1 \leqq \omega_2 \leqq \cdots \leqq \omega_s.$$

A node $\Theta = (A,B,C;X,Y)$ for which the state space X admits a decomposition $X = X_1 \oplus X_2 \oplus X_3 \oplus X_4$ with the properties described above will be called a <u>node with centralized singularities (relative to the contour Γ and the points $\varepsilon_1,\varepsilon_2$)</u>. Note that the operators A_2,A_3 and $A_2^{\times},A_3^{\times}$ appearing in the centers of A and $A^{\times}$ all act on a finite dimensional space and their spectra consist of a single eigenvalue which is either ε_1 or ε_2. It follows that the diagonal elements in the operator matrices (3.2) have no spectrum on Γ, and hence A and $A^{\times}$ have no spectrum on Γ. Thus a node with centralized singularities in a node without spectrum on Γ. Let M, $M^\times$ be the associated pair of spectral subspaces. From the fact that the spectra of A_1 and A_2 are in Ω_Γ^+ and the spectra of A_3 and A_4 in Ω_Γ^-, we conclude that $M = X_1 \oplus X_2$. In a similar way one shows that $M^\times = X_2 \oplus X_4$. It follows that $X_2 = M \cap M^\times$ and X_3 is a complement of $M+M^\times$ in X. In particular, we see that for a node with centralized singularities condition (3.1) is fulfilled. One of the main points in the definition of a node with centralized singularities is given by the following theorem.

THEOREM 3.1. *Let W be the transfer function of a node Θ with centralized singularities, i.e., for the node Θ the formulas (3.2), (3.3) and the properties (i) - (vi) hold true. Put*

$$W_-(\lambda) = I + C_1(\lambda-A_1)^{-1}B_1,$$
$$W_+(\lambda) = I + C_4(\lambda-A_4)^{-1}B_4,$$
$$D(\lambda) = I + C_2(\lambda-A_2)^{-1}B_2 + C_3(\lambda-A_3)^{-1}B_3.$$

Then

$$W(\lambda) = W_-(\lambda)D(\lambda)W_+(\lambda), \qquad \lambda \in \Gamma, \tag{3.8}$$

and the operator function $D(\lambda)$ is of the form

$$D(\lambda) = \Pi_0 + \sum_{j=1}^{t} \left(\frac{\lambda-\varepsilon_1}{\lambda-\varepsilon_2}\right)^{-\alpha_j}\Pi_{-j} + \sum_{j=1}^{s} \left(\frac{\lambda-\varepsilon_1}{\lambda-\varepsilon_2}\right)^{\omega_j}\Pi_j, \tag{3.9}$$

where $\Pi_{-t},\dots,\Pi_{-1},\Pi_1,\dots,\Pi_s$ are mutually disjoint one dimensional projections of Y and $\Sigma_{j=-t}^{s}\Pi_j$ is the identity operator on Y. Furthermore, the factorization (3.8) is a right Wiener-Hopf factorization of W relative to Γ and the points $\varepsilon_1,\varepsilon_2$. In particular, $-\alpha_1,\dots,-\alpha_t$, $\omega_1,\dots,\omega_s$ are the right factorization indices.

PROOF. Let $W(\lambda) = I + C(\lambda-A)^{-1}B$ with A, B and C as in

(3.2) and (3.3), and assume that the properties (i) - (vi) hold true. From the geometrical factorization principle (in Section 1.1 in [BGK1]) and the special form of the operator matrix representations for A and $A^{\times}$ in (3.2) it is readily seen that the factorization (3.8) holds true. Further, the spectral conditions on A_1, $A_1^{\times}$ and A_4, $A_4^{\times}$ (see the properties (i), (ii)) imply that the operator functions W_- and W_+ have the properties which are necessary for the factors in a right Wiener-Hopf factorization. Thus it remains to show that the middle term $D(\lambda)$ in the right hand side of (3.8) has the desired diagonal form. This will involve the properties of the center parts only.

From the special form of the operator matrix for $A-A^{\times}$ one easily sees that

$$(3.10)\qquad B_2C_2 = A_2-A_2^{\times}, \quad B_3C_2 = 0;$$

$$(3.11)\qquad B_2C_3 = 0, \quad B_3C_3 = A_3-A_3^{\times}.$$

We shall need the following identities:

$$(3.12)\qquad (A_2-A_2^{\times})e_{j\alpha_j} = \sum_{\upsilon=0}^{\alpha_j-1} \binom{\alpha_j}{\upsilon+1}(\varepsilon_1-\varepsilon_2)^{\upsilon+1} d_{j,\alpha_j-\upsilon};$$

$$(3.13)\qquad (A_2-A_2^{\times})e_{jk} = 0, \quad k = 1,\dots,\alpha_j-1;$$

$$(3.14)\qquad (A_3-A_3^{\times})g_{jk} = (\varepsilon_2-\varepsilon_1)^k\binom{\omega_j}{k}g_{j1}, \quad k = 1,\dots,\omega_j.$$

One proves these identities by straightforward calculation (cf., e.g., the proof of formula (1.11) in Section II.1 below).

Put $z_j = C_2e_{j\alpha_j}$ for $j = 1,\dots,t$. According to (3.10) we have $B_2z_j = B_2C_2e_{j\alpha_j} = (A_2-A_2^{\times})e_{j\alpha_j}$ and $B_3z_j = 0$. From (3.12) it is clear that the vectors $B_2z_1,\dots,B_2z_t$ are linearly independent. Since rank $B_2 = t$, we may conclude that

$$(3.15)\qquad Y = \operatorname{Ker} B_2 \oplus \operatorname{sp}\{z_1,\dots,z_t\}$$

and the vectors $z_1,\dots,z_t$ are linearly independent. Note that $z_1,\dots,z_t$ are in $\operatorname{Im} C_2$. But rank C_2 is also equal to t. Hence $\operatorname{Im} C_2 = \operatorname{sp}\{z_1,\dots,z_t\}$. We shall see that this implies that

$$(3.16)\qquad C_2 e_{jk} = 0, \quad k = 1,\dots,\alpha_j-1.$$

Indeed, take $1 \leq k \leq \alpha_j-1$. From $(A_2-A_2^{\times})e_{jk} = 0$ it is clear that $B_2C_2e_{jk} = 0$. It follows that $C_2e_{jk} \in \operatorname{Im} C_2 \cap \operatorname{Ker} B_2$, but the latter space consists of the zero element only (cf. (3.15)). From (3.16) and (3.5b) we may conclude that

$$(3.17)\qquad C_2 d_{j\alpha_j} = z_j, \quad C_2 d_{jk} = 0 \quad (k=1,\dots,\alpha_j-1).$$

Next, we note that

$$D(\lambda)z_j = z_j + C_2(\lambda-A_2)^{-1}B_2z_j + C_3(\lambda-A_3)^{-1}B_3z_j$$

$$= z_j + C_2(\lambda-A_2)^{-1}B_2z_j$$

$$= z_j + C_2\{\sum_{\beta=1}^{\infty} \frac{1}{(\lambda-\varepsilon_1)^{\beta}}(A_2-\varepsilon_1)^{\beta-1}(A_2-A_2^{\times})e_{j\alpha_j}\}$$

$$= z_j + C_2\{\sum_{\beta=1}^{\infty} \frac{1}{(\lambda-\varepsilon_1)^{\beta}}[\sum_{\upsilon=\beta-1}^{\alpha_j-1} \binom{\alpha_j}{\upsilon+1}(\varepsilon_1-\varepsilon_2)^{\upsilon}d_{j,\alpha_j-\upsilon+\beta-1}]\}$$

$$= z_j + \sum_{\beta=1}^{\alpha_j-1} \frac{1}{(\lambda-\varepsilon_1)^{\beta}}\binom{\alpha_j}{\beta}(\varepsilon_1-\varepsilon_2)^{\beta}z_j$$

$$= \left(\frac{\lambda-\varepsilon_2}{\lambda-\varepsilon_1}\right)^{\alpha_j}z_j.$$

Put $y_j = \frac{1}{(\varepsilon_1-\varepsilon_2)\omega_j}C_3g_{j1}$ for $j = 1,\dots,s$. Using the identities (3.11) and (3.14) we see that $B_2y_j = 0$ and $B_3y_j = g_{j1}$. It follows that $y_1,\dots,y_s$ are linearly independent vectors in $\operatorname{Ker} B_2$ and, since rank $B_3 = s$, there exists a closed subspace Y_0 of Y such that and $B_2Y_0 = B_3Y_0 = (0)$ (cf. (3.15)). Next, we prove

$$(3.19) \qquad C_3g_{j\upsilon} = (\varepsilon_2-\varepsilon_1)^{\upsilon}\binom{\omega_j}{\upsilon}y_j.$$

Obviously $C_3g_{j\upsilon} \in \operatorname{Im} C_3$. The vectors $y_1,\ldots,y_s$ are also in $\operatorname{Im} C_3$. Since rank C_3 is equal to s, it follows that $C_3g_{j\upsilon} = \Sigma_{k=1}^{s}\gamma_k y_k$. Now

$$B_3C_3g_{j\upsilon} = \sum_{k=1}^{s}\gamma_k B_3y_k = \sum_{k=1}^{s}\gamma_k g_{k1}.$$

On the other hand (using formula (3.14)):

$$B_3C_3g_{j\upsilon} = (A_3-A_3^{\times})g_{j\upsilon} = (\varepsilon_2-\varepsilon_1)^{\upsilon}\binom{\omega_j}{\upsilon}g_{j1}.$$

It follows that $\gamma_k = 0$ for $k \neq j$ and $\gamma_j = (\varepsilon_2-\varepsilon_1)^{\upsilon}\binom{\omega_j}{\upsilon}$, which proves (3.19). Note that

$$\begin{aligned}
D(\lambda)y_j &= y_j + C_2(\lambda-A_2)^{-1}B_2y_j + C_3(\lambda-A_3)^{-1}B_3y_j \\
&= y_j + C_3(\lambda-A_3)^{-1}B_3y_j \\
&= y_j + C_3\Big\{\sum_{\beta=1}^{\infty}\frac{1}{(\lambda-\varepsilon_2)^{\beta}}(A_3-\varepsilon_2)^{\beta-1}g_{j1}\Big\} \\
&= y_j + \sum_{\beta=1}^{\omega_j}\frac{1}{(\lambda-\varepsilon_2)^{\beta}}C_3g_{j\beta} \\
&= y_j + \sum_{\beta=1}^{\omega_j}\frac{1}{(\lambda-\varepsilon_2)^{\beta}}(\varepsilon_2-\varepsilon_1)^{\beta}\binom{\omega_j}{\beta}y_j \\
&= \Big(\frac{\lambda-\varepsilon_1}{\lambda-\varepsilon_2}\Big)^{\omega_j}y_j.
\end{aligned}$$

Obviously, $D(\lambda)y_0 = y_0$ for each $y_0 \in Y_0$.

For $1 \leq j \leq t$ define Π_{-j} to be the projection of Y onto the space spanned by z_j along the space spanned by Y_0, the vectors $y_1,\ldots,y_s$ and the vectors $z_k (k \neq j)$. Similarly, vectors $y_1,\dots,y_s$ and the vectors $z_k (k \neq j)$. Similarly, for $1 \leqq j \leqq s$ define Π_j to be the projection onto $sp\{y_j\}$ along the space spanned by Y_0, the vectors $z_1,\dots,z_t$ and the vectors $y_k (k \neq j)$. Further, we define Π_0 to be the projection of y onto Y_0 along the vectors $z_1,\dots,z_t$ and $y_1,\dots,y_s$. Then the projections $\Pi_{-t},\dots,\Pi_{-1},\Pi_0,\Pi_1,\dots,\Pi_s$ have the desired properties and formula (3.9) holds true.□

The converse of Theorem 3.1 also holds true, that is, if the operator function W is analytic on Γ (and normalized to I at ∞ if ∞ ∈ Γ) and if W admits a right Wiener-Hopf factorization with respect to Γ, then W is the transfer function of a node with centralized singularities. This is proved in [BGK4].

We say that a node $\Theta = (A,B,C;X,Y)$ with centralized singularities is <u>simple</u> if the spaces X_1 and X_4 consist of the zero element only. In that case $X = X_2 \oplus X_3$ with X_2 and X_3 finite dimensional and with respect to the decomposition $X = X_2 \oplus X_3$

$$A = \begin{pmatrix} A_2 & 0 \\ 0 & A_3 \end{pmatrix}, \qquad A^\times = \begin{pmatrix} A_2^\times & 0 \\ 0 & A_3^\times \end{pmatrix},$$

$$B = \begin{pmatrix} B_2 \\ B_3 \end{pmatrix}, \qquad C = [C_2 \quad C_3],$$

where the specified operators have the properties described in (iii) - (vi) above. A simple node with centralized singularities has by definition a finite dimensional state space. Moreover such a node is a minimal node. Note that the proof of Theorem 3.1 shows that the transfer function of a simple node with centralized singularities is the diagonal term in a Wiener-Hopf factorization. In [BGK4] it is proved that, conversely, any minimal realization of the diagonal term in a

Wiener-Hopf factorization is a simple node with centralized singularities.

Let $\Theta = (A,B,C;X,Y)$ be a node with centralized singularities, and assume that Θ is finite dimensional and minimal. Then the factorization (3.8) is a minimal factorization (cf. Theorem 4.8 in [BGK1]). However, between the factors in a Wiener-Hopf factorization there may be pole-zero cancellation and hence, in general, such a factorization is not a minimal one. It follows that a node without spectrum on Γ and satisfying condition (3.1) does not have to be a node with centralized singularities. The best one can hope for is that a dilation with a sufficiently large state space has centralized singularities.

It turns out that the dimension of the state space is not the only obstruction. It may happen that there exists a dilation Θ such that Θ does not have centralized singularities while on the other hand the state space dimension of Θ is as large as one likes. To see this we consider the following example.

EXAMPLE 3.2. Put $W(\lambda) = (\lambda-\varepsilon_1)/(\lambda-a)$, where $a \in \Omega_\Gamma^-$, $a \neq \varepsilon_2$. The scalar function $W(\lambda)$ is the transfer function of the minimal node $\Theta_0 = (a,1,a-\varepsilon_1;\mathbb{C},\mathbb{C})$ and

$$(3.20)\qquad W(\lambda) = \left(\frac{\lambda-\varepsilon_1}{\lambda-\varepsilon_2}\right)\left(\frac{\lambda-\varepsilon_2}{\lambda-a}\right), \qquad \lambda \in \Gamma,$$

is a right Wiener-Hopf factorization of W relative to Γ. Let $A_1: X_1 \to X_1$ be a linear operator on the n-dimensional linear space X_1, and assume that the Jordan normal form of A_1 consists of a single Jordan block with eigenvalue b, say. We shall suppose that $b \notin \Gamma$ and $b \neq a$, $b \neq \varepsilon_1$. Put $X = X_1 \oplus \mathbb{C}$, and consider

$$(3.21)\qquad A = \begin{pmatrix} A_1 & 0 \\ 0 & a \end{pmatrix}, \qquad B = \begin{pmatrix} 0 \\ 1 \end{pmatrix}, \qquad C = \begin{bmatrix} 0 & a-\varepsilon_1 \end{bmatrix}.$$

The node $\Theta = (A,B,C,;X,\mathbb{C})$ is a dilation of the node Θ_0 and the dimension of its state space is n+1. Observe that the associate main operator is given by

$$A^{\times} = \begin{pmatrix} A_1 & 0 \\ 0 & \varepsilon_1 \end{pmatrix}.$$

It follows that A and $A^{\times}$ have no spectrum on Γ.

Since $b \neq a$, any A-invariant subspace Z is of the form $Z = N \oplus L$, where N is an A_1-invariant subspace of X_1 and $L = (0)$ or $L = \mathbb{C}$. From the fact that the Jordan normal form of A_1 consists of a single Jordan block it follows that the invariant subspace structure of A_1 is very rigid and consists of a chain

$$(3.22) \qquad (0) \subset N_1 \subset N_2 \subset \dots \subset N_n = X_1$$

of n+1 subspaces. Since $\varepsilon_1 \neq b$, the invariant subspaces of $A^{\times}$ are of the same form as those of A. Now, let Z and $Z^{\times}$ be nonzero subspaces invariant under A and $A^{\times}$, respectively, and assume that $X = Z \oplus Z^{\times}$. We know that $Z = N \oplus L$ and $Z^{\times} = N^{\times} \oplus L^{\times}$, where N and $N^{\times}$ are members of the chain (3.22). Thus $N \subset N^{\times}$ or $N^{\times} \subset N$. But then the fact that $Z \cap Z^{\times} = (0)$ implies that either $N = (0)$ or $N^{\times} = (0)$. It follows that there are only two possibilities, namely $Z = X_1 \oplus (0)$, $Z^{\times} = (0) \oplus \mathbb{C}$ or $Z = (0) \oplus \mathbb{C}, Z^{\times} = X_1 \oplus (0)$. In the first case the corresponding factorization is $W(\lambda) = 1.W(\lambda)$ and the second case gives $W(\lambda) = W(\lambda).1$. Thus, the geometrical principle of factorization applied to the node Θ yields only trivial factorizations and does not give the Wiener-Hopf factorization (3.20). In this way one sees that the node Θ does not have centralized singularities.

To obtain a dilation with an infinite dimensional state space which does not have centralized singularities, we take for the operator A_1 in (3.21) a unicellular operator (see [GK2], Section I.9) which acts on an infinite dimensional Banach space X_1 and whose spectrum consists of the point b only. For example, we may take $X_1 = L_2[0,1]$ and

$$(A_1 f)(t) = b - 2i \int_t^1 f(s)ds.$$

Then the node $\Theta = (A,B,C;X,Y)$ has an infinite dimensional state space and with arguments similar to the ones used before one sees that Θ does not have centralized singularities.

From the previous example it is clear that not any dilation with a sufficiently large state space has the desired centralized singularities. To obtain the Wiener-Hopf factorization one has to construct a specific dilation. This is exactly the problem which is solved in the next sections. Starting with a node $\Theta = (A,B,C;X,Y)$ which has no spectrum on Γ and for which condition (3.1) is fulfilled, we shall construct explicitly in terms of A, B, C and the spaces M and $M^\times$ a dilation of Θ with centralized singularities. Among other things the following theorem will be proved.

THEOREM 3.3. *Let $\Theta = (A,B,C;X,Y)$ be a node without spectrum on Γ, and let $M,M^\times$ be the pair of spectral subspaces associated with Θ and Γ. Assume*

$$\dim(M \cap M^\times) < \infty, \quad \dim(X/M+M^\times) < \infty,$$

and let K be a complement of $M+M^\times$ in X. Then Θ may be dilated to a node $\hat{\Theta}$ with centralized singularities and with state space

$$\hat{X} = (M \cap M^\times) \oplus (M \cap M^\times) \oplus X \oplus K \oplus K.$$

II. INCOMING CHARACTERISTICS

II.1. Incoming bases

Let $\Theta = (A,B,C;X,Y)$ be a node without spectrum on Γ, and let $M,M^\times$ be the pair of spectral subspaces associated with Θ and Γ. In this section we study complements of $M+M^\times$ in X. Throughout this section we work under the assumption that

$$\text{(1.1)} \qquad \operatorname{codim}(M+M^\times) = \dim(X/M+M^\times) < \infty.$$

Since $M,M^\times$ are operator ranges, the same is true for $M+M^\times$, and hence (1.1) implies that $M+M^\times$ is closed in X.

We begin by considering the sequences of <u>incoming subspaces</u> for Θ. These subspaces are defined by

$$(1.2)\qquad H_j = M + M^\times + \operatorname{Im} B + \operatorname{Im} AB + \dots + \operatorname{Im} A^{j-1}B,$$

where $j = 0,1,2,\dots$. Of course H_0 is meant to be $M+M^\times$. Note that the spaces H_j do not change if in (1.2) the operator A is replaced by $A^\times$. Obviously $H_0 \subset H_1 \subset H_2 \subset \dots$. Since $H_0 = M+M^\times$ has finite codimension in X, not all inclusions are proper. Let ω be the smallest integer such that $H_\omega = H_{\omega+1}$. We claim that $H_\omega = X$. To prove this, first note that H_ω is invariant under A. Thus $H_\omega = M + M^\times + \operatorname{Im}(A|B)$ (cf. formula (1.2) in Ch.I), and the next lemma shows that $H_\omega = X$.

LEMMA 1.1. <u>The space</u> $M + M^\times + \operatorname{Im}(A|B)$ <u>is equal to</u> X.

PROOF. Write $X = X_1 \oplus X_0$, where $X_1 = M + M^\times + \operatorname{Im}(A|B)$. We know that X_1 is invariant under A and $A^\times$. Further, since $A^\times - A = BC$, the compressions of A and $A^\times$ to X_0 coincide. It follows that the operator matrices of A and $A^\times$ with respect to the decomposition $X = X_1 \oplus X_0$ are of the following form:

$$A = \begin{pmatrix} A_1 & * \\ 0 & A_0 \end{pmatrix}, \quad A^\times = \begin{pmatrix} A_1^\times & * \\ 0 & A_0 \end{pmatrix}.$$

Since A has no spectrum on Γ and X_o is finite dimensional, also A_0 has no spectrum on Γ. Let M_0 (resp. $M_0^\times$) be the spectral subspace of A_0 corresponding to the part of $\sigma(A_0)$ inside (resp. outside) Γ. Then $M_0 \subset M + X_1 = X_1$ and $M_0^\times \subset M^\times + X_1 = X_1$, and thus $M_0 = M_0^\times = \{0\}$. But then $\sigma(A_0)$ must be empty, which proves that $X_0 = \{0\}$. □

Let ε be a complex number. Note that the spaces H_j do not change if in (1.2) the operator A is replaced by $A-\varepsilon$.

It follows that

$$(1.3)\qquad H_1 + (A-\varepsilon)H_k = H_{k+1}, \qquad k = 1,2,\dots .$$

We shall use these identities to construct a system of vectors

$$(1.4)\qquad f_{jk}, \quad k = 1,\dots,\omega_j, \qquad j = 1,\dots,s$$

with the following properties:

(1) $1 \leqq \omega_1 \leqq \omega_2 \leqq \dots \leqq \omega_s$;

(2) $(A-\varepsilon)f_{jk} - f_{jk+1} \in M + M^{\times} + \operatorname{Im} B, \quad k = 1,\dots,\omega_j,$ where by definition $f_{j,\omega_j+1} = 0$;

(3) the vectors f_{jk}, $\min\{p \mid \omega_p \geqq k\} \leqq j \leqq s$, form a basis for H_k modulo H_{k-1}.

Such a system of vectors we shall call an <u>incoming basis</u> for Θ (with respect to the operator $A-\varepsilon$).

For all incoming bases for Θ the integers s and $\omega_1,\dots,\omega_s$ are the same and independent of the choice of ε. In fact

$$(1.5)\qquad s = \dim(H_1/H_0).$$

$$(1.6)\qquad \omega_j = \#\{k \mid s-\dim(H_k/H_{k-1}) \leqq j-1\}.$$

Both formulas are clear from condition (3). We call $\omega_1,\dots,\omega_s$ the <u>incoming indices</u> of the node Θ. Observe that $\omega_s = \omega$. The incoming indices are also given by the following identity:

$$(1.7)\qquad \{j \mid 1 \leqq j \leqq s,\ \omega_j = k\} = \dim(H_k/H_{k-1}) - \dim(H_{k+1}/H_k).$$

Note that for an incoming basis $\{f_{jk}\}_{k=1,j=1}^{\omega_j \; s}$ the following holds:

(3a) the vectors f_{jk}, $k = 1,\ldots,\omega_j$, $j = 1,\ldots,s$ form a basis for a complement of $M+M^{\times}$;

(3b) the vectors $f_{11},\ldots,f_{s1}$ form a basis of $M + M^{\times} + \operatorname{Im} B$ modulo $M+M^{\times}$.

Conversely, if (1.4) is a system of vectors such that (1), (2), (3a) and (3b) are satisfied, then the system (1.4) is an incoming basis for Θ. This is readily seen by using (1.3).

We now come to the construction of an incoming basis, which is based on (1.3) and uses a method employed in [GKS], Section 1.6. Put

$$s_k = \dim(H_1/H_0) - \dim(H_{k+1}/H_k), \quad k = 0,1,\ldots,\omega.$$

Obviously $s_\omega = \dim(H_1/H_0) = s$. Fix $k \in \{1,\ldots,\omega\}$. From (1.3) it is clear that $A-\varepsilon$ induces a surjective linear transformation

$$[A-\varepsilon]: H_k/H_{k-1} \to H_{k+1}/H_k.$$

This implies that $s_k \geqq s_{k-1}$ and $\dim \operatorname{Ker}[A-\varepsilon] = s_k - s_{k-1}$.

Assume $[A-\varepsilon][f] = 0$, where $[f]$ denotes the class in H_k/H_{k-1} containing f. Then there exists $g \in [f]$ such that $(A-\varepsilon)g \in H_1$. Indeed from $[A-\varepsilon][f] = 0$ it follows that $(A-\varepsilon)f \in H_k$ and so by (1.3) there exists $h \in H_{k-1}$ such that $(A-\varepsilon)(f-h) \in H_1$. The vector $g = f-h$ has the good properties. Using this result we can choose vectors f_{jk}, $s_{k-1}+1 \leqq j \leqq s_k$, in H_k, linearly independent modulo H_{k-1}, such that $[f_{s_{k-1}+1}],\ldots,[f_{s_k}]$ is a basis of $\operatorname{Ker}[A-\varepsilon]$ and

$$(A-\varepsilon)f_{jk} \in H_1, \quad s_{k-1}+1 \leqq j \leqq s_k.$$

Now let $f_{s_k+1,k+1},\ldots,f_{s,k+1}$ be a basis for H_{k+1} modulo

H_k. According to (1.3) there exist vectors $f_{jk}, s_k+1 \leqq j \leqq s$ in H_k such that

$$(A-\varepsilon)f_{jk} - f_{jk+1} \in H_1, \quad s_k+1 \leqq j \leqq s.$$

Clearly, the vectors f_{jk}, $s_{k-1}+1 \leqq j \leqq s$, form a basis of H_k modulo H_{k-1}. Repeating the construction for k-1 instead of k it is clear how one can get in a finite number of steps an incoming basis. Note that $\omega_j = k$ whenever $s_{k-1}+1 \leqq j \leqq s_k$.

Let the system of vectors (1.4) be an incoming basis for Θ with respect to $A-\varepsilon$. Put.

(1.8) $$K = \text{span}\{f_{jk} \mid k = 1,\ldots,\omega_j,\ j = 1,\ldots,s\}.$$

Clearly, $X = (M+M^\times) \oplus K$. Define $T: K \to K$ by

(1.9) $$(T-\varepsilon)f_{jk} = f_{jk+1}, \quad k = 1,\ldots,\omega_j,$$

where, as before, $f_{j,\omega_j+1} = 0$. We call T the <u>incoming operator</u> for Θ associated with the incoming basis (1.4). Note that with respect to the basis (1.4) the matrix of T has Jordan normal form with ε as the only eigenvalue and the sizes of the Jordan blocks are $\omega_1,\ldots,\omega_s$.

The next proposition shows how a given incoming basis with a parameter ε may be transformed into an incoming basis with a different complex parameter.

PROPOSITION 1.2. *Let ε_1 and ε_2 be complex numbers, and let the system* (1.4) *be an incoming basis for* Θ *with respect to* $A-\varepsilon_1$. *Put*

$$g_{jk} = \sum_{\upsilon=0}^{k-1} \binom{\upsilon+\omega_j-k}{\upsilon} (\varepsilon_2-\varepsilon_1)^{\upsilon} f_{j,k-\upsilon}. \tag{1.10}$$

Then $\{g_{jk}\}_{k=1,\,j=0}^{\omega_j,\ s}$ *is an incoming basis for* Θ *with respect to the operator* $A-\varepsilon_2$. *Further, if* T_1 *and* T_2 *are the incoming operators associated with the incoming bases* (1.4) *and* (1.10), *respectively, then* T_1 *and* T_2 *act on the same space* K *and*

$$(T_1-T_2)g_{jk} = -(\varepsilon_2-\varepsilon_1)^k \binom{\omega_j}{k} g_{j1} \tag{1.11}$$

for $k = 1,\dots,\omega_j$, $j = 1,\dots,s$.

PROOF. For $1 \leq k \leq \omega_j$ we have $g_{jk}-f_{jk} \in H_{k-1}$. It follows that condition (3) of an incoming basis holds for the vectors (1.10). To check condition (2), let us write $f \equiv g$ whenever $f-g \in H_1$. We have

$$(A-\varepsilon_2)g_{j\omega_j} = (A-\varepsilon_1)g_{j\omega_j} - (\varepsilon_2-\varepsilon_1)g_{j\omega_j}$$

$$\equiv \sum_{\upsilon=0}^{\omega_j-1} (\varepsilon_2-\varepsilon_1)^{\upsilon}(A-\varepsilon_1)f_{j,\omega_j-\upsilon} - \sum_{\upsilon=0}^{\omega_j-1} (\varepsilon_2-\varepsilon_1)^{\upsilon+1} f_{j,\omega_j-\upsilon}$$

$$\equiv \sum_{\upsilon=1}^{\omega_j-1} (\varepsilon_2-\varepsilon_1)^{\upsilon} f_{j,\omega_j-\upsilon+1} - \sum_{\upsilon=1}^{\omega_j} (\varepsilon_2-\varepsilon_1)^{\upsilon} f_{j,\omega_j-\upsilon+1}$$

$$\equiv -(\varepsilon_2-\varepsilon_1)^{\omega_j} f_{j1} \equiv 0.$$

This shows that $(A-\varepsilon_2)g_{j\omega_j} \in H_1$. Next, take $1 \leq k \leq \omega_j-1$. Then

$$(A-\varepsilon_2)g_{jk} = (A-\varepsilon_1)g_{jk} - (\varepsilon_2-\varepsilon_1)g_{jk}$$

$$\equiv \sum_{\upsilon=0}^{k-1} \binom{\upsilon+\omega_j-k}{\upsilon}(\varepsilon_2-\varepsilon_1)^{\upsilon} f_{j,k-\upsilon+1} - \sum_{\upsilon=0}^{k-1}\binom{\upsilon+\omega_j-k}{\upsilon}(\varepsilon_2-\varepsilon_1)^{\upsilon+1} f_{j,k-\upsilon}$$

$$\equiv f_{j,k+1} + \sum_{\upsilon=1}^{k}\binom{\upsilon+\omega_j-k}{\upsilon}(\varepsilon_2-\varepsilon_1)^{\upsilon} f_{j,k-\upsilon+1} +$$

$$- \sum_{\upsilon=1}^{k}\binom{\upsilon+\omega_j-(k+1)}{\upsilon-1}(\varepsilon_2-\varepsilon_1)^{\upsilon} f_{j,k-\upsilon+1}$$

$$\equiv f_{j,k+1} + \sum_{\upsilon=1}^{k}\binom{\upsilon+\omega_j-(k+1)}{\upsilon}(\varepsilon_2-\varepsilon_1)^{\upsilon} f_{j,k+1-\upsilon} \equiv g_{j,k+1}.$$

Thus $(A-\varepsilon_2)g_{jk}-g_{jk+1} \in H_1$. Hence the system (1.10) is an incoming basis.

Now consider the corresponding incoming operators. Clearly, T_1 and T_2 act on the same space. We have

$$(T_1-T_2)g_{jk} = (T_1-\varepsilon_1)g_{jk}-(T_2-\varepsilon_2)g_{jk}-(\varepsilon_2-\varepsilon_1)g_{jk}.$$

Take $k = \omega_j$. Then $(T_2-\varepsilon_2)g_{j\omega_j} = 0$, and one finds that

$$(T_1-T_2)g_{j\omega_j} = (T_1-\varepsilon_1)g_{j\omega_j} - (\varepsilon_2-\varepsilon_1)g_{j\omega_j}$$

$$= \sum_{\upsilon=0}^{\omega_j-2} (\varepsilon_2-\varepsilon_1)^{\upsilon} f_{j,\omega_j-\upsilon+1} - (\varepsilon_2-\varepsilon_1)g_{j\omega_j} = -(\varepsilon_2-\varepsilon_1)^{\omega_j} f_{j,1}.$$

Since $f_{j1} = g_{j1}$, this proves (1.11) for $k = \omega_j$.

Finally, take $1 \leqq k \leqq \omega_j-1$. We have

$$(T_1-T_2)g_{jk} = \sum_{\upsilon=0}^{k-1}\binom{\upsilon+\omega_j-k}{\upsilon}(\varepsilon_2-\varepsilon_1)^{\upsilon} f_{j,k-\upsilon+1} +$$

$$- \sum_{\upsilon=0}^{k}\binom{\upsilon+\omega_j-k-1}{\upsilon}(\varepsilon_2-\varepsilon_1)^{\upsilon} f_{j,k+1-\upsilon} - (\varepsilon_2-\varepsilon_1)g_{jk}$$

$$= -\binom{\omega_j-1}{k}(\varepsilon_2-\varepsilon_1)^k f_{j1} - (\varepsilon_2-\varepsilon_1)g_{jk} +$$

$$+ \sum_{\upsilon=1}^{k-1}\binom{\upsilon+\omega_j-k-1}{\upsilon-1}(\varepsilon_2-\varepsilon_1)^{\upsilon} f_{j,k+1-\upsilon}$$

$$= -\binom{\omega_j-1}{k}(\varepsilon_2-\varepsilon_1)^k f_{j1} - (\varepsilon_2-\varepsilon_1)g_{jk} +$$

$$+ \sum_{\upsilon=0}^{k-2}\binom{\upsilon+\omega_j-k}{\upsilon}(\varepsilon_2-\varepsilon_1)^{\upsilon+1} f_{j,k-\upsilon}$$

$$= -\binom{\omega_j-1}{k}(\varepsilon_2-\varepsilon_1)^k f_{j1} - \binom{\omega_j-1}{k-1}(\varepsilon_2-\varepsilon_1)^k f_{j1} = -\binom{\omega_j}{k}(\varepsilon_2-\varepsilon_1)^k f_{j1}.$$

Since $f_{j1} = g_{j1}$, the proposition is proved. □

If the vectors of two incoming bases $\{f_{jk}\}_{k=1,j=1}^{\omega_j \ \ s}$ and $\{g_{jk}\}_{k=1,j=1}^{\omega_j \ \ s}$ are related by (1.10), then reversely

$$f_{jk} = \sum_{\upsilon=0}^{k-1}\binom{\upsilon+\omega_j-k}{\upsilon}(\varepsilon_1-\varepsilon_2)^{\upsilon} g_{j,k-\upsilon}. \tag{1.12}$$

This is proved by direct checking. It follows that (1.11) may be replaced by

$$(T_1-T_2)f_{jk} = (\varepsilon_1-\varepsilon_2)^k\binom{\omega_j}{k}f_{j1}. \tag{1.13}$$

We conclude this section with a remark about the definition of a node with centralized singularities as given in the preceding section. Let $\Theta = (A,B,C;X,Y)$ be such a node, and let $\{f_{jk}\}_{k=1,j=1}^{\omega_j \ \ s}$ and $\{g_{jk}\}_{k=1,j=1}^{\omega_j \ \ s}$ be the bases of X_3 introduced in property (v) of a node with centralized singularities. Since the operator B_3 appearing in property (vi) of a node with centralized singularities has rank s, one checks easily that the basis $\{f_{jk}\}_{k=1,j=1}^{\omega_j \ \ s}$ is an incoming basis for Θ relative to the operator $A-\varepsilon_1$ and $A_3^{\times}$ is the corresponding incoming operator. Similarly, $\{g_{jk}\}_{k=1,j=1}^{\omega_j \ \ s}$ is an incoming basis for Θ relative to the operator $A-\varepsilon_2$ and A_3 is the corresponding incoming operator. The formulas (3.7a) and (3.7b) in Ch.I tell us that the two bases are related as in (1.10) and (1.12).

II.2. Feedback operators related to incoming bases

In this section we continue the study of incoming bases. Again $\Theta = (A,B,C;X,Y)$ is a node without spectrum on the contour Γ and $M,M^{\times}$ is the pair of spectral subspaces associated with Θ and Γ. We shall assume that

$$\dim(M \cap M^{\times}) < \infty, \quad \dim(X/M+M^{\times}) < \infty.$$

Throughout this section ε_1 and ε_2 are two fixed complex numbers, $\varepsilon_1 \in \Omega_\Gamma^+$ and $\varepsilon_2 \in \Omega_\Gamma^-$. A triple

$$(2.1) \qquad \mathcal{D}_{in} = \left(\{f_{jk}\}_{k=1,\,j=1}^{\omega_j\;\;s}, \{g_{jk}\}_{k=1,\,j=1}^{\omega_j\;\;s}, \{y_j\}_{j=1}^{s}\right)$$

is called a <u>triple of associated incoming data</u> for the node Θ (with respect to the contour Γ and the points $\varepsilon_1,\varepsilon_2$) if $\{f_{jk}\}_{k=1,\,j=1}^{\omega_j\;\;s}$ and $\{g_{jk}\}_{k=1,\,j=1}^{\omega_j\;\;s}$ are incoming bases for Θ with respect to the operators $A-\varepsilon_1$ and $A-\varepsilon_2$, respectively, and the following identities hold true:

$$(2.2) \qquad g_{jk} = \sum_{\upsilon=0}^{k-1} \binom{\upsilon+\omega_j-k}{\upsilon} (\varepsilon_2-\varepsilon_1)^{\upsilon} f_{j,k-\upsilon}, \qquad k = 1,\ldots,\omega_j;$$

$$(2.3) \qquad f_{j1}-By_j = g_{j1}-By_j \in M+M^{\times}, \qquad j = 1,\ldots,s.$$

The construction of a triple of incoming data is readily understood from the results of Section II.1. Indeed, if one starts with an incoming basis $\{f_{jk}\}_{k=1,\,j=1}^{\omega_j\;\;s}$ for Θ with respect to the operator $A-\varepsilon_1$, then according to Proposition 1.2 formula (6.2) defines an incoming basis $\{g_{jk}\}_{k=1,\,j=1}^{\omega_j\;\;s}$ for Θ with respect to the operator $A-\varepsilon_2$. Further, we know that the vectors $f_{11},\ldots,f_{s1}$ form a basis of $M+M^{\times}+\operatorname{Im} B$ modulo $M+M^{\times}$; thus there exist vectors $y_1,\ldots,y_s$ in Y such that $f_{j1}-By_j \in M+M^{\times}$ for $j = 1,\ldots,s$. Since $f_{j1} = g_{j1}$, we see that (2.3) holds true. Note that the vectors $y_1,\ldots,y_s$ form a basis of Y modulo $B^{-1}[M+M^{\times}]$; in particular

(2.4) $$Y = \mathrm{sp}\{y_1,\ldots,y_s\} \oplus B^{-1}[M+M^{\times}].$$

With a triple of associated incoming data $\mathcal{D}_{in}$ is related in a natural way a complement of $M+M^{\times}$, namely the space K spanned by the vectors f_{jk}, $k = 1,\ldots,\omega_j$, $j = 1,\ldots,s$. Of course, cf. (2.2), the space K is also spanned by the vectors g_{jk}, $k = 1,\ldots,\omega_j$, $j = 1,\ldots,s$. The incoming operator corresponding to the first incoming basis in $\mathcal{D}_{in}$ is denoted by T_1 and the incoming operator corresponding to the second incoming basis in $\mathcal{D}_{in}$ is denoted by T_2; both T_1 and T_2 act on the subspace K.

Two operators F_1,F_2: $K \to Y$ are called a <u>pair of feedback operators corresponding to</u> $\mathcal{D}_{in}$ if

(2.5) $$Ax - T_1x + BF_1x \in M+M^{\times} \quad (x \in K);$$

(2.6) $$A^{\times}x - T_2x - BF_2x \in M+M^{\times} \quad (x \in K);$$

(2.7) $$(C+F_1 + F_2)f_{jk} = \binom{\omega_j}{k}(\varepsilon_1-\varepsilon_2)^k y_j.$$

From (1.10) one easily deduces that instead of (2.7) we may write

(2.8) $$(C + F_1 + F_2)g_{jk} = -\binom{\omega_j}{k}(\varepsilon_2-\varepsilon_1)^k y_j.$$

The term "feedback operator" is taken from systems theory, where in general this term refers to an operator from the state space X into the input space Y (cf. [KFA,W]).

The construction of a pair of feedback operators F_1,F_2 proceeds as follows. First, recall that $Af_{jk} - T_1f_{jk}$ is an element of $M+M^{\times}$ + Im B. So for some $u_{jk} \in Y$ we have $Af_{jk} - T_1f_{jk} + Bu_{jk} \in M+M^{\times}$. Now define F_1: $K \to Y$ by setting $F_1f_{jk} = u_{jk}$. Then F_1 satisfies (2.5). Next, we choose F_2: $K \to Y$ such that (2.7) holds. This defines F_2 uniquely, and it remains to show that (2.6) is satisfied. According to (1.13) and (2.7) we have

$$A^{\times}f_{jk} - T_2 f_{jk} - BF_2 f_{jk} =$$

$$= Af_{jk} - T_1 f_{jk} + BF_1 f_{jk} + (T_1 - T_2)f_{jk} - B(C+F_1+F_2)f_{jk}$$

$$= Af_{jk} - T_1 f_{jk} + BF_1 f_{jk} + (\varepsilon_1 - \varepsilon_2)^k \binom{\omega_j}{k}(f_{j1} - By_j),$$

which belongs to $M+M^{\times}$ by (2.5) and (2.3). This completes the construction of the operators F_1, F_2.

A pair of feedback operators F_1, F_2 corresponding to a triple of associated incoming data $\mathcal{D}_{in}$ is said to be an improved pair of feedback operators if the following additional properties hold true:

(2.9) $$Z_1 x := Ax - T_1 x + BF_1 x \in M \quad (x \in K);$$

(2.10) $$Z_2 x := A^{\times}x - T_2 x - BF_2 x \in M^{\times} \quad (x \in K);$$

(2.11) $$Z_1 x = Z_2 x = 0 \text{ for } x \in K \cap \operatorname{Ker}(C+F_1+F_2).$$

This last condition is automatically fulfilled whenever $M \cap M^{\times} = (0)$. To see this we subtract (2.10) from (2.9) which yields the following identity

(2.12) $$B(C+F_1+F_2)x = Z_1 x + (T_1 - T_2)x - Z_2 x \quad (x \in K).$$

Now, if $M \cap M^{\times} = (0)$, then $X = M \oplus K \oplus M^{\times}$, and hence the three terms in the right hand side of (2.12) are zero whenever $(C+F_1+F_2)x = 0$. In particular, (2.11) holds true.

Any triple of associated incoming data $\mathcal{D}_{in}$ can be transformed into a new triple of associated incoming data which has an improved pair of feedback operators. To see this start with the triple $\mathcal{D}_{in}$ given by (2.1), and let F_1, F_2 be an arbitrary pair of feedback operators for this triple. Since $M \cap M^{\times}$ is finite dimensional, one can construct closed subspaces L and $L^{\times}$ such that

(2.13) $M+M^{\times} = L \oplus L^{\times}, \quad L \subset M, \quad L^{\times} \subset M^{\times}.$

So, according to formulas (2.5) and (2.6) there exist operators $V_1, V_2: K \to L$ and $V_1^{\times}, V_2^{\times}: K \to L^{\times}$ such that

(2.14) $Ax - T_1x + BF_1x = V_1x + V_1^{\times}x \quad (x \in K);$

(2.15) $A^{\times}x - T_1x - BF_2x = V_2x + V_2^{\times}x \quad (x \in K).$

Next we use that ε_1 is inside Γ. So the spectra of T_1 and of the restriction $A^{\times}|M^{\times}$ are disjoint. Hence there exists a unique operator $U_1: K \to M^{\times}$ such that

(2.16) $U_1T_1x - A^{\times}U_1x = V_1^{\times}x \quad (x \in K).$

Similarly, using that ε_2 is outside Γ, there exists a unique operator $U_2: K \to M$ such that

(2.17) $U_2T_2x - AU_2x = V_2x \quad (x \in K).$

Put

(2.18) $f'_{jk} = f_{jk} + U_1f_{jk} + U_2f_{jk}, \quad g'_{jk} = g_{jk} + U_1g_{jk} + U_2g_{jk};$

(2.19) $F_1'f'_{jk} = (F_1-CU_1)f_{jk}, \quad F_2'f'_{jk} = (F_2-CU_2)f_{jk}.$

LEMMA 2.1. *The triple*

$$\mathcal{D}'_{in} = (\{f'_{jk}\}_{k=1,\, j=1}^{\omega_j \quad s}, \{g'_{jk}\}_{k=1,\, j=1}^{\omega_j \quad s}, \{y_j\}_{j=1}^{s})$$

is a triple of associated incoming data and F_1', F_2' *is an improved pair of feedback operators corresponding to* $\mathcal{D}'_{in}$.

PROOF. For $x \in K$ we have

$$A(x+U_1x+U_2x) = Ax + AU_1x + AU_2x =$$

$$= T_1x+U_1T_1x+U_2T_1x-BF_1x+V_1x+V_1^{\times}x+(A^{\times}U_1x-U_1T_1x) +$$

$$+ BCU_1x+(AU_2x-U_2T_2x) + U_2(T_2-T_1)x$$

$$= T_1x+U_1T_1x+U_2T_1x-B(F_1-CU_1)x + (V_1-V_2)x + U_2(T_2-T_1)x.$$

Recall that V_1,V_2 and U_2 map K into M. It follows that

$$(A-\varepsilon_1)f'_{jk} - f'_{j,k+1} =$$

$$= (A-\varepsilon_1)(f_{jk}+U_1f_{jk}+U_2f_{jk}) +$$

$$-\{(T_1-\varepsilon_1)f_{jk} + U_1(T_1-\varepsilon_1)f_{jk} + U_2(T_1-\varepsilon_1)f_{jk}\} =$$

$$= - B(F_1-CU_1)f_{jk} + (V_1-V_2)f_{jk} + U_2(T_2-T_1)f_{jk},$$

which is an element of $M+\mathrm{Im}B$. Since $f'_{jk}-f_{jk}$ is an element of $M+M^{\times}$, it is clear that the vectors f'_{jk}, $k = 1,\ldots,\omega_j$, $j = 1,\ldots,s$, form an incoming basis for Θ with respect to the operator $A-\varepsilon_1$. Further, F'_1 is a feedback operator which has the property laid down in (2.9).

Next one computes that

$$A^{\times}(x+U_1x+U_2x) = T_2x + U_1T_2x + U_2T_2x +$$

$$+ B(F_2-CU_2)x + (V_2^{\times}-V_1^{\times})x + U_1(T_1-T_2)x$$

for each $x \in K$. Since $V_1^{\times},V_2^{\times}$ and U_1 map K into $M^{\times}$, we may conclude that

$$A^{\times}(x+U_1x+U_2x) - (T_2x+U_1T_2x+U_2T_2x) +$$

$$- B(F_2-CU_2)x \in M^{\times} + \mathrm{Im}\, B \qquad (x \in K).$$

In particular, $(A-\varepsilon_2)g'_{jk} - g'_{jk+1} \in M^{\times} + \mathrm{Im}B$, where $g'_{j\omega_j+1} = 0$. Since $g'_{jk}-g_{jk} \in M+M^{\times}$, it is clear that the vectors g'_{jk}, $k = 1,\ldots,\omega_j$, $j = 1,\ldots,s$, form an incoming basis for Θ with respect to the operator $A-\varepsilon_2$. Note that F'_2 is a feedback operator which has the property laid down in (2.10).

Since g_{jk} and f_{jk} are related through formula (2.2), this formula remains true if g_{jk} is replaced by g'_{jk} and f_{jk} by f'_{jk}. From $f_{j1}-f'_{j1} \in M+M^{\times}$ and $g_{j1}-g'_{j1} \in M+M^{\times}$ we conclude that (2.3) also holds with f_{j1} replaced by f'_{j1} and g_{j1} by g'_{j1}. So we have proved that the triple $\mathcal{D}'_{in}$ is a triple of associated incoming data. Since

$$(2.20)\qquad (C+F'_1+F'_2)f'_{jk} = (C+F_1+F_2)f_{jk},$$

it is clear that the operators F'_1,F'_2 form a pair of feedback operators corresponding to $\mathcal{D}'_{in}$.

It remains to establish the analogue of (2.11) for F'_1,F'_2. Take $a \in K$, and consider $b = a + U_1a + U_2a$. Assume $(C+F'_1+F'_2)b = 0$. According to (2.20) this implies that $(C+F_1+F_2)a = 0$. Subtracting (2.14) from (2.15) for $x = a$ yields

$$(V_2-V_1)a + (T_2-T_1)a + (V_2^{\times}-V_1^{\times})a = 0.$$

Note that $(V_2-V_1)a \in L$, $(T_2-T_1)a \in K$ and $(V_2^{\times}-V_1^{\times})a \in L^{\times}$. From our choice of L and $L^{\times}$ we know that $X = L \oplus K \oplus L^{\times}$. Thus $(V_2-V_1)a = 0$, $(T_2-T_1)a = 0$ and $(V_2^{\times}-V_1^{\times})a = 0$. But then it follows that

$$A(a+U_1a+U_2a) - (T_1a+U_1T_1a+U_2T_1a) + B(F_1-CU_1)a = 0;$$

$$A^{\times}(a+U_1a+U_2a) - (T_2a+U_1T_2a+U_2T_2a) - B(F_2-CU_2)a = 0.$$

The last two identities form the analogue of (2.11) for F'_1,F'_2. Thus F'_1,F'_2 is an improved pair of feedback operators.□

II.3. Factorization with non-negative indices

In this section we assume that $M \cap M^{\times} = (0)$, and thus

$$X = M \oplus K \oplus M^{\times}. \tag{3.1}$$

We prove that a transfer function with this property admits a right Wiener-Hopf factorization with non-negative indices. In fact, using the incoming characteristics and the associated feedback operators we can describe explicitly the factors in the Wiener-Hopf factorization including the diagonal terms.

THEOREM 3.1. *Let W be the transfer function of a node $\Theta = (A,B,C;X,Y)$ without spectrum on the contour Γ, and assume that*

$$M \cap M^{\times} = (0), \qquad \dim(X/M+M^{\times}) < \infty, \tag{3.2}$$

where $M, M^{\times}$ is the pair of spectral subspaces associated with Θ and Γ. Then W admits a right Wiener-Hopf factorization with respect to Γ,

$$W(\lambda) = W_-(\lambda)D(\lambda)W_+(\lambda), \qquad \lambda \in \Gamma, \tag{3.3}$$

with factors given by

$$W_-(\lambda) = I + (C+F_1P_2)(P_1+P_2)[\lambda-(A-BCP_3+BF_1P_2)]^{-1}B, \tag{3.4}$$

$$W_-(\lambda)^{-1} = I - (C+F_1P_2)(P_1+P_2)(\lambda-A^{\times})^{-1}B, \tag{3.5}$$

$$W_+(\lambda) = I + (C+F_2P_2)(P_2+P_3)(\lambda-A)^{-1}B, \tag{3.6}$$

$$W_+(\lambda)^{-1} = I - (C+F_2P_2)(P_2+P_3)[\lambda-(A^{\times}+BCP_1-BF_2P_2)]^{-1}B, \tag{3.7}$$

$$D(\lambda)y = \begin{cases} \left(\frac{\lambda-\varepsilon_1}{\lambda-\varepsilon_2}\right)^{\omega_j} y_j, & y = y_j \quad (j=1,\dots,s), \\ y \quad , & y \in B^{-1}[M+M^{\times}]. \end{cases} \tag{3.8}$$

Here $F_1,F_2\colon K \to Y$ *is an improved pair of feedback operators corresponding to a triple*

$$\mathcal{D}_{in} = \left(\{g_{jk}\}_{k=1,\,j=1}^{\omega_j\quad s},\ \{g_{jk}\}_{k=1,\,j=1}^{\omega_j\quad s},\{y_j\}_{j=1}^{s}\right)$$

of associated incoming data for the node Θ (*relative to* Γ *and the points* $\varepsilon_1,\varepsilon_2$), *and for* $i = 1,2,$ *and* 3 *the operator* P_i *stands for the projection of* X *onto the* i-*th space in the decomposition* $M \oplus K \oplus M^\times$ *along the other spaces in this decomposition.*

PROOF. Note that

$$(3.9)\qquad A^\times + B(C+F_1P_2)(P_1+P_2) = A - BCP_3 + BF_1P_2,$$

$$(3.10)\qquad A - B(C+F_2P_2)(P_2+P_3) = A^\times + BCP_1 - BF_2P_2.$$

Let $T_1,T_2\colon K \to K$ be the incoming operators for Θ corresponding to the incoming data $\mathcal{D}_{in}$, and let $Z_1\colon K \to M$ and $Z_2\colon K \to M^\times$ be the operators defined by (2.9) and (2.10), respectively. Write A and $A^\times$ as 3×3 operator matrices relative to the decomposition $X = M \oplus K \oplus M^\times$:

$$(3.11)\qquad A = \begin{pmatrix} A_{11} & A_{12} & A_{13} \\ 0 & A_{22} & A_{23} \\ 0 & A_{32} & A_{33} \end{pmatrix},\quad A^\times = \begin{pmatrix} A^\times_{11} & A^\times_{12} & 0 \\ A^\times_{21} & A^\times_{22} & 0 \\ A^\times_{31} & A^\times_{32} & A^\times_{33} \end{pmatrix}.$$

Then, relative to the decomposition $X = M \oplus K \oplus M^\times$,

$$(3.12)\qquad A - BCP_3 + BF_1P_2 = \begin{pmatrix} A_{11} & Z_1 & 0 \\ 0 & T_1 & 0 \\ 0 & 0 & A^\times_{33} \end{pmatrix},$$

$$(3.13) \qquad A^{\times} + BCP_1 - BF_2P_2 = \begin{pmatrix} A_{11} & 0 & 0 \\ 0 & T_2 & 0 \\ 0 & Z_2 & A_{33}^{\times} \end{pmatrix}.$$

We see that as A and $A^{\times}$, the operators $A - BCP_3 + BF_1P_2$ and $A^{\times} + BCP_1 - BF_2P_2$ have no spectrum on Γ. But then we can use (3.9) and (3.10) to show that indeed for each $\lambda \in \Gamma$ the right hand side of (3.5) is the inverse of the right hand side of (3.4) and the right hand side of (3.7) is the inverse of the right hand side of (3.6).

Next, we compute $W_-(\lambda)^{-1}W(\lambda)W_+(\lambda)^{-1}$ for $\lambda \in \Gamma$. Take $\lambda \in \Gamma$. Then

$$W_-(\lambda)^{-1}W(\lambda) = W_-(\lambda)^{-1} + W_-(\lambda)^{-1}C(\lambda-A)^{-1}B$$

$$= W_-(\lambda)^{-1} + C(\lambda-A)^{-1}B + $$
$$+ (C+F_1P_2)(P_1+P_2)(\lambda-A^{\times})^{-1}(A^{\times}-A)(\lambda-A)^{-1}B$$

$$= I + C(\lambda-A)^{-1}B - (C+F_1P_2)(P_1+P_2)(\lambda-A)^{-1}B$$

$$= I + (CP_3-F_1P_2)(\lambda-A)^{-1}B.$$

Next

$$W_-(\lambda)^{-1}W(\lambda)W_+(\lambda)^{-1} = W_-(\lambda)^{-1}W(\lambda) +$$
$$- (C+F_2P_2)(P_2+P_3)[\lambda-(A^{\times}+BCP_1-BF_2P_2]^{-1}B +$$
$$(CP_3-F_1P_2)(\lambda-A)^{-1}(A^{\times}+BCP_1-BF_2P_2-A)\cdot$$
$$\cdot\ [\lambda-(A^{\times}+BCP_1-BF_2P_2)]^{-1}B$$

$$= I - (C+F_2P_2)(P_2+P_3)[\lambda-(A^{\times}+BCP_1-BF_2P_2)]^{-1}B$$

$$+ (CP_3-F_1P_2)[\lambda-(A^{\times}+BCP_1-BF_2P_2)]^{-1}B$$

$$= I - (C + F_1+F_2)P_2[\lambda-(A^{\times}+BCP_1-BF_2P_2)]^{-1}B.$$

From (3.13) we see that

$$P_2[\lambda-(A^{\times}+BCP_1-BF_2P_2)]^{-1} = P_2(\lambda-T_2)^{-1}P_2,$$

and thus

$$W_-(\lambda)^{-1}W(\lambda)W_+(\lambda)^{-1} = I - (C+F_1+F_2)P_2(\lambda-T_2)^{-1}P_2B$$

$$= \begin{cases} (\frac{\lambda-\varepsilon_1}{\lambda-\varepsilon_2})^{\omega_j}y_j, & y = y_j \ (j=1,\dots,s), \\ \lambda & y \in B^{-1}[M+M^{\times}]. \end{cases}$$

Hence $W_-(\lambda)^{-1}W(\lambda)W_+(\lambda)^{-1} = D(\lambda)$ for $\lambda \in \Gamma$ and (3.3) is proved.

Put

$$A_o = \begin{pmatrix} A_{22} & A_{23} \\ A_{32} & A_{33} \end{pmatrix}, \quad A_o^{\times} = \begin{pmatrix} A_{11}^{\times} & A_{12}^{\times} \\ A_{21}^{\times} & A_{22}^{\times} \end{pmatrix}.$$

Since M is the spectral subspace of A corresponding to the part of $\sigma(A)$ in Ω_Γ^+ the operator A_o has its spectrum in Ω_Γ^-. In a similar way $\sigma(A_o^{\times}) \subset \Omega_\Gamma^+$. Next, observe that

$$(P_2+P_3)(\lambda-A)^{-1} = (P_2+P_3)(\lambda-A)^{-1}(P_2+P_3)$$

$$= (P_2+P_3)(\lambda-A_o)^{-1}(P_2+P_3), \quad \lambda \in \Gamma.$$

It follows that $W_+(\cdot)$ has an analytic continuation defined on the closure of Ω_Γ^+. From

$$(P_1+P_2)(\lambda-A^{\times})^{-1} = (P_1+P_2)(\lambda-A^{\times})^{-1}(P_1+P_2)$$

$$= (P_1+P_2)(\lambda-A_o^{\times})^{-1}(P_1+P_2), \quad \lambda \in \Gamma,$$

we may conclude that $W_-(\cdot)$ has an analytic continuation defined on the closure of Ω_Γ^-. Using (3.12) (resp. (3.13)) it is easy to see that $W_-(\cdot)$ (resp. $W_+(\cdot)^{-1}$) has an analytic continuation defined on the closure of Ω_Γ^- (resp. Ω_Γ^+). It follows that (3.3) is a right Wiener-Hopf factorization and the theorem is proved.□

III. OUTGOING CHARACTERISTICS

III.1. Outgoing bases

Continuing the discussion of the preceding section, we now assume (instead of (1.1)in Ch.II)) that

$$(1.1)\qquad \dim(M \cap M^\times) < \infty.$$

Under this assumption we shall investigate the structure of $M \cap M^\times$.

We begin by considering the sequence of outgoing subspaces of Θ. These subspaces are defined by

$$(1.2)\qquad K_j = M \cap M^\times \cap \operatorname{Ker} C \cap \operatorname{Ker} CA \cap \dots \cap \operatorname{Ker} CA^{j-1},$$

where $j = 0,1,2,\dots$ .For $j = 0$ the right hand side of (1.2) is interpreted as $M \cap M^\times$. The spaces K_j do not change when in (1.2) the operator A is replaced by $A^\times$. Clearly $K_0 \supset K_1 \supset K_2 \supset \dots$ with not all inclusions proper because of (1.1). Let α be the smallest non-negative integer such that $K_\alpha = K_{\alpha+1}$. We claim that $K_\alpha = (0)$. To see this, note that $K_\alpha = M \cap M^\times \cap \operatorname{Ker}(C|A)$ and apply the following lemma.

LEMMA 1.1. The space $M \cap M^\times \cap \operatorname{Ker}(C|A) = (0)$.

PROOF. Put $Z = M \cap M^\times \cap \operatorname{Ker}(C|A)$. Then Z is a finite dimensional subspace of X, invariant under both A and $A^\times$. Suppose Z is non-trivial, and let μ be an eigenvalue of the restriction $A|Z$ of A to Z. Since A and $A^\times$ coincide on Z, we have that μ is an eigenvalue of $A^\times|Z$ too. But then $\mu \in \sigma(A|M) \cap \sigma(A^\times|M^\times)$, which contradicts the fact that $\sigma(A|M)$ and $\sigma(A^\times|M^\times)$ are disjoint. □

Let ε be a complex number. The spaces K_j do not change if in (1.2) the operator A is replaced by $A-\varepsilon$. Hence

(1.3) $$K_j \cap (A-\varepsilon)^{-1}K_j = K_{j+1}, \quad j = 0,1,2,\ldots .$$

We shall use these identities to construct a system of vectors

(1.4) $$d_{jk}, \quad k = 1,\ldots,\alpha_j; \quad j = 1,\ldots,t$$

with the following properties

(1) $\alpha_1 \geqq \alpha_2 \geqq \cdots \geqq \alpha_t \geqq 1$;

(2) $(A-\varepsilon)d_{jk} = d_{j,k+1}$, $k = 1,\ldots,\alpha_j-1$, $j = 1,\ldots,t$;

(3) the vectors d_{jk}, $k = 1,\ldots,\alpha_j-m$, $\alpha_j \geqq m+1$, for a basis of K_m.

Such a system of vectors we shall call an <u>outgoing basis</u> for Θ (with respect to the operator $A-\varepsilon$).

For all outgoing bases for Θ the integers t and $\alpha_1,\ldots,\alpha_t$ are the same and independent of the choice of ε. In fact one sees from (3) that

(1.5) $$t = \dim(K_0/K_1),$$

(1.6) $$\alpha_j = \#\{m \mid \dim(K_{m-1}/K_m) \geqq j\}, \quad j = 1,\ldots,t.$$

We call $\alpha_1,\ldots,\alpha_t$ the <u>outgoing indices</u> of the node Θ. Observe that $\alpha_1 = \alpha$. The outgoing indices are also determined by

(1.7) $$\#\{j \mid 1 \leqq j \leqq t, \alpha_j = k\} = \dim(K_{p-1}/K_p) - \dim(K_p/K_{p+1}).$$

For an outgoing basis $\{d_{jk}\}_{k=1,j=1}^{\alpha_j\ \ t}$ the following holds:

(3a) the vectors d_{jk}, $k = 1,\dots,\alpha_j$, $j=1,\dots,t$, form a basis of $M \cap M^{\times}$;

(3b) the vectors d_{jk}, $k = 1,\dots,\alpha_j - 1$, $\alpha_j \geqq 2$, form a basis of $M \cap M^{\times} \cap \operatorname{Ker} C$.

Conversely, if (1.4) is a system of vectors such that (1), (2), (3a) and (3b) are satisfied, then the system (1.4) is an outgoing basis for Θ. This is readily seen by using (1.3).

Now, let us make an outgoing basis. The construction is based on (1.3) and uses a method employed in [GKS], Section I.6. Define

$$t_m = \dim(K_0/K_1) - \dim(K_m/K_{m+1}), \quad m = 0,1,\dots,\alpha.$$

Obviously $t_\alpha = \dim(K_0/K_1) = t$. Putting $n_j = t_\alpha - t_{\alpha-j} = t - t_{\alpha-j}$, we have $n_j = \dim(K_{\alpha-j}/K_{\alpha-j+1})$ and

$$n_1 + \dots + n_j = \dim K_{\alpha-j}, \quad j = 1,\dots,\alpha.$$

In particular, $n_1 = \dim K_{\alpha-1}$. Let $d_{11},\dots,d_{n_1,1}$ be a basis of $K_{\alpha-1}$. For $i = 1,\dots,n_1$ write $d_{i2} = (A-\varepsilon)d_{i1}$. From (1.3) it is clear that $d_{12},\dots,d_{n_1,2}$ are vectors in $K_{\alpha-2}$, linearly independent modulo $K_{\alpha-1}$. But then $n_2 = \dim(K_{\alpha-2}/K_{\alpha-1}) \geqq n_1$, and we can choose vectors $d_{n_1+1,1},\dots,d_{n_2,1}$ such that $d_{11},\dots,d_{n_2,1},d_{12},\dots,d_{n_1,2}$ form a basis of $K_{\alpha-2}$. Put

$$d_{i3} = (A-\varepsilon)d_{i2}, \quad i = 1,\dots,n_1,$$

$$d_{i2} = (A-\varepsilon)d_{i1}, \quad 1 = n_1+1,\dots,n_2.$$

Again using (1.3) we see that $d_{13},\dots,d_{n_1,3},d_{n_1+1,2},\dots,d_{n_2,2}$ are vectors in $K_{\alpha-3}$, linearly independent modulo $K_{\alpha-2}$. It follows that $n_3 \geqq n_2$ and by choosing additional vectors $d_{n_2+1,1},\dots,d_{n_3,1}$ we can produce a basis of $K_{\alpha-3}$. Proceeding in this way one obtains in a finite number of steps an outgoing basis for Θ. Observe that the construction shows that $0 = t_0 \leqq t_1 \leqq \dots \leqq t_\alpha = t$. Also $\alpha_j = k$ whenever $t_{k-1}+1 \leqq j \leqq t_k$.

Let the system of vectors (1.4) be an outgoing basis for Θ with respect to $A-\varepsilon$. Define $S: M \cap M^\times \to M \cap M^\times$ be

$$(1.8) \qquad (S-\varepsilon)d_{jk} = d_{j,k+1}, \quad k = 1,\dots,\alpha_j, \quad j = 1,\dots,t.$$

where $d_{j,\alpha_j+1} = 0$. We call S the <u>outgoing operator</u> for Θ associated with the outgoing basis (1.4). With respect to the basis (1.4) the matrix of S has Jordan normal form with ε as the only eigenvalue. There are t Jordan blocks with sizes $\alpha_1,\dots,\alpha_t$.

The next proposition is the counterpart of Proposition 1.2. in Ch. II.

PROPOSITION 1.2. <u>Let ε_1 and ε_2 be complex numbers, and let the system</u> (1.4) <u>be an outgoing basis for Θ with respect to $A-\varepsilon_1$. Put</u>

$$(1.9) \qquad e_{jk} = \sum_{\upsilon=0}^{k-1} \binom{k-1}{\upsilon} (\varepsilon_1-\varepsilon_2)^\upsilon d_{j,k-\upsilon}.$$

<u>Then $\{e_{jk}\}_{k=1,j=1}^{\alpha_j \quad t}$ is an outgoing basis for Θ with respect to the operator $A-\varepsilon_2$. Further, if S_1 and S_2 are the outgoing operators associated with the outgoing bases</u> (1.4) <u>and</u> (1.9), <u>respectively, then S_1 and S_2 coincide on $M \cap M^\times \cap \operatorname{Ker} C$ and</u>

$$(1.10) \qquad (S_1-S_2)e_{j\alpha_j} = \sum_{\upsilon=0}^{\alpha_j-1} \binom{\alpha_j}{\upsilon+1} (\varepsilon_1-\varepsilon_2)^{\upsilon+1} d_{j,\alpha_j-\upsilon}.$$

PROOF. The proof is analogous to that of Proposition 1.2. in Ch. II. First one shows, by computation, that

$$(A-\varepsilon_2)e_{jk} = e_{j,k+1}, \quad k = 1,\dots,\alpha_j-1, \quad \alpha_j \geqq 2.$$

Clearly (3) is satisfied with d_{jk} replaced by e_{jk}. So the vectors (1.9) form an outgoing basis for Θ with respect to

$A-\varepsilon_2$. The identity (1.10) can be checked by a straightforward calculation. A similar computation yields

$$(1.11)\qquad (S_1-S_2)e_{jk} = 0, \quad k = 1,\dots,\alpha_j-1, \quad \alpha_j \geq 2.$$

In view of (3b), this gives that S_1 and S_2 coincide on $M \cap M^\times \cap \operatorname{Ker} C$, and the proof is complete. □

If the vectors of two outgoing bases $\{d_{jk}\}_{k=1,j=1}^{\omega_j\ \ t}$ and $\{e_{jk}\}_{k=1,j=1}^{\omega_j\ \ t}$ are related by (1.9), then reversely

$$(1.12)\qquad d_{jk} = \sum_{\upsilon=0}^{k-1} \binom{k-1}{\upsilon}(\varepsilon_2-\varepsilon_1)^{\upsilon} e_{j,k-\upsilon}.$$

Hence (1.10) may be replaced by

$$(1.13)\qquad (S_2-S_1)d_{j\alpha_j} = \sum_{\upsilon=0}^{\alpha_j-1} \binom{\alpha_j}{\upsilon+1}(\varepsilon_2-\varepsilon_1)^{\upsilon+1} e_{j,\alpha_j-\upsilon}.$$

As in Section II.1. we conclude this section with a remark about the definition of a node with centralized singularities as given in Section I.3. Let $\Theta = (A,B,C;X,Y)$ be such a node, and let $\{d_{jk}\}_{k=1,j=1}^{\alpha_j\ \ t}$ and $\{e_{jk}\}_{k=1,j=1}^{\alpha_j\ \ t}$ be the bases of X_2 introduced in property (iii) of a node with centralized singularities. In the proof of Theorem 3.1 in Ch.I we have shown (cf. formulas (3.16) and (3.17)) in Ch.I. that

$$Cd_{jk} = Ce_{jk} = 0, \quad k = 1,\dots,\alpha_j-1.$$

Using this together with property (iv) of a node with centralized singularities one easily checks that the basis $\{d_{jk}\}_{k=1,j=1}^{\alpha_j\ \ t}$ is an outgoing basis for Θ relative to the operator $A-\varepsilon_2$ and $A_2^\times$ is the corresponding outgoing operator. Similarly, $\{e_{jk}\}_{k=1,j=1}^{\alpha_j\ \ t}$ is an outgoing basis for Θ relative to the operator A_2 the corresponding outgoing operator. Note that the formulas (3.5a) and (3.5b) in Ch.I tell us that the two bases are related as in (1.9) and (1.12).

III.2. Output injection operators related to outgoing bases

We continue the study of outgoing bases. Again $\Theta = (A,B,C;X,Y)$ is a node without spectrum on Γ, the spaces $M,M^{\times}$ are the spectral subspaces associated with Θ and Γ, and $\varepsilon_1,\varepsilon_2$ are two fixed complex numbers, $\varepsilon_1 \in \Omega_{\Gamma}^{+}$ and $\varepsilon_2 \in \Omega_{\Gamma}^{-}$. Throughout this section we assume that

$$\dim(M \cap M^{\times}) < \infty, \qquad \dim(X/M+M^{\times}) < \infty.$$

A triple

$$(2.1) \qquad \mathcal{D}_{out} = \left(\{d_{jk}\}_{k=1,\,j=1}^{\alpha_j\quad t},\{e_{jk}\}_{k=1,\,j=1}^{\alpha_j\quad t},\{z_j\}_{j=1}^{t}\right)$$

is called a <u>triple of associated outgoing data</u> for the node Θ (with respect to the contour Γ and the points $\varepsilon_1,\varepsilon_2$) if $\{d_{jk}\}_{k=1,\,j=1}^{\alpha_j\quad t}$ and $\{e_{jk}\}_{k=1,\,j=1}^{\alpha_j\quad t}$ are outgoing basis for Θ with respect to the operators $A-\varepsilon_1$ and $A-\varepsilon_2$, respectively, and the following identities hold true:

$$(2.2) \qquad e_{jk} = \sum_{\upsilon=0}^{k-1}\binom{k-1}{\upsilon}(\varepsilon_1-\varepsilon_2)^{\upsilon}d_{j,k-\upsilon}, \qquad k = 1,\dots,\alpha_j;$$

$$(2.3) \qquad z_j = Cd_{j\alpha_j} = Ce_{j\alpha_j}, \qquad j = 1,\dots,t.$$

The construction of a triple $\mathcal{D}_{out}$ is clear from the results of Section III.1. One starts with an outgoing basis $\{d_{jk}\}_{k=1,\,j=1}^{\alpha_j\quad t}$ for Θ with respect to the operator $A-\varepsilon_1$. Then, according to Proposition 1.2, formula (2.2) defines an outgoing basis $\{e_{jk}\}_{k=1,\,j=1}^{\alpha_j\quad t}$ for Θ with respect to the operator $A-\varepsilon_2$. Since the vectors d_{jk}, $k = 1,\dots,\alpha_j-1$, belong to the kernel of C we have $Cd_{j,\alpha_j} = Ce_{j,\alpha_j}$, and hence the vectors $z_1,\dots,z_t$ are well-defined by (2.3). Note that the vectors $z_1,\dots,z_t$ form a basis of $C[M\cap M^\times]$. Since $BC[M\cap M^\times] \subset M+M^\times$, we conclude that

$$(2.4) \qquad sp\{z_1,\dots,z_t\} = C[M\cap M^\times] \subset B^{-1}[M+M^\times].$$

To a triple of associated outgoing data $\mathcal{D}_{out}$ there corresponds in a natural way two outgoing operators. By S_1 we denote the outgoing operator corresponding to the first outgoing basis in $\mathcal{D}_{out}$, and S_2 denotes the outgoing operator corresponding to the second basis in $\mathcal{D}_{out}$. Both S_1 and S_2 act on the space $M \cap M^\times$.

Two operators $B_1,B_2\colon Y \to M\cap M^\times$ are called a <u>pair of output injection operators corresponding to</u> $\mathcal{D}_{out}$ if for some projection ρ of $M+M^\times$ onto $M\cap M^\times$ the following holds:

$$(2.5) \qquad A^\times x - S_1 x + B_1 Cx \in \operatorname{Ker} \rho \quad (x \in M \cap M^\times);$$

$$(2.6) \qquad Ax - S_2 x - B_2 Cx \in \operatorname{Ker} \rho \quad (x \in M \cap M^\times).$$

Note that all vectors appearing in (2.5) and (2.6) are in $M+M^\times$ and hence belong to the domain of the projection ρ. By substracting (2.5) from (2.6) one obtains that

$$(2.7) \qquad \rho(B-B_1-B_2)Cx = (S_2-S_1)x \quad (x \in M \cap M^\times).$$

To construct such a pair of output injection operators we consider operators $G_1,G_2,G_1^\times,G_2^\times\colon Y \to X$ which have the property that

$$(2.8) \qquad G_1 z_j = (A-\varepsilon_1)d_{j\alpha_j}, \quad G_1^\times z_j = (A^\times-\varepsilon_1)d_{j\alpha_j},$$

$$(2.9) \qquad G_2 z_j = (A-\varepsilon_2)e_{j\alpha_j}, \quad G_2^\times z_j = (A^\times-\varepsilon_2)e_{j\alpha_j}$$

for $j = 1,\dots,t$. Since the vectors $z_1,\dots,z_t$ are linearly independent such operators exist. Note that

$$(2.10)\qquad Ax - S_1x - G_1Cx = 0 \quad (x \in M \cap M^\times);$$

$$(2.11)\qquad Ax - S_2x - G_2Cx = 0 \quad (x \in M \cap M^\times).$$

Furthermore, (2.10) and (2.11) remain true if A is replaced by $A^\times$, G_1 by $G_1^\times$ and G_2 by $G_2^\times$. Now let Π be any projection of X onto $M \cap M^\times$, and let ρ be the restriction of Π to $M+M^\times$. Put

$$(2.12)\qquad B_1 = -\Pi G_1^\times, \quad B_2 = \Pi G_2.$$

Then B_1 and B_2 have the desired properties.

There is a lot of freedom in the choice of the operators B_1, B_2, and the properties of B_1 and B_2 can be improved by specifying the projection ρ and the operators G_1, G_2, $G_1^\times$ and $G_2^\times$. For example one can prove that there exists a pair of output injection operators B_1, B_2 such that

$$(2.13)\qquad A^\times x - S_1\rho x + B_1Cx \in \operatorname{Ker}\rho \quad (x \in M^\times),$$

$$(2.14)\qquad Ax - S_2\rho x - B_2Cx \in \operatorname{Ker}\rho \quad (x \in M),$$

$$(2.15)\qquad \rho(B-B_1-B_2)y = 0, \quad y \in Y_0,$$

where ρ is a projection of $M+M^\times$ onto $M \cap M^\times$ and Y_0 is a complement of $\mathrm{sp}\{z_1,\dots,z_t\}$ in $B^{-1}[M+M^\times]$. Such a pair B_1, B_2 we shall call an <u>improved</u> pair of output injection operators. We shall obtain the existence of an improved pair of output injection operators as a corollary of a more general theorem which concerns the intertwinning of outgoing and incoming data and which will be proved in the next section. The projection ρ appearing in (2.13), (2.14) and (2.15) we shall refer to as <u>the projection corresponding to the improved pair of output injection operators</u> B_1, B_2.

The term output injection operator is taken from systems theory; in general it refers to an operator which maps the output space into the state space. Roughly speaking, the notion of a (resp. an improved) pair of output injection operators may be viewed as the dual of the notion of a (resp. an improved) pair of feedback operators.

III.3. Factorization with non-positive indices

In this section we assume that $M+M^\times = X$, and we prove that a transfer function with this property admits a right Wiener-Hopf factorization with non-positive indices. Moreover, using the outgoing data and the associated output injection operators we can describe explicitly the factors in the Wiener-Hopf factorization

THEOREM 3.1. *Let* W *be the transfer function of a node* $\Theta = (A,B,C;X,Y)$ *without spectrum on the contour* Γ, *and assume that*

$$\dim(M\cap M^\times) < \infty, \quad X = M+M^\times, \tag{3.1}$$

where $M,M^\times$ *is the pair of spectral subspaces associated with* Θ *and* Γ. *Then* W *admits a right Wiener-Hopf factorization with respect to* Γ,

$$W(\lambda) = W_-(\lambda)D(\lambda)W_+(\lambda), \quad \lambda \in \Gamma, \tag{3.2}$$

with factors given by

$$W_-(\lambda) = I + C(\lambda-A)^{-1}(\rho_1+\rho_2)(B-B_1), \tag{3.3}$$

$$W_-(\lambda)^{-1} = I - C[\lambda-(A-(\rho_1+\rho_2)(B-B_1)C]^{-1}(\rho_1+\rho_2)(B-B_1), \tag{3.4}$$

$$W_+(\lambda) = I + C[\lambda-(A^\times+(\rho_2+\rho_3)(B-B_2)C]^{-1}(\rho_2+\rho_3)(B-B_2), \tag{3.5}$$

(3.6) $$W_+(\lambda)^{-1} = I - C(\lambda-A^\times)^{-1}(\rho_2+\rho_3)(B-B_2),$$

(3.7) $$D(\lambda)z = \begin{cases} \left(\frac{\lambda-\varepsilon_2}{\lambda-\varepsilon_1}\right)^{\alpha_j} z_j, & z = z_j \ (j=1,\dots,t), \\ z & z \in Y_o. \end{cases}$$

Here $B_1, B_2 \colon Y \to M \cap M^\times$ _is an improved pair of output injection operators corresponding to a triple_

$$\mathcal{D}_{out} = \left(\{d_{jk}\}_{k=1,j=1}^{\alpha_j \ \ t}, \{e_{jk}\}_{k=1,j=1}^{\alpha_j \ \ t}, \{z_j\}_{j=1}^{t}\right)$$

of associated outgoing data for the node Θ _(relative to_ Γ _and the points_ $\varepsilon_1, \varepsilon_2$_). If_ ρ _is the projection corresponding to the output injection operators_ B_1, B_2_, then_

(3.8) $$X = (M \cap \operatorname{Ker}\rho) \oplus (M \cap M^\times) \oplus (M^\times \cap \operatorname{Ker}\rho),$$

and for $i = 1,2$ _and_ 3 _the operator_ ρ_i _is the projection of_ X _onto the i-th space in the decomposition_ (3.8) _along the other spaces in this decomposition. Further,_ Y_0 _is the subspace in_ (2.15).

PROOF. Write A and $A^\times$ as 3×3 operator matrices relative to the decomposition (3.8):

$$A = \begin{pmatrix} A_{11} & A_{12} & A_{13} \\ A_{21} & A_{22} & A_{23} \\ 0 & 0 & A_{33} \end{pmatrix}, \quad A^\times = \begin{pmatrix} A_{11}^\times & 0 & 0 \\ A_{21}^\times & A_{22}^\times & A_{23}^\times \\ A_{31}^\times & A_{32}^\times & A_{33}^\times \end{pmatrix}.$$

Since M is the spectral subspace of A corresponding to the part of $\sigma(A)$ in Ω_p^+, we have $\sigma(A_{33}) \subset \Omega_\Gamma^-$. Similarly, $\sigma(A_{11}^\times) \subset \Omega_\Gamma^+$.

Let $S_1, S_2 \colon M \cap M^\times \to M \cap M^\times$ be the outgoing operators corresponding to the triple $\mathcal{D}_{out}$. Then relative to the decomposition (3.8)

$$(3.9)\qquad A-(\rho_1+\rho_2)(B-B_1)C=\begin{pmatrix}A_{11}^{\times} & 0 & 0\\ * & S_1 & 0\\ 0 & 0 & A_{33}\end{pmatrix},$$

$$(3.10)\qquad A^{\times}+(\rho_2+\rho_3)(B-B_2)C=\begin{pmatrix}A_{11}^{\times} & 0 & 0\\ 0 & S_2 & *\\ 0 & 0 & A_{33}\end{pmatrix}.$$

To prove (3.9) and (3.10) one uses the formulas (2.13) and (2.14). From (3.9) and (3.10) it is clear that the operators $A-(\rho_1+\rho_2)(B-B_1)C$ and $A^{\times}+(\rho_2+\rho_3)(B-B_2)C$ have no spectrum on Γ. Thus for $\lambda \in \Gamma$ the right hand sides of (3.3) - (3.6) are well-defined. Take $\lambda \in \Gamma$. Let $W_-(\lambda)$ and $W_+(\lambda)$ be defined by the right hand sides of (3.3) and (3.5), respectively. Then it is clear that the operators $W_-(\lambda)$ and $W_+(\lambda)$ are invertible and their inverses are given by the right hand sides of (3.4) and (3.6), respectively.

Next, we compute $W_-(\lambda)^{-1}W(\lambda)W_+(\lambda)^{-1}$ for $\lambda \in \Gamma$. Take $\lambda \in \Gamma$. Then

$$W_-(\lambda)^{-1}W(\lambda) = W_-(\lambda)^{-1} + C(\lambda-A)^{-1}B +$$

$$+ C[\lambda-(A-(\rho_1+\rho_2)(B-B_1)C]^{-1}[A-(\rho_1+\rho_2)(B-B_1)C-A](\lambda-A)^{-1}B$$

$$= W_-(\lambda)^{-1}+C[\lambda-(A-(\rho_1+\rho_2)(B-B_1)C]^{-1}B$$

$$= I + C[\lambda-(A-(\rho_1+\rho_2)(B-B_1)C]^{-1}(\rho_3B+\rho_2B_1).$$

Thus

$$W_-(\lambda)^{-1}W(\lambda)W_+(\lambda)^{-1} = W_-(\lambda)^{-1}W(\lambda)-C(\lambda-A^{\times})^{-1}(\rho_2+\rho_3)(B-B_2) +$$

$$-C[\lambda-(A-(\rho_1+\rho_2)(B-B_1)C]^{-1}\cdot$$

$$\cdot[A-(\rho_1+\rho_2)(B-B_1)-A^{\times}](\lambda-A^{\times})^{-1}(\rho_2+\rho_3)(B-B_2)$$

$$= W_-(\lambda)^{-1}W(\lambda) - C[\lambda-(A-(\rho_1+\rho_2)(B-B_1)C]^{-1}(\rho_2+\rho_3)(B-B_2)$$

$$= I - C[\lambda-(A-(\rho_1+\rho_2)(B-B_1)C]^{-1}\rho_2(B-B_1-B_2).$$

Now use (3.9) and $\rho_2 = \rho$. One sees that

$$W_-(\lambda)^{-1}W(\lambda)W_+(\lambda)^{-1} = I - C\rho[\lambda-S_1]^{-1}\rho(B-B_1-B_2).$$

$$= \begin{cases} \left(\frac{\lambda-\varepsilon_2}{\lambda-\varepsilon_1}\right)^{\alpha_j} z_j, & z = z_j,\ j = 1,\dots,t, \\ z, & z \in Y_0. \end{cases}$$

Here we used (2.15) and the fact that (cf. (2.7) and (2.3))

$$\rho(B-B_1-B_2)z_j = (S_2-S_1)e_{j\alpha_j},\quad j = 1,\dots,t.$$

It follows that $W_-(\lambda)^{-1}W(\lambda)W_+(\lambda)^{-1} = D(\lambda)$ for $\lambda \in \Gamma$ and (3.2) is proved.

Since $\mathrm{Im}(\rho_1+\rho_2) = M$, the operator function $W_-(.)$ has an analytic continuation defined on the closure of Ω_Γ^-. From $\mathrm{Im}(\rho_2+\rho_3) = M^\times$ it follows that $W_+(.)^{-1}$ has an analytic continuation defined on the closure of Ω_Γ^+. Introduce

$$A_o = \begin{pmatrix} A_{11}^\times & 0 \\ * & S_1 \end{pmatrix}, \quad A_o = \begin{pmatrix} S_2 & * \\ 0 & A_{33} \end{pmatrix}.$$

Obviously, $\sigma(A_o) \subset \Omega_\Gamma^+$ and $\sigma(A_o^\times) \subset \Omega_\Gamma^-$. From (3.4), (3.5), (3.9) and (3.10) one sees that

$$W_-(\lambda)^{-1} = I - C(\rho_1+\rho_2)(\lambda-A_o)^{-1}(\rho_1+\rho_2)(B-B_1),$$

$$W_+(\lambda) = I + C(\rho_2+\rho_3)(\lambda-A_o^\times)^{-1}(\rho_2+\rho_3)(B-B_2).$$

But then it is clear that $W_-(.)^{-1}$ and $W_+(.)$ have analytic continuations defined on the closures of Ω_Γ^- and Ω_Γ^+, respectively. Thus (3.2) is a right Wiener-Hopf factorization and the theorem is proved. □

For the factorization formulas in Theorem 3.2 it is important to know the construction of an improved pair of output injection operators. For the general case we postponed this construction to Section IV.1. If $X = M+M^\times$, which is the case considered here, then the construction of an improved pair of output injection operators is less involved and goes as follows.

Let

$$\mathcal{D}_{out} = \left(\{d_{jk}\}_{k=1,}^{\alpha_j}{}_{j=1}^{t}, \{e_{jk}\}_{k=1,}^{\alpha_j}{}_{j=1}^{t}, \{z_j\}_{j=1}^{t}\right),$$

and let Y_0 be a closed linear complement of span $\{z_1,\ldots,z_t\}$ in Y. Since $\dim(M\cap M^\times) < \infty$ and $X = M+M^\times$, there exist bounded linear operators $H: Y_0 \to M$ and $H^\times: Y_0 \to M^\times$ such that $By = Hy - H^\times y$ for each $y \in Y$. Let G_1, G_2, $G_1^\times$ and $G_2^\times$ be the linear operators from Y into X defined by (2.8), (2.9) and

$$\text{(3.11)} \qquad G_1y = G_2y = Hy, \; G_1^\times = G_2^\times y = H_y^\times, \qquad y \in Y_0.$$

It follows that

$$B = G_1 - G_1^\times = G_2 - G_2^\times,$$

$$\operatorname{Im} G_\upsilon \subset M, \; \operatorname{Im}, G_\upsilon^\times \subset M^\times \quad (\upsilon = 1,2).$$

Consider the operators

$$K = A - G_2C|M: M \to M,$$

$$K^\times = A^\times - G_1^\times C|M^\times: M^\times \to M^\times.$$

Let N be a closed linear complement of $M\cap M^\times$ in M. The 2×2 operator matrix representation of K relative to the decomposition $M = N \oplus (M \cap M^\times)$ has the following form

$$(3.12) \qquad K = \begin{pmatrix} A_o^\times & 0 \\ * & S_2 \end{pmatrix}.$$

Here $A_o^\times = \eta A^\times | N$, where η is the projection of X onto N along $M^\times$. To prove (3.12) one uses (2.11) and the fact that $A - G_2C = A^\times - G_2^\times C$. From the definition of $A_o^\times$ it is clear that $\sigma(A_o^\times) \subset \Omega_\Gamma^+$, and hence $M \cap M^\times$ is the spectral subspace of K corresponding to the part of $\sigma(K)$ inside Ω_p^-. In an analogous way one shows that $M \cap M^\times$ is the spectral subspace of $K^\times$ corresponding to the past of $\sigma(K^\times)$ inside Ω_Γ^+. Define

$$\rho(x) = \begin{cases} x - \frac{1}{2\pi i} \int_\Gamma (\lambda - K)^{-1} x d\lambda, & x \in M, \\ \frac{1}{2\pi i} \int_\Gamma (\lambda - K^\times)^{-1} x d\lambda, & x \in M^\times. \end{cases}$$

Clearly, ρ is a projection of X onto $M \cap M^\times$. Put $B_1 = -\rho G_1^\times$ and $B_2 = \rho G_2$. Then (2.13)-(2.15) hold true, and hence B_1, B_2 is an improved pair of output injection operators associated to $\mathcal{D}_{out}$.

IV. MAIN RESULTS

IV.1. <u>Interwinning relations for incoming and outgoing data</u>

Let F_1, F_2: $K \to Y$ be an improved pair of feedback operators corresponding to a triple of associated incoming data $\mathcal{D}_{in}$. As before, T_1, T_2: $K \to K$ denote the incoming operators associated with $\mathcal{D}_{in}$. Further, $\mathcal{D}_{out}$ is a triple of associated outgoing data with outgoing operators S_1, S_2: $M \cap M^\times \to M \cap M^\times$. We say that an improved pair B_1, B_2: $Y \to M \cap M^\times$ of output injection operators corresponding to $\mathcal{D}_{out}$ is <u>coupled</u> to the pair F_1, F_2 if there exist operators $\tau, \tau^\times$: $K \to M \cap M^\times$ such that the following interwinning relations hold true:

(1) $\tau T_1 x - S_2\tau x = -\rho Z_1 x + B_2(C+F_1)x$,

(2) $\tau T_2 x - S_2\tau x = - B_2F_2x$,

(3) $\tau^\times T_2 x - S_1\tau^\times x = -\rho Z_2 x - B_1(C+F_2)x$,

(4) $\tau^\times T_1 x - S_1\tau^\times x = B_1F_1x$

for $x \in K$. Here ρ is a projection of $M + M^\times$ onto $M\cap M^\times$ corresponding to the pair B_1, B_2, and the operators Z_1, Z_2: $K \to M+M^\times$ are defined by (2.9) and (2.10)in Ch.II, respectively.

THEOREM 1.1. <u>Given an improved pair</u> F_1, F_2 <u>of feedback operators corresponding to a triple</u> $\mathcal{D}_{in}$, <u>any triple</u> $\mathcal{D}_{out}$ <u>has an improved pair</u> B_1, B_2 <u>of output injection operators coupled to the pair</u> F_1, F_2.

PROOF. Let $\mathcal{D}_{in}$ and $\mathcal{D}_{out}$ be given by (2.1) in Ch.II and (2.1) in Ch.III, respectively. By applying (2.4) in Ch.II and (2.4) in Ch.III we see that there exists a closed subspace Y_0 of finite codimension in Y such that

$$(1.1) \qquad Y = sp\{z_1,\dots,z_t\} \oplus Y_0 \oplus sp\{y_1,\dots,y_s\},$$

$$(1.2) \qquad B^{-1}[M+M^\times] = sp\{z_1,\dots,z_t\} \oplus Y_0.$$

Since $M\cap M^\times$ is a finite dimensional subspace of $M+M^\times$, there exist bounded linear operators H: $Y_0 \to M$ and $H^\times$: $Y_0 \to M^\times$ such that $By = Hy - H^\times y$ for each $y \in Y_0$.

Now, let us return to the operators G_1, G_2, $G_1^\times$ and $G_2^\times$ appearing in (2.8) and (2.9) in Ch.III. So far the action of these operators is prescribed on the vectors $z_1,\dots,z_t$ only. Using the operators H and $H^\times$ introduced in the preceding paragraph we fix the action of the operators G_1, G_2, $G_1^\times$ and $G_2^\times$ on $B^{-1}[M+M^\times]$ by setting

$$G_1y = G_2y = Hy,\quad G_1^\times y = G_2^\times y = H^\times y,\quad y \in Y_0. \tag{1.3}$$

Next, recall that (cf. formula (2.7) in Ch.II).

$$\mathrm{sp}\{y_1,\dots,y_s\} = (C+F_1+F_2)K. \tag{1.4}$$

Thus each $y \in \mathrm{sp}\{y_1,\dots,y_s\}$ can be written as $y = (C+F_1+F_2)x$ for some $x \in K$. Further, by subtracting (2.10) from (2.9) in Ch.II one sees that

$$B(C+F_1+F_2)x = Z_1x + (T_1-T_2)x - Z_2x,\quad x \in K. \tag{1.5}$$

We use these connections to extend the definitions of the operators G_1, G_2, $G_1^\times$, $G_2^\times$ to all of Y by setting

$$G_1y = Z_1x,\qquad G_1^\times y = -(T_1-T_2)x + Z_2x, \tag{1.6}$$

$$G_2y = Z_1x + (T_1-T_2)x,\quad G_2^\times y = Z_2x, \tag{1.7}$$

for $y = (C+F_1+F_2)x$, $x \in K$. Since the pair of feedback operators F_1,F_2 is an improved pair, (2.11) in Ch.II, holds true and hence for a given y the definitions of (1.6) and (1.7) do not depend on the special choice of x.

For G_1, G_2, $G_1^\times$ and $G_2^\times$ defined in this way the following holds true:

$$B = G_1 - G_1^\times = G_2 - G_2^\times; \tag{1.8}$$

$$\mathrm{Im}\, G_1 \subset M,\qquad \mathrm{Im}\, G_1^\times \subset K+M^\times; \tag{1.9}$$

$$\mathrm{Im}\, G_2 \subset M+K,\quad \mathrm{Im}\, G_2^\times \subset M^\times. \tag{1.10}$$

Let η be a projection of X onto $M \cap M^\times$ such that

$\eta x = 0$ for $x \in K$. Consider

$$V_1x = \begin{cases} \eta(Ax - S_1\eta x - G_1Cx), & x \in M^{\times}, \\ \eta(Ax - G_1Cx - G_1F_2x), & x \in K, \\ 0, & x \in M; \end{cases}$$

$$V_2x = \begin{cases} \eta(A^{\times}x - S_2\eta x - G_2^{\times}Cx), & x \in M, \\ \eta(A^{\times}x - G_2^{\times}Cx - G_2^{\times}F_1x), & x \in K, \\ 0, & x \in M^{\times}. \end{cases}$$

Because of (2.10) and (2.11) in Ch.III the vectors V_1x and V_2x are well-defined for $x \in M \cap M^{\times}$. Observe that $V_1, V_2: X \to M \cap M^{\times}$ are bounded linear operators. So there exist unique operators $U_1, U_2: X \to M \cap M^{\times}$ such that

(1.11) $\quad S_1U_1 - U_1A = V_1, \quad U_1x = 0 \quad (x \in M);$

(1.12) $\quad S_2U_2 - U_2A^{\times} = V_2, \quad U_2x = 0 \quad (x \in M^{\times}).$

Introduce the following operators

(1.13) $\quad B_1 = -\eta G_1^{\times} + U_1B, \; B_2 = \eta G_2 + U_2B.$

We claim that B_1, B_2 is an improved pair of output injection operators coupled to the pair F_1, F_2.

To see this, put

(1.14) $\quad \rho x = (\eta + U_1 + U_2)x, \quad x \in M+M^{\times}.$

Note that ρ is a projection of $M+M^{\times}$ onto $M\cap M^{\times}$. Let us check that formulas (2.13), (2.14) and (2.15) in Ch.III hold true. Take $x \in M^{\times}$. Then

$$\rho(A^{\times}x-S_1\rho x+B_1C)x =$$

$$= \eta A^{\times}x + U_1A^{\times}x - S_1\eta x - S_1U_1x - \eta G_1^{\times}Cx + U_1BCx$$

$$= \eta(Ax-S_1\eta x-G_1Cx) + U_1Ax - S_1U_1x = 0,$$

which proves (2.13) in Ch.III. Similarly, for $x \in M$

$$\rho(Ax-S_2\rho x-B_2Cx) =$$

$$= \eta Ax + U_2Ax - S_2\eta x - S_2U_2x - \eta G_2Cx - U_2BCx$$

$$= \eta(A^{\times}x-S_2\eta x-G_2^{\times}Cx) + U_2A^{\times}x-S_2U_2x = 0,$$

and hence (2.14) in Ch.III holds true. To prove (2.15) in Ch.III, take y in Y_0. Then

$$\rho(By-B_1y-B_2y) =$$

$$= \eta By+U_1By+U_2By-B_1y-B_2y$$

$$= \eta Hy-\eta H^{\times}y+U_1By+U_2By-B_1y-B_2y$$

$$= (\eta G_2+U_2B)y + (-\eta G_1^{\times}+U_1B)y - B_1y-B_2y = 0.$$

So (2.15) in Ch.III holds true. It follows that B_1, B_2 is an improved pair of output injection operators with ρ as the corresponding projection.

Next we prove the coupling. Define $\tau, \tau^{\times}: K \to M \cap M^{\times}$ by setting

$$(1.15) \qquad \tau x + U_2x, \quad \tau^{\times}x = U_1x \quad (x \in K).$$

With ρ, τ and $\tau^{\times}$ defined by (1.14) and (1.15), respectively, we have to check the properties (1) - (4) listed in the beginning of this section. Take $x \in K$. Note that

$$S_2\tau x = S_2U_2x = U_2A^\times x + V_2x$$

$$= (\eta+U_2)A^\times x - \eta G_2^\times Cx - \eta G_2^\times F_1x$$

$$= (\eta+U_2)Ax - (\eta G_2+U_2B)Cx - \eta G_2^\times F_1x$$

$$= (\eta+U_2)Ax - B_2Cx - \eta G_2^\times F_1x.$$

Recall that $Z_1x = Ax - T_1x + BF_1x \in M$ (cf. formula (2.9) in Ch.II). It follows that

$$\tau T_1x - S_2\tau x = U_2T_1x - (\eta+U_2)Ax + B_2Cx + \eta G_2^\times F_1x$$

$$= -(\eta+U_2)Z_1x + (\eta+U_2)BF_1x + B_2Cx + \eta G_2^\times F_1x$$

$$= -\rho Z_1x + (\eta G_2+U_2B)F_1x + B_2Cx$$

$$= -\rho Z_1x + B_2(C+F_1)x.$$

So (1) holds true. To prove (2), note that

$$\tau T_2x - S_2\tau x = U_2T_2x - (\eta+U_2)A^\times x + \eta G_2^\times Cx + \eta G_2^\times F_1x$$

$$= -(\eta+U_2)Z_2x - (\eta+U_2)BF_2x + \eta G_2^\times(C+F_1)x$$

$$= -(\eta+U_2)\{G_2^\times(C+F_1+F_2)x + BF_2x - G_2^\times(C+F_1)x\}$$

$$= -(\eta+U_2)(BF_2x+G_2^\times F_2x)$$

$$= -(\eta B+\eta G_2^\times)F_2x - U_2BF_2x + - B_2F_2x.$$

Next, note that

$$S_1\tau^\times x = S_1U_1x = U_1Ax + V_1x = (\eta+U_1)Ax - \eta G_1Cx - \eta G_1F_2x$$

$$= (\eta+U_1)A^\times x - (\eta G_1^\times-U_1B)Cx - \eta G_1F_2x$$

$$= (\eta+U_1)A^\times x + B_1Cx - \eta G_1F_2x.$$

Recall that $Z_2x = A^{\times}x - T_2x - BF_2x \in M^{\times}$ (cf. formula (2.10) in Ch.II). It follows that

$$\tau^{\times}T_2x - S_1\tau^{\times}x = U_1T_2x - (\eta+U_1)A^{\times}x - B_1Cx + \eta G_1F_1x$$

$$= -(\eta+U_1)Z_2x - (\eta+U_1)BF_2x - B_1Cx + \eta G_1F_2x$$

$$= -\rho Z_2x + (\eta G_1^{\times}-U_1B)F_2x - B_1Cx$$

$$= -\rho Z_2x - B_1(C+F_2)x,$$

which shows that (3) holds true. Finally, to prove (4) we note that

$$\tau^{\times}T_1x - S_1\tau^{\times}x = U_1T_1x - (\eta+U_1)Ax + \eta G_1(C+F_2)x$$

$$= -(\eta+U_1)Z_1x + (\eta+U_1)BF_1x + (\eta+U_1)G_1(C+F_2)x$$

$$= -(\eta+U_1)\{G_1(C+F_1+F_2)x-BF_1x-G_1(C+F_2)x\}$$

$$= -(\eta+U_1)(G_1F_1x-BF_1x) = -\eta G_1^{\times}F_1x + U_1BF_1x = B_1F_1x.$$

IV.2. Dilation to a node with centralized singularities

In this section $\Theta = (A,B,C;X,Y)$ is a node without spectrum on the contour Γ and $M,M^{\times}$ is the pair of spectral subspaces associated with Θ and Γ. We assume that

$$\dim(M \cap M^{\times}) < \infty, \qquad \dim(X/M+M^{\times}) < \infty.$$

Let $F_1,F_2\colon K \to Y$ be an improved pair of feedback operators corresponding to a triple of associated incoming data $\mathcal{D}_{in}$ of

Θ. As before, $T_1, T_2: K \to K$ denote the incoming operators associated with $\mathcal{D}_{in}$. Further, $\mathcal{D}_{out}$ is a triple of associated outgoing data with outgoing operators $S_1, S_2: M \cap M^\times \to M \cap M^\times$, and $B_1, B_2: Y \to M \cap M^\times$ is an improved pair of output injection operators for $\mathcal{D}_{out}$ which is coupled to the pair F_1, F_2. Put

$$\hat{X} = (M \cap M^\times) \oplus (M \cap M^\times) \oplus X \oplus K \oplus K,$$

and introduce the following operators

$$\hat{A} = \begin{pmatrix} S_1 & 0 & 0 & 0 & 0 \\ 0 & S_2 & B_2C & B_2F_1 & B_2F_2 \\ 0 & 0 & A & BF_1 & 0 \\ 0 & 0 & 0 & T_1 & 0 \\ 0 & 0 & 0 & 0 & T_2 \end{pmatrix}, \quad \hat{B} = \begin{pmatrix} B_1 \\ B_2 \\ B \\ 0 \\ 0 \end{pmatrix}, \tag{2.1}$$

$$\hat{C} = [0 \quad 0 \quad C \quad F_1 \quad F_2]. \tag{2.2}$$

THEOREM 2.1. <u>The node $\hat{\Theta} = (\hat{A}, \hat{B}, \hat{C}; \hat{X}, Y)$ is a dilation of Θ and $\hat{\Theta}$ is a node with centralized singularities.</u>

To prove Theorem 2.1 we have to show that $\hat{X}$ admits a decomposition

$$\hat{X} = \hat{X}_1 \oplus \hat{X}_2 \oplus \hat{X}_3 \oplus \hat{X}_4 \tag{2.3}$$

such that the operator matrices of $\hat{A}$, $\hat{A}^\times$, $\hat{B}$ and $\hat{C}$ relative to this decomposition have certain special properties (see Section I.3). To describe the spaces appearing in (2.3) let ρ be a projection of $M+M^\times$ onto $M \cap M^\times$ corresponding to the pair B_1, B_2. Further, let $\tau, \tau^\times: K \to M \cap M^\times$ be such that the coupling properties (1) - (4) listed in the beginning of Section IV.1 are fulfilled. Then

$$\hat{X}_1 = \left\{ \begin{pmatrix} 0 \\ \rho m+\tau x \\ m+x \\ x \\ 0 \end{pmatrix} \middle| x \in K,\ m \in M \right\},$$

$$\hat{X}_2 = \left\{ \begin{pmatrix} u \\ u \\ u \\ 0 \\ 0 \end{pmatrix} \middle| u \in M \cap M^{\times} \right\},$$

$$\hat{X}_3 = \left\{ \begin{pmatrix} 0 \\ 0 \\ x \\ x \\ x \end{pmatrix} \middle| x \in K \right\},$$

$$\hat{X}_4 = \left\{ \begin{pmatrix} \rho m^{\times}+\tau^{\times} x \\ 0 \\ m^{\times}+x \\ 0 \\ x \end{pmatrix} \middle| x \in K,\ m^{\times} \in M^{\times} \right\}.$$

To see that formula (2.3) holds true with $\hat{X}_1, \hat{X}_2, \hat{X}_3$ and $\hat{X}_4$ defined as above, put $N = M \cap \operatorname{Ker}\rho$ and $N^{\times} = M^{\times} \cap \operatorname{Ker}\rho$. Obviously,

$$X = N \oplus (M \cap M^{\times}) \oplus K \oplus N. \tag{2.4}$$

For $i = 1,2,3,4$ we let Q_i denote the projection of X onto the i-th space in the decomposition (2.4) along the other spaces in this decomposition. Now consider

$$\begin{pmatrix} x_1 \\ x_2 \\ x_0 \\ x_3 \\ x_4 \end{pmatrix} = \begin{pmatrix} 0 \\ \rho m+\tau a \\ m+a \\ a \\ 0 \end{pmatrix} + \begin{pmatrix} u \\ u \\ u \\ 0 \\ 0 \end{pmatrix} + \begin{pmatrix} 0 \\ 0 \\ b \\ b \\ b \end{pmatrix} + \begin{pmatrix} \rho m^{\times}+\tau^{\times} c \\ 0 \\ m^{\times}+c \\ 0 \\ c \end{pmatrix} \tag{2.5}$$

Here $m \in M$, $m^{\times} \in M^{\times}$, a, b and c are in K and $u \in M \cap M^{\times}$, and these elements we have to find. Clearly

$$x_1 = u + \rho m^{\times} + \tau^{\times} c,$$

$$x_2 = \rho m + \tau a + u,$$

$$x_0 = m + u + a + b + c + m^{\times},$$

$$x_3 = a + b,$$

$$x_4 = b + c.$$

Thus

$$x_0 = (m-\rho m) + (\rho m+u+\rho m^{\times}) + (a+b+c) + (m^{\times}-\rho m^{\times}).$$

Observe that $m-\rho m \in N$ and $m^{\times}-\rho m^{\times} \in N^{\times}$. Further $\rho m+u+\rho m^{\times} \in M\cap M^{\times}$ and $a+b+c \in K$. But then one easily checks that the following identities hold true:

$$b = x_3 + x_4 - Q_3 x_0,$$

$$a = Q_3 x_0 - x_4,$$

$$c = Q_3 x_0 - x_3,$$

$$u = x_1 + x_2 - Q_2 x_0 - \tau a - \tau^{\times} c,$$

$$m = Q_1 x_0 + x_2 - u - \tau a,$$

$$m^{\times} = Q_4 x_0 + x_1 - u - \tau^{\times} c.$$

Thus given the left hand side of (2.5), the elements in the right hand side of (2.5) are uniquely determined. It follows that (2.3) holds true.

Let $\hat{Q}_j$ be the projection of $\hat{X}$ onto the space $\hat{X}_j$ along the spaces $\hat{X}_i (i\neq j)$.

LEMMA 2.2. *One has*

$$\hat{Q}_2\hat{B}y = \begin{pmatrix} u(y) \\ u(y) \\ u(y) \\ 0 \\ 0 \end{pmatrix}, \qquad \hat{Q}_3\hat{B}y = \begin{pmatrix} 0 \\ 0 \\ b(y) \\ b(y) \\ b(y) \end{pmatrix},$$

<u>where</u>

$$(2.6) \qquad u(y) = \begin{cases} (S_1-S_2)x & \underline{\text{for}}\ y = Cx,\ x \in M \cap M^\times, \\ 0 & \underline{\text{for}}\ y \in Y_0 \oplus \mathrm{sp}\{y_1,\dots,y_s\}; \end{cases}$$

$$(2.7) \qquad b(y) = \begin{cases} 0 & \underline{\text{for}}\ y \in \mathrm{sp}\{z_1,\dots,z_t\} \oplus Y_0, \\ (T_2-T_1)x & \underline{\text{for}}\ y = (C+F_1+F_2)x,\ x \in K. \end{cases}$$

<u>In particular</u>, rank $\hat{Q}_2\hat{B} = s$ <u>and</u> rank $\hat{Q}_3\hat{B} = t$.

PROOF. In (2.6) and (2.7) the vectors $y_1,\dots,y_s$ and $z_1,\dots,z_t$ come from the given triples $\mathcal{D}_{in}$ and $\mathcal{D}_{out}$, respectively. The space Y_0 is as in (2.15) in Ch.III. So

$$(2.8) \qquad Y = \mathrm{sp}\{z_1,\dots,z_t\} \oplus Y_0 \oplus \mathrm{sp}\ \{y_1,\dots,y_s\}.$$

From the proof of the decomposition (2.3) it is clear that

$$u(y) = B_1y + B_2y - Q_2By - \tau Q_3By - \tau^\times Q_3By,$$

$$b(y) = -\ Q_3By.$$

Take $y \in \mathrm{sp}\{z_1,\dots,z_t\} \oplus Y_0 = B^{-1}[M+M^\times]$. Then $By \in M+M^\times$. So $Q_2By = \rho By$ and $Q_3By = 0$. From the last equality, the first equality in (2.7) is clear. Also, we see that $u(y) = \rho(B_1+B_2-B)y$. Hence, by formula (2.7) in Ch. III, the first equality in (2.6) holds true. Also, by (2.15) in Ch. III we have $u(y) = 0$ for $y \in Y_0$.

Next take $y = (C+F_1+F_2)x$ with $x \in K$. So y is an arbitrary element of $\mathrm{sp}\{y_1,\dots,y_s\}$. By subtracting (2.10) from (2.9) in Ch. II, we see that

$$(2.9) \qquad By = Z_1x + (T_1-T_2)x - Z_2x.$$

Since $Z_1x \in M$ and $Z_2x \in M^\times$, this implies that

$$(2.10)\qquad Q_2By = \rho(Z_1x-Z_2x),\quad Q_3B = (T_1-T_2)x.$$

From the second equality in (2.10) the second equality in (2.7) is clear. From the properties (1) - (4) for the coupling operators (see the beginning of Section IV.1) we see that

$$\begin{aligned} B_1y + B_2y &= (B_1+B_2)(C+F_1+F_2)x \\ &= \rho Z_1x + \tau(T_1-T_2)x - \rho Z_2x + \tau^\times(T_1-T_2)x \\ &= Q_2By + \tau Q_3By + \tau^\times Q_3By. \end{aligned}$$

But then $u(y) = 0$, and (2.6) is proved too.

From (1.10) and (1.11) in Ch.III we know that $\operatorname{rank}(S_1-S_2)$ is equal t, and hence the same is true for rank $\hat{Q}_2\hat{B}$. Also, by (1.11) in Ch. II, $\operatorname{rank}(T_1-T_2) = s$, which implies that $\hat{Q}_3\hat{B} = s$.□

From (2.4) in Ch.III and (2.8) in Ch.II it is clear that

$$(2.11)\qquad \operatorname{rank} \hat{C}\hat{Q}_2 = t,\quad \operatorname{rank} \hat{C}\hat{Q}_3 = s.$$

As usual we write $\hat{A}^\times = \hat{A} - \hat{B}\hat{C}$. Note that

$$\hat{A}^\times = \begin{pmatrix} S_1 & 0 & -B_1C & -B_1F_1 & -B_1F_2 \\ 0 & S_2 & 0 & 0 & 0 \\ 0 & 0 & A^\times & 0 & -BF_2 \\ 0 & 0 & 0 & T_1 & 0 \\ 0 & 0 & 0 & 0 & T2 \end{pmatrix}.$$

LEMMA 2.3. <u>The spaces X_1 and X_4 are invariant under</u> $\hat{A}$ and $\hat{A}^\times$, <u>respectively. More precisely for</u> $x \in K$, $m \in M$ <u>and</u> $m^\times \in M^\times$ <u>we have</u>

$$(2.12)\qquad \hat{A}\begin{pmatrix} 0 \\ \rho m+\tau x \\ m+x \\ x \\ 0 \end{pmatrix} = \begin{pmatrix} 0 \\ \rho(Am+Z_1x)+\tau T_1x \\ Am+Z_1x+T_1x \\ T_1x \\ 0 \end{pmatrix},$$

$$(2.13)\qquad \hat{A}^{\times}\begin{pmatrix} \rho m^{\times}+\tau^{\times}x \\ 0 \\ m^{\times}+x \\ 0 \\ x \end{pmatrix} = \begin{pmatrix} \rho(A^{\times}m^{\times}+Z_2x)+\tau^{\times}T_2x \\ 0 \\ A^{\times}m^{\times}+Z_2x+T_2x \\ 0 \\ T_2x \end{pmatrix},$$

<u>where</u> $Z_1x = Ax - T_1x + BF_1x$ <u>and</u> $Z_2x = A^{\times}x - T_2x - BF_2x$.

PROOF. To prove (2.12) we only have to show that

$$(2.14)\quad S_2\rho m + S_2\tau x + B_2Cm + B_2Cx + B_2F_1x = \rho Am + \rho Z_1x + \tau T_1x.$$

We use the properties (1) and (2) of the coupling operators and formula (2.14) in Ch.III. So

$$\begin{aligned} &S_2\rho m + S_2\tau x + B_2Cm + B_2Cx + B_2F_1x = \\ &= \rho Am + \tau T_2x + B_2(C+F_1+F_2)x \\ &= \rho Am + \rho Z_1x + \tau T_1x, \end{aligned}$$

and (2.14) is proved.

To prove (2.13) one has to check that

$$(2.15)\qquad S_1\rho m^{\times}+S_1\tau^{\times}x-B_1Cm^{\times}-B_1Cx-B_1F_2x = \rho A^{\times}m^{\times}+\rho Z_2x+\tau^{\times}T_2x.$$

But this is easily done by applying the properties (3) and (4) of the coupling operators and formula (2.13) in Ch.III. □

LEMMA 2.4. <u>For</u> $u \in M \cap M^{\times}$ <u>one has</u>

$$(2.16)\qquad \hat{A}\begin{pmatrix} u \\ u \\ u \\ 0 \\ 0 \end{pmatrix} = \begin{pmatrix} S_1u \\ S_1u \\ S_1u \\ 0 \\ 0 \end{pmatrix} + \begin{pmatrix} 0 \\ \rho(Au-S_1u) \\ Au-S_1u \\ 0 \\ 0 \end{pmatrix},$$

$$(2.17)\qquad \hat{A}^{\times}\begin{pmatrix} u \\ u \\ u \\ 0 \\ 0 \end{pmatrix} = \begin{pmatrix} S_2u \\ S_2u \\ S_2u \\ 0 \\ 0 \end{pmatrix} + \begin{pmatrix} \rho(A^{\times}u-S_2u) \\ 0 \\ A^{\times}u-S_2u \\ 0 \\ 0 \end{pmatrix}.$$

In particular,

$$(2.18)\qquad \hat{A}\hat{X}_2 \subset \hat{X}_1 + \hat{X}_2, \qquad \hat{A}^{\times}\hat{X}_2 \subset \hat{X}_2 + \hat{X}_4.$$

PROOF. To prove (2.16) one has to check that $\rho Au = S_2u + B_2Cu$ for $u \in M \cap M^{\times}$. But this is clear from (2.6) in Ch.III. To check (2.17) one applies formula (2.5) in Ch.III in a similar way.□

LEMMA 2.5. *For* $x \in K$ *one has*

$$(2.19)\qquad \hat{A}\begin{pmatrix} 0 \\ 0 \\ x \\ x \\ x \end{pmatrix} = \begin{pmatrix} 0 \\ 0 \\ T_2x \\ T_2x \\ T_2x \end{pmatrix} + \begin{pmatrix} 0 \\ \rho Z_1x+\tau(T_1-T_2)x \\ Z_1x+(T_1-T_2)x \\ (T_1-T_2)x \\ 0 \end{pmatrix},$$

$$(2.20)\qquad \hat{A}^{\times}\begin{pmatrix} 0 \\ 0 \\ x \\ x \\ x \end{pmatrix} = \begin{pmatrix} 0 \\ 0 \\ T_1x \\ T_1x \\ T_1x \end{pmatrix} + \begin{pmatrix} \rho Z_2x+\tau\,(T_2-T_1)x \\ 0 \\ Z_2x+(T_2-T_1)x \\ 0 \\ (T_2-T_1)x \end{pmatrix}.$$

In particular,

(2.21) $\quad \hat{A}\hat{X}_3 \subset \hat{X}_1 + \hat{X}_3, \qquad \hat{A}^{\times}\hat{X}_3 \subset \hat{X}_3 + \hat{X}_4.$

PROOF. To prove (2.19) one applies properties (1) and (2) of the coupling operators and formula (2.9) in Ch.II. For (2.20) we need properties (3) and (4) of the coupling operators and formula (2.10) in Ch.II. □

From the expressions of $\hat{A}$ and $\hat{A}^{\times}$ it is clear that both operators have no spectrum on Γ. By $\hat{P}$ (resp. $\hat{P}^{\times}$) we denote the Riesz projection of $\hat{A}$ (resp. $\hat{A}^{\times}$) corresponding to the part of the spectrum in Ω_{Γ}^{+}. We have

$$\hat{P} = \begin{pmatrix} I & 0 & 0 & 0 & 0 \\ 0 & 0 & P_1 & P_2 & 0 \\ 0 & 0 & P & P_3 & 0 \\ 0 & 0 & 0 & I & 0 \\ 0 & 0 & 0 & 0 & 0 \end{pmatrix}, \quad I-\hat{P}^{\times} = \begin{pmatrix} 0 & 0 & P_1^{\times} & 0 & P_2^{\times} \\ 0 & I & 0 & 0 & 0 \\ 0 & 0 & I-P^{\times} & 0 & P_3^{\times} \\ 0 & 0 & 0 & 0 & 0 \\ 0 & 0 & 0 & 0 & I \end{pmatrix}.$$

Here P (resp. $P^{\times}$) is the Riesz projection corresponding to the part of $\sigma(A)$ (resp. $\sigma(A^{\times})$) in Ω_{Γ}^{+}.

LEMMA 2.6. <u>The following identities hold true</u>:

(2.22) $\quad P_1 = \rho P, \qquad P_1^{\times}x = \rho(I-P^{\times});$

(2.23) $\quad P_2x = \tau x - P_1x, \qquad P_2^{\times}x = \tau^{\times}x - P_1^{\times}x \quad (x \in K);$

(2.24) $\quad P_3x = (I-P)x, \qquad P_3^{\times}x = P^{\times}x \quad (x \in K).$

PROOF. Since $\hat{P}^2 = \hat{P}$, we have

(2.25) $\quad P_1P = P_1, \quad PP_3 = 0.$

Further, $\hat{A}\hat{P} = \hat{P}\hat{A}$ yields the following identities:

(2.26) $P_1A - S_2P_1 = B_2CP,$

(2.27) $AP_3 - P_3T_1 = -(I-P)BF_1,$

(2.28) $S_2P_2 - P_2T_1 = P_1BF_1 - B_2F_1 - B_2CP_3.$

According to formula (2.14) in Ch.III we have

(2.29) $(\rho P)A - S_2(\rho P) = B_2CP.$

Since the spectra of $A|M$ and S_2 are disjoint and $P_1P = P_1$, we see from (2.26) and (2.29) that $P_1 = \rho P$. Next, we use (2.9) in Ch.II to show that

(2.30) $A(I-P)x - (I-P)T_1x = -(I-P)BF_1x \quad (x \in K).$

Further, observe that the spectra of $A|\mathrm{Ker}\, P$ and T_1 are disjoint. So, by comparing (2.27) and (2.30) we may conclude that $I - P = (I-P)P_3$. But $PP_3 = 0$. Hence $P_3 = I-P$. Now take $x \in K$ and rewrite (2.28) as follows:

$$\begin{aligned} S_2P_2x - P_2T_1x &= \rho PBF_1x - B_2F_1x - B_2C(I-P)x \\ &= -B_2(C+F_1)x + \rho PBF_1x + B_2CPx \\ &= -B_2(C+F_1)x + \rho PBF_1x + \rho PAx - S_2\rho Px. \end{aligned}$$

It follows that

$$S_2(P_2+\rho P)x-(P_2+\rho P)T_1x = -B_2(C+F_1)x+\rho(Ax-T_1x+BF_1x).$$

Now compare this last identity to property (1) of the coupling operators and use that the spectra of T_1 and S_2 are disjoint. One sees that $\tau x = P_2x + \rho Px$ for $x \in K$, and hence $P_2x = \tau x - P_1x$ for $x \in K$. We have proved the first equalities in (2.22), (2.23) and (2.24).

From $(I-\hat{P}^\times)^2 = I-\hat{P}^\times$ we see that

(2.31) $\quad P_1^\times P^\times = 0, \quad P^\times P_3^\times = P_3^\times.$

Further, $\hat{A}^\times(I-\hat{P}^\times) = (I-\hat{P}^\times)\hat{A}^\times$ yields the following identities:

(2.23) $\quad P_1^\times A^\times - S_1 P_1^\times = - B_1 C(I-P^\times),$

(2.33) $\quad A^\times P_3^\times - P_3^\times T_2 = P^\times BF_2,$

(2.34) $\quad S_1 P_2^\times - P_2^\times T_2 = - P_1^\times BF_2 + B_1 F_2 + B_1 CP_3^\times.$

By comparing (2.32) with (2.13) in Ch.III and using disjointness of spectra we see that $P_1^\times = \rho(I-P^\times)$. In a similar way we obtain from (2.10) in Ch.II and (2.33) that $P_3^\times = P^\times x$ for $x \in K$. Next take $x \in K$ and rewrite (2.34) as follows:

$$\begin{aligned} S_1 P_2^\times x - P_2^\times T_2 x &= - \rho(I-P^\times)BF_2 x + B_1 F_2 x + B_1 CP^\times x \\ &= B_1(C+F_2)x - \rho(I-P^\times)BF_2 x - B_1 C(I-P^\times)x \\ &= B_1(C+F_2)x - \rho(I-P^\times)BF_2 x + \rho A^\times(I-P^\times)x - S_1\rho(I-P^\times)x. \end{aligned}$$

It follows that

$$\begin{aligned} &S_1(P_2^\times+\rho(I-P^\times))x - (P_2^\times+\rho(I-P^\times))T_2 x \\ &= B_1(C+F_2)x + \rho(A^\times x - T_2 x - BF_2 x). \end{aligned}$$

Next compare this to property (3) of the coupling operators. Since the spectra of S_1 and T_2 are disjoint, it follows that $\tau^\times x = P_2^\times x + \rho(I-P^\times)x$ for $x \in K$, and hence the second equality in (2.23) holds true.□

LEMMA 2.7. <u>The node $\hat{\Theta}$ has no spectrum on Γ and for the spectral subspaces $\hat{M},\hat{M}^\times$ associated with Θ and Γ one has</u>

(2.35) $\quad \hat{M} = \hat{X}_1 \oplus \hat{X}_2, \quad \hat{M}^\times = \hat{X}_2 \oplus \hat{X}_4.$

PROOF. According to Lemma 2.6 we have

$$\hat{P}\begin{pmatrix} x_1 \\ x_2 \\ x_0 \\ x_3 \\ x_4 \end{pmatrix} = \begin{pmatrix} x_1 \\ x_1 \\ x_1 \\ 0 \\ 0 \end{pmatrix} + \begin{pmatrix} 0 \\ \rho(PX_0-Px_3-x_1)+\tau x_3 \\ Px_0-Px_3-x_1+x_3 \\ x_3 \\ 0 \end{pmatrix},$$

from which the first identity in (2.35) is clear. To prove the second identity in (2.35) one uses

$$(I-P^{\times})\begin{pmatrix} x_1 \\ x_2 \\ x_0 \\ x_3 \\ x_4 \end{pmatrix} = \begin{pmatrix} x_2 \\ x_2 \\ x_2 \\ 0 \\ 0 \end{pmatrix} + \begin{pmatrix} \rho m^{\times}+\tau^{\times}x_4 \\ 0 \\ m^{\times}+x_4 \\ 0 \\ x_4 \end{pmatrix},$$

where $m^{\times} = (I-P^{\times})x_0 - (I-P^{\times})x_4 - x_2$.□

PROOF OF THEOREM 2.1. From Lemmas 2.3, 2.4 and 2.5 it is clear that relative to the decomposition (2.3) the operators $\hat{A}$ and $\hat{A}^{\times}$ can be written in the following form:

$$\hat{A} = \begin{pmatrix} \hat{A}_1 & * & * & * \\ 0 & \hat{S}_1 & 0 & * \\ 0 & 0 & \hat{T}_2 & * \\ 0 & 0 & 0 & \hat{A}_4 \end{pmatrix}, \quad \hat{A}^{\times} = \begin{pmatrix} \hat{A}_1^{\times} & 0 & 0 & 0 \\ * & \hat{S}_2 & 0 & 0 \\ * & 0 & \hat{T}_1 & 0 \\ * & * & * & \hat{A}_4^{\times} \end{pmatrix}.$$

Properties (iv) and (vi) of a node with centralized singularities hold true for $\hat{\Theta}$ because of Lemma 2.2 and formula (2.11). By formula (2.16) and (2.17) the operators $\hat{S}_1$ and $\hat{S}_2$ have the desired action and hence property (iii) holds true. Similarly, property (v) is satisfied by the formulas (2.19) and (2.20). In both cases the desired bases in $\hat{X}_2$ and $\hat{X}_3$ are induced by the outgoing and incoming bases of $M \cap M^{\times}$ and K, respectively. It remains to prove the spectral properties (i) and (ii).

From the first identity in (2.35) it is clear that

$$\sigma\left[\begin{pmatrix} \hat{A}_1 & * \\ 0 & \hat{S}_1 \end{pmatrix}\right] \subset \Omega_\Gamma^+, \qquad \sigma\left[\begin{pmatrix} \hat{T}_2 & * \\ 0 & \hat{A}_4 \end{pmatrix}\right] \subset \Omega_\Gamma^-.$$

Since $\sigma(\hat{S}_1) \subset \Omega_\Gamma^+$ and $\sigma(\hat{T}_2) \subset \Omega_\Gamma^-$, we may conclude that $\sigma(\hat{A}_1) \subset \Omega_\Gamma^+$ and $\sigma(\hat{A}_4) \subset \Omega_\Gamma^-$. Similarly, using the second identity in (2.35), one shows that $\hat{A}_1^{\times}$ and $\hat{A}_4^{\times}$ have the desired spectral properties.□

IV.3. Main Theorem and corollaries

THEOREM 3.1. Let W be the transfer function of a node $\Theta = (A,B,C;X,Y)$ without spectrum on the contour Γ, and assume that

$$\dim(M \cap M^{\times}) < \infty, \qquad \dim(X/M+M^{\times}) < \infty,$$

where $M, M^{\times}$ is the pair of spectral subspaces associated with Θ and Γ. Then W admits a right Wiener-Hopf factorization with respect to Γ,

$$(3.1) \qquad W(\lambda) = W_-(\lambda)D(\lambda)W_+(\lambda), \qquad \lambda \in \Gamma,$$

with factors given by

$$\begin{aligned} W_-(\lambda) = I &+ C(\lambda-A)^{-1}(Q_1B+Q_2B+\tau^{\times}Q_3B-B_1) \\ &+ C(\lambda-A)^{-1}Z_1(\lambda-T_1)^{-1}Q_3B \\ &+ (C+F_1)(\lambda-T_1)^{-1}Q_3B, \end{aligned}$$

$$W_-(\lambda)^{-1} = I - (C+F_1Q_3)(Q_1+Q_3)(\lambda-A^\times)^{-1}B - C(\lambda-S_1)^{-1}V_1(Q_1+Q_3)(\lambda-A^\times)^{-1}B - C(\lambda-S_1)^{-1}(Q_2B+\tau^\times Q_3B-B_1),$$

$$W_+(\lambda) = I + (C+F_2Q_3)(Q_3+Q_4)(\lambda-A)^{-1}B + C(\lambda-S_2)^{-1}V_2(Q_3+Q_4)(\lambda-A)^{-1}B + C(\lambda-S_2)^{-1}(Q_2B-B_2+\tau Q_3B),$$

$$W_+(\lambda)^{-1} = I - C(\lambda-A^\times)^{-1}(Q_2B-B_2+\tau Q_3B+Q_4B) - C(\lambda-A^\times)^{-1}Z_2(\lambda-T_2)^{-1}Q_3B - (C+F_2)(\lambda-T_2)^{-1}Q_3B,$$

$$(3.2)\qquad D(\lambda)y = \begin{cases} \left(\frac{\lambda-\varepsilon_1}{\lambda-\varepsilon_2}\right)^{\omega_j} y_j, & y = y_j \quad (j=1,\ldots,s), \\ y \quad , & y \in Y_0, \\ \left(\frac{\lambda-\varepsilon_2}{\lambda-\varepsilon_1}\right)^{\alpha_j} z_j, & y = z_j \quad (j=1,\ldots,t). \end{cases}$$

Here $T_1, T_2\colon K \to K$ are incoming operators corresponding to a triple

$$\mathcal{D}_{in} = \left(\{f_{jk}\}_{k=1,\ j=1}^{\omega_j \quad s}, \{g_{jk}\}_{k=1,\ j=1}^{\omega_j \quad s}, \{y_j\}_{j=1}^{s}\right)$$

of associated incoming data for Θ (relative to Γ and the points $\varepsilon_1, \varepsilon_2$), and $F_1, F_2\colon K \to Y$ is an improved pair of feedback operators corresponding to $\mathcal{D}_{in}$. The operators $S_1, S_2\colon M\cap M^\times \to M\cap M^\times$ are outgoing operators corresponding to a triple

$$\mathcal{D}_{out} = \left(\{d_{jk}\}_{k=1,\ j=1}^{\alpha_j \quad t}, \{e_{jk}\}_{k=1,\ j=1}^{\alpha_j \quad t}, \{z_j\}_{j=1}^{t}\right)$$

of associated outgoing data for Θ, and $B_1, B_2\colon Y \to M\cap M^\times$ is an improved pair of output injection operators corresponding to

$\mathcal{D}_{out}$, *which is coupled to the pair* F_1, F_2 *by coupling operators* $\tau, \tau^\times : K \to M \cap M^\times$. *If* ρ *is the projection corresponding to the output injection operators* B_1, B_2, *then*

(3.3) $$X = (M \cap \operatorname{Ker}\rho) \oplus (M \cap M^\times) \oplus K \oplus (M^\times \cap \operatorname{Ker}\rho),$$

and for $i = 1,2,3$ *and* 4 *the operator* Q_i *stands for the projection of* X *onto the* i-th *space in the decomposition* (3.3) *along the other spaces in this decomposition. Further*

(3.4) $$Z_1 x = Ax - T_1 x + BF_1 x, \qquad x \in K,$$

(3.5) $$Z_2 x = A^\times x - T_2 x - BF_2 x, \qquad x \in K,$$

(3.6) $$V_1 x = Q_2 A^\times x + (B_1 - \tau^\times Q_3 B)(C + F_1 Q_3)x, \qquad x \in \operatorname{Im}(Q_1 + Q_3),$$

(3.7) $$V_2 x = Q_2 A x + (\tau Q_3 B - B_2)(F_2 Q_3 + C)x, \qquad x \in \operatorname{Im}(Q_3 + Q_4).$$

Finally, Y_0 *is a subspace of* Y *such that*

$$Y = \operatorname{sp}\{z_1, \ldots, z_t\} \oplus Y_0 \oplus \operatorname{sp}\{y_1, \ldots, y_s\}.$$

PROOF. Let $\hat{\Theta} = (\hat{A}, \hat{B}, \hat{C}; \hat{X}, Y)$ be the dilation of Θ constructed in the previous section. According to Theorem 2.1, the node $\hat{\Theta}$ is a node with centralized singularities. So, by Theorem 3.1 in Ch.I, the transfer function of $\hat{\Theta}$ admits a right Wiener-Hopf factorization. But $\hat{\Theta}$ and Θ have the same transfer function, and thus W admits a right Wiener-Hopf factorization relative to Γ. From (3.9) and the proof of Theorem 3.1 in Ch.I it is clear that the factorization can be made in such a way that the middle term $D(\lambda)$ in the right hand side of (3.1) has the desired form (3.2). Theorem 3.1 in Ch.I applied to $\hat{\Theta}$ also yields explicit formulas for the factors W_- and W_+ in (3.1), namely

(3.8) $$W_-(\lambda) = I + \hat{C}\hat{Q}_1(\lambda - \hat{A}_1)^{-1}\hat{Q}_1\hat{B},$$

(3.9) $$W_+(\lambda) = I + \hat{C}\hat{Q}_4(\lambda-\hat{A}_4)^{-1}\hat{Q}_4\hat{B},$$

where $\hat{Q}_1$ and $\hat{Q}_4$ are the projections of $\hat{X}$ introduced in the previous section. We shall use these identities to get the desired expressions for $W_-(\lambda)^{\pm 1}$ and $W_+(\lambda)^{\pm 1}$.

Without further explanation we shall use the notation introduced in Section 2. Consider the node $\Theta_- = (A_-,B_-,C_-;M\oplus K,Y)$, where

$$A_-: M \oplus K \to M \oplus K, \quad A_- = \begin{pmatrix} A_M & Z_1 \\ 0 & T_1 \end{pmatrix},$$

$$B_-: Y \to M \oplus K, \quad B_- = \begin{pmatrix} (Q_1+Q_2)B-B_1+\tau^\times Q_3B \\ Q_3B \end{pmatrix},$$

$$C_-: M \oplus K \to Y, \quad C_- = [C_M \quad C_K+F_1].$$

Here A_M denotes the restriction of A to M considered as an operator from M into M, and for any subspace L of X the operator C_L denotes the restriction of C to L. The node Θ_- is isomorphic to the node $\hat{\Theta}_1 = (\hat{A}_1,\hat{Q}_1\hat{B},\hat{C}\hat{Q}_1;\hat{X}_1,Y)$. Indeed, define

$$J: M \oplus K \to \hat{X}_1, \quad J\begin{bmatrix} m \\ a \end{bmatrix} = \begin{pmatrix} 0 \\ \rho m+\tau a \\ m+a \\ a \\ 0 \end{pmatrix}.$$

Then J is an invertible bounded linear operator and one checks easily that J is a node similarly between Θ_- and $\hat{\Theta}_1$. It follows that Θ_- and $\hat{\Theta}_1$ have the same transfer function, and thus $W_-(\lambda) = I + C_-(\lambda-A_-)^{-1}B_-$. Now write $(\lambda-A_-)^{-1}$ as a 2×2

operator matrix and compute the matrix product $C_-(\lambda-A_-)^{-1}B_-$. One finds the desired expression for $W_-(\lambda)$.

To get the right formula for $W_-(\lambda)^{-1}$ we first compute $A_-^\times = A_- - B_-C_-$. We have

$$(3.10)\qquad A_-^\times = \begin{matrix} U_M^\times & U_K^\times - S_1\tau^\times \\ Q_3A^\times(Q_1+Q_2) & Q_3A^\times Q_3 \end{matrix},$$

where for any subspace L of X the operator $U_L^\times$ is defined by

$$U_L^\times x = (Q_1+Q_2+\tau^\times Q_3)A^\times x + B_1Cx, \qquad x \in L.$$

To see this, take $m \in M$. Then $Am \in M$, and hence $Am = (Q_1+Q_2)Am$ and $Q_3Am = 0$. It follows that

$$Am - (Q_1+Q_2)BCm + B_1Cm + \tau^\times Q_3BCm =$$

$$= (Q_1+Q_2+\tau^\times Q_3)A^\times m + B_1Cm = U_M^\times m.$$

Further, $-Q_3BCm = Q_3A^\times m = Q_3A^\times(Q_1+Q_2)m$. Next, take $x \in K$. Note that $Z_1x \in M$. So $(Q_1+Q_2)Z_1x = Z_1x$ and $Q_3Z_1x = 0$. In particular $Q_3Ax + Q_3BF_1 = Q_3T_1x$. Further, by property (4) of the coupling operators, $B_1F_1x - \tau^\times Q_3T_1x = -S_1\tau^\times x$. Using these identities one gets

$$\begin{aligned} &Z_1x - (Q_1+Q_2)B(C+F_1)x + B_1(C+F_1)x - \tau^\times Q_3B(C+F_1)x \\ &= (Q_1+Q_2)(Ax-T_1x+BF_1x) - (Q_1+Q_2)BCx + B_1Cx \\ &\quad - (Q_1+Q_2)BF_1x + B_1F_1x - \tau^\times Q_3BCx - \tau^\times Q_3BF_1x \\ &= (Q_1+Q_2+\tau^\times Q_3)A^\times x + B_1Cx - \tau^\times Q_3Ax + B_1F_1x - \tau^\times Q_3BF_1x \\ &= (Q_1+Q_2+\tau^\times Q_3)A^\times x + B_1Cx - \tau^\times Q_3T_1x + B_1F_1x \\ &= U_K^\times x - S_1\tau^\times x. \end{aligned}$$

Finally, since $Q_3Z_1x = 0$, we have $Q_3Ax = T_1x - Q_3BF_1x$, and thus

$$T_1x - Q_3B(C+F_1)x = Q_3A^{\times}Q_3x.$$

This prove (3.10).

Note that $M = (M \cap M^{\times}) \oplus \operatorname{Im}Q_1$. Thus

$$M \oplus K = (M \cap M^{\times}) \oplus \operatorname{Im}(Q_1+Q_2). \tag{3.11}$$

Hence we may write $A_-^{\times}$ as a 2×2 operator matrix with respect to the decomposition given by the right hand side of (3.11); we have

$$A_-^{\times} = \begin{pmatrix} S_1 & V_1 \\ 0 & (Q_1+Q_3)A^{\times}(Q_1+Q_3) \end{pmatrix}, \tag{3.12}$$

where V_1 is given by (3.6). Take $u \in M \cap M^{\times}$. Then

$$Q_2A_-^{\times}u = Q_2U_M^{\times}u = Q_2(A^{\times}u+B_1Cu) = Q_2(A^{\times}u-S_1u+B_1Cu) + S_1u,$$

and, by formula (2.13) in Ch.III, we see that $Q_2A_-^{\times}u = S_1u$. Since $A^{\times}u \in M^{\times}$, we have

$$Q_1A_-^{\times}u = Q_1U_M^{\times}u = Q_1A^{\times}u = 0,$$

$$Q_3A_-^{\times}u = Q_3A^{\times}u = 0.$$

Next, take $x \in \operatorname{Im}Q_1$. Then $Ax \in M$, and thus $Q_3Ax = 0$. It follows that

$$Q_2A_-^{\times}x = Q_2U_M^{\times}x = (Q_2+\tau^{\times}Q_3)A^{\times}x + B_1Cx$$

$$= Q_2A^{\times}x + (B_1-\tau^{\times}Q_3B)Cx = V_1x.$$

Further,

$$(Q_1+Q_3)(A_-^{\times}x = Q_1U_M^{\times}x + Q_3A^{\times}x = Q_1A^{\times}x + Q_3A^{\times}x = (Q_1+Q_3)A^{\times}x.$$

Finally, take $a \in \operatorname{Im}Q_3 = K$. Then $Q_3Aa = T_1a - Q_3BF_1a$ by formula (2.9) in Ch.II. Further, according property (4) of the coupling operators, we have $\tau^\times T_1a - S_1\tau^\times a = B_1F_1a$. It follows that

$$Q_2A_-^\times a = (Q_2+\tau^\times Q_3)A^\times a + B_1Ca - S_1\tau^\times a$$

$$= Q_2A^\times a + (B_1-\tau^\times Q_3B)Ca + \tau^\times Q_3Aa - S_1\tau^\times a$$

$$= Q_2A^\times a + (B_1-\tau^\times Q_3B)Ca + \tau^\times T_1a - S_1\tau^\times a - \tau^\times Q_3BF_1a$$

$$= Q_2A^\times a + (B_1-\tau^\times Q_3B)(C+F_1)a = V_1a.$$

Since $(Q_1+Q_3)A_-^\times a = (Q_1+Q_3)Aa$, the representation (3.12) is proved.

With respect to the decomposition $(M\cap M^\times) \oplus \operatorname{Im}(Q_1+Q_3)$ the operators B_- and C_- have the following matrix representations:

$$(3.13)\qquad B_- = \begin{pmatrix} Q_2B-B_1+\tau^\times Q_3B \\ (Q_1+Q_3)B \end{pmatrix},$$

$$(3.14)\qquad C_- = [\,C_{M\cap M^\times} \quad C(Q_1+Q_3)+F_1Q_3\,].$$

Since $W_-(\lambda) = I + C_-(\lambda-A_-)^{-1}B_-$, we have $W_-(\lambda)^{-1} = I-C_-(\lambda-A_-^\times)^{-1}B_-$. Now compute $(\lambda-A_-^\times)^{-1}$ as a 2×2 operator matrix, using the representation (3.12). Next, using the formulas (3.13) and (3.14), make the matrix product $C_-(\lambda-A_-^\times)^{-1}B_-$, and observe that $(Q_1+Q_3)(\lambda-A^\times)^{-1}(Q_1+Q_3) = (Q_1+Q_3)(\lambda-A^\times)^{-1}$. In this way one finds the desired formula for $W_-(\lambda)^{-1}$.

Next, we study in more detail the function W_+. Define $A_+ : M^\times \oplus K \to M^\times \oplus K$ by setting

$$A_+ = \begin{pmatrix} U_M^\times & U_K-S_2\tau \\ Q_3A(Q_2+Q_4) & Q_3AQ_3 \end{pmatrix}.$$

Here for any subspace L of X the operator U_L is defined by

$$U_L x = (Q_2+\tau Q_3+Q_4)Ax - B_2Cx, \qquad x \in L.$$

Further, $B_+: Y \to M^\times \oplus K$ and $C_+: M^\times \oplus K \to Y$ are defined by

$$(3.15) \qquad B_+ = \begin{pmatrix} (Q_2+Q_4)B-B_2+\tau Q_3B \\ Q_3B \end{pmatrix},$$

$$(3.16) \qquad C_+ = [C_{M^\times} \quad C_K+F_2].$$

It is not difficult to check that the node $\Theta_+ = (A_+,B_+,C_+;M^\times\oplus K,Y)$ is similar to the node $\hat{\Theta}_4 = (\hat{A}_4,\hat{Q}_4B,C\hat{Q}_4;\hat{X}_4,Y)$ with the node similarity given by

$$J^\times: M^\times \oplus K \to \hat{X}_4, \quad J^\times \begin{pmatrix} m^\times \\ c \end{pmatrix} = \begin{pmatrix} \rho m^\times+\tau^\times c \\ 0 \\ m^\times+c \\ 0 \\ c \end{pmatrix},$$

which is an invertible bounded linear operator. It follows that Θ_+ and $\hat{\Theta}_4$ have the same transfer function, and thus $W_+(\lambda) = I + C_+(\lambda-A_+)^{-1}B_+$.

To get the function W_+ in the desired form, we use the fact that $M^\times \oplus K = (M \cap M^\times) \oplus \mathrm{Im}(Q_3+Q_4)$. It turns out that with respect to the decomposition $(M \cap M^\times) \oplus \mathrm{Im}(Q_3+Q_4)$ the operator A_+ has the following form:

$$(3.17) \qquad A_+ = \begin{pmatrix} S_2 & V_2 \\ 0 & (Q_3+Q_4)A(Q_3+Q_4) \end{pmatrix},$$

where V_2 is given by (3.7). To see this, take $u \in M \cap M^\times$. Then

$$Q_2A_+u = Q_2Au - B_2Cu = Q_2(Au-S_2u-B_2Cu) + S_2u,$$

and thus, using formula (2.14) in Ch.III, we see that $Q_2A_+u = S_2u$. Since $Au \in M = \mathrm{Im}(Q_1+Q_2)$, we have

$$Q_3A_+u = Q_3A(Q_2+Q_4)u = Q_3Au = 0,$$
$$Q_4A_+u = Q_4U_{M^\times}u = Q_4Au = 0.$$

Next, take $x \in \mathrm{Im}Q_4$. Then $A^\times x \in M^\times$, and thus $Q_3A^\times x = 0$. It follows that

$$\begin{aligned} Q_2A_+x &= Q_2U_{M^\times}x = (Q_2+\tau Q_3)Ax - B_2Cx \\ &= Q_2Ax + \tau Q_3BCx - B_2Cx = V_2x. \end{aligned}$$

Further,

$$(Q_3+Q_4)A_+x = Q_4U_{M^\times}x + Q_3Ax = Q_4Ax + Q_3Ax = (Q_3+Q_4)Ax.$$

Finally, take $a \in \mathrm{Im}\ Q_3 = K$. Then $Q_3A^\times a = T_2a + Q_3BF_2a$. Further, by property (2) of the coupling operators, we have $\tau T_2a - S_2\tau a = -B_2F_2a$. It follows that

$$\begin{aligned} Q_2A_+a &= Q_2U_Ka-S_2\tau a = (Q_2+\tau Q_3)Aa - B_2Ca - S_2\tau a \\ &= Q_2Aa + (\tau Q_3B-B_2)Ca + \tau Q_3A^\times a - S_2\tau a \\ &= Q_2Aa + (\tau Q_3B-B_2)Ca + \tau T_2a - S_2\tau a + \tau Q_3BF_2a \\ &= Q_2a + (\tau Q_3B-B_2)(C+F_2)a = V_2a. \end{aligned}$$

Since $(Q_3+Q_4)A_+a = (Q_3+Q_4)Aa$ the representation (3.17) is proved.

With respect to the decomposition $(M \cap M^\times) \oplus \mathrm{Im}(Q_3+Q_4)$ the operators B_+ and C_+ have the following matrix representations:

(3.18) $$B_+ = \begin{pmatrix} Q_2B-B_2+\tau Q_3B \\ (Q_3+Q_4)B \end{pmatrix},$$

(3.19) $$C_+ = \begin{pmatrix} C_{M\cap M^\times} & C(Q_3+Q_4)+F_2Q_3 \end{pmatrix}.$$

Now compute $(\lambda-A_+)^{-1}$, using the 2×2 matrix representation (3.17). Next, using the expressions (3.18) and (3.19), make the matrix product $C_+(\lambda-A_+)^{-1}B_+$, and observe that $(Q_3+Q_4)(\lambda-A)^{-1}(Q_3+Q_4) = (Q_3+Q_4)(\lambda-A)^{-1}$. In this way one finds the desired formula for $W_+(\lambda)$.

To get the desired formula for $W_+(\lambda)^{-1}$ we compute $A_+^\times = A_+ - B_+C_+$. With respect to the decomposition $M^\times \oplus K$ the operator $A_+^\times$ has the following 2×2 operator matrix representation:

(3.20) $$A_+^\times = \begin{pmatrix} A^\times_{M^\times} & Z_2 \\ 0 & T_2 \end{pmatrix}.$$

Here $A^\times_{M^\times}$ is the restriction of $A^\times$ to $M^\times$ considered as an operator from $M^\times$ into $M^\times$. To prove (3.20) one applies (2.13) and the fact that $A_+^\times(J^\times)^{-1} = (J^\times)^{-1}\hat{A}_4^\times$. Now compute $(\lambda-A_+^\times)^{-1}$, using the representation (3.20). Next, using (3.15) and (3.16), make the matrix product $C_+(\lambda-A_+^\times)^{-1}B_+$. In this way one finds the desired formula for $W_+(\lambda)^{-1}$.□

COROLLARY 3.2. *Let W be the transfer function of a node $\Theta = (A,B,C;X,Y)$ without spectrum on the contour Γ, and assume that*

$$\dim(M\cap M^\times) < \infty, \quad \dim(X/M+M^\times) < \infty,$$

where $M,M^\times$ is the pair of spectral subspaces associated with Θ and Γ. Put

$$s = \dim\frac{M+M^\times+\mathrm{Im}B}{M+M^\times}, \qquad t = \dim\frac{M\cap M^\times}{M\cap M^\times\cap \mathrm{Ker}C},$$

$$\omega_j = \#\left\{k \,\middle|\, s - \dim\frac{M+M^{\times}+\mathrm{Im}B+\ldots+\mathrm{Im}A^{k-1}B}{M+M^{\times}+\mathrm{Im}B+\ldots+\mathrm{Im}A^{k-2}B} \leqq j-1\right\},$$

$$\alpha_j = \#\left\{m \,\middle|\, \dim\frac{M\cap M^{\times}\cap\mathrm{Ker}C\cap\ldots\cap\mathrm{Ker}CA^{m-2}}{M\cap M^{\times}\cap\mathrm{Ker}C\cap\ldots\cap\mathrm{Ker}CA^{m-1}} \geqq j\right\}.$$

<u>Then</u> $-\alpha_1 \leqq -\alpha_2 \leqq \cdots \leqq -\alpha_t \leqq \omega_1 \leqq \omega_2 \leqq \cdots \leqq \omega_s$ <u>are the right factorization indices of</u> W <u>with respect to</u> Γ.

PROOF. Apply Theorem 3.1. Formula (3.2) gives the factorization indices. They are determined by the formulas (1.5), (1.6) in Ch.II and (1.5), (1.6) in Ch.III, which are exactly the formulas given above.□

The particular case when $X = M \oplus M^{\times}$ is also covered by Theorem 3.1 Note that $X = M \oplus M^{\times}$ means that $M \cap M^{\times} = (0)$ and $X = M+M^{\times}$. So, if $X = M \oplus M^{\times}$, then by Theorem 3.1 the function W admits a Wiener-Hopf factorization relative to Γ and all factorization indices are zero (cf. Corollary 3.2), and thus the factorization is a canonical one. Further, in this case, the projections Q_2 and Q_3 are both zero and $Q_4 = I-Q_1$ is equal to the projection of X along M onto $M^{\times}$ (i.e., $Q_4 = \Pi$, where Π is as in Theorem 2.1 in Ch.I). Inserting this information in the formulas for $W_-(\lambda)^{\pm 1}$ and $W_+(\lambda)^{\pm 1}$ yields the formulas for canonical factorization given in Theorem 2.1 in Ch.I.

We conclude with a few remarks about the case when W is an $m\times m$ rational matrix function which is analytic at ∞ and has the value I at ∞. Assume W has no poles and zeros on Γ. Then W appears as the transfer function of a finite dimensional node $\Theta = (A,B,C;\mathbb{C}^n,\mathbb{C}^m)$ such that A and $A^{\times}$ have no eigenvalues on Γ. Since the state space of Θ is finite dimensional, it follows that the condition

$$\dim(M\cap M^{\times}) < \infty, \quad \dim(X/M+M^{\times}) < \infty$$

is fulfilled. Thus W admits a Wiener-Hopf factorization relative to Γ, and we can apply Theorem 3.1 to get explicit formulas for the factors. In this case the diagonal term $D(\lambda)$ may be written as an m×m diagonal matrix whose diagonal entries are given by

$$\left(\frac{\lambda-\varepsilon_1}{\lambda-\varepsilon_2}\right)^{-\alpha_1},\ldots,\left(\frac{\lambda-\varepsilon_1}{\lambda-\varepsilon_2}\right)^{-\alpha_t},1,\ldots,1,\left(\frac{\lambda-\varepsilon_1}{\lambda-\varepsilon_2}\right)^{\omega_1},\ldots,\left(\frac{\lambda-\varepsilon_1}{\lambda-\varepsilon_2}\right)^{\omega_s}.$$

To get $D(\lambda)$ as a diagonal matrix the factors W_- and W_+ have to be multiplied by suitable constant invertible operators.

The factors W_- and W_+ produced in the proof of Theorem 3.1 are transfer functions of nodes with state spaces $M \oplus K$ and $M^\times \oplus K$, respectively. Note that $M \oplus K$ and $M^\times \oplus K$ are subspaces of the state space of Θ. It follows that in the rational case the McMillan degree $\delta(W_-)$ of W_- (i.e., the minimal dimension of the state space of a node for W_- (see, e.g., [BGK1], Section 4.2)) is less than the McMillan degree $\delta(W)$ of W. Similarly, $\delta(W_+) \leq \delta(W)$. This leads to the following corollary.

COROLLARY 3.3. *An m×m rational matrix function W with no poles and zeros on the contour Γ admits a Wiener-Hopf factorization with respect to Γ with factors W_- and W_+ that are rational matrix functions with McMillan degrees not exceeding the McMillan degree of W.*

PROOF. By applying a suitable Möbius transformation we may assume that W is analytic at ∞ and has the value I at ∞. But then, as explained above, the corollary is a consequence of (the proof of) Theorem 3.1. □

REFERENCES

[BGK1] Bart, H., Gohberg, I., Kaashoek, M.A.: *Minimal factorization of matrix and operator functions.*

Operator Theory: Advances and Applications, Vol.1. Basel-Boston-Stuttgart, Birkhäuser Verlag, 1979.

[BGK2] Bart, H. Gohberg, I., Kaashoek, M.A.: Wiener-Hopf factorization of analytic operator functions and realization. Rapport 231, Wiskundig Seminarium, Vrije Universiteit, Amsterdam, 1983.

[BGK3] Bart, H., Gohberg, I., Kaashoek, M.A.: Wiener-Hopf factorization and realization. In: Mathematical Theory of Networks and Systems, Proceedings of the MTNS-83 International Symposium, Beer Sheva, Israel, Lecture Notes in Control and Information Sciences, no. 58 (Ed. P. Fuhrmann), Springer Verlag, Berlin, 1984, pp.42-62.

[BGK4] Bart, H., Gohberg, I., Kaashoek, M.A.: Invariants for Wiener-Hopf equivalence of analytic operator functions. This volume.

[Br] Brodskii, M.S.: Triangular and Jordan representation of linear operators. Transl. Math. Monographs, Vol.32, Providence, R.I., Amer. Math. Soc., 1970.

[CG] Clancey, K., Gohberg, I.: Factorization of matrix functions and singular integral operators. Operator Theory: Advances and Applications, Vol.3, Basel-Boston-Stuttgart, Birkhäuser Verlag, 1981.

[GF] Gohberg, I.C., Feldman, I.A.: Convolution equations and projection methods for their solution. Transl. Math. Monographs, Vol.41, Providence, R.I., Amer. Math. Soc., 1974.

[GK1] Gohberg, I.C., Krein, M.G.: Systems of integral equations on a half line with kernels depending on the difference of arguments. Amer. Math. Soc. Transl. (2) 14, 217-287 (1960).

[GK2] Gohberg, I.C., Krein, M.G.: Theory and applications of Volterra operators in Hilbert space. Transl. Math. Monographs, Vol.24, Providence, R.I., Amer. Math. Soc., 1970.

[GKr] Gohberg, I., Krupnik, N.: Einführung in die Theorie der eindimensionalen singulären Integraloperatoren.

Basel-Boston-Stuttgart, Birkhäuser Verlag, 1979.

[GKS] Gohberg, I., Kaashoek, M.A., Van Schagen, F.: Similarity of operator blocks and canonical forms. I. General results, feedback equivalence and Kronecker indices. Integral Equations and Operator Theory 3, 350-396 (1980).

[GL] Gohberg, I.C., Leiterer, J.: A criterion for factorization of operator functions with respect to a contour. Sov. Math. Dokl. 14, No.2, 425-429 (1973).

[K] Krein, M.G.: Integral equations on the half line with a kernel depending on the difference of the arguments. Amer. Math. Soc. Transl. (2)22, 163-288 (1962).

[KFA] Kalman, R.E., Falb, P.F., Arbib, M.A.: Topics in mathematical systems theory. New York, McGraw-Hill, 1969.

[KR] Kaashoek, M.A., Ran, A.C.M.: Symmetric Wiener-Hopf factorization of selfadjoint rational matrix functions and realization. This volume.

[W] Wonham, W.H.: Linear Multivariable Control. Berlin, Springer Verlag, 1974.

H. Bart,
Econometrisch Instituut
Erasmus Universiteit
Postbus 1738
3000 DR Rotterdam
The Netherlands

I. Gohberg
Dept of Mathematical Sciences
The Raymond and Beverly Sackler
Faculty of Exact Sciences
Tel-Aviv University
Ramat-Aviv. Israel

M.A. Kaashoek
Subfaculteit Wiskunde en
Informatica
Vrije Universiteit
Postbus 7161
1007 MC Amsterdam
The Netherlands

Operator Theory:
Advances and Applications, Vol. 21

INVARIANTS FOR WIENER-HOPF EQUIVALENCE OF ANALYTIC OPERATOR FUNCTIONS

H. Bart, I. Gohberg, M.A. Kaashoek

Necessary conditions for Wiener-Hopf equivalence are established in terms of the incoming and outgoing subspaces associated with realizations of the given analytic operator functions. Other results about the behaviour of the incoming and outgoing subspaces under certain elementary operations are also included.

1. Introduction and main result

Let W_1 and W_2 be continuous functions on a positively oriented contour Γ in the complex plan $\mathbb{C}$ and assume that their values are invertible bounded linear operators on a complex Banach space Y. The functions W_1 and W_2 are called (right) Wiener-Hopf equivalent with respect to Γ if W_1 and W_2 are related in the following way:

$$(1.1)\qquad W_1(\lambda) = W_-(\lambda)W_2(\lambda)W_+(\lambda), \qquad \lambda \in \Gamma,$$

where W_- and W_+ are analytic on the outer domain Ω_Γ^- (which contains ∞) and the inner domain Ω_Γ^+ of Γ, respectively, both W_- and W_+ are continuous up to the boundary and their values are invertible operators on Y. In case that Y is finite dimensional and the functions W_1 and W_2 are Hölder continuous on the contour it is well-known (see e.g:, [CG]) that W_1 and W_2 are right Wiener-Hopf equivalent if and only if W_1 and W_2 have the same right factorization indices.

Next let us consider the case when the functions W_1 and W_2 are analytic on Γ and appear as transfer functions of nodes $\Theta_1 = (A_1,B_1,C;X_1,Y)$ and $\Theta_2 = (A_2,B_2,C_2;X_2,Y)$ without spectrum on Γ (see [BGK4], Ch.I), that is, W_1 and W_2 are given in the realized form

$$W_1(\lambda) = I + C_1(\lambda-A_1)^{-1}B_1, \qquad \lambda \in \Gamma,$$

$$W_2(\lambda) = I + C_2(\lambda-A_2)^{-1}B_2, \qquad \lambda \in \Gamma,$$

where $A_\upsilon: X_\upsilon \to X_\upsilon$, $B_\upsilon: Y \to X_\upsilon$ and $C_\upsilon: X_\upsilon \to Y$ are bounded linear operators, the space X_υ is a complex Banach space and the operators A_υ and $A_\upsilon^{\times} := A_\upsilon - B_\upsilon C_\upsilon$ have no spectrum on the contour Γ. Now, let M_υ and $M_\upsilon^{\times}$ be the spectral subspaces

$$M_\upsilon = \operatorname{Im}\Big(\frac{1}{2\pi i}\int_\Gamma (\lambda-A_\upsilon)^{-1}d\lambda\Big), \tag{1.2}$$

$$M_\upsilon^{\times} = \operatorname{Ker}\Big(\frac{1}{2\pi i}\int_\Gamma (\lambda-A_\upsilon^{\times})^{-1}d\lambda\Big), \tag{1.3}$$

and assume in addition that

$$\dim(M_\upsilon \cap M_\upsilon^{\times}) < \infty, \qquad \operatorname{codim}(M_\upsilon + M_\upsilon^{\times}) < \infty \qquad (\upsilon=1,2). \tag{1.4}$$

Then it follows from Corollary 3.2 in Ch.IV of [BGK4] that W_1 and W_2 are right Wiener-Hopf equivalent if and only if

$$\begin{aligned}\dim\big(M_1\cap M_1^{\times}\cap \operatorname{Ker}C_1\cap\ldots\cap\operatorname{Ker}C_1A_1^{m-1}\big) &= \\ = \dim\big(M_2\cap M_2\cap \operatorname{Ker}C_2\cap\ldots\cap\operatorname{Ker}C_2A_2^{m-1}\big)&\end{aligned} \tag{1.5}$$

and

$$\begin{aligned}\operatorname{codim}(M_1+M_1^{\times}+\operatorname{Im}B_1+\ldots+\operatorname{Im}A_1^{m-1}B_1) &= \\ = \operatorname{codim}(M_2+M_2^{\times}+\operatorname{Im}B_2+\ldots+\operatorname{Im}A_2^{m-1}B_2)&\end{aligned} \tag{1.6}$$

for $m = 0,1,2,\ldots$. (Actually, the results of [BGK4] apply to a sligthly more general situation involving contours on the Riemann sphere $\mathbb{C} \cup \{\infty\}$.)

In general, when the finite dimensionality conditions in (1.4) are not fulfilled, then conditions (1.5) and (1.6) do not give much information and one may expect that they have to be replaced by the existence of certain isomorphisms between the subspaces (or their complements) involved. The next theorem, which is the main result of this paper, guarantees the existence of such isomorphisms.

THEOREM 1.1. *For* $\upsilon = 1,2$, *let*

$$W_\upsilon(\lambda) = I + C_\upsilon(\lambda - A_\upsilon)^{-1}B_\upsilon$$

be the transfer function of a node $\Theta_\upsilon = (A_\upsilon, B_\upsilon, C_\upsilon; X_\upsilon Y)$ *without spectrum on the contour* Γ. *Suppose* W_1 *and* W_2 *are Wiener-Hopf equivalent with respect to* Γ. *Let the Riesz projections* $P_1, P_1^\times$ *and the bounded linear operators* ψ, Φ: $X_1 \to X_2$ *be given by*

$$P_1 = \frac{1}{2\pi i}\int_\Gamma (\lambda - A_1)^{-1} d\lambda, \quad P_1^\times = \frac{1}{2\pi i}\int_\Gamma (\lambda - A_1^\times)^{-1} d\lambda,$$

$$\Psi = \frac{1}{2\pi i}\int_\Gamma (\lambda - A_2^\times)^{-1} B_2 W_-(\lambda)^{-1} C_1 (\lambda - A_1)^{-1} d\lambda,$$

$$\Phi = \frac{1}{2\pi i}\int_\Gamma (\lambda - A_2^\times)^{-1} B_2 W_-(\lambda)^{-1} C_1 (\lambda - A_1)^{-1} (I - P_1) P_1^\times d\lambda,$$

where W_- *is as in* (1.1). *Then, for* $m = 0,1,2,\dots$

$$\Psi[M_1 \cap M_1^\times \cap \mathrm{Ker}C_1 \cap \dots \cap \mathrm{Ker}C_1 A_1^{m-1}] = M_2 \cap M_2^\times \cap \mathrm{Ker}C_2 \cap \dots \cap \mathrm{Ker}C_2 A_2^{m-1},$$

$$\mathrm{Ker}\Psi \cap M_1 \cap M_1^\times \cap \mathrm{Ker}C_1 \cap \dots \cap \mathrm{Ker}C_1 A_1^{m-1} = (0),$$

$$M_1 + M_1^\times + \mathrm{Im}B_1 + \dots + \mathrm{Im}A_1^{m-1}B_1 = \Phi^{-1}[M_2 + M_2^\times + \mathrm{Im}B_2 + \dots + \mathrm{Im}A_2^{m-1}B_2],$$

$$\mathrm{Im}\Phi + M_2 + M_2^\times + \mathrm{Im}B_2 + \dots + \mathrm{Im}A_2^{m-1}B_2 = X_2.$$

Here $M_1, M_1^\times$, M_2 and $M_2^\times$ *are the spectral subspaces given by* (1.2) *and* (1.3), *i.e.*, $M_\upsilon = \mathrm{Im}P_\upsilon$ *and* $M_2^\times = \mathrm{Im}P_\upsilon^\times$ $(\upsilon=1,2)$.

We do not know whether in general the existence of appropriate isomorphims between the subspaces (or their complements) implies Wiener-Hopf equivalence. There are indications (see Theorem III.2.1 in [GKS] which deals with Wiener-Hopf equivalence of operator polynomials) that a converse statement is true under the additional condition that the subspaces are complemented.

Theorem 1.1 is proved in Section 7 and to prove it we use results from Sections 2,3,5 and 6. In Section 4 we describe the behaviour of the spaces

$$(1.7)\qquad H_m = M + M^\times + \operatorname{Im}B + \operatorname{Im}AB + \dots + \operatorname{Im}A^{m-1}B,$$

$$(1.8)\qquad K_m = M \cap M^\times \cap \operatorname{Ker}C \cap \operatorname{Ker}CA \cap \dots \cap \operatorname{Ker}CA^{m-1}.$$

under the operation of dilation. Along the way we obtain a proof of the only if part of Theorem 2.2 in Ch.I of [BGK4]. For the type of contour considered here, this theorem can be formulated as follows.

THEOREM 1.2. _Let_ $W(\lambda) = I + C(\lambda-A)^{-1}B$ _be the transfer function of a node_ $\Theta = (A,B,C;X,Y)$ _without spectrum on the contour_ Γ, _and let_ M _and_ $M^\times$ _be the spectral subspaces_

$$(1.9)\qquad M = \operatorname{Im}\left(\frac{1}{2\pi i}\int_\Gamma (\lambda-A)^{-1}d\lambda\right),$$

$$(1.10)\qquad M^\times = \operatorname{Ker}\left(\frac{1}{2\pi i}\int_\Gamma (\lambda-A^\times)^{-1}d\lambda\right).$$

Then W _admits a (right) Wiener-Hopf factorization with respect to_ Γ _if and only if_

$$\dim(M\cap M^\times) < \infty, \qquad \operatorname{codim}(M+M^\times) < \infty.$$

The if part of this result is proved in [BGK4]. In fact, the setting there is slightly more general than the one considered here. We shall comment on this after the proof of the only if part, which will be given in Section 6.

Generally speaking, notation and terminology are as in [BGK4], the paper preceding the present one in this volume. For the convenience of the reader, we recall a few of the main definitions, thereby making some of the notions already used above more precise. A <u>node</u> (or a <u>system</u>) is a quintet $\Theta = (A,B,C;X,Y)$, where $A: X \to X$, $B: Y \to X$ and $C: X \to Y$ are (bounded) linear operators acting between complex Banachspaces X and Y. In other words $A \in L(X)$, $B \in L(Y,X)$ and $C \in L(X,Y)$. The node is said to be <u>finite dimensional</u> if both X and Y are finite dimensional spaces; otherwise it is called <u>infinite dimensional</u>. The spectrum of the operator A is denoted by $\sigma(A)$ and the operator function

$$W(\lambda) = I + C(\lambda - A)^{-1}B,$$

defined and analytic on the complement of $\sigma(A)$, is called the <u>transfer function</u> of Θ. We say that Θ <u>has no spectrum on</u> (the given contour) Γ if A and $A^{\times} := A-BC$ have no spectrum on Γ. In that case one can introduce the spectral subspaces (1.9) and (1.10). The pair $M,M^{\times}$ obtained this way is called the <u>pair of spectral subspaces associated with</u> Θ <u>and</u> Γ. The spaces H_m given by (1.7) are called the <u>incoming subspaces</u>, the spaces K_m given by (1.8) the <u>outgoing subspaces</u> for Θ. Note that $H_0 = M + M^{\times}$ and $K_0 = M \cap M^{\times}$. (For slightly more general definitions involving contours on the Riemann sphere, see [BGK4]).

An earlier version of the present paper appeared in the report [BGK2]; the main results were also announced in [BGK3].

2. Simple nodes with centralized singularities

In this section we study operator functions having the diagonal form of the middle factor in a Wiener-Hopf factorization. As in Section I.2 of [BGK4], we fix complex numbers ε_1 in Ω_Γ^+ and ε_2 in Ω_Γ^-.

Let Y be a complex Banach space and consider the operator function

$$(2.1)\qquad D(\lambda) = \Pi_0 + \sum_{j=1}^{t} \left(\frac{\lambda-\varepsilon_1}{\lambda-\varepsilon_2}\right)^{-\alpha_j}\Pi_{-j} + \sum_{j=1}^{s} \left(\frac{\lambda-\varepsilon_1}{\lambda-\varepsilon}\right)^{\omega_j}\Pi_j,$$

Here $\Pi_{-t},\ldots,\Pi_{-1},\Pi_1,\ldots,\Pi_s$ are mutually disjoint one dimensional projections of Y, $\sum_{j=-t}^{s} \Pi_j$ is the identity operator I_Y on Y, $\alpha_1,\ldots,\alpha_t$, $\omega_1,\ldots,\omega_s$ are positive integers and $-\alpha_1 \leqq \cdots \leqq -\alpha_t < 0 < \omega_1 < \cdots < \omega_s$ (cf. Section I.3 of [BGK4], formula (3.9)). We allow for t or s to be zero. This simply means that in (2.1) the first or second sum does not appear.

For $j = 1,\ldots,t$, let $A_j^-: \mathbb{C}^{\alpha_j} \to \mathbb{C}^{\alpha_j}$, $B_j^-: \mathbb{C} \to \mathbb{C}^{\alpha_j}$ and $C_j^-: \mathbb{C}^{\alpha_j} \to \mathbb{C}$ be given by

$$A_j^- = \begin{pmatrix} \varepsilon_1 & 0 & \cdots & \cdots & 0 \\ 1 & \varepsilon_1 & & & \vdots \\ 0 & 1 & \ddots & & \vdots \\ \vdots & & \ddots & \ddots & 0 \\ 0 & \cdots & 0 & 1 & \varepsilon_1 \end{pmatrix}, \qquad B_j^- = \begin{pmatrix} \binom{\alpha_j}{\alpha_j}(\varepsilon_1-\varepsilon_2)^{-1} \\ \binom{\alpha_j}{\alpha_j-1}(\varepsilon_1-\varepsilon_2)^{-2} \\ \vdots \\ \binom{\alpha_j}{1}(\varepsilon_1-\varepsilon_2)^{-\alpha_j} \end{pmatrix},$$

$$C_j^- = (\varepsilon_1-\varepsilon_2)^{\alpha_j+1}[0\ \ldots\ 0\ \ 1].$$

Then the transfer function of the node $\Theta_j^- = (A_j^-,B_j^-,C_j^-;\mathbb{C}^{\alpha_j},\mathbb{C})$ has the form $((\lambda-\varepsilon_1)/(\lambda-\varepsilon_2))^{-\alpha_j}$. Identifying $\mathbb{C}$ with the one-dimensional subspace $\operatorname{Im}\Pi_{-j}$ of Y, we obtain a node $\Theta_j^- = (A_j^-,B_j^-,C_j^-;\mathbb{C}^{\alpha_j},\operatorname{Im}\Pi_{-j})$ whose transfer function is

$$D_j^-(\lambda) = \left(\frac{\lambda-\varepsilon_1}{\lambda-\varepsilon_2}\right)^{-\alpha_j} I_{\mathrm{Im}\Pi_{-j}}.$$

Clearly the node Θ_j^- has no spectrum on the given contour Γ and $\mathbb{C}^{\alpha_j},\mathbb{C}^{\alpha_j}$ is the pair of spectral subspaces associated with Θ_j^- and Γ. More generally, the outgoing subspaces of Θ_j^- are

$$\mathbb{C}^{\alpha_j}, \mathbb{C}^{\alpha_j-1}\times(0),\ldots,\mathbb{C}\times(0)^{\alpha_j-1},(0)^{\alpha_j},(0)^{\alpha_j},\ldots$$

and the incoming subspaces of Θ_j^- are all equal to $\mathbb{C}^{\alpha_j}$. It follows that Θ_j^- has no incoming indices and only one outgoing index, namely α_j. In fact, the standard basis of $\mathbb{C}^{\alpha_j}$ is an outgoing basis for Θ_j^- with respect to the operator $A_j^- - \varepsilon_1$ (cf. [BGK4]).

Next, for $j = 1,\ldots,s$, let $A_j^+ : \mathbb{C}^{\omega_j} \to \mathbb{C}^{\omega_j}$, $B_j^+ : \mathbb{C} \to \mathbb{C}^{\omega_j}$ and $C_j^+ : \mathbb{C}^{\omega_j} \to \mathbb{C}$ be given by

$$A_j^+ = \begin{pmatrix} \varepsilon_2 & 0 & \cdot & \cdot & \cdot & \cdot & 0 \\ 1 & \varepsilon_2 & & & & & \cdot \\ 0 & 1 & & & & & \cdot \\ \vdots & & \cdot & & & & \cdot \\ \vdots & & & \cdot & & & \cdot \\ \vdots & & & & \cdot & & 0 \\ 0 & \cdot & \cdot & \cdot & 0 & 1 & \varepsilon_2 \end{pmatrix}, \qquad B_j^+ = (\varepsilon_2-\varepsilon_1)^{\omega_j+1}\begin{pmatrix} 1 \\ 0 \\ \cdot \\ \cdot \\ \cdot \\ 0 \\ 0 \end{pmatrix},$$

$$C_j^+ = \left[\binom{\omega_j}{1}\frac{1}{(\varepsilon_2-\varepsilon_1)^{\omega_j}} \quad \cdots \quad \binom{\omega_j}{\omega_j-1}\frac{1}{(\varepsilon_2-\varepsilon_1)^2} \quad \binom{\omega_j}{\omega_j}\frac{1}{(\varepsilon_2-\varepsilon_1)}\right].$$

Then the transfer function of the node $\Theta_j^+ = (A_j^+,B_j^+,C_j^+;\mathbb{C}^{\omega_j},\mathbb{C})$ has the form $((\lambda-\varepsilon_1)/(\lambda-\varepsilon_2))^{\omega_j}$. Identifying $\mathbb{C}$ with the one-dimensional subspace $\mathrm{Im}\,\Pi_j$ of Y, we obtain a node $\Theta_j^+ = (A_j^+,B_j^+,C_j^+;\mathbb{C}^{\omega_j},\mathrm{Im}\Pi_j)$ whose transfer function is

$$D_j^+(\lambda) = \left(\frac{\lambda-\varepsilon_1}{\lambda-\varepsilon_2}\right)^{\omega_j} I_{\mathrm{Im}\Pi_j}.$$

This node has no spectrum on Γ and $(0)^{\omega_j},(0)^{\omega_j}$ is the pair of spectral subspaces associated with Θ_j^+ and Γ. More generally, the incoming subspaces of Θ_j^+ are

$$(0)^{\omega_j}, \mathbb{C}\times(0)^{\omega_j-1}, \dots, \mathbb{C}^{\omega_j-1}\times(0), \mathbb{C}^{\omega_j}, \mathbb{C}^{\omega_j}, \dots$$

and the outgoing subspaces of Θ_j^+ are all equal to $(0)^{\omega_j}$. It follows that Θ_j^+ has no outgoing indices and only one incoming index, namely ω_j. In fact, the standard basis of $\mathbb{C}^{\omega_j}$ is an incoming basis for Θ_j^+ with respect to the operator $A_j^+-\varepsilon_2$ (cf.[BGK4]).

Now let us return to (2.1). Note that $D(\lambda)$ can be written as

$$D(\lambda) = \Pi_0 + \sum_{j=1}^{t} D_j^-(\lambda)\Pi_{-j} + \sum_{j=1}^{s} D_j^+(\lambda)\Pi_j.$$

It follows that D is the transfer function of the "direct sum" of the nodes $\Theta_1^-,\dots,\Theta_t^-,\Theta_1^+,\dots,\Theta_s^+$ and the trivial node $\Theta_0 = (0,0,0;(0),\mathrm{Im}\pi_0)$ whose transfer function is $I_{\mathrm{Im}\Pi_0}$. More, precisely, D is the transfer function of the node

$\Theta_D = (A_D,B_D,C_D;X_D,Y)$, where $X_D = \mathbb{C}^{\alpha_1+\dots+\alpha_t+\omega_1+\dots+\omega_s}$ and

$$A_D = \begin{pmatrix} A_1^- & & & & & \\ & \ddots & & & & \\ & & A_t^- & & & \\ & & & A_1^+ & & \\ & & & & \ddots & \\ & & & & & A_s^+ \end{pmatrix} : X_D \to X_D,$$

$$B_D = \begin{pmatrix} B_1^- & & & & & \\ & \ddots & & & & \\ & & B_t^- & & 0 & \\ & & & B_1^+ & & \\ & & & & \ddots & \\ & & & & & B_s^+ \end{pmatrix} : Y \to X_D,$$

$$C_D = \begin{pmatrix} C_1^- & & & & & \\ & \ddots & & & & \\ & & C_t^- & & & \\ & & 0 & & & \\ & & & C_1^+ & & \\ & & & & \ddots & C_s^+ \end{pmatrix} : X_D \to Y.$$

Here the matrix representations are taken with respect to the decompositions

$$X_D = \mathbb{C}^{\alpha_1} \oplus \ldots \oplus \mathbb{C}^{\alpha_t} \oplus \mathbb{C}^{\omega_1} \oplus \ldots \oplus \mathbb{C}^{\omega_s},$$

$$Y = \operatorname{Im} \Pi_{-1} \oplus \ldots \oplus \operatorname{Im} \Pi_{-t} \oplus \operatorname{Im} \Pi_0 \oplus \operatorname{Im} \Pi_1 \oplus \ldots \oplus \operatorname{Im} \Pi_s.$$

Clearly Θ_D has no spectrum on Γ and for the pair $M_D, M_D^\times$ of spectral subspaces associated with Θ_D and Γ, we have

$$M_D = M_D^\times = \mathbb{C}^{\alpha_1} \oplus \ldots \oplus \mathbb{C}^{\alpha_t} \oplus (0)^{\omega_1} \oplus \ldots \oplus (0)^{\omega_s}.$$

More generally, we can get the outgoing and incoming subspaces of Θ_D by talking the direct sums of the corresponding spaces of the nodes $\Theta_1^-,\ldots,\Theta_t^-,\Theta_1^+,\ldots,\Theta_s^+$. In a similar way an outgoing basis for Θ_D (with respect to the operator $A_D-\varepsilon_1$) and an incoming basis for Θ_D (with respect to the operator $A_D-\varepsilon_2$) can be obtained. It follows that

$$(2.2) \qquad t = \dim \frac{M_D \cap M_D^\times}{M_D \cap M_D^\times \cap \operatorname{Ker} C_D}, \qquad s = \dim \frac{M_D + M_D^\times + \operatorname{Im} B_D}{M_D + M_D^\times},$$

$\alpha_1,\ldots,\alpha_t$ are the outgoing indices of Θ_D and $\omega_1,\ldots,\omega_s$ are the incoming indices of Θ_D. The last two assertions can also be formulated as

$$(2.3) \quad \alpha_j = \#\left\{m \,\middle|\, \dim \frac{M_D \cap M_D^\times \cap \operatorname{Ker} C_D \cap \ldots \cap \operatorname{Ker} C_D A_D^{m-2}}{M_D \cap M_D^\times \cap \operatorname{Ker} C_D \cap \ldots \cap \operatorname{Ker} C_D A_D^{m-1}} \geqq j\right\}, \quad j = 1,\ldots,t,$$

$$(2.4) \quad \omega_j = \#\left\{m \,\middle|\, \dim \frac{M_D + M_D^\times + \operatorname{Im} B_D + \ldots + \operatorname{Im} A_D^{m-1} B_D}{M_D + M_D^\times + \operatorname{Im} B_D + \ldots + \operatorname{Im} A_D^{m-2} B_D} \geqq s-j+1\right\}, \quad j = 1,\ldots,s.$$

Note also that

$$(2.5) \qquad \dim(M_D \cap M_D^\times) = \alpha_1 + \dots + \alpha_t,$$

$$(2.6) \qquad \dim(X_D / M_D + M_D^\times) = \omega_1 + \dots + \omega_s.$$

In particular $\dim(M_D \cap M_D^\times)$ and $\dim(X_D/M_D + M_D^\times)$ are finite.

The node Θ_D constructed above is clearly minimal. Also it is a simple node with centralized singularities in the sense of Section I.3 of [BGK4]. The following result was already announced in the discussion after the proof of Theorem 3.1 in Ch.I of [BGK4].

PROPOSITION 2.1. _Let_ D _be as above and let the node_ Θ _be a minimal realization of_ D _on a neighbourhood of_ ∞. _Then_ Θ _is a simple node with centralized singularities._

PROOF. Since the transfer functions of Θ and Θ_D coincide on a neighbourhood of ∞, the nodes Θ and Θ_D are quasi-similar (cf. [H]). Now the state space of Θ_D is finite dimensional, so in fact Θ and Θ_D are similar. As we observed, Θ_D is a node with centralized singularities. It follows that Θ is also of this type. □

For completeness we recall from Section I.3 of [BGK4] that the transfer function of a simple node with centralized singularities has the form of the middle term in a Wiener-Hopf factorization, i.e., it admits a representation as in (2.1)

3. Multiplication by plus and minus terms

Part of the next theorem was already announced in the discussion ensuing the proof of Theorem 3.1 in Ch.I of [BGK4]. For simplicity we restrict ourselves here to the case when Γ is a positively oriented contour in the (finite) complex plane. The general case was treated in [BGK2].

THEOREM 3.1. *Let* W *be an operator function defined and analytic on an open neighbourhood of the contour* Γ *and with values in* $L(Y)$, *where* Y *is a complex Banach space. Then, given a (right) Wiener-Hopf factorization*

$$(3.1)\qquad W(\lambda) = W_-(\lambda)D(\lambda)W_+(\lambda), \qquad \lambda \in \Gamma$$

relative to the contour Γ *and the points* $\varepsilon_1, \varepsilon_2$, *with the middle term* D *in the form* (2.1), *i.e.*,

$$D(\lambda) = \Pi_0 + \sum_{j=1}^{t} \left(\frac{\lambda-\varepsilon_1}{\lambda-\varepsilon_2}\right)^{-\alpha_j}\Pi_{-j} + \sum_{j=1}^{s} \left(\frac{\lambda-\varepsilon_1}{\lambda-\varepsilon_2}\right)^{\omega_j}\Pi_j,$$

$(-\alpha_1 \leqq -\alpha_2 \leqq \cdots \leqq -\alpha_t < 0 < \omega_1 \leqq \omega_2 \leqq \cdots \leqq \omega_s)$, *there exists a node* $\Theta = (A,B,C;X,Y)$ *with the following properties:*

(a) Θ *is a node with centralized singularities relative to the contour* Γ *and the points* $\varepsilon_1, \varepsilon_2$ *(in particular* Θ *has no spectrum on* Γ*)*;

(b) *On the contour* Γ *the transfer function of* Θ *coincides with* W, *i.e.*,

$$W(\lambda) = I_Y + C(\lambda I_X - A)^{-1}B, \qquad \lambda \in \Gamma;$$

(c) *If* $M, M^\times$ *is the pair of spectral subspaces associated with* Θ *and* Γ, *then*

$$\dim(M \cap M^\times) < \infty, \qquad \operatorname{codim}(M + M^\times) < \infty;$$

(d) *The number* t *of negative factorization indices in the Wiener-Hopf factorization* (3.1) *and the negative factorization indices* $-\alpha_1, \ldots, -\alpha_t$ *themselves are given by*

$$t = \dim \frac{M \cap M^\times}{M \cap M^\times \cap \operatorname{Ker} C},$$

$$\alpha_j = \#\left\{m \,\middle|\, \dim \frac{M \cap M^\times \cap \operatorname{Ker} C \cap \ldots \cap \operatorname{Ker} CA^{m-2}}{M \cap M^\times \cap \operatorname{Ker} C \cap \ldots \cap \operatorname{Ker} CA^{m-1}} \geqq j\right\},$$

$$j = 1, \ldots, t;$$

in other words the absolute values of the negative factorization indices are just the outgoing indices of Θ;

(e) *The number* s *of positive factorization indices in the Wiener-Hopf factorization* (3.1) *and the positive factorization indices* $\omega_1,\dots,\omega_s$ *themselves are given by*

$$s = \dim \frac{M + M^{\times} + \mathrm{Im} B}{M + M^{\times}},$$

$$\omega_j = \left\{m \,\middle|\, \dim \frac{M + M^{\times} + \mathrm{Im} B + \dots + \mathrm{Im} A^{m-1} B}{M + M^{\times} + \mathrm{Im} B + \dots + \mathrm{Im} A^{m-2} B} \geq s-j+1\right\},$$

$$j = 1,\dots,s;$$

in other words the positive factorization indices are just the incoming indices of Θ.

The condition imposed on W can also be formulated as follows: On (a neighbourhood of) Γ, the function W can be written as the transfer function of a node (cf. [BGK1], Section 2.3).

In order to simplify the proof of Theorem 3.1 and motivated by the definition of Wiener-Hopf factorization, given in Section I.2 of [BGK4], we introduce the following terminology. An operator function W is called a *plus (minus) function* (with respect to the contour Γ) if W is analytic on Ω_Γ^+ (Ω_Γ^-), continuous on the closure of Ω_Γ^+ (Ω_Γ^-) and has invertible values. If W is a plus (minus) function, the same is true for W^{-1} (given by ($W^{-1}(\lambda) = W(\lambda)^{-1}$).

A related concept can be defined for nodes. We call a node $\Theta = (A,B,C;X,Y)$ a *plus (minus) node* (with respect to Γ) if the spectra of A and $A^{\times}$ are contained in Ω_Γ^- (Ω_Γ^+). In particular such a node has no spectrum on Γ. The transfer function of a plus (minus) node is obviously a plus (minus) function. In fact it is even analytic on an open neighbourhood (in $\mathbb{C} \cup \{\infty\}$) of the closure of Ω_Γ^+ (Ω_Γ^-). Conversely, suppose W is analytic and takes invertible values on an open neighbourhood U of the closure of Ω_Γ^+. In the case considered here where Ω_Γ^+ is bounded, we may assume that U is bounded too, and then W can be written (on U) as the transfer function of a plus node (cf. [BGK1], Ch.II). Similar remarks hold when Ω_Γ^+ is replaced by Ω_Γ^-, provided $W(\infty) = I_Y$.

We shall now consider products of the type $\Theta = \Theta_-\Theta_0\Theta_+$, where Θ_+ is a plus node, Θ_- is a minus node and Θ_0 is a node without spectrum on Γ. Recall from [BGK1] that, if $\Theta_- = (A_-,B_-,C_-;X_-,Y)$, $\Theta_0 = (A_0,B_0,C_0;X_0Y)$ and $\Theta_+ = (A_+,B_+,C_+;X_+,Y)$, the product $\Theta = \Theta_-\Theta_0\Theta_+$ is given by $\Theta = (A,B,C;X,Y)$, where $X = X_- \oplus X_0 \oplus X_+$ and

$$A = \begin{pmatrix} A_- & B_-C_0 & B_-C_+ \\ 0 & A & B_0C_+ \\ 0 & 0 & A_+ \end{pmatrix} : X_- \oplus X_0 \oplus X_+ \to X_- \oplus X_0 \oplus X_+, \tag{3.2}$$

$$B = \begin{pmatrix} B_- \\ B_0 \\ B_+ \end{pmatrix} : Y \to X_- \oplus X_0 \oplus X_+, \tag{3.3}$$

$$C = [C_- \quad C_0 \quad C_+]: X_- \oplus X_0 \oplus X_+ \to Y. \tag{3.4}$$

Since Θ_-, Θ_0 and Θ_+ have no spectrum on Γ, the operator A has no spectrum on Γ. The same is true for the operator $A^\times$, which can be written as

$$A^\times = \begin{pmatrix} A_-^\times & 0 & 0 \\ -B_0C_- & A_0^\times & 0 \\ -B_+C_- & -B_+C_0 & A_+^\times \end{pmatrix} : X_- \oplus X_0 \oplus X_+ \to X_- \oplus X_0 \oplus X_+. \tag{3.5}$$

So $\Theta = \Theta_-\Theta_0\Theta_+$ also has no spectrum on Γ.

It is convenient to introduce the following notation. If E_-, E_0 and E_+ are subspaces of X_-, X_0 and X_+, respectively, then, by definition,

$$\begin{pmatrix} E_- \\ E_0 \\ E_+ \end{pmatrix} = E_- \oplus E_0 \oplus E_+ .$$

THEOREM 3.2. *Suppose* $\Theta_-, \Theta_0, \Theta_+$ *are as above and* $\Theta = \Theta_-\Theta_0\Theta_+$.

(1) *Let* $H_{00}, H_{01}, H_{02}, \ldots$ *be the incoming subspaces of* Θ_0. *Then the incoming subspaces* $H_0, H_1, H_2, \ldots$ *of* Θ *are given by*

$$H_m = \begin{pmatrix} X_- \\ H_{0m} \\ X_+ \end{pmatrix}, \qquad m = 0,1,2,\ldots ;$$

(2) *Let* $K_{00}, K_{01}, K_{02}, \ldots$ *be the outgoing subspaces of* Θ_0. *Then the outgoing subspaces* $K_0, K_1, K_2, \ldots$ of Θ *are given by*

$$K_m = \begin{pmatrix} (0) \\ K_{0m} \\ (0) \end{pmatrix}, \qquad m = 0,1,2,\ldots .$$

Note that all these subspaces are well-defined because Θ_0 and Θ have no spectrum on Γ.

PROOF. First consider the case $m = 0$. We then need to show that

$$(3.6) \qquad M+M^\times = \begin{pmatrix} X_- \\ M_0+M_0^\times \\ X_+ \end{pmatrix}, \qquad M\cap M^\times = \begin{pmatrix} (0) \\ M_0\cap M_0^\times \\ (0) \end{pmatrix},$$

where $M_0, M_0^\times$ is the pair of spectral subspaces associated with Θ_0 and $M, M^\times$ is the pair of spectral subspaces associated with Θ. Write $M_0 = \operatorname{Im}P_0$, $M = \operatorname{Im}P$, where P_0 is the Riesz projection corresponding to the part of $\sigma(A_0)$ lying in Ω_Γ^+ and P is the Riesz projection corresponding to the part of $\sigma(A)$ lying in Ω_Γ^+. Since $\sigma(A_-) \subset \Omega_\Gamma^+$ and $\sigma(A_+) \subset \Omega_\Gamma^-$, the projection P has the form

$$(3.7)\qquad P = \begin{pmatrix} I & P_1 & P_2 \\ 0 & P_0 & P_3 \\ 0 & 0 & 0 \end{pmatrix} : X_- \oplus X_0 \oplus X_+ \to X_- \oplus X_0 \oplus X_-.$$

The fact that $P^2 = P$ implies certain relationships between the operators appearing in (3.7) and these can be used to show that

$$P = \begin{pmatrix} I & P_1 & P_2 \\ 0 & I & P_3 \\ 0 & 0 & I \end{pmatrix} \cdot \begin{pmatrix} I & 0 & 0 \\ 0 & P_0 & 0 \\ 0 & 0 & I \end{pmatrix} \cdot \begin{pmatrix} I & P_2 & P_2 \\ 0 & I & P_3 \\ 0 & 0 & I \end{pmatrix},$$

while, moreover, the first and last factor in the right hand side of this identity are each others inverse. From this we see that

$$(3.8)\qquad M = \begin{pmatrix} I & P_1 & P_2 \\ 0 & I & P_3 \\ 0 & 0 & I \end{pmatrix} \begin{pmatrix} X_- \\ M_0 \\ (0) \end{pmatrix} = \begin{pmatrix} X_- \\ M_0 \\ (0) \end{pmatrix}.$$

In the same way one proves that

$$(3.9)\qquad M^\times = \begin{pmatrix} (0) \\ M_0^\times \\ X_+ \end{pmatrix}.$$

Combining (3.8) and (3.9), one immediately gets (3.6).

We have proved the desired result now for $m = 0$. For arbitrary m, it is obtained by an induction argument based on the special form of the operator A, B and C and the identities (3.6). □

COROLLARY 3.3. *Suppose* Θ_-, Θ_0 *and* Θ_+ *are as above and* $\Theta = \Theta_-\Theta_0\Theta_+$. *For* $j = 1,2$, *let* $M, M_j^\times$ *be the pair of spectral subspaces associated with* Θ_j *and* Γ. *Then*

(3.10) $\dim(M_0 \cap M_0^\times) < \infty, \quad \dim(X_0/M_0+M_0^\times) < \infty$

if and only if

(3.11) $\dim(M \cap M^\times) < \infty, \quad \dim(X/M+M^\times) < \infty,$

and in that case the incoming (outgoing) indices of Θ_0 *and* Θ *are the same.*

PROOF. The equivalence of (3.10) and (3.11) is an immediate consequence of (3.6). If (3.10) or (3.11) is satisfied, then there are incoming indices and outgoing indices associated with Θ_0 and Θ. From formula (1.6) in Ch.II of [BGK4] formula (1.6) in Ch.III of [BGK4] and Theorem 3.2 above, we see that the incoming (outgoing) indices of Θ_0 and Θ are the same. □

If (3.10) or (3.11) is satisfied, there exist incoming (outgoing) bases for Θ_0 and Θ. Given an incoming (outgoing) basis for Θ_0 one can easily construct an incoming (outgoing) basis for Θ. Just map the basis for Θ_0, which consists of vectors in X_0, into X by the natural injection of X_0 into $X = X_- \oplus X_0 \oplus X_+$.

We are now ready to prove Theorem 3.1.

PROOF OF THEOREM 3.1. First suppose that W_- and W_+ can be written as the transfer function of a minus node $\Theta_- = (A_-,B_-,C_-;X_-,Y)$ and a plus node $\Theta_+ = (A_+,B_+,C_+;X_+,Y)$, respectively. Let $\Theta_D = (A_D,B_D,C_D;X_D,Y)$ be the realization of D constructed in Section 2. Then Θ_D is a simple node with centralized singularities and (2.2) - (2.6) hold. Put $\Theta = \Theta_-\Theta_D\Theta_+$ and write $\Theta = (A,B,C;X,Y)$. Then A, B, C and $A^\times$ are given by (3.2) - (3.5) with A_0, B_0, C_0 and X_0 replaced by A_D, B_D, C_D and X_D, respectively. From this it is clear that Θ is a node with centralized singularities. For $\lambda \in \Gamma$, the transfer function of Θ has the value $W_-(\lambda)D(\lambda)W_+(\lambda) = W(\lambda)$. So on Γ the transfer function of Θ coincides with W. Let $M,M^\times$ be the pair of spectral subspaces associated with Θ and Γ. From (2.5),

(2.6) and Corollary 3.3 we see that $\dim(M\cap M^{\times}) < \infty$ and $\dim(X/M+M^{\times}) < \infty$. Also the assertions (d) and (e) hold true because of (2.2), (2.3), (2.4) and Theorem 3.2.

Next, assume that there exists an invertible bounded linear operator S on Y such that the functions $\tilde{W}_-(\lambda) = S^{-1}W_-(\lambda)$ and $\tilde{W}_+(\lambda) = W_+(\lambda)S$ can be written as the transfer function of a minus node $\tilde{\Theta}_-$ and a plus node $\tilde{\Theta}_+$, respectively. Consider the factorization

$$(3.12) \qquad \tilde{W}(\lambda) = \tilde{W}_-(\lambda)D(\lambda)\tilde{W}_+(\lambda), \qquad \lambda \in \Gamma,$$

where $\tilde{W}(\lambda) = S^{-1}W(\lambda)S$. This is a Wiener-Hopf factorization of the type considered in the first paragraph of this proof. Put $\tilde{\Theta} = \tilde{\Theta}_-\tilde{\Theta}_D\tilde{\Theta}_+$, where Θ_D is again as in Section 2. As we saw above $\tilde{\Theta}$ is a realization (on Γ) of $\tilde{W}$ with the desired properties. Write $\tilde{\Theta} = (\tilde{A},\tilde{B},\tilde{C};X,Y)$ and define $\Theta = (A,B,C;X,Y)$ by

$$A = \tilde{A}, \qquad B = \tilde{B}S^{-1}, \qquad C = S\tilde{C}.$$

Then $A^{\times} = \tilde{A}^{\times}$ and one easily verifies that (a) - (e) are satisfied. Actually, the incoming (outgoing) subspaces for Θ and the incoming (outgoing) subspaces for $\tilde{\Theta}$ are exactly the same.

It remains to prove that an operator S with the properties mentioned above can always be found. For this, we argue as follows. Recall that ∞ belongs to Ω_{Γ}^{-}. For S we take the (invertible) operator $W_-(\infty)$. Define $\tilde{W}_-$, $\tilde{W}_+$ and $\tilde{W}$ as in the second paragraph of the proof. Then $\tilde{W}_-(\infty) = I_Y$ and we have the Wiener-Hopf factorization (3.12). Clearly

$$\tilde{W}_-(\lambda) = \tilde{W}(\lambda)\tilde{W}_+(\lambda)^{-1}D(\lambda)^{-1}, \qquad \lambda \in \Gamma.$$

Inspecting both sides of this identity and using the specific properties of $\tilde{W}_-,\tilde{W}_+$, D and the fact that $\tilde{W}$ is analytic on a neighbourhood of Γ, we see that $\tilde{W}_-$ has an analytic extension, again denoted by $\tilde{W}_-$, to a neighbourhood U of the closure $\Omega_\Gamma^- \cup \Gamma$ of Ω_Γ^-. By taking U sufficiently small, we may assume that W_- takes only invertible values on U. But then $\tilde{W}_-$ can be written (on U) as the transfer function of a minus node (cf. [BGK1], Chapter II, especially Theorem 2.5 and Corollary 2.7). Writing (3.12) in the form

$$\tilde{W}_+(\lambda) = D(\lambda)^{-1}\tilde{W}_-(\lambda)^{-1}\tilde{W}(\lambda), \qquad \lambda \in \Gamma,$$

we see that $\tilde{W}_+$ has an analytic extension, again denoted by $\tilde{W}_+$, to a neighbourhood V of the closure $\Omega_\Gamma^+ \cup \Gamma$ of Ω_Γ^+ and V can be chosen in such a way that $\tilde{W}_+$ takes only invertible values on V. Since Ω_Γ^+ is bounded, we may assume that V is bounded and it follows from the results discussed in [BGK1], Chapter II that $\tilde{W}_+$ can be written (on V) as the transfer function of a plus node. This completes the proof. □

4. Dilation

Suppose $\Theta_0 = (A_0, B_0, C_0; X_0, Y)$ is a node and $\Theta = (A, B, C; X, Y)$ is a dilation of Θ_0. This means that the state space X of Θ admits a topological decomposition $X = X_1 \oplus X_0 \oplus X_2$ relative to which the operators A, B and C have the form

$$A = \begin{pmatrix} A_1 & * & * \\ 0 & A_0 & * \\ 0 & 0 & A_2 \end{pmatrix} : X_1 \oplus X_0 \oplus X_2 \to X_1 \oplus X_0 \oplus X_2,$$

$$B = \begin{pmatrix} * \\ B_0 \\ 0 \end{pmatrix} : Y \to X_1 \oplus X_0 \oplus X_2,$$

$$C = [0 \quad C_0 \quad *] : X_1 \oplus X_0 \oplus X_2 \to Y.$$

Clearly $\sigma(A) \subset \sigma(A_1) \cup \sigma(A_0) \cup \sigma(A_2)$. In general this inclusion may be strict, but in the finite dimensional case (to which we shall restrict ourselves later) it is not. On the complement of $\sigma(A_1) \cup \sigma(A_0) \cup \sigma(A_2)$, the transfer functions of Θ_0 and Θ coincide.

From now on we shall assume that Θ_0 has no spectrum on the contour Γ and, in addition, that A_1 and A_2 have no spectrum on Γ. Then A has no spectrum on Γ and the same is true for $A^\times$ because $A^\times$ has the form

$$A^\times = \begin{pmatrix} A_1 & * & * \\ 0 & A_0^\times & * \\ 0 & 0 & A_2 \end{pmatrix} : X_1 \oplus X_0 \oplus X_2 \to X_1 \oplus X_0 \oplus X_2.$$

So the conditions imposed on Θ_0 and A_1, A_2 imply that Θ has no spectrum on Γ too. In the finite dimensional case the converse is also true.

THEOREM 4.1. _Let_ $\Theta_0 = (A_0, B_0, C_0; X_0, Y)$ _and its dilation_ $\Theta = (A,B,C;X,Y)$ _be as above. Then there exists an invertible bounded linear operator_ $S: X \to X$ _such that the following holds:_

(1) _Let_ $H_{00}, H_{01}, H_{02}, \ldots$ _be the incoming subspaces of_ Θ_0. _Then the incoming subspaces_ $H_0, H_1, H_2, \ldots$ _of_ Θ _are given by_

$$H_m = S \begin{pmatrix} X_1 \\ H_{0m} \\ X_2 \end{pmatrix}, \quad m = 0,1,2,\ldots ;$$

(2) _Let_ $K_{00}, K_{01}, K_{02}, \ldots$ _be the outgoing subspaces of_ Θ_0. _Then the outgoing subspaces_ $K_0, K_1, K_2, \ldots$ _of_ Θ _are given by_

$$K_m = S \begin{pmatrix} (0) \\ K_{0m} \\ (0) \end{pmatrix}, \quad m = 0,1,2,\ldots .$$

Here we use the same notation as in Theorem 3.2. The incoming and outgoing subspaces mentioned in the theorem all exist because Θ_0 and Θ have no spectrum on Γ.

PROOF. Let $M_0, M_0^\times$ and $M, M^\times$ be the pairs of spectral subspaces associated with Θ_0 and Θ, respectively. Write $M_0 = \operatorname{Im} P_0$, $M_0^\times = \operatorname{Ker} P_0^\times$, $M = \operatorname{Im} P$ and $M^\times = \operatorname{Ker} P^\times$, where $P_0, P_0^\times$, P and $P^\times$ are the appropriate Riesz projections. With respect to the decomposition $X = X_1 \oplus X_0 \oplus X_2$, the projections P and $P^\times$ have the form

$$(4.1) \qquad P = \begin{pmatrix} P_1 & R_1 & R_2 \\ 0 & P_0 & R_3 \\ 0 & 0 & P_2 \end{pmatrix} : X_1 \oplus X_0 \oplus X_2 \to X_1 \oplus X_0 \oplus X_2,$$

$$(4.2) \qquad P^\times = \begin{pmatrix} P_1 & R_1^\times & R_2^\times \\ 0 & P_0^\times & R_3^\times \\ 0 & 0 & P_2 \end{pmatrix} : X_1 \oplus X_0 \oplus X_2 \to X_1 \oplus X_0 \oplus X_2.$$

This is clear from the matrix representations of A and $A^\times$. The fact that P and $P^\times$ are projections implies certain relationships between the operators appearing in (4.1) and (4.2). Taking advantage of these relationships one can check that with

$$S_1 = (I-P_1)R_1P_0 - P_1R_1^\times(I-P_0^\times),$$
$$S_2 = (I-P_1)R_2P_2 - P_1R_2^\times(I-P_2),$$
$$S_3 = (I-P_0)R_3P_2 - P_0^\times R_3^\times(I-P_2),$$

the following identity holds

$$\begin{pmatrix} P_1 & R_1 & R_2 \\ 0 & P_0 & R_3 \\ 0 & 0 & P_2 \end{pmatrix} \cdot \begin{pmatrix} I & -R_1^\times(I-P_0^\times)P_0-R_1 & R_1(I-P_0)R_3P_2-R_2 \\ 0 & I & -R_3 \\ 0 & 0 & I \end{pmatrix} =$$

$$= \begin{pmatrix} I & S_1 & S_2 \\ 0 & I & S_3 \\ 0 & 0 & I \end{pmatrix} \cdot \begin{pmatrix} P_1 & 0 & 0 \\ 0 & P_0 & 0 \\ 0 & 0 & P_2 \end{pmatrix} .$$

From this we see that

$$(4.3) \qquad M = \begin{pmatrix} I & S_1 & S_2 \\ 0 & I & S_3 \\ 0 & 0 & I \end{pmatrix} \begin{pmatrix} \operatorname{Im} P_1 \\ M_0 \\ \operatorname{Im} P_2 \end{pmatrix} .$$

By symmetry (replace P by $I-P^{\times}$ and $P^{\times}$ by $I-P$), one also has

$$(4.4) \qquad M^{\times} = \begin{pmatrix} I & S_1 & S_2 \\ 0 & I & S_3 \\ 0 & 0 & I \end{pmatrix} \begin{pmatrix} \operatorname{Ker} P_1 \\ M_0^{\times} \\ \operatorname{Ker} P_2 \end{pmatrix} .$$

Combining (4.3) and (4.4), one gets

$$(4.5) \quad M+M^{\times} = \begin{pmatrix} I & S_1 & S_2 \\ 0 & I & S_3 \\ 0 & 0 & I \end{pmatrix} \begin{pmatrix} X_1 \\ M_0+M_0^{\times} \\ X_2 \end{pmatrix}, \quad M\cap M^{\times} = \begin{pmatrix} I & S_1 & S_2 \\ 0 & I & S_3 \\ 0 & 0 & I \end{pmatrix} \begin{pmatrix} (0) \\ M_0\cap M_0^{\times} \\ (0) \end{pmatrix} .$$

For S we now take the invertible operator

$$S = \begin{pmatrix} I & S_1 & S_2 \\ 0 & I & S_3 \\ 0 & 0 & I \end{pmatrix} : X_1 \oplus X_0 \oplus X_2 \to X_1 \oplus X_0 \oplus X_2 .$$

Then the desired holds true for $m = 0$. For arbitrary m, it is obtained by a straightforward argument based on the special form of the operators A, B and C and the identities (4.5). □

As in the proof of Theorem 4.1, let $M_0, M_0^{\times}$ and $M, M^{\times}$ be the pairs of spectral subspaces associated with Θ_0 and Θ, respectively. Then clearly $\dim(M_0\cap M_0^{\times}) < \infty$, $\dim(X_0/M_0+M_0^{\times}) < \infty$ if and only if $\dim(M\cap M^{\times}) < \infty$, $\dim(X/M+M^{\times}) < \infty$, and in that case the incoming (outgoing) indices of Θ_0 and Θ are the same (cf.

Corollary 3.3). This result can also be explained on the level of incoming (outgoing) bases. Given an incoming (outgoing) basis for Θ_0 one can construct an incoming (outgoing) basis for Θ and vice versa. The construction uses the operator S introduced in the proof of Theorem 4.1. We omit the details.

Now let us restrict ourselves to the finite dimensional case. Let W be a rational m×m matrix function having the value I_m at ∞. Suppose W has neither poles nor zeros on the contour Γ. Then W can be written (in several ways) as the transfer function of a finite dimensional node Θ without spectrum on Γ. Associated with such a node are incoming and outgoing spaces. Theorem 1.1 implies the validity of formulas (1.5) and (1.6). In other words, the dimensions of the outgoing spaces and the codimension of the incoming spaces are "spectral characteristics" of the transfer function W: they do not depend on the choice of the realization Θ for W. In the finite dimensional case considered here, the proof of this fact can be given as follows.

Let Θ_1 and Θ_2 be finite dimensional nodes without spectrum on the contour Γ, both having the rational m×m matrix function as their transfer function. Then Θ_1 is the dilation of a minimal node Θ_{10} and Θ_2 is the dilation of a minimal node Θ_{20}. The transfer functions of Θ_{10} and Θ_{20} coincide on Γ and hence they coincide (as rational functions) on the whole Riemann sphere. But then Θ_{10} and Θ_{20} are similar by the well-known state space isomorphism theorem. So we may assume that Θ_1 and Θ_2 are dilations of one single minimal node Θ_0. The desired result is now immediate from Theorem 4.1.

5. Spectral characteristics of transfer functions: outgoing spaces

In this section we investigate the situation where the transfer functions of (possibly infinite dimensional) nodes coincide on the given (positively oriented) contour Γ in the (finite) complex plane.

THEOREM 5.1. *Let* $\Theta_1 = (A_1,B_1,C_1;X_1,Y)$ *and* $\Theta_2 = (A_2,B_2,C_2;X_2,Y)$ *be nodes without spectrum on the contour* Γ. *Suppose that the transfer functions of* Θ_1 *and* Θ_2 *coincide on* Γ. *Then the bounded linear operator* $\Psi: X_1 \to X_2$, *given by*

$$\Psi = \frac{1}{2\pi i}\int_\Gamma (\lambda-A_2^\times)^{-1}B_2C_1(\lambda-A_1)^{-1}d\lambda, \tag{5.1}$$

maps the outgoing subspaces of Θ_1 *one-to-one onto the corresponding outgoing subspaces of* Θ_2. *In other words, for* $m = 0,1,2,\ldots$, *the operator* Ψ *maps* $M_1\cap M_1^\times\cap \operatorname{Ker}C_1 \cap\ldots\cap \operatorname{Ker}C_1A_1^{m-1}$ *one-to-one onto* $M_2\cap M_2^\times\cap\operatorname{Ker}C_2 \cap\ldots\cap \operatorname{Ker}C_2A_2^{m-1}$. *Here* $M_1,M_1^\times$ *and* $M_2,M_2^\times$ *are the pairs of spectral subspaces associated with* Γ *and* Θ_1,Θ_2, *respectively.*

It is interesting to note that when $\Theta_1 = \Theta_2$, the operator Ψ acts on $M_1 \cap M_1^\times$ as the identity operator.

PROOF. The proof requires several steps. We begin by showing that Ψ maps $M_1 \cap M_1^\times$ into $M_2 \cap M_2^\times$.

Take $x \in M_1 \cap M_1^\times$. Recall that $M_j = \operatorname{Im}P_j$ and $M_j^\times = \operatorname{Ker}P_j^\times$, where the Riesz projections P_j and $P_j^\times$ are given by

$$P_j = \frac{1}{2\pi i}\int_\Gamma (\lambda-A_j)^{-1}d\lambda, \qquad P_j^\times = \frac{1}{2\pi i}\int_\Gamma (\lambda-A_j^\times)^{-1}d\lambda.$$

Here $j = 1,2$. So $x \in M_1 \cap M_1^\times$ means that $x = P_1x = (I-P_1^\times)x$. But then

$$\Psi x = \frac{1}{2\pi i}\int_\Gamma (I-P_2^\times)(\lambda-A_2^\times)^{-1}B_2C_1(\lambda-A_1)^{-1}xd\lambda + $$
$$+ \frac{1}{2\pi i}\int_\Gamma P_2^\times(\lambda-A_2^\times)^{-1}B_2C_1(\lambda-A_1)^{-1}P_1xd\lambda.$$

Consider the latter integral. The functions $P_2^\times(\lambda-A_2^\times)^{-1}$ and $(\lambda-A_1)^{-1}P_1$ both have an analytic extension to a neighbourhood of $\Omega_\Gamma^- \cup \Gamma$ vanishing at ∞. Hence the integral vanishes. So

$$\Psi x = \frac{1}{2\pi i}\int_\Gamma (I-P_2^\times)(\lambda-A_2^\times)^{-1}B_2C_1(\lambda-A_1)^{-1}xd\lambda,$$

which implies that $\Psi x \in \operatorname{Im}(I-P_2^\times) = \operatorname{Ker} P_2^\times = M_2^\times$.

For $\lambda \in \Gamma$, let $W(\lambda) = I + C_1(\lambda-A_1)^{-1}B_1$. By hypothesis $W(\lambda) = I + C_2(\lambda-A_2)^{-1}B_2$ too. Moreover, W takes invertible values on Γ, so we can write

$$\Psi = \frac{1}{2\pi i}\int_\Gamma (\lambda-A_2^\times)^{-1}B_2W(\lambda)^{-1}C_1(\lambda-A_1)^{-1}d\lambda.$$

Using that $B_2C_2 = A_2-A_2^\times$ and $B_1C_1 = A_1-A_1^\times$, one easily verifies that

$$(\lambda-A_2^\times)^{-1}B_2W(\lambda) = (\lambda-A_2)^{-1}B_2, \qquad \lambda \in \Gamma,$$

$$W(\lambda)C_1(\lambda-A_1^\times)^{-1} = C_1(\lambda-A_1)^{-1}, \qquad \lambda \in \Gamma.$$

It follows that Ψ can also be written as

$$\Psi = \frac{1}{2\pi i}\int_\Gamma (\lambda-A_2)^{-1}B_2C_1(\lambda-A_1^\times)^{-1}d\lambda.$$

But then

$$\Psi x = \frac{1}{2\pi i}\int_\Gamma P_2(\lambda-A_2)^{-1}B_2C_1(\lambda-A_1^\times)^{-1}x\,d\lambda +$$

$$+ \frac{1}{2\pi i}\int_\Gamma (I-P_2)(\lambda-A_2)^{-1}B_2C_1(\lambda-A_1^\times)^{-1}(I-P_1^\times)x\,d\lambda.$$

The latter integral again vanishes. Indeed, the functions $(I-P_2)(\lambda-A_2)^{-1}$ and $(\lambda-A_1^\times)^{-1}(I-P_1^\times)$ have analytic extensions to a neighbourhood of $\Omega_\Gamma^+ \cup \Gamma$, and so we can apply Cauchy's theorem. Thus

$$\Psi x = \frac{1}{2\pi i}\int_\Gamma P_2(\lambda-A_2)^{-1}B_2C_1(\lambda-A_1)^{-1}x\,d\lambda,$$

and in particular $\Psi x \in \operatorname{Im} P_2 = M_2$. We conclude that $\Psi x \in M_2 \cap M_2^\times$.

From now on we shall consider Ψ as a bounded linear operator from $M_1 \cap M_1^\times$ into $M_2 \cap M_2^\times$. In an analogous way the expression

$$(5.2)\qquad \Phi y = \frac{1}{2\pi i}\int_\Gamma (\mu-A_1)^{-1}B_1C_2(\mu-A_2^\times)^{-1}y\,d\mu, \qquad y \in M_2 \cap M_2^\times$$

defines a bounded linear operator from $M_2 \cap M_2^\times$ into $M_1 \cap M_1^\times$. We shall prove that Ψ and Φ are each others inverse.

Take $\lambda,\mu \in \Gamma$. If $\lambda \neq \mu$ one can use the resolvent equation to show that

$$C_1(\lambda-A_1)^{-1}(\mu-A_1)^{-1}B_1 = \frac{1}{\lambda-\mu}[C_1(\mu-A_1)^{-1}B_1 - C_1(\lambda-A_1)^{-1}B_1]$$
$$= \frac{1}{\lambda-\mu}[W(\mu) - W(\lambda)] = C_2(\lambda-A_2)^{-1}(\mu-A_2)^{-1}B_2.$$

By continuity, it follows that the first and last term in these identities are also equal for $\lambda = \mu$. Take $y \in M_2 \cap M_2^\times$ and put $x = \Phi y$. Then $x \in M_1 \cap M_1^\times$. Combining (5.1), (5.2) and the identity established above, we get

$$\psi\Phi y =$$
$$= \left(\frac{1}{2\pi i}\right)^2\int_\Gamma\int_\Gamma (\lambda-A_2^\times)^{-1}B_2C_1(\lambda-A_1)^{-1}(\mu-A_1)^{-1}B_1C_2(\mu-A_2^\times)^{-1}y\,d\mu\,d\lambda$$
$$= \left(\frac{1}{2\pi i}\right)^2\int_\Gamma\int_\Gamma (\lambda-A_2^\times)^{-1}B_2C_2(\lambda-A_2)^{-1}(\mu-A_2)^{-1}B_2C_2(\mu-A_2^\times)^{-1}y\,d\mu\,d\lambda$$
$$= \left(\frac{1}{2\pi i}\right)^2\int_\Gamma\int_\Gamma (\lambda-A_2^\times)^{-1}(A_2-A_2^\times)(\lambda-A_2)^{-1}(\mu-A_2)^{-1}(A_2-A_2^\times)(\mu-A_2^\times)^{-1}y\,d\mu\,d\lambda$$
$$= \left(\frac{1}{2\pi i}\right)^2\int_\Gamma\int_\Gamma [(\lambda-A_2)^{-1}-(\lambda-A_2^\times)^{-1}].[(\mu-A_2)^{-1}-(\mu-A_2^\times)^{-1}]y\,d\mu\,d\lambda$$
$$= \left(\frac{1}{2\pi i}\int_\Gamma [(\lambda-A_2)^{-1}-(\lambda-A_2^\times)^{-1}]d\lambda\right).\left(\frac{1}{2\pi i}\int_\Gamma [(\mu-A_2)^{-1}-(\mu-A_2^\times)^{-1}]d\mu\right)y$$
$$= (P_2-P_2^\times)(P_2-P_2^\times)y$$
$$= y.$$

So $\Psi\Phi$ is the identity operator on $M_2 \cap M_2^\times$ and, by symmetry, $\Psi\Phi$ is the identity operator on $M_1 \cap M_1^\times$.

It is clear now that Ψ maps $M_1 \cap M_1^\times$ one-to-one onto $M_2 \cap M_2^\times$. Our next step is to show that Ψ maps

$M_1 \cap M_1 \cap \operatorname{Ker} C_1$ into $M_2 \cap M_2^\times \cap \operatorname{Ker} C_2$ and

$$\Psi A_1 x = A_2 \Psi x, \qquad x \in M_1 \cap M_1^\times \cap \operatorname{Ker} C_1. \tag{5.3}$$

Take $x \in M_1 \cap M_1^\times \cap \operatorname{Ker} C_1$. Then

$$\begin{aligned} C_2 \Psi x &= \frac{1}{2\pi i} \int_\Gamma C_2(\lambda - A_2)^{-1} B_2 C_1 (\lambda - A_1^\times)^{-1} x \, d\lambda \\ &= \frac{1}{2\pi i} \int_\Gamma [W(\lambda) - I] C_1 (\lambda - A_1^\times)^{-1} x \, d\lambda \\ &= \frac{1}{2\pi i} \int_\Gamma W(\lambda) C_1 (\lambda - A_1^\times)^{-1} x \, d\lambda - \frac{1}{2\pi i} \int_\Gamma C_1 (\lambda - A_1^\times)^{-1} x \, d\lambda \\ &= \frac{1}{2\pi i} \int_\Gamma C_1 (\lambda - A_1)^{-1} x \, d\lambda - \frac{1}{2\pi i} \int_\Gamma C_1 (\lambda - A_1^\times)^{-1} x \, d\lambda \\ &= C_1 P_1 x - C_1 P_1^\times x \\ &= C_1 x = 0. \end{aligned}$$

Thus $\Psi x \in \operatorname{Ker} C_2$. Since $\Psi x \in M_2 \cap M_2^\times$ too, we have $\Psi x \in M_2 \cap M_2^\times \cap \operatorname{Ker} C_2$.

In order to prove (5.3), we argue as follows. From $C_1 x = 0$ and $x \in M_1 \cap M_1^\times$ it is clear that $A_1 x = A_1^\times x \in M_1 \cap M_1^\times$. Further

$$\begin{aligned} \psi A_1 x &= \frac{1}{2\pi i} \int_\Gamma (\lambda - A_2)^{-1} B_2 C_1 (\lambda - A_1^\times)^{-1} A_1^\times x \, d\lambda \\ &= \frac{1}{2\pi i} \int_\Gamma (\lambda - A_2)^{-1} B_2 C_1 (\lambda - A_1^\times)^{-1} (A_1^\times - \lambda + \lambda) x \, d\lambda \\ &= \frac{1}{2\pi i} \int_\Gamma \lambda (\lambda - A_2)^{-1} B_2 C_1 (\lambda - A_1^\times)^{-1} x \, d\lambda - \frac{1}{2\pi i} \int_\Gamma (\lambda - A_2)^{-1} B_2 C_1 x \, d\lambda \\ &= \frac{1}{2\pi i} \int_\Gamma \lambda (\lambda - A_2)^{-1} B_2 C_1 (\lambda - A_1^\times)^{-1} x \, d\lambda \\ &= \frac{1}{2\pi i} \int_\Gamma (\lambda - A_2 + A_2)(\lambda - A_2)^{-1} B_2 C_1 (\lambda - A_1^\times)^{-1} x \, d\lambda \\ &= \frac{1}{2\pi i} \int_\Gamma B_2 C_1 (\lambda - A_1^\times)^{-1} x \, d\lambda + \frac{1}{2\pi i} \int_\Gamma A_2 (\lambda - A_2)^{-1} B_2 C_1 (\lambda - A_1^\times)^{-1} x \, d\lambda \end{aligned}$$

$$= B_2C_1P_1^{\times}x + A_2\left(\frac{1}{2\pi i}\int_{\Gamma}(\lambda-A_2)^{-1}B_2C_1(\lambda-A_1^{\times})^{-1}x\,d\lambda\right)$$

$$= A_2\psi x.$$

Here we used that $P_1^{\times}x = 0$.

For $j = 1,2$, let $K_{j0}, K_{j1}, K_{j2}, \ldots$ be the outgoing subspaces of Θ_j. Thus

$$K_{jm} = M_j \cap M_j^{\times} \cap \operatorname{Ker} C_j \cap \ldots \cap \operatorname{Ker} C_jA_j^{m-1}.$$

We proved already that ψ maps K_{10} one-to-one onto K_{20} and K_{11} into K_{21}. Assume that for some non-negative integer n the inclusion $\psi[K_{1n}] \subset K_{2n}$ is correct, and take $x \in K_{1,n+1}$. Then $x \in K_{11}$. Since $A_1K_{11} \subset K_{10} = M_1 \cap M_1^{\times}$, we also have $A_1x \in K_{1n}$. Thus $\psi x \in K_{21}$ and $A_2\psi x = \psi A_1 x \in \psi[K_{1n}]$. According to the induction hypothesis $\psi[K_{1n}] \subset K_{2n}$, so $A_2\psi x \in K_{2n}$. But then $\psi x \in K_{2,n+1}$. We conclude that $\psi[K_{1m}] \subset K_{2m}$ for all m. Analogously $\Phi[K_{2m}] \subset K_{1m}$ for all m. Here Φ is the operator given by (5.2). Since Φ and ψ are each others inverse, it follows that $\psi[K_{1m}] = K_{2m}$, and the proof is complete.□

6. Spectral characteristics of transfer functions: incoming spaces

In this section we continue the investigation started in Section 5.

THEOREM 6.1. *Let* $\Theta_1 = (A_1,B_1,C_1;X_1,Y)$ *and* $\Theta_2 = (A_2,B_2,C_2;X_2,Y)$ *be nodes without spectrum on the contour* Γ. *Suppose that the transfer functions of* Θ_1 *and* Θ_2 *coincide on* Γ. *Let the Riesz projections* $P_1, P_1^{\times}$ *and the bounded linear operator* $\Phi\colon X_1 \to X_2$ *be given by*

$$P_1 = \frac{1}{2\pi i}\int_{\Gamma}(\lambda-A_1)^{-1}d\lambda, \qquad P_1^{\times} = \frac{1}{2\pi i}\int_{\Gamma}(\lambda-A_1)^{-1}d\lambda,$$

$$\Phi = \frac{1}{2\pi i}\int_{\Gamma}(\lambda-A_2^{\times})^{-1}B_2C_1(\lambda-A_1)^{-1}(I-P_1)P_1^{\times}d\lambda.$$

Then, for $m = 0,1,2,\ldots,$

$$M_1+M_1^\times+\mathrm{Im}B_1+\ldots+\mathrm{Im}A_1^{m-1}B_1 = \Phi^{-1}[M_2+M_2^\times+\mathrm{Im}B_2+\ldots+\mathrm{Im}A_2^{m-1}B_2],$$

$$X_2 = \mathrm{Im}\Phi + M_2 + M_2^\times + \mathrm{Im}\ B_2 + \ldots + \mathrm{Im}\ A_2^{m-1}B_2.$$

Here $M_1, M_1^\times$ *and* $M_2, M_2^\times$ *are the pairs of spectral subspaces associated with* Γ *and* Θ_1, Θ_2, *respectively.*

Note that $\Phi = \psi(I-P_1)P_1^\times$, where ψ is as in Theorem 5.1. This is clear from formula (5.1). By Cauchy's theorem Φ can also be written as

(6.1) $$\Phi = \frac{1}{2\pi i}\int_\Gamma P_2^\times(\lambda-A_2^\times)^{-1}B_2C_1(\lambda-A_1)^{-1}(I-P_1)P_1^\times d\lambda.$$

If $\Theta_1 = \Theta_2$, then $\Phi = -P_1^\times(I-P_1)P_1^\times$.

PROOF. We begin by showing that Φ maps H_{1m} into H_{2m}. Here $H_{j0}, H_{j1}, H_{j2},\ldots$ are the incoming subspaces of Θ_j, i.e.

$$H_{jm} = M_j+M_j^\times+\mathrm{Im}\ B_j+\ \ldots\ +\mathrm{Im}\ A_j^{m-1}B_j =$$
$$= M_j+M_j^\times+\mathrm{Im}B_j+\ \ldots\ +\mathrm{Im}(A_j^\times)^{m-1}B_j.$$

By P_2 and $P_2^\times$ we denote the Riesz projections

$$P_2 = \frac{1}{2\pi i}\int_\Gamma(\lambda-A_2)^{-1}d\lambda,\qquad P_2^\times = \frac{1}{2\pi i}\int_\Gamma(\lambda-A_2^\times)^{-1}d\lambda.$$

Recall that $M_j = \mathrm{Im}P_j$ and $M_j^\times = \mathrm{Im}\ P_j^\times$ $(j=1,2)$.

Let m be a non-negative integer and take $x \in H_{1m}$. Write x in the form

$$x = P_1u + (I-P_1^\times)v + B_1y_0+\ldots+(A_1^\times)^{m-1}B_1y_{m-1}.$$

Then $P_1^\times x = P_1^\times P_1u + P_1^\times B_1y_0 + \ldots + P_1^\times(A_1^\times)^{m-1}B_1y_{m-1}$. For $\lambda \in \Gamma$, put

$$f(\lambda) = C_1(\lambda-A_1)^{-1}(I-P_1)P_1^{\times}x + \sum_{k=0}^{m-1}\lambda^k y_k.$$

Clearly f is analytic on a neighbourhood of $\Omega_\Gamma^+ \cup \Gamma$ and

$$\frac{1}{2\pi i}\int_\Gamma P_2^{\times}(\lambda-A_2^{\times})^{-1}B_2 f(\lambda)d\lambda =$$

$$= \frac{1}{2\pi i}\int_\Gamma P_2^{\times}(\lambda-A_2^{\times})^{-1}B_2C_1(\lambda-A_1)^{-1}(I-P_1)P_1^{\times}x d\lambda +$$

$$+ \sum_{k=0}^{m-1}\left(\frac{1}{2\pi i}\int_\Gamma \lambda^k P_2^{\times}(\lambda-A_2^{\times})^{-1}B_2 y_k d\lambda\right)$$

$$= \Phi x + \sum_{k=0}^{m-1} P_2^{\times}(A_2^{\times})^k B_2 y_k.$$

So in order to prove that $\Phi x \in H_{2m}$, it suffices to show that

$$\frac{1}{2\pi i}\int_\Gamma P_2^{\times}(\lambda-A_2^{\times})^{-1}B_2 f(\lambda)d\lambda \in M_2+M_2^{\times}. \tag{6.2}$$

As we shall see, this requires a considerable effort.

First of all,

$$\frac{1}{2\pi i}\int_\Gamma (\lambda-A_1^{\times})^{-1}B_1 f(\lambda)d\lambda = \frac{1}{2\pi i}\int_\Gamma P_1^{\times}(\lambda-A_1^{\times})^{-1}B_1 f(\lambda)d\lambda$$

$$= \sum_{k=0}^{m-1}\left(\frac{1}{2\pi i}\int_\Gamma \lambda^k P_1^{\times}(\lambda-A_1^{\times})^{-1}B_1 y_k d\lambda\right) +$$

$$+ \frac{1}{2\pi i}\int_\Gamma P_1^{\times}(\lambda-A_1^{\times})^{-1}B_1C_1(\lambda-A_1)^{-1}(I-P_1)P_1^{\times}P_1 u d\lambda +$$

$$+ \sum_{k=0}^{m-1}\left(\frac{1}{2\pi i}\int_\Gamma P_1^{\times}(\lambda-A_1^{\times})^{-1}B_1C_1(\lambda-A_1)^{-1}(I-P_1)P_1^{\times}(A_1^{\times})^k B_1 y_k\right)d\lambda$$

$$= \sum_{k=0}^{m-1} P_1^{\times}(A_1^{\times})^k B_1 y_k +$$

$$+ \frac{1}{2\pi i}\int_\Gamma P_1^{\times}(\lambda-A_1^{\times})^{-1}(A_1-A_1^{\times})(\lambda-A_1)^{-1}(I-P_1)P_1^{\times}P_1 u d\lambda +$$

$$+ \sum_{k=0}^{m-1}\left(\frac{1}{2\pi i}\int_\Gamma P_1^{\times}(\lambda-A_1^{\times})^{-1}(A_1-A_1^{\times})(\lambda-A_1)^{-1}(I-P_1)P_1^{\times}(A_1^{\times})^k B_1 y_k d\lambda\right)$$

$$= \sum_{k=0}^{m-1} P_1^{\times}(A_1^{\times})^k B y_k + \frac{1}{2\pi i}\int_\Gamma P_1^{\times}[(\lambda-A_1)^{-1}-(\lambda-A_1^{\times})^{-1}](I-P_1)P_1^{\times}P_1 u d\lambda +$$

$$+ \sum_{k=0}^{m-1}\Big(\frac{1}{2\pi i}\int_\Gamma P_1^\times[(\lambda-A_1)^{-1}-(\lambda-A_1^\times)^{-1}](I-P_1)P_1^\times(A_1^\times)^k B_1 y_k d\lambda\Big)$$

$$= \sum_{k=0}^{m-1} P_1^\times(A_1^\times)^k By_k - P_1^\times(I-P_1)P_1^\times P_1 u - \sum_{k=0}^{m-1} P_1^\times(I-P_1)P_1^\times(A_1^\times)^k B_1 y_k$$

$$= P_1^\times P_1 P_1^\times P_1 u - P_1^\times P_1 u + \sum_{k=0}^{m-1} P_1^\times P_1 P_1^\times(A_1^\times)^k B_1 y_k = P_1^\times z,$$

where $z \in \operatorname{Im} P_1 = M_1$.

Secondly, we introduce another auxiliary function. For $\lambda \in \Gamma$, put

$$g(\lambda) = W(\lambda)^{-1}f(\lambda) + \frac{1}{2\pi i}\int_\Gamma C_1(\lambda-A_1^\times)^{-1}(\mu-A_1^\times)^{-1}B_1 f(\mu)d\mu +$$

$$+ C_1(\lambda-A_1^\times)^{-1}(I-P_1^\times)z.$$

Here $W(\lambda) = I + C_1(\lambda-A_1)^{-1}B_1 = I + C_2(\lambda-A_2)^{-1}B_2$. Then

$$W(\lambda)g(\lambda) - \frac{1}{2\pi i}\int_\Gamma C_1(\lambda-A_1)^{-1}(\mu-A_1)^{-1}B_1 g(\mu)d\mu =$$

$$= f(\lambda) + \frac{1}{2\pi i}\int_\Gamma W(\lambda)C_1(\lambda-A_1^\times)^{-1}(\mu-A_1^\times)^{-1}B_1 f(\mu)d\mu +$$

$$+ W(\lambda)C_1(\lambda-A_1^\times)^{-1}(I-P_1^\times)z +$$

$$- \frac{1}{2\pi i}\int_\Gamma C_1(\lambda-A_1)^{-1}(\mu-A_1)^{-1}B_1 W(\mu)^{-1}f(\mu)d\mu +$$

$$-\Big(\frac{1}{2\pi i}\Big)^2\int_\Gamma\int_\Gamma C_1(\lambda-A_1)^{-1}(\mu-A_1)^{-1}B_1 C_1(\mu-A_1^\times)^{-1}(\upsilon-A_1^\times)^{-1}B_1 f(\upsilon)d\upsilon d\mu +$$

$$- \frac{1}{2\pi i}\int_\Gamma C_1(\lambda-A_1)^{-1}(\mu-A_1)^{-1}B_1 C_1(\mu-A_1^\times)^{-1}(I-P_1^\times)z d\mu$$

$$= f(\lambda) + \frac{1}{2\pi i}\int_\Gamma C_1(\lambda-A_1)^{-1}(\mu-A_1^\times)^{-1}B_1 f(\mu)d\mu + C_1(\lambda-A_1)^{-1}(I-P_1^\times)z +$$

$$- \frac{1}{2\pi i}\int_\Gamma C_1(\lambda-A_1)^{-1}(\mu-A_1^\times)^{-1}B_1 f(\mu)d\mu +$$

$$-\left(\frac{1}{2\pi i}\right)^2 \int_\Gamma \int_\Gamma C_1(\lambda-A_1)^{-1}(\mu-A_1)^{-1}(A_1-A_1^\times)(\mu-A_1^\times)^{-1}(\upsilon-A_1^\times)^{-1}B_1 f(\upsilon)d\upsilon d\mu +$$

$$-\frac{1}{2\pi i}\int_\Gamma C_1(\lambda-A_1)^{-1}(\mu-A_1)^{-1}(A_1-A_1^\times)(\mu-A_1^\times)^{-1}(I-P_1^\times)zd\mu$$

$$= f(\lambda) + \frac{1}{2\pi i}\int_\Gamma C_1(\lambda-A_1)^{-1}(\mu-A_1^\times)^{-1}B_1 f(\mu)d\mu + C_1(\lambda-A_1)^{-1}(I-P_1^\times)z +$$

$$-\frac{1}{2\pi i}\int_\Gamma C_1(\lambda-A_1)^{-1}(\mu-A_1^\times)^{-1}B_1 f(\mu)d\mu +$$

$$-\left(\frac{1}{2\pi i}\right)^2 \int_\Gamma \int_\Gamma C_1(\lambda-A_1)^{-1}[(\mu-A_1)^{-1}-(\mu-A_1^\times)^{-1}](\upsilon-A_1^\times)^{-1}B_1 f(\upsilon)d\mu d\upsilon +$$

$$-\frac{1}{2\pi i}\int_\Gamma C_1(\lambda-A_1)^{-1}[(\mu-A_1)^{-1}-(\mu-A_1^\times)^{-1}](I-P_1^\times)zd\mu$$

$$= f(\lambda) + C_1(\lambda-A_1)^{-1}(I-P_1^\times)z - \frac{1}{2\pi i}\int_\Gamma C_1(\lambda-A_1)^{-1}P_1(\upsilon-A_1^\times)^{-1}B_1 f(\upsilon)d\upsilon +$$

$$+\frac{1}{2\pi i}\int_\Gamma C_1(\lambda-A_1)^{-1}P_1^\times(\upsilon-A_1^\times)^{-1}B_1 f(\upsilon)d\upsilon - C_1(\lambda-A_1)^{-1}P_1(I-P_1^\times)z$$

$$= f(\lambda) - C_1(\lambda-A_1)^{-1}(P_1z-P_1P_1^\times z-z+P_1^\times z) +$$

$$-\frac{1}{2\pi i}\int_\Gamma C_1(\lambda-A_1)^{-1}P_1(\upsilon-A_1^\times)^{-1}B_1 f(\upsilon)d\upsilon +$$

$$+\frac{1}{2\pi i}\int_\Gamma C_1(\lambda-A_1)^{-1}(\upsilon-A_1^\times)^{-1}B_1 f(\mu)d\mu$$

$$= f(\lambda) - C_1(\lambda-A_1)^{-1}(I-P_1)P_1^\times z +$$

$$+\frac{1}{2\pi i}\int_\Gamma C_1(\lambda-A_1)^{-1}(I-P_1)(\upsilon-A_1^\times)^{-1}B_1 f(\upsilon)d\upsilon$$

$$= f(\lambda) - C_1(\lambda-A_1)^{-1}(I-P_1)P_1^\times z +$$

$$+ C_1(\lambda-A_1)^{-1}(I-P_1)\left(\frac{1}{2\pi i}\int_\Gamma (\upsilon-A_1^\times)^{-1}B_1 f(\upsilon)d\upsilon\right) = f(\lambda).$$

As a third step in the proof of (6.2), we show that

$$(6.3)\qquad \int_\Gamma (I-P_2)(\lambda-A_2)^{-1}B_2 g(\lambda)d\lambda = 0.$$

From the resolvent equation it is clear that

$$(6.4)\quad C_1(\lambda-A_1^\times)^{-1}(\mu-A_1^\times)^{-1}B_1 = C_2(\lambda-A_2^\times)^{-1}(\mu-A_2^\times)^{-1}B_2, \qquad \lambda,\mu \in \Gamma$$

(cf. the proof of Theorem 5.1), and hence

$$g(\lambda) = W(\lambda)^{-1}f(\lambda) + \frac{1}{2\pi i}\int_\Gamma C_2(\lambda-A_2^\times)^{-1}(\mu-A_2^\times)^{-1}B_2 f(\mu)d\mu + + C_1(\lambda-A_1^\times)^{-1}(I-P_1^\times)z.$$

It follows that

$$\int_\Gamma (I-P_2)(\lambda-A_2)^{-1}B_2 g(\lambda)d\lambda =$$
$$= \int_\Gamma (I-P_2)(\lambda-A_2)^{-1}B_2 W(\lambda)^{-1}f(\lambda)d\lambda +$$
$$+ \frac{1}{2\pi i}\int_\Gamma\int_\Gamma (I-P_2)(\lambda-A_2)^{-1}B_2C_2(\lambda-A_2^\times)^{-1}(\mu-A_2^\times)^{-1}B_2 f(\mu)d\mu d\lambda +$$
$$+ \int_\Gamma (I-P_2)(\lambda-A_2)^{-1}B_2C_1(\lambda-A_1^\times)^{-1}(I-P_1^\times)zd\lambda.$$

The latter integral is zero by Cauchy's theorem. So

$$\int_\Gamma (I-P_2)(\lambda-A_2)^{-1}B_2 g(\lambda)d\lambda =$$
$$= \int_\Gamma (I-P_2)(\lambda-A_2^\times)^{-1}B_2 f(\lambda)d\lambda +$$
$$+ \frac{1}{2\pi i}\int_\Gamma\int_\Gamma (I-P_2)[(\lambda-A_2)^{-1}-(\lambda-A_2^\times)^{-1}](\mu-A_2^\times)^{-1}B_2 f(\mu)d\lambda d\mu$$
$$= \int_\Gamma (I-P_2)(\mu-A_2^\times)^{-1}(I-P_2^\times)B_2 f(\mu)d\mu,$$

which vanishes again by Cauchy's theorem.

We are now ready to prove (6.2). Recall from the proof of Theorem 5.1 that

$$C_1(\lambda-A_1)^{-1}(\mu-A_1)^{-1}B_1 = C_2(\lambda-A_2)^{-1}(\mu-A_2)^{-1}B_2, \qquad \lambda,\mu \in \Gamma$$

(cf. the identity (6.4)). Using this and the expression for $f(\lambda)$ in terms of $g(\lambda)$ established above, we obtain

$$\frac{1}{2\pi i}\int_\Gamma P_2^\times(\lambda-A_2^\times)^{-1}B_2f(\lambda)d\lambda =$$

$$= \frac{1}{2\pi i}\int_\Gamma P_2^\times(\lambda-A_2^\times)^{-1}B_2W(\lambda)g(\lambda)d\lambda +$$

$$-\left(\frac{1}{2\pi i}\right)^2\int_\Gamma\int_\Gamma P_2^\times(\lambda-A_2^\times)^{-1}B_2C_1(\lambda-A_1)^{-1}(\mu-A_1)^{-1}B_1g(\mu)d\mu d\lambda$$

$$= \frac{1}{2\pi i}\int_\Gamma P_2^\times(\lambda-A_2^\times)^{-1}B_2W(\lambda)g(\lambda)d\lambda +$$

$$-\left(\frac{1}{2\pi i}\right)^2\int_\Gamma\int_\Gamma P_2^\times(\lambda-A_2^\times)^{-1}B_2C_2(\lambda-A_2)^{-1}(\mu-A_2)^{-1}B_2g(\mu)d\mu d\lambda$$

$$= \frac{1}{2\pi i}\int_\Gamma P_2^\times(\lambda-A_2)^{-1}B_2g(\lambda)d\lambda +$$

$$-\left(\frac{1}{2\pi i}\right)^2\int_\Gamma\int_\Gamma P_2^\times[(\lambda-A_2)^{-1}-(\lambda-A_2^\times)^{-1}](\mu-A_2)^{-1}B_2g(\mu)d\lambda d\mu$$

$$= \frac{2}{2\pi i}\int_\Gamma P_2^\times(\lambda-A_2)^{-1}B_2g(\lambda)d\lambda - \frac{1}{2\pi i}\int_\Gamma P_2^\times P_2(\mu-A_2)^{-1}B_2g(\mu)d\mu$$

$$= \frac{1}{2\pi i}\int_\Gamma (2P_2^\times-P_2^\times P_2)(\lambda-A_2)^{-1}B_2g(\lambda)d\lambda$$

$$= \frac{1}{2\pi i}\int_\Gamma [P_2^\times P_2+2P_2^\times(I-P_2)](\lambda-A_2)^{-1}B_2g(\lambda)d\lambda.$$

Hence, by formula (6.3),

$$\frac{1}{2\pi i}\int_\Gamma P_2^\times(\lambda-A_2^\times)^{-1}B_2f(\lambda)d\lambda = P_2^\times P_2\left(\frac{1}{2\pi i}\int_\Gamma (\lambda-A_2)^{-1}B_2g(\lambda)d\lambda\right).$$

The desired result (6.2) is now clear from the fact that $\operatorname{Im} P_2^\times P_2 \subset \operatorname{Im} P_2 + \operatorname{Ker} P_2^\times = M_2 + M_2^\times$.

We have proved that Φ maps H_{1m} into H_{2m}. In other words $H_{1m} \subset \Phi^{-1}[H_{2m}]$. In order to establish the reverse inclusion, we introduce the operator

$$\psi = \frac{1}{2\pi i}\int_\Gamma (\lambda-A_1^\times)^{-1}B_1C_2(\lambda-A_2)^{-1}(I-P_2)P_2^\times d\lambda.$$

Note that ψ is obtained from Φ by interchanging the roles of Θ_1 and Θ_2. So, by symmetry, we have $H_{2m} \subset \psi^{-1}[H_{1m}]$. Assume now that

$$(6.5) \qquad x - \psi\Phi x \in M_1 + M_1^\times, \quad x \in X_1.$$

Then we can argue as follows. Take $x \in \Phi^{-1}[H_{2m}]$. Then $\Phi x \in H_{2m} \subset \psi^{-1}[H_{1m}]$, and so $\psi\Phi x \in H_{1m}$. We conclude that $x = x - \psi\Phi x + \psi\Phi x \in M_1 + M_1^\times + H_{1m} = H_{2m}$.

Next we prove (6.5). Take $x \in X_1$. From (6.1) it is clear that $\Phi x \in \operatorname{Im} P_2^\times$. Thus

$$\psi\Phi x = \frac{1}{2\pi i} \int_\Gamma (\lambda - A_1^\times)^{-1} B_1 C_2 (\lambda - A_2)^{-1} (I - P_2) \Phi x d\lambda.$$

Observe that

$$\Phi x = \frac{1}{2\pi i} \int_\Gamma (\lambda - A_2^\times)^{-1} B_2 W(\lambda) W(\lambda)^{-1} C_1 (\lambda - A_1)^{-1} (I - P_1) P_1^\times x d\lambda$$

$$= \frac{1}{2\pi i} \int_\Gamma (\lambda - A_2)^{-1} B_2 C_1 (\lambda - A_1^\times)^{-1} (I - P_1) P_1^\times x d\lambda,$$

and hence

$$\Psi\Phi x = \left(\frac{1}{2\pi i}\right)^2 \int_\Gamma\int_\Gamma (\lambda - A_1^\times)^{-1} B_1 C_2 (\lambda - A_2)^{-1} (I - P_2)(\mu - A_2)^{-1} B_2 \cdot$$

$$\cdot C_1 (\mu - A_1^\times)^{-1})(I - P_1) P_1^\times x d\mu d\lambda.$$

As we shall see presently,

$$(6.6) \qquad \begin{aligned} &C_1(\lambda - A_1)^{-1}(I - P_1)(\mu - A_1)^{-1} B_1 = \\ &= C_2(\lambda - A_2)^{-1}(I - P_2)(\mu - A_2)^{-1} B_2 \quad \lambda, \mu \in \Gamma. \end{aligned}$$

It follows that $\psi\Phi x$ must be equal to

$$\left(\frac{1}{2\pi i}\right)^2 \int_\Gamma\int_\Gamma [(\lambda - A_1)^{-1} - (\lambda - A_1^\times)^{-1}](I - P_1)[(\mu - A_1)^{-1} - (\mu - A_2^\times)^{-1}](I - P_1) P_1^\times x d\mu d\lambda$$

$$= (P_1 - P_1^\times)(I - P_1)(P_1 - P_1^\times) P_1^\times x = x - (I - P_1^\times)(I - P_1 P_1^\times)x - P_1 P_1^\times x.$$

So $x - \psi\Phi x \in \operatorname{Im} P_1 + \operatorname{Im}(I-P_1^\times) = M_1 + M_1^\times$.

To see that (6.6) is true, recall that the transfer functions of Θ_1 and Θ_2 coincide a Γ. By using the resolvent equation a number of times, one gets

$$C_1(\lambda-A_1)^{-1}(\upsilon-A_1)^{-1}(\mu-A_1)^{-1}B_1 = C_2(\lambda-A_2)^{-1}(\upsilon-A_2)^{-1}(\mu-A_2)^{-1}B_2, \quad \lambda,\mu,\upsilon \in \Gamma.$$

Formula (6.6) appears by integrating this expression over Γ (integration variable υ).

The first statement in Theorem 5.1, namely that $H_{1m} = \Phi^{-1}[H_{2m}]$, has now been proved. So we turn to the second assertion holding that $X_2 = \operatorname{Im}\Phi + H_{2m}$. Since $H_{20}, H_{21}, \ldots$ is an increasing sequence, it suffices to show that $X_2 = \operatorname{Im}\Phi + M_2 + M_2^\times$. But this is true indeed, because

$$(6.7) \qquad Y - \psi\Phi y \in M_2 + M_2^\times, \quad y \in X_2.$$

To see that (6.7) holds, consider (6.5) and interchange the roles of Θ_1 and Θ_2. This proves the theorem.□

We are now ready to prove the only if part of Theorem 1.2.

PROOF of THEOREM 1.2 (only if part). Assume W admits a (right) Wiener-Hopf factorization with respect to the given contour Γ. Then, by Theorem 3.1, there exists a node $\tilde{\Theta}$ without spectrum on Γ such that the transfer function of $\tilde{\Theta}$ coincides on Γ with W, while in addition

$$(6.8) \qquad \dim(\tilde{M}\cap\tilde{M}^\times) < \infty, \quad \operatorname{codim}(\tilde{M}+\tilde{M}^\times) < \infty.$$

Here $\tilde{M}, \tilde{M}^\times$ is the pair of spectral subspaces associated with $\tilde{\Theta}$ and Γ. From Theorem 5.1 it is clear that

$\dim(M \cap M^{\times}) = \dim(\tilde{M} \cap \tilde{M}^{\times})$ and $\operatorname{codim}(M+M^{\times}) = \operatorname{codim}(\tilde{M}+\tilde{M}^{\times})$. Thus (6.8) holds with $\tilde{M}$ and $\tilde{M}^{\times}$ replaced by M and $M^{\times}$, respectively.□

For the case dealt with in this paper, where Γ ia a positively oriented contour in the finite complex plane, Theorem 1.2 is the same as Theorem 2.2 in Section I.2 of [BGK4]. As was mentioned in Section 1, the situation considered in [BGK4] is slightly more general. However, it can be easily reduced to the case treated here. Indeed, let W, Θ, Γ, M and $M^{\times}$ be as in Theorem 2.2 in Section I.2 of [BGK4]. A small change in the contour Γ does not affect the fact that Θ has no spectrum on Γ. Neither does it change the incoming and outgoing subspaces of Θ. So we may assume that $\infty \notin \Gamma$, i.e., that Γ is a contour in the finite complex plane $\mathbb{C}$. But then either the inner domain Ω_{Γ}^{+} or the outer domain Ω_{Γ}^{-} of Γ is bounded. The case when Ω_{Γ}^{+} is covered by Theorem 1.2. For the case when Ω_{Γ}^{-} is bounded, the desired follows by changing the orientation of the contour Γ and applying Theorem 1.2 to the node $\Theta^{\times} = (A^{\times}, B, -C; X, Y)$.

7. Spectral characteristics and Wiener-Hopf equivalence

The aim of this section is to prove Theorem 1.1, the main result of this paper.

PROOF OF THEOREM 1.1. There is no loss of generality in assuming that the functions W_- and W_+ appearing in the Wiener-Hopf equivalence (1.1) can be written as the transfer function of a minus node Θ_- and a plus node Θ_+, respectively. This can be seen with an argument analogous to that used in the last part of the proof of Theorem 3.1.

Write $\Theta_- = (A_-, B_-, C_-; X_-, Y)$, $\Theta_+ = (A_+, B_+, C_+; X_+, Y)$ and define $\Theta = (A, B, C; X, Y)$ as the product of Θ_-, Θ_2 and Θ_+, i.e., $\Theta = \Theta_-\Theta_2\Theta_+$. Then A, B, C and $A^{\times}$ are given by (3.2) - (3.5) with A_0, B_0, C_0 and X_0 replaced by A_2, B_2, C_2 and X_2, respectively. Clearly, Θ is a node without spectrum on Γ and, for $\lambda \in \Gamma$, the transfer function of Θ has the value $W_-(\lambda)W_2(\lambda)W_+(\lambda) = W_1(\lambda)$. So on Γ the transfer functions of Θ_1 and Θ coincide. This enables us to apply Theorems 5.1 and 6.1.

Let $\Psi_0 : X_1 \to X$ and $\Phi_0 : X_1 \to X$ be given by

$$\Psi_0 = \frac{1}{2\pi i} \int_\Gamma (\lambda - A^\times)^{-1} B C_1 (\lambda - A_1)^{-1} d\lambda,$$

$$\Phi_0 = \frac{1}{2\pi i} \int_\Gamma (\lambda - A^\times)^{-1} B C_1 (\lambda - A_1)^{-1} (I - P_1) P_1^\times d\lambda.$$

Then ψ_0 and Φ_0 are bounded linear operators and, for $m = 0,1,2,\ldots,$

$$\Psi_0[K_{1m}] = K_m, \quad \operatorname{Ker}\Psi_0 \cap K_{1m} = (0),$$

$$H_{1m} = \Phi_0^{-1}[H_m], \quad \operatorname{Im}\Phi_0 + H_m = X.$$

Here the $K_{1m}(K_m)$ are the outgoing spaces of $\Theta_1(\Theta)$ and the $H_{1m}(H_m)$ are the incoming spaces of $\Theta_1(\Theta)$. From Theorem 3.2 we know that

$$K_m = \begin{pmatrix} (0) \\ K_{2m} \\ (0) \end{pmatrix}, \qquad H_m = \begin{pmatrix} X_- \\ H_{2m} \\ X_+ \end{pmatrix},$$

where the K_{2m} and H_{2m} are the outgoing and incoming spaces of Θ_2, respectively. So, if J is the canonical projection of X onto X_2, i.e.,

$$J = [0 \quad I \quad 0] : X_- \oplus X_2 \oplus X_+ \to X_2,$$

then $K_{2m} = J[K_m]$ and $H_m = J^{-1}[H_{2m}]$. Thus

$$(J \circ \Psi_0)[K_{1m}] = K_{2m}, \qquad \operatorname{Ker}(J \circ \Psi_0) \cap K_{1m} = (0),$$

$$H_{1m} = (J \circ \Phi_0)^{-1}[H_{2m}], \qquad \operatorname{Im}(J \circ \Phi_0) + H_{2m} = X_2.$$

An easy computation, based on the special form of the operators $A^\times$ and B, shows that $J \circ \Psi_0 = \Psi$ and $J \circ \Phi_0 = \Phi$, with Ψ and Φ as in the theorem, and the proof is complete. □

The operators Ψ and Φ appearing in Theorem 1.1 can also be written as

$$\Psi = \frac{1}{2\pi i}\int_\Gamma (\lambda-A_2)^{-1}B_2W_+(\lambda)C_1(\lambda-A_1^\times)^{-1}d\lambda,$$

$$\Phi = \frac{1}{2\pi i}\int_\Gamma (\lambda-A_2)^{-1}B_2W_+(\lambda)C_1(\lambda-A_1)^{-1}(I-P_1)P_1^\times d\lambda,$$

where W_+ is as in (1.1). The proof is based on (7.1) and the identities $(\lambda-A_2^\times)^{-1}B_2W_2(\lambda) = (\lambda-A_2)^{-1}B_2$ and $W_1(\lambda)^{-1}C_1(\lambda-A_1)^{-1} = C_1(\lambda-A_1^\times)^{-1}$ (cf. the proof of Theorem 5.1).

REFERENCES

[BGK1] Bart, H., Gohberg, I., Kaashoek, M.A.: Minimal factorization of matrix and operator functions. Operator Theory: Advances and Applications, Vol.1. Basel-Boston-Stuttgart, Birkhäuser Verlag, 1979.

[BGK2] Bart, H., Gohberg, I., Kaashoek, M.A.: Wiener-Hopf factorization of analytic operator functions and realization. Rapport 231, Wiskundig Seminarium, Vrije Universiteit, Amsterdam, 1983.

[BGK3] Bart, H., Gohberg, I., Kaashoek, M.A.: Wiener-Hopf factorization and realization. In: Mathematical Theory of Networks and Systems, Proceedings of the MTNS-83 International Symposium, Beer Sheva, Israel, Lecture Notes in Control and Information Sciences, no. 58 (Ed. P. Fuhrmann), Springer Verlag, Berlin, 1984, pp. 42-62.

[BGK4] Bart, H., Gohberg, I., Kaashoek, M.A.: Explicit Wiener-Hopf factorization and realization. This volume.

[CG] Clancey, K., Gohberg, I.: Factorization of matrix functions and singular integral operators. Operator Theory: Advances and Applications, Vol.3, Basel-Boston-Stuttgart, Birkhäuser Verlag, 1981.

[GKS] Gohberg, I., Kaashoek, M.A., Van Schagen, F.: Similarity of operator blocks and canonical forms. II. Infinite dimensional case and Wiener-Hopf factorization. In: Topics in Modern Operator Theory. Operator Theory: Advances and Applications, Vol.2. Basel-Boston-Stuttgart, Birkhäuser Verlag, 1981, pp. 121-170.

[H] Helton, J.W.: Systems with infinite-dimensional state space: the Hilbert space approach. Proc. IEEE 64 (1), 145-160 (1976).

H. Bart
Econometrisch Instituut
Erasmus Universiteit
Postbus 1738
3000Dr Rotterdam
The Netherlands

I. Gohberg
Dept of Mathematical Sciences
The Raymond and Beverly Sackler
Faculty of Exact Sciences
Tel-Aviv University
Ramat-Aviv Israel

M.A. Kaashoek
Subfaculteit Wiskunde en
Informatica
Vrije Universiteit
Postbus 7161
1007 MC Amsterdam
The Netherlands

Operator Theory:
Advances and Applications, Vol. 21

MULTIPLICATION BY DIAGONALS AND REDUCTION TO CANONICAL FACTORIZATION

H. Bart, I. Gohberg, M.A. Kaashoek

The calculus of matrix functions in realized form is developed further. Special attention is paid to the operation of multiplications by diagonal matrix functions. The analysis yields a reduction of matrix functions to functions that admit canonical factorization.

1. INTRODUCTION.

Let W be an m×m rational matrix function without poles or zeros on a given contour Γ and having the value I_m at ∞. In this paper we analyse the effect on realizations of W when W is multiplied on the left by a (Wiener-Hopf) diagonal, i.e., a matrix function of the form

$$(1.1)\qquad D(\lambda) = \begin{pmatrix} \left(\frac{\lambda-\varepsilon_1}{\lambda-\varepsilon_2}\right)^{\upsilon_1} & & \\ & \ddots & \\ & & \left(\frac{\lambda-\varepsilon_1}{\lambda-\varepsilon_2}\right)^{\upsilon_m} \end{pmatrix},$$

where ε_1 is in the inner domain Ω_Γ^+ of Γ and ε_2 is in the outer domain Ω_Γ^- of Γ.

The main result is Theorem 3.1. It provides a procedure for choosing the exponents $\upsilon_1,\dots,\upsilon_m$ in (1.1) in such a way that the function $D(\lambda)W(\lambda)$ admits a canonical Wiener-Hopf factorization. As a corollary, a theorem originally proved by I.S. Chibotaru (see [Ch]; cf. also [CG], Section VI.2) is obtained. The main result in Section 2 is a general theorem describing the pair of spectral subspaces associated with a product of two nodes.

Throughout the paper we shall assume that Γ is a positively oriented contour in the (finite) complex plane $\mathbb{C}$ (and hence Ω_Γ^+ is bounded). From the last paragraph of Section 6 in [BGK4], one sees that this assumption does not affect the generality of the results.

We shall use the notation and terminology of [BGK3] and [BGK4]. For the convenience of the reader we recall a few of the main definitions. A <u>node</u> (or a <u>system</u>) is a quintet $\Theta = (A,B,C;X,Y)$, where $A\colon X \to X$, $B\colon Y \to X$ and $C\colon X \to Y$ are (bounded linear) operators acting between the complex Banach spaces X and Y. In other words, $A \in L(X)$, $B \in L(Y,X)$ and $C \in L(X,Y)$. The node Θ is called <u>finite dimensional</u> if both X and Y are finite dimensional. The space X is called the <u>state space</u> of Θ. The <u>transfer function</u> of Θ is the function

$$W_\Theta(\lambda) = I + C(\lambda - A)^{-1}B.$$

If Θ is finite dimensional, W_Θ is (or may be identified with) a rational matrix function having the value I at ∞. Conversely, every rational (square) matrix funtion having the value I at ∞ can be written as the transfer function of a finite dimensional node. We say that Θ <u>has no spectrum on the contour</u> Γ if A has no spectrum on Γ and the same holds true for the operator A-BC, which henceforth will be denoted by $A^\times$. In that case one can introduce the spectral subspaces

$$M = \operatorname{Im}\Big(\frac{1}{2\pi i}\int_\Gamma (\lambda - A)^{-1}d\lambda\Big), \qquad M^\times = \operatorname{Ker}\Big(\frac{1}{2\pi i}\int_\Gamma (\lambda - A^\times)^{-1}d\lambda\Big).$$

The pair $M, M^\times$ is called the <u>pair of spectral subspaces associated with</u> Θ <u>and</u> Γ.

An earlier version of this paper appeared in [BGK2].

2. Spectral pairs associated with products of nodes

The following theorem concerns the pair of spectral subspaces associated with a product of nodes (cf. [BGK1]). Similar results hold for the incoming and outgoing subspaces as introduced in [BGK3]. A special case of the theorem (multiplication by minus or plus nodes) was already obtained in Section 3 of [BGK4].

THEOREM 2.1. *Suppose* $\Theta_1 = (A_1,B_1,C_1;X_1,Y)$ *and* $\Theta_2 = (A_2,B_2,C_2;X_2,Y)$ *are nodes without spectrum on the contour* Γ. *For* $j = 1,2$, *let* $M_j, M_j^\times$ *be the pair of spectral subspaces associated with* Γ *and* Θ_j. *Also, let* $M, M^\times$ *be the pair of spectral subspaces associated with* Γ *and the product* $\Theta = \Theta_1\Theta_2$ *of* Θ_1 *and* Θ_2. *Then*

$$(2.1)\qquad M = \begin{pmatrix} I & R & M_1 \\ 0 & I & M_2 \end{pmatrix}, \qquad M^\times = \begin{pmatrix} I & 0 & M_1^\times \\ R^\times & I & M_2^\times \end{pmatrix},$$

where $R: X_2 \to X_1$ *and* $R^\times: X_1 \to X_2$ *are the bounded linear operators given by*

$$(2.2)\qquad R = \frac{1}{2\pi i}\int_\Gamma (\lambda-A_1)^{-1}B_1C_2(\lambda-A_2)^{-1}d\lambda,$$

$$(2.3)\qquad R^\times = \frac{1}{2\pi i}\int_\Gamma (\lambda-A_2^\times)^{-1}B_2C_1(\lambda-A_1^\times)^{-1}d\lambda.$$

The notation used in (2.1) was introduced in Section 3 of [BGK4].

PROOF. Recall from [BGK1], Section 1.1 that $\Theta = (A,B,C;X,Y)$, where $X = X_1 \oplus X_2$ and

$$A = \begin{pmatrix} A_1 & B_1C_2 \\ 0 & A_2 \end{pmatrix}, \qquad B = \begin{pmatrix} B_1 \\ B_2 \end{pmatrix}, \qquad C = \begin{pmatrix} C_1 & C_2 \end{pmatrix}.$$

Also $A^\times = A-BC$ has the representation

$$A^{\times} = \begin{pmatrix} A_1^{\times} & 0 \\ -B_2C_1 & A_2^{\times} \end{pmatrix}.$$

From this it is clear that Θ has no spectrum on Γ, and so M and $M^{\times}$ are well-defined.

Write $M = \operatorname{Im} P$, $M^{\times} = \operatorname{Ker} P^{\times}$, $M_1 = \operatorname{Im} P_1$, $M_1^{\times} = \operatorname{Ker} P_1^{\times}$, $M_2 = \operatorname{Im} P_2$ and $M_2^{\times} = \operatorname{Ker} P_2^{\times}$, where $P, P^{\times}, P_1, P_1^{\times}, P_2$ and $P_2^{\times}$ are the appropriate Riesz projections. Using the integral representation of these projections, one easily sees that

$$P = \begin{pmatrix} P_1 & P \\ 0 & P_2 \end{pmatrix}, \quad P^{\times} = \begin{pmatrix} P_1^{\times} & 0 \\ -R^{\times} & P_2^{\times} \end{pmatrix},$$

where $R\colon X_2 \to X_1$ and $R^{\times}\colon X_1 \to X_2$ are given by (2.2) and (2.3), respectively. Since P and $P^{\times}$ are projections, we have $R = P_1R+RP_2$ and $R^{\times} = R^{\times}P_1^{\times}+P_2^{\times}R^{\times}$. It follows that

$$P = \begin{pmatrix} I & R \\ 0 & I \end{pmatrix} \begin{pmatrix} P_1 & 0 \\ 0 & P_2 \end{pmatrix} \begin{pmatrix} I & R \\ 0 & I \end{pmatrix},$$

$$I-P^{\times} = \begin{pmatrix} I & 0 \\ R^{\times} & I \end{pmatrix} \begin{pmatrix} I-P_1^{\times} & 0 \\ 0 & I-P_2^{\times} \end{pmatrix} \begin{pmatrix} I & 0 \\ R^{\times} & I \end{pmatrix}.$$

The identities (2.1) are now clear from $M = \operatorname{Im} P$, $M^{\times} = \operatorname{Im}(I-P^{\times})$ and $M_j = \operatorname{Im} P_j$, $M_j^{\times} = \operatorname{Im}(I-P_j^{\times})$, $j = 1,2$. □

REMARK 2.2. Let Θ_1, Θ_2, etc. be as in Theorem 2.1 and suppose in addition that $\sigma(A_1) \subset \Omega_\Gamma^-$ and $\sigma(A_1^{\times}) \subset \Omega_\Gamma^+$. Then $P_1 = 0$ and $P_1^{\times} = I$. So in this case the identities (2.1) simplify to

$$M = \begin{pmatrix} I & R \\ 0 & I \end{pmatrix} \begin{pmatrix} (0) \\ M_2 \end{pmatrix}, \quad M^{\times} = \begin{pmatrix} (0) \\ M_2^{\times} \end{pmatrix}.$$

This we shall use in the next section where for Θ_1 we take a minimal node whose transfer function has the diagonal form (i.1) with all exponents $\upsilon_1,\ldots,\upsilon_m$ non-negative.

3. Multiplication by diagonals

In this section we restrict ourselves to the finite dimensional case and we study the effect of multiplication (on the left) by an m × m diagonal matrix D of the form (1.1), i.e.,

$$(3.1)\qquad D(\lambda) = \begin{pmatrix} \left(\frac{\lambda-\varepsilon_1}{\lambda-\varepsilon_2}\right)^{\upsilon_1} & & & \\ & \cdot & & \\ & & \cdot & \\ & & & \left(\frac{\lambda-\varepsilon_1}{\lambda-\varepsilon_2}\right)^{\upsilon_m} \end{pmatrix}.$$

Here ε_1 lies in the (bounded) inner domain Ω_Γ^+ of the given contour Γ and ε_2 lies in the outer domain Ω_Γ^- of Γ.

For $k = 1,\ldots,m$, let $\Pi_k: \mathbb{C}^m \to \mathbb{C}$ be the canonical projection of $\mathbb{C}^m$ onto the k-th coordinate space. In other words Π_k assigns to each vector in $\mathbb{C}^m$ its k-th coordinate. The next theorem is the main result of this chapter. For the definition of outgoing bases, see [BGK3].

THEOREM 3.1. *Let* $\Theta = (A,B,C;X,\mathbb{C}^m)$ *be a finite dimensional node without spectrum on the contour* Γ, *let* $M,M^\times$ *be the pair of spectral subspaces associated with* Γ *and* Θ, *and let* $\{e_{jk}\}_{k=1,j=1}^{\alpha_j \; t}$ *be an outgoing basis for* Θ *with respect to the operator* $A-\varepsilon_2$. *Then the restriction of* $A-\varepsilon_2$ *to* M *is an invertible operator on* M *and the complex* $t \times m$ *matrix*

$$(3.2)\qquad \left(\Pi_k C(A-\varepsilon_2)^{-1} e_{j1}\right)_{j=1,k=1}^{t \quad m}$$

has rank t. *Suppose* $s_1,\ldots,s_t$ *are integers among* $1,\ldots,m$ *such that*

(3.3) $$\det\left(\Pi_{s_k} C(A-\varepsilon_2)^{-1} e_{j1}\right)_{j,k=1}^{p} \neq 0, \quad p = 1,\dots,t.$$

Further, assume that the exponents $\upsilon_1,\dots,\upsilon_m$ in the diagonal entries of D, given by (3.1), are non-negative and satisfy

(3.4) $$\upsilon_{s_k} \geqq \alpha_k, \quad k = 1,\dots,t.$$

Then, if $\tilde{M},\tilde{M}^{\times}$ is the pair of spectral subspaces associated with Γ and the product node $\tilde{\Theta} = \Theta_D\Theta$, where Θ_D is an arbitrary finite dimensional realization of D without spectrum on Γ, the following identities hold:

(3.5) $$\tilde{M} \cap \tilde{M}^{\times} = (0),$$

(3.6) $$\dim(\tilde{X}/\tilde{M}+\tilde{M}^{\times}) = \dim(X/M+M^{\times}) + \sum_{k=1}^{t} (\upsilon_{s_k} - \alpha_k) + \sum_{\substack{j=1,\dots,m \\ j\neq s_1,\dots,s_t}} \upsilon_j .$$

In particular $\dim(\tilde{X}/\tilde{M}+\tilde{M}^{\times}) \geqq \dim(X/M+M^{\times})$, equality occurring if and only if

(3.7) $$\begin{aligned} \upsilon_{s_k} &= \alpha_k, \quad && k = 1,\dots,t, \\ \upsilon_j &= 0, && j = 1,\dots,m;\ j \neq s_1,\dots,s_t . \end{aligned}$$

Here $\tilde{X}$ is the state space of the node $\tilde{\Theta}$.

The fact that the matrix (3.2) has rank t implies that there do exist integers $s_1,\dots,s_t$ among $1,\dots,m$ such that (3.3) is satisfied.

PROOF. The spectrum of the restriction A_M of A to M lies in the inner domain Ω_Γ^+ of Γ. Since $\varepsilon_2 \in \Omega_\Gamma^-$, it follows that $A_M-\varepsilon_2$ is invertible. Suppose $a_1,\dots,a_t$ are complex numbers such that

$$\sum_{j=1}^{t} a_j C(A_M-\varepsilon_2)^{-1} e_{j1} = 0.$$

Put $u = a_1e_{11}+\dots+a_te_{t1}$. Then $u \in M \cap M^{\times}$ and $Cv = 0$, where

$v = (A_M-\varepsilon_2)^{-1}u \in M$. Observe that $(A^\times-\varepsilon_2)v = (A-BC-\varepsilon_2)v = (A-\varepsilon_2)v = u$. So

$$v = \frac{(\lambda-A^\times)v}{\lambda-\varepsilon_2} + \frac{u}{\lambda-\varepsilon_2}, \qquad \lambda \neq \varepsilon_2.$$

Let $P^\times$ be the Riesz projection corresponding to $A^\times$ and Γ. Then

$$P^\times v = \frac{1}{2\pi i}\int_\Gamma (\lambda-A^\times)^{-1}v\,d\lambda = \frac{1}{2\pi i}\int_\Gamma \frac{v}{\lambda-\varepsilon_2}d\lambda + \frac{1}{2\pi i}\int_\Gamma \frac{(\lambda-A^\times)^{-1}u}{\lambda-\varepsilon_2}d\lambda.$$

The first integral in this sum vanishes because $\varepsilon_2 \in \Omega_\Gamma^-$; the second because its integrand is analytic on a neighbourhood of $\Omega_\Gamma^+ \cup \Gamma$ too. Thus $v \in M \cap M^\times \cap \operatorname{Ker} C$. Now $M \cap M^\times \cap \operatorname{Ker} C$ is spanned by the vectors e_{jk} with $k = 1,\dots,\alpha_j-1$, $j = 1,\dots,t$ ($\alpha_j \geqq 2$). It follows that $u = (A-\varepsilon_2)v$ is a linear combination of the vectors $(A-\varepsilon_2)e_{jk} = e_{j,k+1}$, where k and j are subject to the same restrictions. On the other hand $u = a_1e_{11}+ \dots + a_te_{t1}$, and we may conclude that $a_1 = \dots = a_t = 0$. This proves that the vectors $C(A_M-\varepsilon_2)^{-1}e_{11},\dots,C(A_M-\varepsilon_2)^{-1}e_{t1}$ are linearly independent in $\mathbb{C}^m$. In other words the matrix (3.2) has rank t.

Next we deal with the second part of the theorem. In view of the results obtained in Sections 5 and 6 of [BGK4], it suffices to consider one single realization Θ_D of D without spectrum on the contour Γ. For Θ_D we now take a special (minimal) realization of D which is constructed as follows (cf. Section 2 in [BGK4]). For $j = 1,\dots,m$, define

$$A_j = \begin{pmatrix} \varepsilon_2 & 0 & \cdot & \cdot & \cdot & 0 \\ 1 & \varepsilon_2 & \cdot & \cdot & \cdot & 0 \\ 0 & 1 & \cdot & & & \cdot \\ \cdot & \cdot & & \cdot & & \cdot \\ \cdot & \cdot & \cdot & \cdot & \cdot & 0 \\ 0 & 0 & \cdot & \cdot & 1 & \varepsilon_2 \end{pmatrix} : \mathbb{C}^{\upsilon_j} \to \mathbb{C}^{\upsilon_j},$$

$$B_j = \begin{pmatrix} 1 \\ 0 \\ \cdot \\ \cdot \\ \cdot \\ 0 \end{pmatrix} : \mathbb{C} \to \mathbb{C}^{\upsilon_j},$$

$$C_j = [\binom{\upsilon_j}{1}(\varepsilon_2-\varepsilon_1)\ \binom{\upsilon_j}{2}(\varepsilon_2-\varepsilon_1)^2 \dots \binom{\upsilon_j}{\upsilon_j}(\varepsilon_2-\varepsilon_1)^{\upsilon_j}]: \mathbb{C}^{\upsilon_j} \to \mathbb{C}.$$

Then $\Theta_j = (A_j, B_j, C_j; \mathbb{C}^{\upsilon_j}, \mathbb{C})$ is a (minimal) realization of the scalar function $((\lambda-\varepsilon_1)/(\lambda-\varepsilon_2)^{\upsilon_j})$. The (minimal) realization $\Theta_D = (A_D, B_D, C_D; X_D, \mathbb{C}^m)$ of D is now obtained by putting

(3.8) $$X_D = \mathbb{C}^{\upsilon_1} \oplus \dots \oplus \mathbb{C}^{\upsilon_m},$$

$$A_D = \begin{pmatrix} A_1 & & & & \\ & A_2 & & & \\ & & \ddots & & \\ & & & & A_m \end{pmatrix} : X_D \to X_D,$$

$$B_D = \begin{pmatrix} B_1 & & & & \\ & B_2 & & & \\ & & \ddots & & \\ & & & & B_m \end{pmatrix} : \mathbb{C}^m \to X_D,$$

$$C_D = \begin{pmatrix} C_1 & & & & \\ & C_2 & & & \\ & & \ddots & & \\ & & & & C_m \end{pmatrix} : X_D \to \mathbb{C}^m.$$

In other words Θ_D is the "direct sum" of the nodes $\Theta_1, \dots, \Theta_m$. Note that Θ_D has no spectrum on Γ. In fact $\sigma(A_D) \subset \{\varepsilon_2\} \subset \Omega_\Gamma^-$ and $\sigma(A_D^\times) \subset \{\varepsilon_1\} \subset \Omega_\Gamma^+$.

Write $\tilde{\Theta} = \Theta_D\Theta = (\tilde{A}, \tilde{B}, \tilde{C}; \tilde{X}, Y)$. Then $\tilde{\Theta}$ is a node without spectrum on Γ. According to Remark 2.2 the pair $\tilde{M}, \tilde{M}^\times$ of spectral subspaces associated with $\tilde{\Theta}$ and Γ can be expressed as

$$(3.9)\qquad \widetilde{M} = \begin{pmatrix} I & R \\ 0 & I \end{pmatrix} \begin{pmatrix} (0) \\ M \end{pmatrix}, \qquad \widetilde{M}^{\times} = \begin{pmatrix} (0) \\ M^{\times} \end{pmatrix},$$

where $R: X \to X_D$ is defined by

$$R = \frac{1}{2\pi i} \int_{\Gamma} (\lambda - A_D)^{-1} B_D C(\lambda - A)^{-1} d\lambda.$$

With respect to the decomposition (3.8), the operator R has the form

$$(3.10)\qquad R = \begin{pmatrix} R_1 \\ \cdot \\ \cdot \\ \cdot \\ R_m \end{pmatrix} : X \to \mathbb{C}^{\upsilon_1} \oplus \dots \oplus \mathbb{C}^{\upsilon_m},$$

with $R_k: X \to \mathbb{C}^{\upsilon_k}$ given by

$$R_k = \frac{1}{2\pi i} \int_{\Gamma} (\lambda - A_k)^{-1} B_k \Pi_k C(\lambda - A)^{-1} d\lambda.$$

Taking into account the definition of A_k and B_k, we get

$$(\lambda - A_k)^{-1} B_k \Pi_k C(\lambda - A)^{-1} = \begin{pmatrix} \dfrac{\Pi_k C(\lambda - A)^{-1}}{\lambda - \varepsilon_2} \\ \vdots \\ \dfrac{\Pi_k C(\lambda - A)^{-1}}{(\lambda - \varepsilon)^{\upsilon_k}} \end{pmatrix} : X \to \mathbb{C}^{\upsilon_k}.$$

Observe now that

$$\frac{1}{2\pi i} \int_{\Gamma} \frac{(\lambda - A)^{-1}}{(\lambda - \varepsilon_2)^{\mu}} d\lambda = (A_M - \varepsilon_2)^{-\mu} P,$$

where P is the Riesz projection corresponding to A and the contour Γ, considered here as an operator from X into $M = \operatorname{Im} P$. It follows that

$$(3.11)\qquad R_K = \begin{bmatrix} \Pi_k C(A_M-\varepsilon_2)^{-1}P \\ \vdots \\ \Pi_k C(A_M-\varepsilon_2)^{-\upsilon_k}P \end{bmatrix} : X \to \mathbb{C}^{\upsilon_k}, \quad k = 1,\dots,m.$$

Here C is viewed as an operator from M into $\mathbb{C}^m$. Combining (3.10) and (3.11) one gets a complete description of R.

Until now we only used that the integers $\upsilon_1,\dots,\upsilon_m$ are non-negative. Next we are going to employ the assumption (3.4), where $s_1,\dots,s_t$ are such that (3.3) is satisfied. Since $\tilde{M}$ and $\tilde{M}^\times$ are given by (3.9), the statement $\tilde{M} \cap \tilde{M}^\times = (0)$ is equivalent to the assertion $M \cap M^\times \cap \mathrm{Ker}\, R = (0)$.

Take $x \in M \cap M^\times \cap \mathrm{Ker}\, R$. Then in particular $Rx = 0$. In view of (3.10) and (3.11) this means that

$$(3.12)\qquad \Pi_k C(A_M-\varepsilon_2)^{-\mu}x = 0, \quad \mu = 1,\dots,\upsilon_k; \quad k = 1,\dots,m.$$

Also $x \in M \cap M^\times$, and so x can be written as a linear combination of the vectors e_{jk}:

$$(3.13)\qquad x = \sum_{\substack{k=1,\dots,\alpha_j \\ j=1,\dots,t}} \lambda_{jk}e_{jk}.$$

From (3.4) we see that $\upsilon_{s_1},\dots,\upsilon_{s_t}$ are strictly positive. Thus (3.12) implies

$$(3.14)\qquad \Pi_{s_k} C(A_M-\varepsilon_2)^{-1}x = 0, \quad k = 1,\dots,t.$$

Observe now that

$$C(A_M-\varepsilon_2)^{-1}x = \sum_{j=1}^{t} \lambda_{j1}C(A_M-\varepsilon_2)^{-1}e_{j1} + \sum_{\substack{k=2,\dots,\alpha_j \\ j=1,\dots,t}} \lambda_{jk}Ce_{j,k-1}$$

$$= \sum_{j=1}^{t} \lambda_{j1}C(A_M-\varepsilon_2)^{-1}e_{j1}.$$

Together with (3.14) this gives

$$\sum_{j=1}^{t} \lambda_{j1}\Pi_{s_k} C(A_M-\varepsilon_2)^{-1} e_{j1} = 0, \quad k = 1,\ldots,t.$$

But then it follows from (3.3) that $\lambda_{11} = \ldots = \lambda_{t1} = 0$. If $\alpha_1 = 1$ (and hence $\alpha_1 = \ldots = \alpha_t = 1$), we may conclude that $x = 0$. If $\alpha_1 \geqq 2$, we continue as follows.

Suppose $\alpha_1 \geqq \ldots \geqq \alpha_q \geqq 2 \geqq \alpha_{q+1} \geqq \ldots \geqq \alpha_t$. Then (3.13) can be written as

$$x = \sum_{\substack{k=2,\ldots,\alpha_j \\ j=1,\ldots,q}} \lambda_{jk} e_{jk}.$$

From (3.4) we see that the $\upsilon_{s_1},\ldots,\upsilon_{s_q}$ are all at least 2. Hence, by (3.12),

$$\Pi_{s_k} C(A_M-\varepsilon_2)^{-2} x = 0, \quad k = 1,\ldots,q.$$

Using that

$$C(A_M-\varepsilon_2)^{-2} x = \sum_{j=1}^{q} \lambda_{j2} C(A_M-\varepsilon_2)^{-1} e_{j1} + \sum_{\substack{k=3,\ldots,\alpha_j \\ j=1,\ldots,q}} \lambda_{jk} C e_{j,k-2}$$

$$= \sum_{j=1}^{q} \lambda_{j2} C(A_M-\varepsilon_2)^{-1} e_{j1},$$

we now obtain

$$\sum_{j=1}^{q} \lambda_{j2}\Pi_{s_k} C(A_M-\varepsilon_2)^{-1} e_{j1} = 0, \quad k = 1,\ldots,q.$$

In view of (3.3) this implies $\lambda_{12} = \ldots = \lambda_{q2} = 0$. Proceeding in this way (or, if one prefers, by a formal finite induction argument) we see that all complex numbers λ_{jk} are zero. Hence $x = 0$ and we have proved (3.5).

In order to establish (3.6) we choose subspaces N and $N^{\times}$ of X such that $M = (M \cap M^{\times}) \oplus N$ and $M^{\times} = (M \cap M^{\times}) \oplus N^{\times}$. Then $M+M^{\times} = N \oplus (M \cap M^{\times}) \oplus N^{\times}$ and so

$$\dim(X/M+M^\times) = \dim X - \dim N^\times - \dim(M\cap M^\times) - \dim N.$$

Next we observe that

$$(3.15)\quad \tilde{M}+\tilde{M}^\times = \begin{pmatrix} I & R \\ 0 & I \end{pmatrix}\begin{pmatrix} (0) \\ N \end{pmatrix} \oplus \begin{pmatrix} I & R \\ 0 & I \end{pmatrix}\begin{pmatrix} (0) \\ M\cap M^\times \end{pmatrix} \oplus \begin{pmatrix} (0) \\ M\cap M^\times \end{pmatrix} \oplus \begin{pmatrix} (0) \\ N^\times \end{pmatrix}.$$

Indeed, from (3.9) it is clear that $M+M^\times$ is the sum of the four spaces appearing in the right hand side of (3.15). To see that the sum is direct, assume $w \in N$, $x,y \in M \cap M^\times$, $z \in N^\times$ and

$$\begin{pmatrix} Rw+Rx \\ w+x+y+z \end{pmatrix} = \begin{pmatrix} 0 \\ 0 \end{pmatrix}.$$

Then $w = z = 0$, $x = -y$ and $Rx = 0$. So $x \in M \cap M^\times \cap \operatorname{Ker} R$. But $M \cap M^\times \cap \operatorname{Ker} R = (0)$. Hence $x = y = 0$ too.

The identity (3.15) implies

$$\dim(\tilde{X}/\tilde{M}+\tilde{M}^\times) = \dim X + \dim X_D - \dim N - \dim N^\times - 2\dim(M\cap M^\times).$$

It follows that

$$\begin{aligned}\dim(\tilde{X}/\tilde{M}+\tilde{M}^\times) &= \dim(X/M+M^\times) + \dim X_D - \dim(M\cap M^\times) \\ &= \dim(X/M+M^\times) + \sum_{k=1}^{m} \upsilon_k - \sum_{j=1}^{t} \alpha_j \\ &= \dim(X/M+M^\times) + \sum_{k=1}^{t} (\upsilon_{s_k} - \alpha_k) + \sum_{\substack{j=1,\dots,m \\ j\neq s_1,\dots,s_t}} \upsilon_j,\end{aligned}$$

and the proof is complete. ||

Theorem 3.1 concerns a situation where the (non-negative) integers $\upsilon_1,\dots,\upsilon_m$ appearing in (3.1) are chosen in a special way (cf. (3.4) and (3.7)). It is also possible to analyse other cases, for instance the case when all the υ_k are equal to one single positive integer υ. This comes down to multiplication (from the left) by $((\lambda-\varepsilon_1)/(\lambda-\varepsilon_2))^{\upsilon}I_m$. The analysis is similar to that given in the proof of Theorem 3.1. Further one can replace left multiplication by multiplication from the right.

We refrain from developing the details here. Neither shall we discuss the case where all exponents $\upsilon_1,\dots,\upsilon_m$ are non-positive. Instead we present a simple application of Theorem 3.1.

Let W be a rational m × m matrix function having no poles or zeros on the contour Γ and with the value I_m at ∞. Then W admits a (right) Wiener-Hopf factorization relative to the contour Γ (and the points $\varepsilon_1,\varepsilon_2$). In the finite dimensional case considered here the Wiener-Hopf factorization can be written in the form

$$(3.16)\qquad W(\lambda) = W_-(\lambda)\begin{pmatrix} \left(\frac{\lambda-\varepsilon_1}{\lambda-\varepsilon_2}\right)^{\kappa_1} & & \\ & \ddots & \\ & & \left(\frac{\lambda-\varepsilon_1}{\lambda-\varepsilon_2}\right)^{\kappa_m}\end{pmatrix} W_+(\lambda),\qquad \lambda\in\Gamma,$$

where W_- is a minus function, W_+ is a plus function and $\kappa_1 \leqq \kappa_2 \leqq \dots \leqq \kappa_m$. The integers $\kappa_1,\dots,\kappa_m$, some of which may be zero now, are unique. We refer to them again as the <u>factorization indices</u> of W. For the definition of minus and plus functions, see Section 3 in [BGK4]. Of course W_- and W_+ are (meant to be) m×m matrix functions.

In the situation of Theorem 3.1., the numbers $-\alpha_1,\dots,-\alpha_t$ are precisely the negative factorization indices of $W_\Theta(\lambda) = I_m + C(\lambda-A)^{-1}B$. This fact enables us to prove the following result due to I.S. Chibotaru [Ch].

COROLLARY 3.2. <u>Let</u> $\kappa_1,\dots,\kappa_m$ <u>be the factorization indices of</u> W <u>relative to the contour</u> Γ. <u>Then, for a suitable permutation</u> $\hat\kappa_1,\dots,\hat\kappa_m$ <u>of</u> $\kappa_1,\dots,\kappa_m$, <u>there exists a factorization</u>

<u>of the form</u>

$$(3.17) \qquad W(\lambda) = \begin{pmatrix} \left(\frac{\lambda-\varepsilon_1}{\lambda-\varepsilon_2}\right)^{\hat{\kappa}_1} & & \\ & \ddots & \\ & & \left(\frac{\lambda-\varepsilon_1}{\lambda-\varepsilon_2}\right)^{\hat{\kappa}_m} \end{pmatrix} \hat{W}_-(\lambda)\hat{W}_+(\lambda), \qquad \lambda \in \Gamma,$$

where $\hat{W}_-$ is a minus function relative to Γ and $\hat{W}_+$ is a plus function with respect to Γ.

Again $\hat{W}_-$ adn $\hat{W}_+$ are m×m matrix functions. The type of factorization appearing in (3.17) was considered for the first time in a paper by G.D. Birkkoff [B]. In contrast with how the situation is for Wiener-Hopf factorizations (3.16), the indices $\hat{\kappa}_1,\dots,\hat{\kappa}_m$ in (3.17) are not unique. Our proof will reflect this fact.

PROOF. We may assume that all factorization indices $\kappa_1,\dots,\kappa_m$ are strictly negative. If necessary, this can be achieved by multiplying W by a suitable negative power of $(\lambda-\varepsilon_1)/(\lambda-\varepsilon_2)$.

Write W as the transfer function of a finite dimensional node $\Theta = (A,B,C;X,\mathbb{C}^m)$ without spectrum on the contour Γ. Let $M,M^\times$ be the pair of spectral subspaces associated with Θ and let $\{e_{jk}\}_{k=1,j=1}^{\alpha_j\ \ t}$ be an outgoing basis for Θ with respect to the operator $A-\varepsilon_2$. The assumption that all factorization indices are strictly negative implies that $M+M^\times = X$, $t = m$ and $\kappa_j = -\alpha_j$, $j = 1,\dots,m$. This is clear from the results obtained in [BGK3] and [BGK4]. Choose $s_1,\dots,s_m$ and $\upsilon_1,\dots,\upsilon_m$ as in Theorem 3.1. To be more specific, $s_1,\dots,s_m$ is a permutation of $1,\dots,m$ and (3.3) and (3.7) are satisfied (with $t = m$). But then

$$\upsilon_{s_j} = -\kappa_j, \qquad j = 1,\dots,m,$$

and so $-\upsilon_1,\dots,-\upsilon_m$ is a permutation of $\kappa_1,\dots,\kappa_m$.

Let D be given by (3.1) with $\upsilon_1,\dots,\upsilon_m$ as indicated above, and choose a finite dimensional realization Θ_D of D having no spectrum on Γ. For instance one can take Θ_D as in the proof of Theorem 3.1. If $\tilde{M},\tilde{M}^{\times}$ is the pair of spectral subspace associated with the product node $\tilde{\Theta} = \Theta_D\Theta$, then (3.5) and (3.6) are satisfied. Now $M+M^{\times} = X$ and it follows that $\tilde{X} = \tilde{M} \oplus \tilde{M}^{\times}$. But then Theorem 2.1 in Ch.I of [BGK3] guarantees that the transfer function $\tilde{W}$ of $\tilde{\Theta}$ admits a canonical (right) Wiener-Hopf factorization with respect to the contour Γ, i.e., a factorization for which all factorization indices are zero. Since $\tilde{W}$ is the product of D and W, we may conclude that W admits a factorization of the form described in the corollary. Indeed, for $\hat{\kappa}_1,\dots,\hat{\kappa}_m$ one can just take the integers $-\upsilon_1,\dots,-\upsilon_m$ (in this order). □

The diagonal factor in (3.17) is on the left. By the same type of argument one can obtain a factorization where it is on the right. Also there are factorizations in which the roles of $\hat{W}_-$ and $\hat{W}_+$ are interchanged. We omit the details.

REFERENCES

[BGK1] Bart, H., Gohberg, I., Kaashoek, M.A.: Minimal factorization of matrix and operator functions. Operator Theory: Advances and Applications, Vol.1. Basel-Boston-Stuttgart, Birkhäuser Verlag, 1979.

[BGK2] Bart, H., Gohberg, I., Kaashoek, M.A.: Wiener-Hopf factorization of analytic operator functions and realization. Rapport 231, Wiskundig Seminarium, Vrije Universiteit, Amsterdam, 1983.

[BGK3] Bart, H., Gohberg, I., Kaashoek, M.A.: Explicit Wiener-Hopf factorization and realization. This volume.

[BGK4] Bart, H., Gohberg, I., Kaashoek, M.A.: Invariants for Wiener-Hopf equivalence of analytic operator functions. This volume.

[B] Birkhoff, G.D.: A theorem on matrices of analytic functions. Math. Ann. 74, 122-133 (1913).

[Ch] Chibotaru, I.S.: The reduction of systems of Wiener-Hopf equations to systems with vanishing indices. Bull. Akad. Stiince RSS Moldoven 8, 54-66 (1967) [Russian].

[CG] Clancey, K., Gohberg, I.: Factorization of matrix functions and singular integral operators. Operator Theory: Advances and Applications., Vol.3, Basel-Boston-Stuttgart, Birkhäuser Verlag, 1981.

H. Bart
Econometrisch Instituut
Erasmus Universiteit
Postbus 1738
3000 DR Rotterdam
The Netherlands

I. Gohberg
Dept of Mathematical Sciences
The Raymond and Beverly Sackler
Faculty of Exact Sciences
Tel-Aviv University
Ramat-Avid Israel

M.A. Kaashoek
Subfaculteit Wiskunde en
Informatica
Vrije Universiteit
Postbus 7161
1007 MC Amsterdam
The Netherlands

Operator Theory:
Advances and Applications, Vol. 21

SYMMETRIC WIENER-HOPF FACTORIZATION OF SELFADJOINT RATIONAL MATRIX FUNCTIONS AND REALIZATION

M.A. Kaashoek and A.C.M. Ran [1)]

Explicit formulas for a symmetric Wiener-Hopf factorization of a selfadjoint rational matrix function are constructed. The formulas are given in terms of realizations that are selfadjoint with respect to a certain indefinite inner product. The construction of the formulas is based on the method of Wiener-Hopf factorization developed in [2].

0. INTRODUCTION AND SUMMARY

<u>0.1 Introduction</u>. According to a relative recent result in the theory of Wiener-Hopf factorization (see [7]) any $m \times m$ selfadjoint matrix function $W(\lambda)$ which is continuous and has a nonzero determinant on the extended real line, admits a factorization of the form

$$(0.1)\qquad W(\lambda) = W_+(\bar{\lambda})^* \Sigma(\lambda) W_+(\lambda), \quad -\infty \le \lambda \le \infty,$$

where $W_+(\lambda)$ is analytic in the open upper half plane and continuous up to the real line, $\det W_+(\lambda) \neq 0$ for $\operatorname{Im}\lambda \ge 0$ (including $\lambda = \infty$) and $\Sigma(\lambda)$ is a selfadjoint block matrix of the following type:

$$(0.2)\qquad \Sigma(\lambda) = \begin{pmatrix} & & & & & & \left(\frac{\lambda-i}{\lambda+i}\right)^{\kappa_1} I_{m_1} \\ & & & & & \iddots & \\ & & & & \left(\frac{\lambda-i}{\lambda+i}\right)^{\kappa_r} I_{m_r} & & \\ & & & \begin{matrix} I_p & 0 \\ 0 & -I_q \end{matrix} & & & \\ & & \left(\frac{\lambda+i}{\lambda-i}\right)^{\kappa_r} I_{m_r} & & & & \\ & \iddots & & & & & \\ \left(\frac{\lambda+i}{\lambda-i}\right)^{\kappa_1} I_{m_1} & & & & & & \end{pmatrix}.$$

---

1) Research of second author supported by the Niels Stensen Stichting at Amsterdam .

Here $0 < \kappa_1 < \ldots < \kappa_r$ are positive integers and the non-negative numbers p and q are determined by the equalities $p-q = \text{signature}\, W(\lambda)$ (which does not depend on λ) and $p+q+2(m_1+\ldots+m_r) = m$. We shall call the factorization (0.1) a symmetric (Wiener-Hopf) factorization. A proof of this symmetric factorization theorem may also be found in [4].

In this paper we shall present an explicit construction of the symmetric factorization for the case when the matrix function $W(\lambda)$ is rational. In particular, for a selfadjoint rational matrix function $W(\lambda)$ we shall give explicit formulas for the factor $W_+(\lambda)$, for its inverse $W_+(\lambda)^{-1}$ and for the indices $\kappa_1,\ldots,\kappa_r$ and the numbers p and q.

To obtain our formulas we use the geometric construction of the Wiener-Hopf factorization carried out in [2] (see also [3]). As in [2] the starting point for the construction is a realization of $W(\lambda)$, i.e., an expression of $W(\lambda)$ in the form

$$(0.3) \qquad W(\lambda) = D + C(\lambda I_n - A)^{-1}B, \quad -\infty < \lambda < \infty,$$

and the final results are stated in terms of certain operators which we derive from A, B, C and D and certain invariant subspaces of A and $A - BD^{-1}C$. In order to obtain the symmetric factorization (0.1) by using the construction of [2] it is necessary to develop further the construction given in [2] and to modify it such that it reflects the symmetry of the factorization.

The selfadjointness of the matrix function $W(\lambda)$ allows one (see [5,8]) to choose the realization (0.3) in such a way that $HA = A^*H$ and $HB = C^*$ for some invertible selfadjoint $n \times n$ matrix H. The indefinite inner product on $\mathbb{C}^n$ induced by the invertible selfadjoint operator H will play an essential role in the construction of the symmetric factorization.

This paper is divided into two chapters. In Chapter I we review and modify the construction of the Wiener-Hopf factorization of an arbitrary rational matrix function given in [2]. This first chapter is organized in such a way that the results for the selfadjoint case, which are derived in the second chapter, appear by specifying further the operators constructed in Chapter I.

0.2 Summary. In this subsection $W(\lambda)$ is an $m \times m$ selfadjoint rational matrix function which does not have poles and zeros on the real line including infinity. To construct a symmetric factorization of $W(\lambda)$ we use the fact that $W(\lambda)$ can be represented (see [5,8]; also Chapter 2 in [1]) in the form

$$(0.4) \qquad W(\lambda) = D + C(\lambda - A)^{-1}B, \quad -\infty \le \lambda \le \infty,$$

where A, B, C and D are matrices of sizes $n \times n$, $m \times n$, $n \times m$ and $m \times m$, respectively, D is selfadjoint and invertible, A has no eigenvalues on the real line and for some invertible selfadjoint $n \times n$ matrix H the following indentities hold true:

$$(0.5) \qquad HA = A^*H, \quad HB = C^*.$$

The fact that $W(\lambda)$ has no zeros on the real line implies (see [1]) that like A the matrix $A^\times := A - BD^{-1}C$ has also no eigenvalues on the real line. The interplay between the spectral properties of A and $A^\times$ will be an essential feature of our construction of a symmetric factorization.

First we consider the case of canonical factorization, when the middle term $\Sigma(\lambda)$ in (0.1) is a constant signature matrix (i.e., the numbers $m_1, \ldots, m_r$ in (0.2) are all zero). Let M be the subspace of $\mathbb{C}^n$ spanned by the eigenvectors and generalized eigenvectors of A corresponding to eigenvalues in the upper half plane, and let $M^\times$ be the space spanned by the eigenvectors and generalized eigenvectors of $A^\times$ corresponding to eigenvalues in the lower half plane. The identities (0.5) imply that $HM = M^\perp$ and $HM^\times = (M^\times)^\perp$. Thus $H(M \cap M^\times) = (M + M^\times)^\perp$, and it follows that

$$\dim M \cap M^\times = \operatorname{codim} M + M^\times = \dim \frac{\mathbb{C}^n}{M + M^\times}.$$

In particular, $M \cap M^\times = (0)$ if and only if $\mathbb{C}^n$ is the direct sum of M and $M^\times$. The next theorem is a corollary of our main factorization theorem.

THEOREM 0.1. *The rational matrix function* $W(\lambda)$ *admits a symmetric canonical Wiener-Hopf factorization if and only if* $M \cap M^\times = (0)$, *and in that case such a factorization is given by*

$$(0.6) \qquad W(\lambda) = [E + ED^{-1}C\Pi(\bar{\lambda}-A)^{-1}B]^*\Sigma[E + ED^{-1}C\Pi(\lambda-A)^{-1}B],$$

where Π *is the projection of* $\mathbb{C}^n = M \oplus M^\times$ *along* M *onto* $M^\times$ *and* Σ *is a constant signature matrix which is congruent to* D, *the congruency being given by the invertible matrix* E (*i.e.*, $D = E^*\Sigma E$).

The case of non-canonical factorization (i.e., $M \cap M^\times \neq (0)$) is much more involved. First we choose subspaces $N \subset M$ and $N^\times \subset M^\times$ such that N (resp. $N^\times$) is a direct complement of $M \cap M^\times$ in M (resp. $M^\times$). The identities (0.5) allow us to construct a direct complement K of $M + M^\times$ in $\mathbb{C}^n$ such that $(HK)^\perp = N \oplus K \oplus N^\times$. In particular,

(0.7) $\mathbb{C}^n = N \oplus M \cap M^\times \oplus K \oplus N^\times.$

For $i = 1,2,3,4$ let Q_i be the projection onto the i-th subspace in the right hand side of (0.7) along the other subspaces in this direct sum decomposition of $\mathbb{C}^n$. The projections Q_i are related as follows

(0.8) $HQ_1 = Q_4^* H, \quad HQ_2 = Q_3^* H.$

In $M \cap M^\times$ one can choose (see [2], Section I.5) bases $\{d_{jk}\}_{k=1,j=1}^{\alpha_j \ \ s}$ and $\{e_{jk}\}_{k=1,j=1}^{\alpha_j \ \ s}$ such that

(i) $1 \le \alpha_1 \le \dots \le \alpha_s$;

(ii) $(A-i)d_{jk} = d_{j,k+1}$ for $k = 1,\dots,\alpha_j-1$;

(ii)' $(A+i)e_{jk} = e_{j,k+1}$ for $k = 1,\dots,\alpha_j-1$;

(iii) $\{d_{jk}\}_{k=1\ ,\ j=1}^{\alpha_j-1 \ \ s}$ and $\{e_{jk}\}_{k=1\ ,\ j=1}^{\alpha_j-1 \ \ s}$ are bases of $M \cap M^\times \cap \operatorname{Ker} C$;

and furthermore the vectors d_{jk} and e_{jk} are connected by

(0.9) $$e_{jk} = \sum_{\nu=0}^{k-1} \binom{k-1}{\nu} (2i)^\nu d_{j,k-\nu}.$$

THEOREM 0.2. *Let* $W(\lambda) = (W_+(\bar{\lambda}))^* \Sigma(\lambda) W_+(\lambda)$ *be a symmetric factorization. Then the indices* $\kappa_1,\dots,\kappa_r$ *appearing in the description* (0.2) *of the middle term are precisely the distinct numbers in the sequence* $\alpha_1,\dots,\alpha_s$ *and the numbers* $m_1,\dots,m_r$ *in* (0.2) *are determined by*

(0.10) $m_j = \#\{i \mid \alpha_i = \kappa_j\}, \quad j = 1,\dots,r.$

In particular, $m_1 + \dots + m_r = s$.

From (0.9) one sees that $Cd_{j\alpha_j} = Ce_{j\alpha_j}$ for $j = 1,\dots,s$. Put $z_j = D^{-1}Cd_{j\alpha_j} = D^{-1}Ce_{j\alpha_j}$. Then

$$Bz_j = BD^{-1}Cd_{j\alpha_j} = (A - A^\times)d_{j\alpha_j} \in M + M^\times, \quad j = 1,\dots,s,$$

and thus $z_1,\dots,z_s$ is a linearly independent set of vectors in $B^{-1}(M + M^\times) \subset C^m$. We shall prove that $z_1,\dots,z_s$ can be extended to a basis $z_1,\dots,z_m$ of C^m such that $z_1,\dots,z_{m-s}$ is a basis of $B^{-1}(M + M^\times)$ and

$$
(0.11)\qquad [\langle Dz_j, z_k\rangle]_{j,k=1}^{m} = \begin{pmatrix} & & & & & & I_{m_1} \\ & & & & & \cdot^{\cdot^{\cdot}} & \\ & & & & I_{m_r} & & \\ & & & \begin{matrix} I_p & 0 \\ 0 & -I_q \end{matrix} & & & \\ & & I_{m_r} & & & & \\ & \cdot^{\cdot^{\cdot}} & & & & & \\ I_{m_1} & & & & & & \end{pmatrix}
$$

where the positive numbers $m_1,\dots,m_r$ are given by (0.10).

We use the bases $\{d_{jk}\}$ and $\{e_{jk}\}$ to introduce the so-called <u>outgoing operators</u> (see [2], Section I.5) $S_1, S_2 : M \cap M^{\times} \to M \cap M^{\times}$, as follows:

$$(S_1 - i)d_{jk} = d_{j,k+1}, \quad k = 1,\dots,\alpha_j,$$

$$(S_2 + i)e_{jk} = e_{j,k+1}, \quad k = 1,\dots,\alpha_j,$$

where, by definition, $d_{jk} = e_{jk} = 0$ for $k = \alpha_j+1$. Thus relative to the basis $\{d_{jk}\}$ (resp. $\{e_{jk}\}$) the operator S_1 (resp. S_2) has a Jordan normal form with i (resp. $-i$) as the only eigenvalue and with blocks of sizes $\alpha_1,\dots,\alpha_s$. With S_1, S_2 we associate operators $G_1, G_2 : \mathbb{C}^m \to M \cap M^{\times}$ by setting

$$Q_2(A^{\times} - S_1 + G_1 D^{-1} C)x = 0, \quad x \in M \cap M^{\times};$$

$$Q_2(A - S_2 - G_2 D^{-1} C)x = 0, \quad x \in M \cap M^{\times};$$

$$Q_2(B - G_1 - G_2)z_j = (S_2 - S_1)d_{j\alpha_j}, \quad j = 1,\dots,s;$$

$$Q_2(B - G_1 - G_2)z_j = 0, \quad j = s+1,\dots,m.$$

Next, we define operators $T_1, T_2 : K \to K$ and $F_1, F_2 : K \to \mathbb{C}^m$ by

$$Q_3 T_1 Q_3 = H^{-1}(Q_2 S_2 Q_2)^{*} H, \quad F_1 Q_3 = -(Q_2 G_2)^{*} H,$$

$$Q_3 T_2 Q_3 = H^{-1}(Q_2 S_1 Q_2)^{*} H, \quad F_2 Q_3 = -(Q_2 G_1)^{*} H.$$

Note that T_1 has i as its only eigenvalue and $-i$ is the only eigenvalue of T_2. The next theorem is our final result.

THEOREM 0.3. *The selfadjoint rational matrix function* $W(\lambda) = D + C(\lambda - A)^{-1}B$ *admits the following symmetric factorization*

$$W(\lambda) = W_+(\bar{\lambda})^* \begin{pmatrix} & & & & & & \left(\frac{\lambda-i}{\lambda+i}\right)^{\kappa_1} I_{m_1} \\ & & & & & \cdot^{\cdot^{\cdot}} & \\ & & & & \left(\frac{\lambda-i}{\lambda+i}\right)^{\kappa_r} I_{m_r} & & \\ & & & \begin{matrix} I_p & 0 \\ 0 & -I_q \end{matrix} & & & \\ & & \left(\frac{\lambda+i}{\lambda-i}\right)^{\kappa_r} I_{m_r} & & & & \\ & \cdot^{\cdot^{\cdot}} & & & & & \\ \left(\frac{\lambda+i}{\lambda-i}\right)^{\kappa_1} I_{m_1} & & & & & & \end{pmatrix} W_+(\lambda),$$

where $\kappa_1,\dots,\kappa_r$ *are the distinct numbers in the sequence* $\alpha_1,\dots,\alpha_s$, *the number* m_j *is equal to the number of times the index* k_j *appears in the sequence* $\alpha_1,\dots,\alpha_s$, *the non-negative numbers* p *and* q *are determined by* $p-q =$ signature D *and* $p-q = m-2s$, *and the factor* $W_+(\lambda)$ *and its inverse* $W_+(\lambda)^{-1}$ *are given by*

$$\begin{aligned} W_+(\lambda) &= E + ED^{-1}(CQ_3 + CQ_4 + F_2Q_3)(\lambda - A)^{-1}B + \\ &\quad + ED^{-1}CQ_2(\lambda - S_2)^{-1}V(\lambda - A)^{-1}B + ED^{-1}CQ_2(\lambda - S_2)^{-1}(Q_2B - G_2), \end{aligned}$$

$$\begin{aligned} W_+(\lambda)^{-1} &= E^{-1} - D^{-1}C(\lambda - A^{\times})^{-1}(Q_2B + Q_4B - G_2)E^{-1} + \\ &\quad - D^{-1}C(\lambda - A^{\times})^{-1}V^{\times}(\lambda - T_2)^{-1}Q_3BE^{-1} + \\ &\quad - D^{-1}(CQ_3 + F_2)(\lambda - T_2)^{-1}Q_3BE^{-1}. \end{aligned}$$

Here E *is the inverse of the matrix* $[z_1,\dots,z_m]$ *and*

$$\begin{aligned} V &= Q_2AQ_4 - G_2D^{-1}CQ_4 + \tfrac{1}{2}Q_2AQ_3 - \tfrac{1}{2}(Q_2B - G_1)D^{-1}(CQ_3 + F_2) + \\ &\quad - (Q_2B - G_1 - G_2)D^{-1}F_1, \end{aligned}$$

$$\begin{aligned} V^{\times} &= Q_4A^{\times}Q_3 - Q_4BD^{-1}F_2 + \tfrac{1}{2}Q_2AQ_3 - \tfrac{1}{2}(Q_2B - G_1)D^{-1}(CQ_3 + F_2) + \\ &\quad - G_1D^{-1}(CQ_3 + F_2). \end{aligned}$$

A somewhat less general version of Theorem 0.3 has appeared in Chapter V of [9]. The fact that $W(\lambda)$ is rational is not essential for our formulas. Theorem 0.3 also holds true for a selfadjoint matrix function which is analytic in a neighbourhood of the real line and admits a representation of the form (0.4), where A is now allowed to be a bounded linear operator acting on an infinite dimensional Hilbert space. Of course in that case it is necessary to assume that $\dim M \cap M^{\times}$ is finite.

I. WIENER-HOPF FACTORIZATION

Throughout this chapter $W(\lambda)$ is a rational $m\times m$ matrix function which does not have poles and zeros on the real line. By definition (see [6,4]) a (right) Wiener-Hopf factorization of $W(\lambda)$ relative to the real line is a representation of $W(\lambda)$ in the form

$$W(\lambda) = W_-(\lambda)\begin{bmatrix} \left(\frac{\lambda-i}{\lambda+i}\right)^{\gamma_1} & & \\ & \ddots & \\ & & \left(\frac{\lambda-i}{\lambda+i}\right)^{\gamma_m}\end{bmatrix} W_+(\lambda), \qquad -\infty \le \lambda \le \infty,$$

where $W_-(\lambda)$ and $W_+(\lambda)$ are rational matrix functions such that $W_+(\lambda)$ (resp. $W_-(\lambda)$) has no poles and zeros in the closed upper (resp. lower) halfplane, including the point infinity. In this chapter we review and modify the construction of the Wiener-Hopf factorization given in [2].

I.1 Realizations with centralized singularities

The first step of the construction in [2] is to represent $W(\lambda)$ in the form

(1.1) $$W(\lambda) = D + C(\lambda - A)^{-1}B, \quad -\infty \le \lambda \le \infty.$$

Here A, B, C and D are matrices of sizes $n\times n$, $n\times m$, $m\times n$ and $n\times m$, respectively, the matrix A has no eigenvalues on the real line and D is invertible. The fact that $W = W(\cdot)$ has no zeros on the real line implies (see [1]) that $A^\times := A - BD^{-1}C$ has no eigenvalues on the real line. Given (1.1) the sextet $\theta = (A,B,C,D;\mathbb{C}^n,\mathbb{C}^m)$ is called a realization of W with main operator A and state space $\mathbb{C}^n$. The matrix $A^\times$ is called the associate main operator of θ. (In what follows we shall often identify a $p\times q$ matrix with its canonical action as an operator from $\mathbb{C}^q$ into $\mathbb{C}^p$.)

In [2] the Wiener-Hopf factorization of W is obtained by using the geometric factorization formulas of Theorem 1.1 in [1]. This requires a realization of W with so-called centralized singularities. The realization $\theta = (A,B,C,D;\mathbb{C}^n,\mathbb{C}^m)$ of W is said to have centralized singularities if the following properties hold true (see [2], Section I.3):

(i) the state space $\mathbb{C}^n$ has a decomposition $\mathbb{C}^n = X_1 \oplus X_2 \oplus X_3 \oplus X_4$ and relative to this decomposition the operators A, $A^\times$, B and C can be written as

$$(1.2)\quad A = \begin{pmatrix} A_1 & * & * & * \\ 0 & A_2 & 0 & * \\ 0 & 0 & A_3 & * \\ 0 & 0 & 0 & A_4 \end{pmatrix}, \quad A^\times = \begin{pmatrix} A_1^\times & 0 & 0 & 0 \\ * & A_2^\times & 0 & 0 \\ * & 0 & A_3^\times & 0 \\ * & * & * & A_4^\times \end{pmatrix},$$

$$B = \begin{pmatrix} B_1 \\ B_2 \\ B_3 \\ B_4 \end{pmatrix}, \qquad C = \begin{pmatrix} C_1 & C_2 & C_3 & C_4 \end{pmatrix},$$

where the entries satisfy the following conditions:

(ii) the eigenvalues of A_1 and $A_1^\times$ are in the upper half plane, the eigenvalues of A_4 and $A_4^\times$ are in the lower half plane;

(iii) $A_2 - i$ and $A_2^\times + i$ have the same nilpotent Jordan form in bases $\{d_{jk}\}_{k=1}^{\alpha_j}{}_{j=1}^{t}$ and $\{e_{jk}\}_{k=1}^{\alpha_j}{}_{j=1}^{t}$, respectively, more precisely:

$$(1.3.i)\quad (A_2 - i)d_{jk} = d_{j,k+1}, \quad k = 1,\dots,\alpha_j \quad (d_{j,\alpha_j+1} = 0),$$

$$(1.3.ii)\quad (A_2^\times + i)e_{jk} = e_{j,k+1}, \quad k = 1,\dots,\alpha_j \quad (e_{j,\alpha_j+1} = 0),$$

where it is assumed that $\alpha_1 \le \dots \le \alpha_t$; and further these two bases are related by

$$(1.4.i)\quad e_{jk} = \sum_{\nu=0}^{k-1} \binom{k-1}{\nu} (2i)^\nu d_{j,k-\nu},$$

$$(1.4.ii)\quad d_{jk} = \sum_{\nu=0}^{k-1} \binom{k-1}{\nu} (-2i)^\nu e_{j,k-\nu};$$

(iv) $\operatorname{rank} B_2 = \operatorname{rank} C_2 = t$;

(v) $A_3^\times - i$ and $A_3 + i$ have the same nilpotent Jordan form in bases $\{f_{jk}\}_{k=1}^{\omega_j}{}_{j=1}^{s}$ and $\{g_{jk}\}_{k=1}^{\omega_j}{}_{j=1}^{s}$, respectively, more precisely:

$$(1.5.i)\quad (A_3^\times - i)f_{jk} = f_{j,k+1}, \quad k = 1,\dots,\omega_j \quad (f_{j,\omega_j+1} = 0),$$

$$(1.5.ii)\quad (A_3 + i)g_{jk} = g_{j,k+1}, \quad k = 1,\dots,\omega_j \quad (g_{j,\omega_j+1} = 0),$$

where it is assumed that $\omega_1 \le \dots \le \omega_s$; and further, these two bases are related by

$$(1.6.i)\quad g_{jk} = \sum_{\nu=0}^{k-1} \binom{\nu+\omega_j-k}{\nu} (-2i)^\nu f_{j,k-\nu},$$

$$(1.6.ii)\quad f_{jk} = \sum_{\nu=0}^{k-1} \binom{\nu+\omega_j-k}{\nu} (2i)^\nu g_{j,k-\nu};$$

(vi) $\operatorname{rank} B_3 = \operatorname{rank} C_3 = s$.

Note that the order of $\alpha_1,\dots,\alpha_t$ is different from the order used in [2]. With these notations one has the following theorem (see Theorem I.3.1 in [2]).

THEOREM 1.1 *Let* $\theta = (A,B,C,D;\mathbb{C}^n,\mathbb{C}^m)$ *be a realization of* W *with centralized singularities. Put*

$$W_-(\lambda) = D + C_1(\lambda - A_1)^{-1}B_1,$$
$$W_+(\lambda) = I + D^{-1}C_4(\lambda - A_4)^{-1}B_4,$$
$$D(\lambda) = I + D^{-1}C_2(\lambda - A_2)^{-1}B_2 + D^{-1}C_3(\lambda - A_3)^{-1}B_3.$$

Then $W(\lambda) = W_-(\lambda)D(\lambda)W_+(\lambda)$, $-\infty \le \lambda \le \infty$, *the factor* $W_+(\lambda)$ *(resp.* $W_-(\lambda)$*) has no poles and zeros in the closed upper (resp. lower) half plane and at infinity, and for some invertible* $m \times m$ *matrix* E

$$ED(\lambda)E^{-1} = \operatorname{diag}\left(\left(\frac{\lambda-i}{\lambda+i}\right)^{-\alpha_1},\dots,\left(\frac{\lambda-i}{\lambda+i}\right)^{-\alpha_t},1,\dots,1,\left(\frac{\lambda-i}{\lambda+i}\right)^{\omega_1},\dots,\left(\frac{\lambda-i}{\lambda+i}\right)^{\omega_s}\right).$$

In particular, modulo a basis transformation in $\mathbb{C}^m$, *the factorization* $W(\lambda) = W_-(\lambda)D(\lambda)W_+(\lambda)$ *is a right Wiener-Hopf factorization of* W *relative to the real line.*

Except for minor changes involving the operator D, the proof of the above theorem can be found in [2]. With Theorem 1.1 the problem to factorize W is reduced to the construction of a realization of W with centralized singularities. Such a construction, which has to start from the representation (1.1), will be carried out in the next sections. First we collect together the necessary data, on the basis of which, by dilating the original realization, we construct a new realization with centralized singularities.

I.2 Incoming data and related feedback operators

In this section $\theta = (A,B,C,D;\mathbb{C}^n,\mathbb{C}^m)$ is a realization of W. So A and $A^\times = A - BD^{-1}C$ have no real eigenvalues. Let P (resp. $P^\times$) be the spectral projection of A (resp. $A^\times$) corresponding to the eigenvalues in the upper half plane. Put $M = \operatorname{Im} P$ and $M^\times = \operatorname{Im}(I - P^\times)$.

We start this section by defining the sequence of incoming subspaces for θ, as follows:

$$H_j = M + M^\times + \operatorname{Im} B + \dots + \operatorname{Im} A^{j-1}B, \quad j = 0,1,2,\dots$$

Here $H_0 = M + M^\times$. Let ε be a complex number. Note that the spaces H_j do not change if A is replaced by either $A - \varepsilon$ or $A^\times - \varepsilon$. It follows that

$$H_1 + (A - \varepsilon)H_k = H_{k+1}.$$

Since $H_n = \mathbb{C}^n$, this identity can be used to construct an incoming basis for $A-\varepsilon$ (see [2], Section I.4). A set of vectors $\{f_{jk} \mid k=1,\dots,\omega_j,\ j=1,\dots,s\}$ is called an incoming basis with respect to $A-\varepsilon$ if

(i) $1 \le \omega_1 \le \dots \le \omega_s$;

(ii) $(A-\varepsilon)f_{jk} - f_{j,k+1} \in M + M^\times + \operatorname{Im} B$, $k=1,\dots,\omega_j$ where $f_{j,\omega_j+1} = 0$ by definition;

(iii) the vectors f_{jk}, $k=1,\dots,\omega_j$, $j=1,\dots,s$ form a basis for a complement of $M+M^\times$;

(iv) the vectors $f_{11},\dots,f_{s1}$ form a basis for $M + M^\times + \operatorname{Im} B$ modulo $M+M^\times$.

The numbers $\omega_1,\dots,\omega_s$ are called the incoming indices; they do not depend on either ε or the choice of the basis.

Let $\{f_{jk}\}_{k=1}^{\omega_j}{}_{j=1}^{s}$ be an incoming basis with respect to $A-\varepsilon$. Denote by K the span of the vectors f_{jk}. Then K is a complement to $M+M^\times$. Connected with this incoming basis is an incoming operator $T: K \to K$ given by $(T-\varepsilon)f_{jk} = f_{j,k+1}$ for $k=1,\dots,\omega_j$ (here, again, $f_{j,\omega_j+1} = 0$). Note that with respect to the basis $\{f_{jk}\}_{k=1}^{\omega_j}{}_{j=1}^{s}$ the operator T has Jordan normal form with ε as the only eigenvalue and Jordan blocks of size $\omega_1,\dots,\omega_s$.

The next proposition shows how a given incoming basis with parameter i may be transformed into an incoming basis with parameter -i (see [2], Proposition I.4.2).

PROPOSITION 2.1. *Let* $\{f_{jk}\}_{k=1}^{\omega_j}{}_{j=1}^{s}$ *be an incoming basis for* θ *with respect to* $A-i$. *Put*

$$g_{jk} = \sum_{\nu=0}^{k-1} \binom{\nu+\omega_j-k}{\nu}(-2i)^\nu f_{j,k-\nu}. \tag{2.1}$$

Then $\{g_{jk}\}_{k=1}^{\omega_j}{}_{j=1}^{s}$ *is an incoming basis for* θ *with respect to* $A+i$. *Further, if* T_1 *and* T_2 *are the incoming operators associated with these incoming bases, respectively, then* T_1 *and* T_2 *satisfy*

$$(T_1 - T_2)g_{jk} = -(-2i)^k\binom{\omega_j}{k}g_{j1} \tag{2.2}$$

for $k=1,\dots,\omega_j$, $j=1,\dots,s$.

By direct checking one proves the following analogue of (2.1):

$$f_{jk} = \sum_{\nu=0}^{k-1} \binom{\nu+\omega_j-k}{\nu}(2i)^\nu g_{j,k-\nu}. \tag{2.3}$$

It follows that (2.2) may be replaced by

$$(T_1 - T_2)f_{jk} = (2i)^k\binom{\omega_j}{k}f_{j1}. \tag{2.4}$$

Now let $\{f_{jk}\}_{k=1}^{\omega_j}{}_{j=1}^{s}$ be an incoming basis for $A-i$ and construct the incoming basis for $A+i$ by (2.1). Next, let $y_1,\dots,y_s$ be vectors in $\mathbb{C}^m$ such that for $j=1,\dots,s$:

$$f_{j1} - By_j = g_{j1} - By_j \in M + M^\times. \tag{2.5}$$

Since the vectors $f_{11},\dots,f_{s1}$ form a basis for $M + M^\times + \operatorname{Im} B$ modulo $M+M^\times$ such a set of vectors $\{y_1,\dots,y_s\}$ exists and is a basis for $\mathbb{C}^m$ modulo $B^{-1}[M+M^\times]$; in particular

$$\mathbb{C}^m = \operatorname{span}\{y_1,\dots,y_s\} \oplus B^{-1}[M+M^\times]. \tag{2.6}$$

In this way we have fixed a set of <u>incoming data</u>:

$$\{f_{jk}\}_{k=1}^{\omega_j}{}_{j=1}^{s}, \quad \{g_{jk}\}_{k=1}^{\omega_j}{}_{j=1}^{s}, \quad \{y_j\}_{j=1}^{s}. \tag{2.7}$$

Let K be the span of the vectors f_{jk}. Two operators $F_1,F_2 : K \to \mathbb{C}^m$ are called a <u>pair of feedback operators</u> corresponding to the incoming data (2.7) if

$$(A - T_1 + BD^{-1}F_1)x \in M + M^\times, \quad x \in K, \tag{2.8}$$

$$(A^\times - T_2 - BD^{-1}F_2)x \in M + M^\times, \quad x \in K, \tag{2.9}$$

$$D^{-1}(C + F_1 + F_2)f_{jk} = \binom{\omega_j}{k}(2i)^k y_j. \tag{2.10}$$

Such a pair of operators can be constructed as follows (see [2], Section I.6). First note that $(A-T_1)f_{jk} \in M + M^\times + \operatorname{Im} B$. So there exist vectors u_{jk} in $\mathbb{C}^m$ such that $(A-T_1)f_{jk} + Bu_{jk} \in M + M^\times$. Define $F_1 : K \to \mathbb{C}^m$ by setting $F_1 f_{jk} = Du_{jk}$. Then F_1 satisfies (2.8). Next, choose F_2 such that (2.10) holds. This defines F_2 uniquely. It remains to show that (2.9) holds. Indeed, using (2.4) and (2.10) we have:

$$\begin{aligned}
&(A^\times - T_2 - BD^{-1}F_2)f_{jk} = \\
&= (A - T_1 + BD^{-1}F_1)f_{jk} + (T_1 - T_2)f_{jk} - BD^{-1}(C + F_1 + F_2)f_{jk} = \\
&= (A - T_1 + BD^{-1}F_1)f_{jk} + (2i)^k\binom{\omega_j}{k}(f_{j1} - By_j) \in M + M^\times.
\end{aligned}$$

I.3. Outgoing data and related output injection operators

In this section we make the same assumptions concerning θ as in the first paragraph of the previous section.

We begin by considering the sequence of <u>outgoing subspaces</u> for θ, given by

$$K_j = M \cap M^\times \cap \operatorname{Ker} C \cap \dots \cap \operatorname{Ker} CA^{j-1}, \quad j = 0,1,2,\dots .$$

Here $K_0 = M \cap M^\times$. The spaces K_j do not change if A is replaced by $A - \varepsilon$ or $A^\times - \varepsilon$. Hence:

$$K_j \cap (A-\varepsilon)^{-1} K_j = K_{j+1}, \quad j = 0,1,2,\ldots .$$

Using this identity one can construct an outgoing basis for $A - \varepsilon$ (see [2], Section I.5). A set of vectors $\{d_{jk} \mid k = 1,\ldots,\alpha_j,\ j = 1,\ldots,t\}$ is called an <u>outgoing basis</u> with respect to $A - \varepsilon$ if

(i) $1 \le \alpha_1 \le \ldots \le \alpha_t$;

(ii) $(A-\varepsilon)d_{jk} = d_{j,k+1}$, $k = 1,\ldots,\alpha_j - 1$;

(iii) the vectors d_{jk}, $k = 1,\ldots,\alpha_j$, $j = 1,\ldots,t$ form a basis for $M \cap M^\times$;

(iv) the vectors d_{jk}, $k = 1,\ldots,\alpha_j - 1$, $j = 1,\ldots,t$ form a basis for K_1.

Note that the order of $\alpha_1,\ldots,\alpha_t$ is different from the order used in [2].

The numbers $\alpha_1,\ldots,\alpha_t$ are called the <u>outgoing indices</u>; they do not depend on either ε or the choice of the outgoing basis.

Connected with the outgoing basis $\{d_{jk}\}_{k=1}^{\alpha_j}{}_{j=1}^{t}$ is an <u>outgoing operator</u> $S : M \cap M^\times \to M \cap M^\times$ given by $(S-\varepsilon)d_{jk} = d_{j,k+1}$ for $k = 1,\ldots,\alpha_j$, $j = 1,\ldots,t$. Here $d_{j,\alpha_j+1} = 0$. The operator S has Jordan normal form with ε as the only eigenvalue and Jordan blocks of size $\alpha_1,\ldots,\alpha_t$ with respect to the basis $\{d_{jk}\}_{k=1}^{\alpha_j}{}_{j=1}^{t}$.

The next proposition is the analogue of Proposition 2.1 (see [2], Proposition 5.2).

PROPOSITION 3.1. *Let* $\{d_{jk}\}_{k=1}^{\alpha_j}{}_{j=1}^{t}$ *be an outgoing basis for* θ *with respect to* $A - i$. *Put*

$$e_{jk} = \sum_{\nu=0}^{k-1} \binom{k-1}{\nu} (2i)^\nu d_{j,k-\nu}. \tag{3.1}$$

Then $\{e_{jk}\}_{k=1}^{\alpha_j}{}_{j=1}^{t}$ *is an outgoing basis for* θ *with respect to* $A + i$. *Further, if* S_1 *and* S_2 *are the outgoing operators associated with these outgoing bases, respectively, then* S_1 *and* S_2 *satisfy*

$$\begin{cases} (S_1 - S_2)e_{j\alpha_j} = \sum_{\nu=0}^{\alpha_j - 1} \binom{\alpha_j}{\nu+1} (2i)^{\nu+1} d_{j,\alpha_j-\nu}, \\ (S_1 - S_2)x = 0 \quad \text{for } x \in M \cap M^\times \cap \operatorname{Ker} C. \end{cases} \tag{3.2}$$

By direct checking one also sees that

$$d_{jk} = \sum_{\nu=0}^{k-1} \binom{k-1}{\nu} (-2i)^\nu e_{j,k-\nu}. \tag{3.3}$$

Hence the first formula in (3.2) may be replaced by

(3.4) $$(S_1 - S_2)d_{j\alpha_j} = -\sum_{\nu=0}^{\alpha_j-1} \binom{\alpha_j}{\nu+1}(-2i)^{\nu+1} e_{j,\alpha_j-\nu}.$$

Let $\{d_{jk}\}_{k=1}^{\alpha_j}{}_{j=1}^{t}$ be an outgoing basis with respect to $A - i$, and let $\{e_{jk}\}_{k=1}^{\alpha_j}{}_{j=1}^{t}$ be the outgoing basis with respect to $A + i$ given by (3.1). Put $z_j = D^{-1}Ce_{j\alpha_j} = D^{-1}Cd_{j\alpha_j}$, $j = 1,\dots,t$. In this way we have fixed a set of <u>outgoing data</u>:

(3.5) $$\{d_{jk}\}_{k=1}^{\alpha_j}{}_{j=1}^{t}, \quad \{e_{jk}\}_{k=1}^{\alpha_j}{}_{j=1}^{t}, \quad \{z_j\}_{j=1}^{t}.$$

Note that the vectors $\{z_j\}_{j=1}^{t}$ form a basis of $D^{-1}C[M \cap M^\times]$. Since $BD^{-1}C[M \cap M^\times] \subset M + M^\times$, we have $z_j \in B^{-1}[M + M^\times]$ for $j = 1,\dots,t$. Let Y_0 be a complement of $\operatorname{span}\{z_j\}_{j=1}^{t}$ in $B^{-1}[M + M^\times]$. Furthermore, choose a complement Y_1 of $B^{-1}[M + M^\times]$ in $\mathbb{C}^m$.

Two operators G_1 and G_2 mapping $\mathbb{C}^m$ into $M \cap M^\times$ are called a <u>pair of output injection operators</u> corresponding to the outgoing data (3.5) if

(3.6) $$\rho(A^\times - S_1 + G_1D^{-1}C)x = 0, \qquad x \in M \cap M^\times,$$

(3.7) $$\rho(A - S_2 - G_2D^{-1}C)x = 0, \qquad x \in M \cap M^\times,$$

(3.8) $$\rho(B - G_1 - G_2)y = 0, \qquad y \in Y_0 \oplus Y_1,$$

(3.9) $$\rho(B - G_1 - G_2)z_j = (S_2 - S_1)d_{j\alpha_j}, \quad j = 1,\dots,t.$$

Here ρ is a projection of $\mathbb{C}^n$ onto $M \cap M^\times$, which we assume to be given in advance. Of course, the definition of G_1 and G_2 does not depend only on the outgoing data (3.5) and the related outgoing operators S_1 and S_2, but also on the choice of the complements Y_0 and Y_1 and the projection ρ.

To construct such a pair of operators, let G_2 be an operator for which $G_2z_j := \rho(A + i)e_{j\alpha_j}$ for $j = 1,\dots,t$. Then G_2 satisfies (3.7). Construct G_1 by (3.8) and (3.9). This determines G_1 completely, and it remains to show that (3.6) holds. To do this first note that (3.9) and (3.2) together imply that

$$\rho((B - G_1 - G_2)D^{-1}Cx - (S_2 - S_1)x) = 0, \quad x \in M \cap M^\times.$$

Subtracting this from (3.7) yields (3.6).

<u>I.4. Dilation to realizations with centralized singularities</u>

In this section we show by construction how an arbitrary realization θ of W may be dilated to a realization with centralized singularities. In the next three paragraphs we fix sets of incoming and outgoing date and the corresponding spaces and operators. On the basis of this information we shall introduce the dilation.

Throughout this section $\theta = (A,B,C,D;\mathbb{C}^n,\mathbb{C}^m)$ is an arbitrary realization of W. Furthermore, $\{e_{jk}\}_{k=1,j=1}^{\alpha_j \; t}$ and $\{d_{jk}\}_{k=1,j=1}^{\alpha_j \; t}$ are outgoing bases for $A+i$ and $A-i$, respectively, and $\{f_{jk}\}_{k=1,j=1}^{\omega_j \; s}$ and $\{g_{jk}\}_{k=1,j=1}^{\omega_j \; s}$ are incoming bases for $A-i$ and $A+i$, respectively. Put $z_j = D^{-1}Ce_{j\alpha_j}$, $j=1,\dots,t$, and let y_j, $j=1,\dots,s$, be such that $f_{j1} - By_j = g_{j1} - By_j \in M + M^{\times}$. In this way we have fixed a set of incoming data:

(4.1) $\qquad \{f_{jk}\}_{k=1,j=1}^{\omega_j \; s}, \quad \{g_{jk}\}_{k=1,j=1}^{\omega_j \; s}, \quad \{y_j\}_{j=1}^{s}$

and a set of outgoing data:

(4.2) $\qquad \{d_{jk}\}_{k=1,j=1}^{\alpha_j \; t}, \quad \{e_{jk}\}_{k=1,j=1}^{\alpha_j \; t}, \quad \{z_j\}_{j=1}^{t}.$

Let T_1, T_2 and S_1, S_2 be the corresponding incoming and outgoing operators, respectively.

The subspace of X spanned by the set of vectors $\{f_{jk}\}_{k=1,j=1}^{\omega_j \; s}$ will be denoted by K. Choose subspaces N and $N^{\times}$ such that $M = N \oplus (M \oplus M^{\times})$ and $M^{\times} = N^{\times} \oplus (M \oplus M^{\times})$. Then X has the following direct sum decomposition:

$$X = N \oplus (M \cap M^{\times}) \oplus K \oplus N^{\times}.$$

Denote by Q_j the projection on the j-th space in this decomposition along the other spaces $(j = 1,2,3,4)$.

Furthermore, let $F_1,F_2 : K \to \mathbb{C}^m$ be a pair of feedback operators corresponding to the incoming data (4.1), and let $G_1,G_2 : \mathbb{C}^m \to M \cap M^{\times}$ be a pair of output injection operators corresponding to the outgoing data (4.2). The operator ρ appearing in the formulas (3.6) - (3.9) (which define the operators G_1,G_2) is chosen to be Q_2 and for the space Y_1 in (3.9) we take the space spanned by the vectors $y_1,\dots,y_t$ (cf. formula (2.6)). The choice of the space Y_0 in (3.9) is not restricted. Put

(4.3) $\qquad \hat{X} = (M \cap M^{\times}) \oplus (M \cap M^{\times}) \oplus \mathbb{C}^n \oplus K \oplus K.$

Consider operators $\hat{A} : \hat{X} \to \hat{X}$, $\hat{B} : \mathbb{C}^m \to \hat{X}$ and $\hat{C} : \hat{X} \to \mathbb{C}^m$ given by

$$\hat{A} = \begin{pmatrix} S_1 & 0 & A_{10} & A_{13} & A_{14} \\ 0 & S_2 & A_{20} & A_{23} & A_{24} \\ 0 & 0 & A & A_{03} & A_{04} \\ 0 & 0 & 0 & T_1 & 0 \\ 0 & 0 & 0 & 0 & T_2 \end{pmatrix}, \qquad \hat{B} = \begin{pmatrix} G_1 \\ G_2 \\ B \\ 0 \\ 0 \end{pmatrix},$$

$$\hat{C} = \begin{pmatrix} 0 & 0 & C & F_1 & F_2 \end{pmatrix}.$$

Also, write for $\hat{A}^{\times} = \hat{A} - \hat{B}D^{-1}\hat{C}$:

$$\hat{A}^{\times} = \begin{pmatrix} S_1 & 0 & A_{10}^{\times} & A_{13}^{\times} & A_{14}^{\times} \\ 0 & S_2 & A_{20}^{\times} & A_{23}^{\times} & A_{24}^{\times} \\ 0 & 0 & A^{\times} & A_{03}^{\times} & A_{04}^{\times} \\ 0 & 0 & 0 & T_1 & 0 \\ 0 & 0 & 0 & 0 & T_2 \end{pmatrix} .$$

Here S_1, S_2, T_1, T_2, F_1, F_2, G_1 and G_2 are defined in the preceding paragraphs. The sextet $\hat{\theta} = (\hat{A},\hat{B},\hat{C},D;\hat{X},\mathbb{C}^m)$ is a dilation of θ and hence again a realization of W. We shall show that for a suitable choice of the operators A_{ij} appearing in the formula for $\hat{A}$ the realization $\hat{\theta}$ has centralized singularities.

THEOREM 4.1. *Define the operators* A_{ij} *as follows. First, set*

(4.4) $\quad A_{10}(Q_1+Q_2) = 0, \quad A_{10}Q_4 = Q_2A^{\times}Q_4 + G_1D^{-1}CQ_4;$

(4.5) $\quad (Q_3+Q_4)A_{04} = 0, \quad Q_1A_{04}Q_3 = -Q_1A^{\times}Q_3 + Q_1BD^{-1}F_2Q_3;$

(4.6) $\quad A_{20}(Q_2+Q_4) = G_2D^{-1}C(Q_2+Q_4), \quad A_{20}Q_1 = Q_2AQ_1;$

(4.7) $\quad (Q_1+Q_3)A_{03} = (Q_1+Q_3)BD^{-1}F_1, \quad Q_4A_{03}Q_3 = -Q_4AQ_3$

Next, put

(4.8) $\quad A_{14} = 0, \quad A_{23} = G_2D^{-1}F_1, \quad A_{13} = \Lambda,$

where Λ *is an arbitrary operator from* K *into* $M \cap M^{\times}$. *Finally, let*

(4.9) $\quad A_{10}Q_3 = -\Lambda, \quad A_{20}Q_3 = Q_2AQ_3 + \Lambda + (Q_2B - G_1 - G_2)D^{-1}F_1;$

(4.10) $\quad Q_2A_{03} = \Lambda + (Q_2B - G_1)D^{-1}F_1;$

(4.11) $\quad Q_2A_{04} = -A_{20}Q_3 + G_2D^{-1}(CQ_3 + F_2), \quad A_{24} = Q_2A_{04}.$

Then $\hat{\theta} = (\hat{A},\hat{B},\hat{C},D;\hat{X},\mathbb{C}^m)$ *is a realization of* W *with centralized singularities.*

To prove Theorem 4.1 we start with an investigation of the spectral properties of $\hat{A}$ and $\hat{A}^{\times}$. Note that whatever the choice is of the operators A_{ij} the operators $\hat{A}$ and $\hat{A}^{\times}$ have no real eigenvalues. Let $\hat{P}$ and $\hat{P}^{\times}$ be the spectral projections of $\hat{A}$ and $\hat{A}^{\times}$, respectively, corresponding to the upper half plane. Then

$$\hat{P} = \begin{pmatrix} I & 0 & P_{10} & P_{13} & P_{14} \\ 0 & 0 & P_{20} & P_{23} & P_{24} \\ 0 & 0 & P & P_{03} & P_{04} \\ 0 & 0 & 0 & I & 0 \\ 0 & 0 & 0 & 0 & 0 \end{pmatrix}, \quad I - \hat{P}^{\times} = \begin{pmatrix} 0 & 0 & P_{10}^{\times} & P_{13}^{\times} & P_{14}^{\times} \\ 0 & I & P_{20}^{\times} & P_{23}^{\times} & P_{24}^{\times} \\ 0 & 0 & (I-P^{\times}) & P_{03}^{\times} & P_{04}^{\times} \\ 0 & 0 & 0 & 0 & 0 \\ 0 & 0 & 0 & 0 & I \end{pmatrix}$$

where P (resp., $P^{\times}$) is the spectral projection on A (resp., $A^{\times}$) corresponding to the upper half plane.

LEMMA 4.2. *Independent of the choice of the operators* A_{ij} *we have:*

$$\hat{M} = \left\{ \begin{pmatrix} u \\ P_{20}m+P_{23}x \\ m+P_{03}x \\ x \\ 0 \end{pmatrix} \;\middle|\; m \in M,\ x \in K,\ u \in M \cap M^{\times} \right\},$$

$$\hat{M}^{\times} = \left\{ \begin{pmatrix} P_{10}^{\times}m^{\times}+P_{14}^{\times}x \\ u \\ m^{\times}+P_{04}^{\times}x \\ 0 \\ x \end{pmatrix} \;\middle|\; m^{\times} \in M,\ x \in K,\ u \in M \cap M^{\times} \right\},$$

$$\hat{M} \cap \hat{M}^{\times} = \left\{ \begin{pmatrix} P_{10}^{\times}a \\ P_{20}a \\ a \\ 0 \\ 0 \end{pmatrix} \;\middle|\; a \in M \cap M^{\times} \right\},$$

$$\hat{M} + \hat{M}^{\times} = \left\{ \begin{pmatrix} x_1 \\ x_2 \\ z+P_{03}x_3+P_{04}^{\times}x_4 \\ x_3 \\ x_4 \end{pmatrix} \;\middle|\; x_1,x_2 \in M \cap M^{\times},\ x_3,x_4 \in K,\ z \in M+M^{\times} \right\}.$$

PROOF. Suppose $\hat{x} = (x_1,x_2,x_0,x_3,x_4)^T \in \operatorname{Im}\hat{P}$. Then $x_4 = 0$. Since $\hat{P}\hat{x} = \hat{x}$ the vector $\hat{x}$ satisfies the identities

$$P_{10}x_0 + P_{13}x_3 = 0,$$

$$P_{20}x_0 + P_{23}x_3 = x_2,$$
$$Px_0 + P_{03}x_3 = x_0.$$

From $\hat{P}^2 = \hat{P}$ one obtains

$$(4.12)\qquad \begin{aligned} &P_{20}P_{03} = 0, \quad P_{10}P = 0, \quad PP_{03} = 0,\\ &P_{10}P_{03} + P_{13} = 0, \; P_{20}P = P_{20}. \end{aligned}$$

So $P_{20}x_0 = P_{20}Px_0$, and putting $m = Px_0$, $x = x_3$ one obtains that $\hat{x}$ has the desired form.

Conversely, for every $m \in M$, $x \in K$ and $u \in M \cap M^\times$ one has

$$\hat{P}\begin{pmatrix} u \\ P_{20}m+P_{23}x \\ m+P_{03}x \\ x \\ 0 \end{pmatrix} = \begin{pmatrix} u+P_{10}m+P_{10}P_{03}x+P_{13}x \\ P_{20}m+P_{20}P_{03}x+P_{23}x \\ Pm+PP_{03}x+P_{03}x \\ x \\ 0 \end{pmatrix} = \begin{pmatrix} u \\ P_{20}m+P_{23}x \\ m+P_{03}x \\ x \\ 0 \end{pmatrix},$$

according to the formulas (4.12). Hence $\hat{M}$ has the desired form.

Using the fact that $\hat{P}^\times$ is a projection one obtains the formula for $\hat{M}^\times$. The formulas for $\hat{M} \cap \hat{M}^\times$ and $\hat{M} + \hat{M}^\times$ now easily follow. □

LEMMA 4.3. *Assume* (4.4) - (4.7) *hold. Then*

(i) $A_{10}P = 0, \quad (I - P)A_{04} = 0;$

(ii) $P^\times A_{03}^\times = 0, \quad A_{20}^\times(I - P^\times) = 0;$

(iii) $P_{10}^\times = Q_2(I - P^\times), \quad P_{20} = Q_2P;$

(iv) $P_{03} = (I - P)Q_3, \quad P_{04}^\times = P^\times Q_3.$

PROOF. From (4.4) and (4.6) it is immediately clear that $A_{10}P = 0$ and $A_{20}^\times(I - P^\times) = 0$, because $\operatorname{Im} P = \operatorname{Im}(Q_1 + Q_2)$, $\operatorname{Im}(I - P^\times) = \operatorname{Im}(Q_2 + Q_4)$. Further, (4.5) implies $\operatorname{Im} A_{04} \subset \operatorname{Im}(Q_1 + Q_2) = \operatorname{Ker}(I - P)$, so $(I - P)A_{04} = 0$. Similarly, $P^\times A_{03}^\times = 0$. This proves (i) and (ii).

From (iii) and (iv) we only show $P_{20} = Q_2P$, the other formulas are obtained similarly. From $\hat{A}\hat{P} = \hat{P}\hat{A}$ one obtains $P_{20}A = S_2P_{20} + A_{20}P$. Since $P_{20}P = P_{20}$, it follows that P_{20} is a solution of the Lijapunov equation.

$$P_{20}PAP - S_2P_{20} = A_{20}P.$$

Now PAP and S_2 have no common eigenvalues, so P_{20} is the unique solution of this equation. We compute $A_{20}P$ from (4.6). Take $x \in M$, then

$$A_{20}x = A_{20}(Q_1 + Q_2)x = Q_2AQ_1x + G_2D^{-1}CQ_2x.$$

According to (3.7) $G_2D^{-1}CQ_2x = Q_2(A - S_2)Q_2x$, so

$$A_{20}x = Q_2Ax - S_2Q_2x.$$

It follows that Q_2P solves the Lijapunov equation mentioned above. Hence $P_{20} = Q_2P$. □

LEMMA 4.4. *Assume* (4.4) - (4.11) *hold. Then*

$$P_{23} = -P_{20}Q_3, \quad P_{14}^{\times} = -P_{10}^{\times}Q_3.$$

PROOF. Take $x \in K$. From $\hat{A}\hat{P} = \hat{P}\hat{A}$ one sees that

$$\text{(4.13)} \quad \begin{aligned} &S_2P_{23} + A_{20}P_{03} + A_{23} = P_{20}A_{03} + P_{23}T_1; \\ &S_2P_{20} + A_{20}P = P_{20}A. \end{aligned}$$

Using these identities and Lemma 4.3 one obtains that $P_{20}Q_3 + P_{23}$ satisfies

$$S_2(P_{20} + P_{23})x - (P_{20} + P_{23})T_1x = Q_2P(A - T_1 + A_{03})x - A_{23}x - A_{20}x$$

for all $x \in K$. We claim that $(A - T_1 + A_{03})x \in M$. Indeed, (4.7) implies $Q_4(A - T_1 + A_{03})x = Q_4(A + A_{03})x = 0$, and since $(A_{03} - BD^{-1}F_1)x \in M^{\times}$ one has $Q_3(A - T_1 + A_{03})x = Q_3(A - T_1 + BD^{-1}F_1)x$ which is zero according to (2.8). Hence

$$\begin{aligned} S_2(P_{20} + P_{23})x - (P_{20} + P_{23})T_1x &= Q_2(A - T_1 + A_{03})x - A_{23}x - A_{20}x = \\ &= Q_2Ax + Q_2A_{03}x - A_{23}x - A_{20}x. \end{aligned}$$

From (4.8), (4.9) and (4.10) one sees that the right hand side of this equation is zero. Hence $P_{20}Q_3 + P_{23}$ satisfies the Lijapunov equation

$$\text{(4.14)} \quad S_2(P_{20}Q_3 + P_{23}) - (P_{20}Q_3 + P_{23})T_1 = 0.$$

Since S_2 and T_1 have no common eigenvalues one obtains $P_{20}Q_3 + P_{23} = 0$.

Similarly, one shows that

$$\text{(4.15)} \quad S_1(P_{10}^{\times}Q_3 + P_{14}^{\times}) - (P_{10}^{\times}Q_3 + P_{14}^{\times})T_2 = 0$$

which implies $P_{10}^{\times}Q_3 + P_{14}^{\times} = 0$. □

To show that $\hat{A}$ and $\hat{A}^{\times}$ have the desired triangular form we have to introduce the spaces X_1, X_2, X_3 and X_4 which appear in the definition of a realization with centralized singularities. This will be done as follows:

$$X_1 = \left\{ \begin{pmatrix} 0 \\ P_{20}m+P_{23}x \\ m+P_{03}x \\ x \\ 0 \end{pmatrix} \;\middle|\; x \in K,\ m \in M \right\},$$

$$X_2 = \left\{ \begin{pmatrix} a \\ a \\ a \\ 0 \\ 0 \end{pmatrix} \;\middle|\; a \in M \cap M^\times \right\}, \qquad X_3 = \left\{ \begin{pmatrix} 0 \\ 0 \\ x \\ x \\ x \end{pmatrix} \;\middle|\; x \in K \right\},$$

$$X_4 = \left\{ \begin{pmatrix} P^\times_{10}m^\times+P^\times_{14}x \\ 0 \\ m^\times+P^\times_{04}x \\ 0 \\ x \end{pmatrix} \;\middle|\; x \in K,\ m^\times \in M^\times \right\}.$$

LEMMA 4.5. *Suppose* (4.4) - (4.11) *hold. Then*

$$\hat{X} = X_1 \oplus X_2 \oplus X_3 \oplus X_4,$$

$$\hat{M} = X_1 \oplus X_2, \qquad \hat{M}^\times = X_2 \oplus X_4.$$

PROOF. (i) From Lemmas 4.2 and 4.3 one sees that $X_2 = \hat{M} \cap \hat{M}^\times$. Clearly, also $X_1 \subset \hat{M}$, $X_4 \subset \hat{M}^\times$. Choose $\hat{x} = (x_1,x_2,x_0,x_3,x_4)^T \in \hat{X}$. Consider the equations

$$\begin{aligned} x_1 &= a + P^\times_{10}m^\times + P^\times_{14}y_2 \\ x_2 &= a + P_{20}m + P_{23}y_1 \\ x_0 &= a + m + P_{03}y_1 + m^\times + P^\times_{04}y_2 + x \\ x_3 &= y_1 + x \\ x_4 &= y_2 + x \end{aligned}$$

where $y_1,y_2,x \in K$, $m \in M$, $m^\times \in M^\times$ and $a \in M \cap M^\times$. With respect to the canonical decomposition $X = N \oplus (M \cap M^\times) \oplus K \oplus N^\times$ we have, using Lemma 4.3:

$$\begin{aligned} x_0 = {} & (I - Q_2)(m - Py_1) + (a + Q_2m + Q_2m^\times - Q_2Py_1 - Q_2(I - P^\times)y_2) + \\ & (x + y_1 + y_2) + (I - Q_2)(m^\times - (I - P^\times)y_2). \end{aligned}$$

So $Q_3x_0 = x + y_1 + y_2$ and $Q_2x_0 = a + Q_2m + Q_2m^\times - Q_2Py_1 - Q_2(I-P^\times)y_2$. From this and

Lemmas 4.3 and 4.4 it follows that

$$
\begin{aligned}
x &= x_3 + x_4 - Q_3 x_0,\\
y_1 &= x_3 - x, \quad y_2 = x_4 - x,\\
a &= x_1 + x_2 - Q_2 x_0,\\
m &= Q_1 x_0 + x_2 - a + P y_1,\\
m^{\times} &= Q_4 x_0 + x_1 - a + (I - P^{\times}) y_2.
\end{aligned}
\tag{4.16}
$$

This shows that $\hat{X} = X_1 \oplus X_2 \oplus X_3 \oplus X_4$, and also the decompositions for $\hat{M}$ and $\hat{M}^{\times}$ are clear. □

PROOF of Theorem 4.1. First we show that with respect to the decomposition $\hat{X} = X_1 \oplus X_2 \oplus X_3 \oplus X_4$ the operators $\hat{A}$ and $\hat{A}^{\times}$ have the desired triangular form. To show that $\hat{A}$ has triangular form we need to show $\hat{A}X_1 \subset X_1$, $\hat{A}X_2 \subset X_1 \oplus X_2$ and $\hat{A}X_3 \subset X_1 \oplus X_3$. Similarly, for the triangular form of $\hat{A}^{\times}$ we have to show $\hat{A}^{\times}X_4 \subset X_4$, $\hat{A}^{\times}X_2 \subset X_2 \oplus X_4$ and $\hat{A}^{\times}X_3 \subset X_3 \oplus X_4$.

Take a vector in X_1 and compute

$$
\hat{A}\begin{pmatrix} 0 \\ P_{20}m+P_{23}x \\ m+P_{03}x \\ x \\ 0 \end{pmatrix} = \begin{pmatrix} A_{10}m+(A_{10}P_{03}+A_{13})x \\ (S_2P_{20}+A_{20})m + (S_2P_{23}+A_{20}P_{03}+A_{23})x \\ Am + (AP_{03}+A_{03})x \\ T_1x \\ 0 \end{pmatrix}.
$$

Now $A_{10}P_{03} + A_{13} = A_{10}(I - P)Q_3 + A_{13} = A_{10}Q_3 + A_{13} = 0$ according to Lemma 4.3 and the formulas (4.8) and (4.9). Hence $A_{10}m + (A_{10}P_{03} + A_{13})x = 0$. Next, using (4.13) and rewriting the second and third coordinates one obtains

$$
\hat{A}\begin{pmatrix} 0 \\ P_{20}m+P_{23}x \\ m+P_{03}x \\ x \\ 0 \end{pmatrix} = \begin{pmatrix} 0 \\ P_{20}(Am-APx+PT_1x+(A-T_1+A_{03})x) + P_{23}T_1x \\ (Am-APx+PT_1x+(A-T_1+A_{03})x) + P_{03}T_1x \\ T_1x \\ 0 \end{pmatrix}.
$$

Since $(A - T_1 + A_{03})x \in M$ for $x \in K$ it follows that $\hat{A}X_1 \subset X_1$. Likewise one shows $\hat{A}^{\times}X_4 \subset X_4$.

Next, for $a \in M \cap M^{\times}$ we have

$$\hat{A}\begin{pmatrix} a \\ a \\ a \\ 0 \\ 0 \end{pmatrix} = \begin{pmatrix} S_1 a \\ S_1 a \\ S_1 a \\ 0 \\ 0 \end{pmatrix} + \begin{pmatrix} 0 \\ A_{20}a + (S_2 - S_1)a \\ (A - S_1)a \\ 0 \\ 0 \end{pmatrix}.$$

Since $A_{20}a + (S_2 - S_1)a = P_{20}(A - S_1)a$ the second term is in X_1. Similarly

$$\hat{A}^{\times}\begin{pmatrix} a \\ a \\ a \\ 0 \\ 0 \end{pmatrix} = \begin{pmatrix} S_2 a \\ S_2 a \\ S_2 a \\ 0 \\ 0 \end{pmatrix} + \begin{pmatrix} P_{10}^{\times}(A^{\times} - S_2)a \\ 0 \\ (A^{\times} - S_2)a \\ 0 \\ 0 \end{pmatrix}.$$

These formulas also yield property (iii) in the definition of a realization with centralized singularities. The bases occuring in property (iii) are induced by the outgoing bases in $M \cap M^{\times}$.

Next, take $x \in K$ and consider

$$(4.17)\qquad \hat{A}\begin{pmatrix} 0 \\ 0 \\ x \\ x \\ x \end{pmatrix} = \begin{pmatrix} 0 \\ 0 \\ T_2 x \\ T_2 x \\ T_2 x \end{pmatrix} + \begin{pmatrix} (A_{10}+A_{13}+A_{14})x \\ (A_{20}+A_{23}+A_{24})x \\ (A-T_1+A_{03}+A_{04}+P(T_1-T_2))x + P_{03}(T_1-T_2)x \\ (T_1-T_2)x \\ 0 \end{pmatrix}.$$

Now $(A_{10} + A_{13} + A_{14})x = 0$. Further $(A - T_1 + A_{03} + A_{04} + P(T_1 - T_2))x \in M$ and, using Lemmas 4.3 and 4.4:

$$P_{20}(A - T_1 + A_{03} + A_{04} + P(T_1 - T_2))x + P_{23}(T_1 - T_2)x =$$
$$= P_{20}(A - T_1 + A_{03} + A_{04})x = Q_2(A + A_{03} + A_{04})x.$$

From (4.8) - (4.11) one easily sees that the latter term equals $(A_{20} + A_{23} + A_{24})x$. It follows that the last term in (4.17) is an element of X_1.

Likewise for $x \in K$ we have

$$(4.18)\qquad \hat{A}^{\times}\begin{pmatrix} 0 \\ 0 \\ x \\ x \\ x \end{pmatrix} = \begin{pmatrix} 0 \\ 0 \\ T_1 x \\ T_1 x \\ T_1 x \end{pmatrix} + \begin{pmatrix} P_{10}^{\times}(A^{\times}-T_2+A_{04}^{\times}+A_{03}^{\times}+(I-P^{\times})(T_2-T_1))x + P_{14}^{\times}(T_2-T_1)x \\ 0 \\ (A^{\times}-T_2+A_{04}^{\times}+A_{03}^{\times}+(I-P^{\times})(T_2-T_1))x + P_{04}^{\times}(T_2-T_1)x \\ 0 \\ (T_2-T_1)x \end{pmatrix}.$$

Formulas (4.17) and (4.18) also yield property (v), where the desired bases in

X_3 are induced by the incoming bases in K.

Further, property (ii) is also clear from the fact that $\hat{M} = X_1 \oplus X_2$, $\hat{M}^\times = X_2 \oplus X_4$. It remains to prove properties (iv) and (vi), which is done in the next lemma. □

LEMMA 4.6. *Suppose* (4.4) - (4.11) *hold, and let* $\hat{Q}_i$ *be the projection of* $\hat{X}$ *onto* X_i *along the spaces* X_j, $j \neq i$. *Then:*

$$\hat{Q}_2\hat{B}y = \begin{pmatrix} u(y) \\ u(y) \\ u(y) \\ 0 \\ 0 \end{pmatrix}, \qquad \hat{Q}_3\hat{B}y = \begin{pmatrix} 0 \\ 0 \\ b(y) \\ b(y) \\ b(y) \end{pmatrix},$$

where

$$u(y) = \begin{cases} (S_1 - S_2)x & \text{for } y = D^{-1}Cx,\ x \in M \cap M^\times, \\ 0 & \text{for } y \in Y_0 \oplus \operatorname{span}\{y_j\}_{j=1}^{s}; \end{cases}$$

$$b(y) = \begin{cases} 0 & \text{for } y \in B^{-1}(M + M^\times), \\ (T_2 - T_1)x & \text{for } y = D^{-1}(C + F_1 + F_2)x,\ x \in K. \end{cases}$$

In particular, rank $\hat{Q}_2\hat{B} = t$ *and* rank $\hat{Q}_3\hat{B} = s$. *Further,*

$$\operatorname{rank} \hat{C}\hat{Q}_2 = \operatorname{rank} C|_{M \cap M^\times} = t,$$

$$\operatorname{rank} \hat{C}\hat{Q}_3 = \operatorname{rank} (C + F_1 + F_2)|_K = s.$$

PROOF. Clearly, rank $\hat{C}\hat{Q}_2 = t$. Further, from (2.10) we obtain rank $\hat{C}\hat{Q}_3 = s$.

Using formulas (4.16) we have $u(y) = (G_1 + G_2 - Q_2By)$ and $b(y) = -Q_3By$. The formula for $u(y)$ is easily obtained from (3.8) and (3.9).

Take $y = D^{-1}(C + F_1 + F_2)x$, $x \in K$. Then

$$By = BD^{-1}(C + F_1 + F_2)x = (A - T_1 + A_{03})x + (T_1 - T_2)x + (A^\times - T_2 + A_{04}^\times)x.$$

Since $(A - T_1 + A_{03})x \in M$ and $(A^\times - T_2 + A_{04}^\times)x \in M^\times$, we have $Q_3By = (T_1 - T_2)x$. Next, take $y \in B^{-1}(M + M^\times)$. Then $Q_3By = 0$. This proves the formula for $b(y)$.

Using (2.4) and (3.2) it easily follows that rank $\hat{Q}_2\hat{B} = t$ and rank $\hat{Q}_3\hat{B} = s$. □

Note that the operator Λ in Theorem 4.1 is completely arbitrary, so we might even have set $\Lambda = 0$. However, the choice of Λ will play an important role in Chapter II, and there the choice $\Lambda = 0$ will not be suitable.

I.5. The final formulas

In this section we apply the factorization formulas of Theorem 1.1 to the dilation $\hat{\theta}$ constructed in Theorem 4.1. This yields the final result of the chapter, which is an improved version of Theorem I.10.1 in [2].

THEOREM 5.1. *Let* $W(\lambda) = D + C(\lambda - A)^{-1}B$ *be a rational matrix function which is regular at infinity, and assume that* A *and* $A^{\times} := A - BD^{-1}C$ *have no real eigenvalues. Then* W *admits a right Wiener-Hopf factorization* $W(\lambda) = W_-(\lambda)D(\lambda)W_+(\lambda)$, $-\infty \le \lambda \le \infty$, *of which the factors are given by:*

$$W_-(\lambda) = D + C(\lambda - A)^{-1}((Q_1 + Q_2)B - G_1) + C(\lambda - A)^{-1}V_-(\lambda - T_1)^{-1}Q_3B + \\ + (CQ_3 + F_1)(\lambda - T_1)^{-1}Q_3B;$$

$$W_-(\lambda)^{-1} = D^{-1} - D^{-1}(C(Q_1 + Q_3) + F_1Q_3)(\lambda - A^{\times})^{-1}BD^{-1} + \\ - D^{-1}CQ_2(\lambda - S_1)^{-1}V_-^{\times}(\lambda - A^{\times})^{-1}BD^{-1} + \\ - D^{-1}CQ_2(\lambda - S_1)^{-1}(Q_2B - G_1)D^{-1};$$

$$W_+(\lambda) = I + D^{-1}(C(Q_3 + Q_4) + F_2Q_3)(\lambda - A)^{-1}B + \\ + D^{-1}CQ_2(\lambda - S_2)^{-1}V_+(\lambda - A)^{-1}B + D^{-1}CQ_2(\lambda - S_2)^{-1}(Q_2B - G_2);$$

$$W_+(\lambda)^{-1} = I - D^{-1}C(\lambda - A^{\times})^{-1}((Q_2 + Q_4)B - G_2) + \\ - D^{-1}C(\lambda - A^{\times})^{-1}V_+^{\times}(\lambda - T_2)^{-1}Q_3B - D^{-1}(CQ_3 + F_2)(\lambda - T_2)^{-1}Q_3B;$$

$$E^{-1}D(\lambda)E = \operatorname{diag}\left(\left(\frac{\lambda-i}{\lambda+i}\right)^{-\alpha_1}, \dots, \left(\frac{\lambda-i}{\lambda+i}\right)^{-\alpha_t}, 1, \dots, 1, \left(\frac{\lambda-i}{\lambda+i}\right)^{\omega_1}, \dots, \left(\frac{\lambda-i}{\lambda+i}\right)^{\omega_s}\right).$$

Here T_1, T_2 *are the incoming operators,* S_1, S_2 *the outgoing operators,* F_1, F_2 *the feedback operators,* G_1, G_2 *the output injection operators and* Q_1, Q_2, Q_3, Q_4 *the projections which were introduced in the first paragraphs of the previous section. Furthermore,*

$$V_- = (Q_1 + Q_2)AQ_3 + (Q_1 + Q_2)BD^{-1}F_1 + \Lambda - G_1D^{-1}F_1,$$
$$V_-^{\times} = Q_2A^{\times}(Q_1 + Q_3) + G_1D^{-1}C(Q_1 + Q_3) + \Lambda Q_3,$$
$$V_+ = Q_2AQ_4 - G_2D^{-1}CQ_4 - \Lambda Q_3 - (Q_2B - G_1 - G_2)D^{-1}F_1,$$
$$V_+^{\times} = Q_4A^{\times}Q_3 - Q_4BD^{-1}F_2 - \Lambda - G_1D^{-1}(CQ_3 + F_2),$$

where Λ *is an arbitrary operator from* K *into* $M \cap M^{\times}$. *Finally,*

$$E = [z_1 \dots z_t \; z_{t+1} \dots z_{m-s} \; y_1 \dots y_s],$$

where $z_1,\dots,z_t$ *and* $y_1,\dots,y_s$ *are the vectors appearing in the sets of outgoing and incoming data* (4.1) *and* (4.2), *respectively, and* $z_{t+1},\dots,z_{m-s}$ *form a basis for a complement* Y_0 *of* $\operatorname{span}\{z_j\}_{j=1}^{t}$ *in* $B^{-1}[M+M^{\times}]$.

PROOF. The idea of the proof is essentially the same as in the proof of Theorem I.10.1 in [2]. We apply Theorems 1.1 and 4.1. So, let $\hat{A}$, $\hat{B}$, $\hat{C}$ and $\hat{A}^{\times} = \hat{A} - \hat{B}D^{-1}\hat{C}$ be as in the previous section, and assume that the entries A_{ij} in $\hat{A}$ are defined as in Theorem 4.1. Furthermore, take X_1, X_2, X_3 and X_4 as in the paragraph preceding Lemma 4.5. Theorem 1.1 applied to the realization $\hat{\theta} = (\hat{A},\hat{B},\hat{C},D;\hat{X},\mathbb{C}^m)$ yields a right Wiener-Hopf factorization $W(\lambda) = W_-(\lambda)D(\lambda)W_+(\lambda)$, $-\infty \le \lambda \le \infty$, with

$$W_-(\lambda) = D + \hat{C}\hat{Q}_1(\lambda - \hat{A}|_{X_1})^{-1}\hat{Q}_1\hat{B},$$
$$W_+(\lambda) = I + D^{-1}\hat{C}\hat{Q}_4(\lambda - \hat{A}|_{X_4})^{-1}\hat{Q}_4\hat{B},$$

where $\hat{Q}_1$ and $\hat{Q}_4$ are the projections introduced in Lemma 4.6. From Lemmas 4.3 and 4.4 it follows that we may write

$$X_1 = \left\{\begin{pmatrix} 0 \\ P_{20}m \\ m+x \\ x \\ 0 \end{pmatrix} \;\middle|\; m \in M,\ x \in K\right\},\quad X_4 = \left\{\begin{pmatrix} P_{10}m^{\times} \\ 0 \\ m^{\times}+x \\ 0 \\ x \end{pmatrix} \;\middle|\; m^{\times} \in M^{\times},\ x \in K\right\}.$$

Define $J : M \oplus K \to X_1$ by

$$J\begin{pmatrix} m \\ x \end{pmatrix} = \begin{pmatrix} 0 \\ P_{20}m \\ m+x \\ x \\ 0 \end{pmatrix}.$$

Then J is invertible, and we have $\hat{Q}_1\hat{B} = JB_-$, $\hat{C}\hat{Q}_1J = C_-$ and $\hat{A}|_{X_1}J = JA_-$, where

$$A_- = \begin{pmatrix} A|_M & (Q_1+Q_2)(A+A_{03})Q_3 \\ 0 & T_1 \end{pmatrix},\quad B_- = \begin{pmatrix} (Q_1+Q_2)B - G_1 \\ Q_3B \end{pmatrix},$$

$$C_- = \begin{pmatrix} C(Q_1+Q_2) & CQ_3+F_1 \end{pmatrix}.$$

It follows that $W_-(\lambda) = D + C_-(\lambda - A_-)^{-1}B_-$. Calculating this gives the formula for W_- with $V_- = (Q_1+Q_2)(A+A_{03})Q_3$. By using the definition of A_{03} (see

Theorem 4.1) one gets the desired expression for V_-.

Further, $W_-(\lambda)^{-1} = D^{-1} - D^{-1}C_-(\lambda - A_-^\times)^{-1}B_-D^{-1}$, where $A_-^\times = A_- - B_-D^{-1}C_-$. Next, one calculates $A_-^\times$, B_- and C_- relative to the decomposition $(M \cap M^\times) \oplus \operatorname{Im}(Q_1 + Q_3)$ $(= M + K)$. This yields

$$A_-^\times = \begin{pmatrix} S_1 & V_-^\times \\ 0 & (Q_1+Q_3)A^\times(Q_1+Q_3) \end{pmatrix}, \quad B_- = \begin{pmatrix} Q_2B - G_1 \\ (Q_1+Q_3)B \end{pmatrix},$$

$$C_- = \begin{pmatrix} CQ_2 & C(Q_1+Q_3) + F_1Q_3 \end{pmatrix}.$$

From these identities the formula for $W_-(\lambda)^{-1}$ is clear.

Analogously, define $J^\times : M^\times \oplus K \to X_4$ by

$$J^\times \begin{pmatrix} m^\times \\ x \end{pmatrix} = \begin{pmatrix} P_{10}^\times m^\times \\ 0 \\ m^\times + x \\ 0 \\ x \end{pmatrix}.$$

Then $J^\times$ is invertible, and $\hat{Q}_4\hat{B} = J^\times B_+$, $\hat{C}\hat{Q}_4J^\times = C_+$ and $\hat{A}^\times|_{X_4}J^\times = J^\times A_+^\times$, where

$$A_+^\times = \begin{pmatrix} A^\times|_{M^\times} & (Q_2+Q_4)(A^\times + A_{04}^\times)Q_3 \\ 0 & T_2 \end{pmatrix}, \quad B_+ = \begin{pmatrix} (Q_2+Q_4)B - G_2 \\ Q_3B \end{pmatrix},$$

$$C_+ = \begin{pmatrix} C(Q_2+Q_4) & CQ_3 + F_2 \end{pmatrix}.$$

Since $W_+(\lambda)^{-1} = I - D^{-1}C_+(\lambda - A_+^\times)^{-1}B_+$, we obtain the formula for $W_+(\lambda)^{-1}$ with $V_+^\times = (Q_2+Q_4)(A^\times + A_{04}^\times)Q_3$. Using the definition of A_{04} (see Theorem 4.1) it follows that

$$\begin{aligned} V_+^\times &= (Q_2+Q_4)(A^\times + A_{04}^\times)Q_3 = (Q_2+Q_4)A^\times Q_3 + (Q_2+Q_4)(A_{04} - BD^{-1}F_2)Q_3 \\ &= Q_4A^\times Q_3 - Q_4BD^{-1}F_2Q_3 + Q_2A^\times Q_3 - Q_2AQ_3 - \Lambda - (Q_2B - G_1 - G_2)D^{-1}F_1 + \\ &\quad + G_2D^{-1}(CQ_3 + F_2) - Q_2BD^{-1}F_2Q_3 \\ &= Q_4A^\times Q_3 - Q_4BD^{-1}F_2Q_3 - Q_2BD^{-1}CQ_3 - \Lambda - (Q_2B - G_1 - G_2)D^{-1}F_1 + \\ &\quad + G_2D^{-1}(CQ_3 + F_2) - Q_2BD^{-1}FQ_3 \\ &= Q_4A^\times Q_3 - Q_4BD^{-1}F_2Q_3 - \Lambda - G_1D^{-1}(CQ_3 + F_2) + \\ &\quad - (Q_2B - G_1 - G_2)D^{-1}(CQ_3 + F_1 + F_2), \end{aligned}$$

which (because of (2.10) and (3.8)) yields the desired formula for $V_+^\times$.

Next, we compute $W_+(\lambda) = I + D^{-1}C_+(\lambda - A_+)^{-1}B_+$. To do this we write $A_+ = A_+^\times + B_+D^{-1}C_+$, B_+ and C_+ as block matrices relative to the decomposition $(M \cap M^\times) \oplus \operatorname{Im}(Q_3 + Q_4)$ $(= M^\times + K)$. We obtain:

$$A_+ = \begin{pmatrix} S_2 & V_+ \\ 0 & (Q_3+Q_4)A(Q_3+Q_4) \end{pmatrix}, \quad B_+ = \begin{pmatrix} Q_2B - G_2 \\ (Q_3+Q_4)B \end{pmatrix},$$

$$C_+ = \begin{pmatrix} CQ_2 & C(Q_3+Q_4) + F_2Q_3 \end{pmatrix}.$$

From these identities the formula for W_+ is clear.

Following the line of arguments of the proof of Theorem I.3.1 in [2] one sees (modulo some changes involving the operator D) that

$$(5.1) \qquad D(\lambda)y = \begin{cases} \left(\frac{\lambda-i}{\lambda+i}\right)^{\omega_j} y_j & \text{for } y = y_j, \quad j = 1,\dots,s; \\ y & \text{for } y \in Y_0; \\ \left(\frac{\lambda-i}{\lambda+i}\right)^{-\alpha_j} z_j & \text{for } y = z_j, \quad j = 1,\dots,t. \end{cases}$$

It follows that $E^{-1}D(\lambda)E$ has the desired diagonal form. □

II. SYMMETRIC WIENER-HOPF FACTORIZATION

In the first two sections of this chapter it is shown that for a selfadjoint rational matrix function the incoming and outgoing data, the incoming and outgoing operators and the corresponding feedback and output injection operators may be chosen in such a way that they are dual to each other in an appropriate sense. On the basis of this duality and Theorem I.5.1 we prove in the third section the main theorems of this paper.

II.1. Duality between incoming and outgoing operators

Throughout this chapter $W(\lambda)$ is an $m \times m$ selfadjoint rational matrix function which does not have poles and zeros on the real line including infinity, and $\theta = (A,B,C,D;\mathbb{C}^n,\mathbb{C}^m)$ is a fixed realization of W which satisfies the following general conditions: D is selfadjoint and invertible, the operators A and $A^\times$ $(:= A - BD^{-1}C)$ have no real eigenvalues, and for some invertible selfadjoint $n \times n$ matrix H we have

$$(1.1) \qquad HA = A^*H, \quad HB = C^*.$$

Such a realization always exists (see [5,8]); in fact, if the state space dimension of θ (i.e., the number n) is minimal among all realizations of W, then the realization θ satifies the above general conditions automatically.

Note that (1.1) implies that $HA^{\times} = (A^{\times})^{*}H$. Let P (resp. $P^{\times}$) be the spectral projection of A (resp. $A^{\times}$) corresponding to the eigenvalues in the upper half plane. Then

$$HP = (I - P)^{*}H, \quad HP^{\times} = (I - P^{\times})^{*}H.$$

Put $M = \operatorname{Im} P$ and $M^{\times} = \operatorname{Im}(I - P^{\times})$. It follows that $HM = M^{\perp}$ and $HM^{\times} = (M^{\times})^{\perp}$. As an easy consequence we have the following identity:

$$(1.2) \qquad [H(M \cap M^{\times})]^{\perp} = M + M^{\times}.$$

Note that (1.2) implies that $\dim(M \cap M^{\times}) = \operatorname{codim}(M + M^{\times})$.

Recall that the incoming and outgoing spaces for θ are the spaces

$$H_j = M + M^{\times} + \operatorname{Im} B + \dots + \operatorname{Im} A^{j-1}B, \quad j = 0,1,2,\dots$$

$$K_j = M \cap M^{\times} \cap \operatorname{Ker} C \cap \dots \cap \operatorname{Ker} CA^{j-1}, \quad j = 0,1,2,\dots$$

respectively. Here $H_0 = M + M^{\times}$, $K_0 = M \cap M^{\times}$. Formula (1.2) gives $(HH_0)^{\perp} = K_0$. Further

$$(H \operatorname{Im} A^{j-1}B)^{\perp} = \operatorname{Ker} B^{*}A^{*j-1}H = \operatorname{Ker} CA^{j-1}.$$

Hence

$$(1.3) \qquad (HH_j)^{\perp} = K_j, \quad j = 0,1,2,\dots .$$

According to [2], Corollary I.10.2 the incoming indices and outgoing indices are determined by:

$$s = \dim H_1/H_0, \quad t = \dim K_0/K_1,$$

$$\omega_j = \#\{k \mid \dim H_k/H_{k-1} \geq s+1-j\},$$

$$\alpha_j = \#\{k \mid \dim K_{k-1}/K_k \geq t+1-j\}.$$

By combining this with (1.3) we get the following proposition.

PROPOSITION 1.1. *The incoming indices* $\omega_1,\dots,\omega_s$ *and the outgoing indices* $\alpha_1,\dots,\alpha_t$ *coincide, that is,* $s = t$ *and* $\omega_j = \alpha_j$, $j = 1,\dots,s$.

Let K be any complement of $M + M^{\times}$ in $\mathbb{C}^n$. Then $\dim K = \dim(M \cap M^{\times})$ (see (1.2)). Further

$$(1.4) \qquad [x,y] = \langle Hx,y \rangle, \quad x \in M \cap M^{\times}, \quad y \in K$$

defines a Hilbert duality between $M \cap M^{\times}$ and K (because of the invertibility of H). We shall use this duality to define incoming data in terms of outgoing data.

LEMMA 1.2. *Let* K *be any complement of* $M+M^\times$ *in* $\mathbb{C}^n$. *Let* $\{d_{jk}\}_{k=1,j=1}^{\alpha_j\ \ s}$ *and* $\{e_{jk}\}_{k=1,j=1}^{\alpha_j\ \ s}$ *be outgoing bases for* A-i *and* A+i, *respectively, and assume that the two bases are related by*

$$(1.5)\qquad d_{jk} = \sum_{\nu=0}^{k-1}\binom{k-1}{\nu}(-2i)^{\nu}e_{j,k-\nu}.$$

Choose in K *a set of vectors* $\{f_{jk}\}_{k=1,j=1}^{\alpha_j\ \ s}$ *satisfying*

$$(1.6)\qquad \langle Hf_{jk},e_{pq}\rangle = \delta_{jp}\cdot\delta_{k,\alpha_p-q+1}.$$

Then $\{f_{jk}\}_{k=1,j=1}^{\alpha_j\ \ s}$ *is an incoming basis for* A-i. *Further, if*

$$(1.7)\qquad g_{jk} = \sum_{\nu=0}^{k-1}\binom{\nu+\alpha_j-k}{\nu}(-2i)^{\nu}f_{j,k-\nu},$$

then $\{g_{jk}\}_{k=1,j=1}^{\alpha_j\ \ s}$ *is an incoming basis for* A+i *and*

$$(1.8)\qquad \langle Hg_{jk},d_{pq}\rangle = \delta_{jp}\cdot\delta_{k,\alpha_p-q+1}.$$

PROOF. First note that because of the duality (1.4) between $M\cap M^\times$ and K there always exists a set of vectors in K satisfying (1.6). Further, $\dim(M\cap M^\times) = \dim K$ implies that the vectors f_{jk} form a basis for K. Also, for $j=1,\dots,s$ the vector $f_{j1}\in(HK_1)^\perp = H_1$. Since $\dim H_1/H_0 = s$, it follows that the vectors $f_{11},\dots,f_{s1}$ form a basis for H_1 modulo H_0.

For $x = \Sigma_{\nu=1}^{s}\Sigma_{\mu=1}^{\alpha_\nu-1}\beta_{\nu\mu}e_{\nu\mu}\in K_1$ we have

$$\langle H((A-i)f_{jk}-f_{j,k+1}),x\rangle = \langle Hf_{jk},(A+i)x\rangle - \langle Hf_{j,k+1},x\rangle =$$
$$= \sum_{\nu=1}^{s}\sum_{\mu=1}^{\alpha_\nu-1}\overline{\beta}_{\nu\mu}\left\{\langle Hf_{jk},e_{\nu,\mu+1}\rangle - \langle Hf_{j,k+1},e_{\nu\mu}\rangle\right\}.$$

From (1.6) it follows that the term between braces is zero, so $(A-i)f_{jk}-f_{j,k+1}\in(HK_1)^\perp = H_1 = M+M^\times+\operatorname{Im}B$. So $\{f_{jk}\mid k=1,\dots,\alpha_j,\ j=1,\dots,s\}$ is an incoming basis for $A-i$.

Let the vectors g_{jk} be given by (1.7). From Proposition I.2.1 we know that they form an incoming basis for $A+i$. It remains to prove formula (1.8). From (1.5) and (1.7) we obtain

$$\langle Hg_{jk},d_{pq}\rangle = \sum_{\ell=0}^{k-1}\sum_{\nu=0}^{q-1}\binom{\ell+\alpha_j-k}{\ell}\binom{q-1}{\nu}.(2i)^{\ell+\nu}(-1)^{\ell}\langle Hf_{j,k-\ell},e_{p,q-\nu}\rangle,$$

which is zero if $p\neq j$. So, let $p=j$. Then, using (1.6) we have

$$\langle Hg_{jk},d_{jq}\rangle = \sum_{\ell=0}^{k-1}\sum_{\nu=0}^{q-1}\binom{\ell+\alpha_j-k}{\ell}\binom{q-1}{\nu}.(2i)^{\ell+\nu}(-1)^{\ell}\delta_{k-\ell,\alpha_j-q+\nu+1}.$$

The Kronecker delta appearing here is non-zero precisely when $\ell+\nu = k-\alpha_j+q-1$. Since $\ell+\nu \ge 0$ this implies $q-1 \ge \alpha_j - k$. Put $\nu = k-\alpha_j+q-1-\ell$ in the formula above. Then $0 \le \nu \le q-1$ gives $0 \le \ell \le k-\alpha_j+q-1$. So

$$<Hg_{jk},d_{jq}> = \sum_{\ell=0}^{k-\alpha_j+q-1} \binom{\ell+\alpha_j-k}{\ell}\binom{q-1}{k-\alpha_j+q-1-\ell}(2i)^{k-\alpha_j+q-1}(-1)^{\ell} =$$

$$= \binom{q-1}{\alpha_j-k}(2i)^{k+q-\alpha_j-1}\sum_{\ell=0}^{k-\alpha_j+q-1}\binom{q-1-\alpha_j+k}{\ell}(-1)^{\ell} = \delta_{k,\alpha_j-q+1}.$$

This proves formula (1.8). □

So far the complement K of $M+M^{\times}$ was arbitrary. We shall now prove that K may be chosen in a special way. Take subspaces N and $N^{\times}$ such that $M = N \oplus (M \cap M^{\times})$ and $M^{\times} = N^{\times} \oplus (M \cap M^{\times})$.

LEMMA 1.3. *There exists a complement* K *of* $M+M^{\times}$ *such that* K *is* H-*neutral and*

$$[H(N \oplus N^{\times})]^{\perp} = K \oplus (M \cap M^{\times}). \tag{1.9}$$

PROOF. Put $L = [H(N \oplus N^{\times})]^{\perp}$. It is easily seen that $M \cap M^{\times} = L \cap [H(M \cap M^{\times})]^{\perp} = L \cap (M+M^{\times})$. Let $x_1,\dots,x_k$ be a basis of $M \cap M^{\times}$. Choose vectors $y_1',\dots,y_k'$ in L such that

$$<Hy_j',x_i> = \delta_{ij}, \quad i,j=1,\dots,k. \tag{1.10}$$

This is possible, since $\mathbb{C}^n$ equipped with the indefinite inner product given by H is a nondegenerate space. Put

$$y_j = y_j' - \tfrac{1}{2}\sum_{\nu=1}^{k} <Hy_j',y_\nu'> x_\nu, \quad j=1,\dots,k, \tag{1.11}$$

and let $K = \text{span}\{y_1,\dots,y_k\}$. Using (1.10) and (1.11) it is easily seen that

$$<Hy_j,y_i> = 0, \quad i,j=1,\dots,n.$$

So K is H-neutral. Note that also $K \subset L$.

Finally, $K \cap (M \cap M^{\times}) = (0)$, since

$$<Hy_j,x_i> = \delta_{ij}, \quad i,j=1,\dots,k,$$

and $\dim L = \text{codim}(N+N^{\times}) = \dim(M \cap M^{\times}) + \text{codim}(M+M^{\times}) = 2\cdot\dim(M \cap M^{\times})$. So $L = K \oplus (M \cap M^{\times})$. □

An H-neutral complement K of $M+M^{\times}$ satisfying (1.9) will be called a <u>canonical complement</u>, and in that case the decomposition

$$\mathbb{C}^n = N \oplus (M \cap M^{\times}) \oplus K \oplus N^{\times} \tag{1.12}$$

is said to be a <u>canonical decomposition</u> for θ.

In what follows Q_i $(i=1,2,3,4)$ is the projection onto the i-th space in the decomposition (1.12) along the other spaces in this decomposition.

LEMMA 1.4. *If* (1.12) *is a canonical decomposition, then*

$$HQ_2 = Q_3^*H, \quad HQ_1 = Q_4^*H. \tag{1.13}$$

PROOF. First note that in a canonical decomposition

$$(HK)^{\perp} = N \oplus K \oplus N^{\times}. \tag{1.14}$$

Using this identity together with (1.2) and $(HM)^{\perp} = M$, $(HM^{\times})^{\perp} = M^{\times}$ it is straightforward to derive (1.13). □

PROPOSITION 1.5. *Let* K *be a canonical complement of* $M + M^{\times}$. *Let* $\{d_{jk}\}_{k=1,j=1}^{\alpha_j\ \ s}$ *and* $\{e_{jk}\}_{k=1,j=1}^{\alpha_j\ \ s}$ *be outgoing bases for* A-i *and* A+i, *respectively, and assume that the two bases are related by* (1.5). *Choose in* K *incoming bases* $\{f_{jk}\}_{k=1,j=1}^{\alpha_j\ \ s}$ *and* $\{g_{jk}\}_{k=1,j=1}^{\alpha_j\ \ s}$ *for* A-i *and* A+i, *respectively, such that* (1.6) *and* (1.7) *are satisfied. Let* S_1, S_2 *be the corresponding outgoing operators and* T_1, T_2 *the corresponding incoming operators. Then*

$$H(Q_3T_1Q_3) = (Q_2S_2Q_2)^*H, \quad H(Q_3T_2Q_3) = (Q_2S_1Q_2)^*H. \tag{1.15}$$

PROOF. First note that Lemma 1.2 guarantees the existence of the incoming bases $\{f_{jk}\}$ and $\{g_{jk}\}$. We shall only prove the first identity in (1.15); the second is proved analogously. Let f_{jk} and e_{pq} be given. Then

$$\langle H(T_1 - i)f_{jk}, e_{pq}\rangle = \langle Hf_{j,k+1}, e_{pq}\rangle = \delta_{jp}\cdot\delta_{k+1,\alpha_p-q+1} \quad (k < \alpha_j).$$

Also

$$\langle Hf_{jk}, (S_2 + i)e_{pq}\rangle = \langle Hf_{jk}, e_{p,q+1}\rangle = \delta_{jp}\cdot\delta_{k,\alpha_p-q} \quad (q < \alpha_p).$$

So these two expressions are equal for $k < \alpha_j$ and $q < \alpha_p$. Suppose $k = \alpha_j$, then $\langle H(T_1 - i)f_{jk}, e_{pq}\rangle = 0$, and using (1.6) one sees that

$$\langle Hf_{j\alpha_j}, (S_2 + i)e_{pq}\rangle = \langle Hf_{j\alpha_j}, e_{p,q+1}\rangle = 0.$$

A similar argument holds in case $q = \alpha_p$. So

$$\langle H(T_1 - i)x, y\rangle = \langle Hx, (S_2 + i)y\rangle$$

for all $x \in K$ and $y \in M \cap M^{\times}$. But then $\langle HT_1Q_3x, Q_2y\rangle = \langle HQ_3x, S_2Q_2y\rangle$ for all x and y in $\mathbb{C}^n$, and we can use (1.13) to get the first equality in (1.15). □

II.2. The basis in $\mathbb{C}^m$ and duality between the feedback operators and the output injection operators

Let $\theta = (A,B,C,D;\mathbb{C}^n,\mathbb{C}^m)$ be a realization of W which satisfies the general conditions stated in the first paragraph of the previous section. We proceed with an analysis of the basis in $\mathbb{C}^m$.

LEMMA 2.1. *Let* K *be any complement of* $M + M^{\times}$ *in* $\mathbb{C}^n$, *and for* $k = 1, \ldots, \alpha_j$ *and* $j = 1, \ldots, s$ *let the vectors* d_{jk}, e_{jk}, f_{jk} *and* g_{jk} *be as in Lemma* 1.2. *For* $j = 1, \ldots, s$ *let* z_j *be given by* $z_j = D^{-1}Ce_{j\alpha_j} = D^{-1}Cd_{j\alpha_j}$. *If a set of vectors* $\{y_1, \ldots, y_s\}$ *satisfies*

$$(2.1) \qquad \langle Dz_k, y_j \rangle = \delta_{jk}, \quad j,k = 1, \ldots, s,$$

then

$$(2.2) \qquad f_{j1} - By_j = g_{j1} - By_j \in M + M^{\times}, \quad j = 1, \ldots, s,$$

and, conversely, if $y_1, \ldots, y_s$ *satisfy* (2.2), *then* (2.1) *holds. Further,* $\langle Dz_j, z \rangle = 0$ *for all* $z \in B^{-1}(M + M^{\times})$ *and* $j = 1, \ldots, s$.

Moreover, if Y_0 *is a subspace of* $B^{-1}(M + M^{\times})$ *such that* $B^{-1}(M + M^{\times}) = Y_0 \oplus \operatorname{span}\{z_1, \ldots, z_s\}$, *then a set of vectors* $\{y_1, \ldots, y_s\}$ *can be chosen such that* (2.1) *holds and*

$$(2.3) \qquad (D \operatorname{span}\{y_1, \ldots, y_s\})^{\perp} = \operatorname{span}\{y_1, \ldots, y_s\} \oplus Y_0.$$

PROOF. Suppose (2.1) holds. Let

$$x = \sum_{i=1}^{s} \sum_{k=1}^{\alpha_i} \beta_{ik} e_{ik} \in M \cap M^{\times}.$$

Then

$$\langle Hx, f_{j1} - By_j \rangle = \beta_{j\alpha_j} - \langle B^{*}Hx, y_j \rangle = \beta_{j\alpha_j} - \sum_{i=1}^{s} \sum_{k=1}^{\alpha_i} \beta_{ik} \langle Ce_{ik}, y_j \rangle =$$

$$= \beta_{j\alpha_j} - \sum_{i=1}^{s} \beta_{i\alpha_i} \langle Ce_{i\alpha_i}, y_j \rangle = \beta_{j\alpha_j} - \sum_{i=1}^{s} \beta_{i\alpha_i} \langle Dz_i, y_j \rangle = 0.$$

So $f_{j1} - By_j \in (H(M \cap M\ ))^{\perp} = M + M^{\times}$.

Conversely, suppose (2.2) holds. Then

$$0 = \langle H(f_{j1} - By_j), e_{k\alpha_k} \rangle = \delta_{jk} - \langle y_j, Ce_{k\alpha_k} \rangle = \delta_{jk} - \langle y_j, Dz_k \rangle.$$

So (2.1) holds.

Let z_j and $z \in B^{-1}(M + M^{\times})$ be given. Then $\langle Dz_j, z \rangle = \langle Ce_{j\alpha_j}, z \rangle = \langle e_{j\alpha_j}, HBz \rangle = \langle He_{j\alpha_j}, Bz \rangle$. Since $Bz \in M + M^{\times}$ and $e_{j\alpha_j} \in M \cap M^{\times} = (H(M + M^{\times}))^{\perp}$, this equals zero.

To prove the last part of the proposition, put $L = (DY_0)^{\perp}$ and apply to L a reasoning analogous to the one used in the proof of Lemma 1.3 (with H replaced by D and $N \oplus N^{\times}$ by Y_0). Then the construction of the vectors $y_1, \ldots, y_s$ proceeds as in the proof of Lemma 1.3. We omit the details. □

Suppose $\{y_1,\ldots,y_s\}$ are constructed such that (2.1) and (2.3) hold. It follows from Lemma 2.1 that for any basis $\{z_{s+1},\ldots,z_{m-s}\}$ in Y_0 the matrix $(\langle Dz_j,z_k\rangle)_{j,k=s+1}^{m-s}$ has the same signature as D. Choose a basis $\{z_{s+1},\ldots,z_{m-s}\}$ in Y_0 such that this matrix has the form $I_p \oplus (-I_q)$, where $p-q = \operatorname{sign} D$ and $p+q+2s = m$.

The different numbers among $\alpha_1,\ldots,\alpha_s$ will be denoted by $\kappa_1,\ldots,\kappa_r$, where we assume $\kappa_1 < \ldots < \kappa_r$. The number of times κ_j appears in $\alpha_1,\ldots,\alpha_s$ will be denoted by m_j. Note that $s = \Sigma_{j=1}^r m_j$.

Now we introduce vectors $z_{m-s+1},\ldots,z_m$ via a renumbering of the vectors $y_1,\ldots,y_s$ as follows

$$z_{m+i-(m_1+\ldots+m_{j+1})} = y_{i+(m_1+\ldots+m_j)},$$

for $i = 1,\ldots,m_{j+1}$ and $j = 0,\ldots,r-1$.

Let S be the matrix $S = [z_1 \cdots z_m]$. Note that S is invertible, and that from the choice of $z_1,\ldots,z_m$ it follows that

$$(2.4)\qquad S^*DS = \begin{pmatrix} & & & & & I_{m_1} \\ & & & & \cdot^{\cdot^{\cdot}} & \\ & & & I_{m_r} & & \\ & & \begin{matrix} I_p & 0 \\ 0 & -I_q \end{matrix} & & & \\ & I_{m_r} & & & & \\ \cdot^{\cdot^{\cdot}} & & & & & \\ I_{m_1} & & & & & \end{pmatrix}$$

The matrix S will be called a <u>congruence</u> matrix of the realization θ.

PROPOSITION 2.2. *Let* $\mathbb{C}^n = N \oplus (M \cap M^\times) \oplus K \oplus N^\times$ *be a canonical decomposition, and let* $G_1,G_2 : \mathbb{C}^m \to M \cap M^\times$ *be a pair of output injection operators (relative to the projection* Q_2 *and) corresponding to the following set of outgoing data*

$$(2.5)\qquad \{d_{jk}\}_{k=1,j=1}^{\alpha_j\ \ s}, \quad \{e_{jk}\}_{k=1,j=1}^{\alpha_j\ \ s}, \quad \{z_j\}_{j=1}^{s}.$$

Choose in K *incoming bases* $\{f_{jk}\}_{k=1,j=1}^{\alpha_j\ \ s}$ *and* $\{g_{jk}\}_{k=1,j=1}^{\alpha_j\ \ s}$ *for* A-i *and* A+i, *respectively, such that* (1.6) *and* (1.7) *are satisfied. Construct vectors* $y_1,\ldots,y_s$ *in* $\mathbb{C}^m$ *such that* (2.1) *and* (2.3) *hold. Then*

$$(2.6)\qquad \{f_{jk}\}_{k=1,j=1}^{\alpha_j\ \ s}, \quad \{g_{jk}\}_{k=1,j=1}^{\alpha_j\ \ s}, \quad \{y_j\}_{j=1}^{s}$$

is a set of incoming data, and the operators $F_1,F_2 : K \to \mathbb{C}^m$ *defined by*

(2.7) $\quad F_1Q_3 = -(Q_2G_2)^*H, \quad F_2Q_3 = -(Q_2G_1)^*H$

form a pair of feedback operators corresponding to the incoming data (2.6).

PROOF. We assume that formulas (I.3.6) up to (I.3.9) hold with $\rho = Q_2$. Since (2.2) holds true, it is clear that (2.6) is a set of incoming data. So we have to prove that the operators F_1, F_2 defined by (2.7) are feedback operators, that is, we have to check that the formulas (I.2.8) up to (I.2.10) are satisfied. Take $x \in K$ and $y \in M \cap M^{\times}$. Then

$$\langle H(A - T_1 + BD^{-1}F_1)x, y\rangle = \langle Hx, Ay\rangle - \langle Hx, S_2y\rangle + \langle F_1x, D^{-1}Cy\rangle =$$
$$= \langle Hx, (A - S_2 - G_2D^{-1}C)y\rangle = 0,$$

because $(A - S_2 - G_2D^{-1}C)y \in \operatorname{Ker} Q_2 = N \oplus K \oplus N^{\times} = (HK)^{\perp}$ (cf. (1.14)). Since $[H(M \cap M^{\times})]^{\perp} = M + M^{\times}$, it follows $(A - T_1 + BD^{-1}F_1)x \in M + M^{\times}$ for all $x \in K$, and (I.2.8) is proved. In a similar way one proves (I.2.9).

To prove (I.2.10), take $y \in \mathbb{C}^m$. Then

$$\langle DD^{-1}(C + F_1 + F_2)f_{jk}, y\rangle = \langle Hf_{jk}, (Q_2B - G_1 - G_2)y\rangle$$

If $y \in (D\operatorname{span}\{y_j\}_{j=1}^{s})^{\perp} = Y_0 \oplus \operatorname{span}\{y_j\}_{j=1}^{s}$, then $(Q_2B - G_1 - G_2)y = 0$ by formula (I.3.8). This implies that $D^{-1}(C + F_1 + F_2)f_{jk} = \Sigma_{\nu=1}^{s} \beta_{jk}^{\nu} y_{\nu}$. Next, note that (2.1) implies that for $\ell = 1, \dots, s$

$$\beta_{jk}^{\ell} = \langle \sum_{\nu=1}^{s} \beta_{jk}^{\nu} y_{\nu}, z_{\ell}\rangle = \langle DD^{-1}(C + F_1 + F_2)f_{jk}, z_{\ell}\rangle =$$
$$= \langle Hf_{jk}, (Q_2B - G_1 - G_2)z_{\ell}\rangle = \langle Hf_{jk}, (S_2 - S_1)d_{\ell\alpha_{\ell}}\rangle,$$

where the last identity follows from (I.3.9). Now, use (I.3.4) and (1.6), and one sees that

$$\beta_{jk}^{\ell} = \sum_{\nu=0}^{\alpha_{\ell}-1} \binom{\alpha_{\ell}}{\nu+1} (2i)^{\nu+1} \langle Hf_{jk}, e_{\ell,\alpha_{\ell}-\nu}\rangle = \binom{\alpha_j}{k} (2i)^k \delta_{\ell j}. \qquad \square$$

II.3. Proof of the main theorems

In this section we prove Theorems 0.1 - 0.3. As before $W(\lambda)$ is a selfadjoint rational $m \times m$ matrix function which is regular on the real line including the point infinity and $\theta = (A,B,C,D;\mathbb{C}^n,\mathbb{C}^m)$ is a realization of W satisfying the general conditions stated in the first paragraph of Section II.1. To construct a symmetric factorization we need a number of auxiliary operators. First choose a canonical decomposition

(3.1) $\quad \mathbb{C}^n = N \oplus (M \cap M^{\times}) \oplus K \oplus N^{\times}$

for θ, and let Q_1, Q_2, Q_3 and Q_4 be the corresponding projections. Next, let

$S_1, S_2 : M \cap M^\times \to M \cap M^\times$ be a pair of outgoing operators and let $G_1, G_2 : ¢^m \to M \cap M^\times$ be a pair of output injection operators associated with a set

$$\{d_{jk}\}_{k=1,j=1}^{\alpha_j \;\; s}, \quad \{e_{jk}\}_{k=1,j=1}^{\alpha_j \;\; s}, \quad \{z_j\}_{j=1}^{s} \tag{3.2}$$

of outgoing data for θ (and the projection Q_2). Further, define operators $T_1, T_1 : K \to K$ and $F_1, F_2 : K \to ¢^m$ by setting

$$Q_3 T_1 Q_3 = H^{-1}(Q_2 S_2 Q_2)^* H, \quad F_1 Q_3 = -(Q_2 G_2)^* H,$$

$$Q_3 T_2 Q_3 = H^{-1}(Q_2 S_1 Q_2)^* H, \quad F_2 Q_3 = -(Q_2 G_1)^* H,$$

where H is the invertible selfadjoint operator associated with θ. Finally, extend the vectors $z_1, \ldots, z_s$ to a basis $z_1, \ldots, z_m$ of $¢^m$ such that $S = [z_1, \ldots, z_m]$ is a congruence matrix for θ (as in Section II.2).

THEOREM 3.1. *A selfadjoint rational* $m \times m$ *matrix function* $W(\lambda)$ *which is regular on the real line and at infinity admits a symmetric Wiener-Hopf factorization* $W(\lambda) = W_+(\bar{\lambda})^* \Sigma(\lambda) W_+(\lambda)$, $-\infty \le \lambda \le \infty$, *of which the factors are given by:*

$$\begin{aligned} W_+(\lambda) &= E + ED^{-1}(C(Q_3 + Q_4) + F_2 Q_3)(\lambda - A)^{-1}B + \\ &\quad + ED^{-1}CQ_2(\lambda - S_2)^{-1}V(\lambda - A)^{-1}B + ED^{-1}CQ_2(\lambda - S_2)^{-1}(Q_2 B - G_2), \end{aligned}$$

$$\begin{aligned} W_+(\lambda)^{-1} &= E^{-1} - D^{-1}C(\lambda - A^\times)^{-1}((Q_2 + Q_4)B - G_2)E^{-1} + \\ &\quad - D^{-1}C(\lambda - A^\times)^{-1}V^\times(\lambda - T_2)^{-1}Q_3 BE^{-1} + \\ &\quad - D^{-1}(CQ_3 + F_2)(\lambda - T_2)^{-1}Q_3 BE^{-1}, \end{aligned}$$

$$\Sigma(\lambda) = \begin{pmatrix} & & & & & & \left(\frac{\lambda-i}{\lambda+i}\right)^{\kappa_1} I_{m_1} \\ & & & & & \cdot^{\cdot^{\cdot}} & \\ & & & & \left(\frac{\lambda-i}{\lambda+i}\right)^{\kappa_r} I_{m_r} & & \\ & & & \begin{matrix} I_p & 0 \\ 0 & -I_q \end{matrix} & & & \\ & & \left(\frac{\lambda+i}{\lambda-i}\right)^{\kappa_r} I_{m_r} & & & & \\ & \cdot^{\cdot^{\cdot}} & & & & & \\ \left(\frac{\lambda+i}{\lambda-i}\right)^{\kappa_1} I_{m_1} & & & & & & \end{pmatrix}.$$

Here E *is the inverse of the congruence matrix* $[z_1 \cdots z_m]$ *mentioned at the end of the first paragraph of this section and*

$$V = Q_2AQ_4 - G_2D^{-1}CQ_4 - \Lambda Q_3 - (Q_2B - G_1 - G_2)D^{-1}F_1,$$
$$V^{\times} = Q_4A^{\times}Q_3 - Q_4BD^{-1}F_2 - \Lambda - G_1D^{-1}(CQ_3 + F_2),$$

where Λ *is an arbitrary operator from* K *into* $M \cap M^{\times}$ *which satisfies the Lijapunov equation*

$$(3.3) \qquad H\Lambda x + \Lambda^* Hx = -HQ_2AQ_3x + H(Q_2B - G_1)D^{-1}(CQ_3 + F_2)x, \quad x \in K.$$

Further, the operators A, B, C, D, Q_2, Q_3, Q_4, S_2, T_2, G_1, G_2, F_1 *and* F_2 *are as in the first paragraph of this section and as usual* $A^{\times} = A - BD^{-1}C$. *Finally,* $\kappa_1,\ldots,\kappa_r$ *are the distinct numbers in the sequence* $\alpha_1,\ldots,\alpha_s$ *of outgoing indices in* (3.2), *the number* m_j *equals the number of times* κ_j *appears in the sequence* $\alpha_1,\ldots,\alpha_s$, *the non-negative numbers* p *and* q *are determined by* $p-q = \operatorname{sign} D$ *and* $p+q = m-2s$.

PROOF. We continue to use the notation introduced in the first paragraph of this section. From Propositions 1.5 and 2.2 it follows that T_1,T_2 is a pair of incoming operators and F_1,F_2 is a pair of feedback operators corresponding to a set of incoming data. So according to Theorem I.5.1 the function W admits a right Wiener-Hopf factorization $W(\lambda) = \widetilde{W}_-(\lambda)\widetilde{D}(\lambda)\widetilde{W}_+(\lambda)$ of which the plus-factor $\widetilde{W}_+(\lambda)$ is precisely equal to $EW_+(\lambda)$. We compute:

$$\begin{aligned} W_+(\overline{\lambda})^* &= E^* + B^*(\lambda - A^*)^{-1}((Q_3^* + Q_4^*)C^* + (F_2Q_3)^*)D^{-1}E^* + \\ &\quad + B^*(\lambda - A^*)^{-1}V^*(Q_2(\overline{\lambda} - S_2)^{-1}Q_2)^*C^*D^{-1}E^* + \\ &\quad + (B^*Q_2^* - (Q_2G_2)^*)(Q_2(\overline{\lambda} - S_2)^{-1}Q_2)^*C^*D^{-1}E^* = \\ &= E^* + C(\lambda - A)^{-1}((Q_2 + Q_1)B - G_1)D^{-1}E^* + \\ &\quad + C(\lambda - A)^{-1}(H^{-1}V^*H)(\lambda - T_1)^{-1}Q_3BD^{-1}E^* + \\ &\quad + (CQ_3 + F_1)(\lambda - T_1)^{-1}Q_3BD^{-1}E^*, \end{aligned}$$

where we have used $C^* = HB$, $HA = A^*H$, $HQ_2 = Q_3^*H$, $HQ_1 = Q_4^*H$ and the definitions of T_1, F_1 and F_2. The next step is to calculate $H^{-1}V^*H$:

$$\begin{aligned} H^{-1}V^*H &= Q_1AQ_3 + Q_1BD^{-1}F_1 + \Lambda + Q_2AQ_3 - (Q_2B - G_1)D^{-1}(CQ_3 + F_2) + \\ &\quad + G_2D^{-1}(CQ_3 + F_1 + F_2) \\ &= (Q_1 + Q_2)AQ_3 + \Lambda - (Q_2B - G_1 - G_2)D^{-1}(CQ_3 + F_2) + G_2D^{-1}F_1 + Q_1BD^{-1}F_1 = \\ &= (Q_1 + Q_2)AQ_3 + \Lambda + (Q_2B - G_1 - G_2)D^{-1}F_1 + G_2D^{-1}F_1 + Q_1BD^{-1}F_1 = \end{aligned}$$

$$= (Q_1 + Q_2)AQ_3 + \Lambda + (Q_1 + Q_2)BD^{-1}F_1 - G_1D^{-1}F_1.$$

Here we used the definition of Λ and the fact that $(Q_2B - G_1 - G_2)D^{-1}(CQ_3 + F_1 + F_2) = 0$. So $H^{-1}V^*H$ is equal to the operator V_- introduced in Theorem I.5.1, and we may conclude that $W_+(\overline{\lambda})^*(E^*)^{-1}D = \widetilde{W}_-(\lambda)$. But then

$$W(\lambda) = W_+(\overline{\lambda})^*\Sigma(\lambda)W_+(\lambda)$$

with $\Sigma(\lambda) = (E^*)^{-1}D\widetilde{D}(\lambda)E^{-1}$. From (2.4) and (I.5.1) it follows that $\Sigma(\lambda)$ has the desired form. The formula for $W_+(\lambda)^{-1}$ is a direct consequence of Theorem I.5.1. □

Note that the Lijapunov equation (3.3) has

$$\Lambda = -\tfrac{1}{2}Q_2AQ_3 + \tfrac{1}{2}(Q_2B - G_1)D^{-1}(CQ_3 + F_2)$$

as one of its solutions. By inserting this solution Λ in the expressions for V and $V^\times$ one obtains the formulas of Theorem 0.3. Theorems 0.1 and 0.2 are immediate corollaries of Theorem 0.3.

We conclude with a remark about the dilation in Theorem I.4.1. Let $\theta = (A,B,C,D;\mathbb{C}^n,\mathbb{C}^m)$ be as in the first paragraph of this section, and as in Theorem I.4.1. Consider the dilation $\hat{\theta} = (\hat{A},\hat{B},\hat{C},D;\hat{X},\mathbb{C}^m)$. Assume that the operators Q_1, Q_2, Q_3, Q_4, S_1, S_2, T_1, T_2, G_1, G_2, F_1 and F_2 appearing in the definition of the operators $\hat{A}$, $\hat{B}$ and $\hat{C}$ are as in the beginning of this section. Further, assume that the operator Λ in Theorem I.4.1 (which we can choose freely) is a solution of equation (3.3). Then the dilation $\hat{\theta}$ has additional symmetry properties. In fact, on $\hat{X}$ a selfadjoint operator $\hat{H}$ can be defined by setting

$$\left\langle \hat{H}\begin{pmatrix} x_1 \\ x_2 \\ x_0 \\ x_3 \\ x_4 \end{pmatrix}, \begin{pmatrix} y_1 \\ y_2 \\ y_0 \\ y_3 \\ y_4 \end{pmatrix} \right\rangle = \langle Hx_0, y_0\rangle - \sum_{j=1}^{4} \langle Hx_j, y_{5-j}\rangle,$$

and one can show that $\hat{H}\hat{A} = (\hat{A})^*\hat{H}$ and $\hat{H}\hat{B} = (\hat{C})^*$. Further, for the spaces X_1, X_2, X_3 and X_4 in Section I.4 the following identities hold true:

$$(\hat{H}X_1)^\perp = X_1 \oplus X_2 \oplus X_3,$$

$$(\hat{H}X_2)^\perp = X_1 \oplus X_2 \oplus X_4,$$

$$(\hat{H}X_3)^\perp = X_1 \oplus X_3 \oplus X_4,$$

$$(\hat{H}X_4)^{\perp} = X_2 \oplus X_3 \oplus X_4.$$

Also the bases in X_2 and X_3 are related by identities of the type (1.6) and (1.8) (replacing H by $\hat{H}$). It follows that (in the sense of [9], Section V.1) $\hat{\theta}$ is a realization with selfadjoint centralized singularities, which one can use to prove Theorem 3.1 directly, instead of employing Theorem I.5.1.

REFERENCES

1. Bart, H., Gohberg, I. and Kaashoek, M.A.: Minimal factorization of matrix and operator functions. Operator Theory: Advances and Applications, Vol 1; Birkhäuser Verlag, Basel, 1979.

2.† Bart, H., Gohberg, I. and Kaashoek, M.A.: Wiener-Hopf factorization of analytic operator functions and realization. Rapport nr. 231, Wiskundig Seminarium der Vrije Universiteit, Amsterdam, 1983.

3. Bart, H., Gohberg, I. and Kaashoek, M.A.: Wiener-Hopf factorization and realization. In: Mathematical Theory of Networks and Systems (Ed. P. Fuhrmann), Lecture Notes in Control and Information Sciences, Vol 58, Springer Verlag, Berlin etc., 1984.

4. Clancey, K. and Gohberg, I.: Factorization of matrix functions and singular integral operators. Operator Theory: Advances and Applications, Vol 3; Birkhäuser Verlag, Basel, 1981.

5. Fuhrmann, P.A.: On symmetric rational transfer functions. Linear Algebra and Applications, 50 (1983), 167-250.

6. Gohberg, I.C. and Krein, M.G.: Systems of integral equations on a half line with kernels depending on the difference of the arguments. Uspehi Mat Nauk 13 (1958) no. 2 (80), 3-72 (Russian) = Amer. Math. Soc. Transl. (2) 14 (1960), 217-287.

7. Nikolaĭčuk, A.M. and Spitkovskiĭ, I.M.: On the Riemann boundary-value problem with hermitian matrix. Dokl. Akad. Nauk SSSR 221 (1975) No. 6. English translation: Soviet Math. Dokl. 16 (1975) No. 2., 533-536.

8. Ran, A.C.M.: Minimal factorization of selfadjoint rational matrix functions. Integral Equations and Operator Theory 5 (1982), 850-869.

9. Ran, A.C.M.: Semidefinite invariant subspaces, stability and applications. Ph-D thesis Vrije Universiteit, Amsterdam, 1984.

† The results from [2] which are used in the present paper can also be found in the sixth article in this volume.

Department of Mathematics & Computer Science, Vrije Universiteit
Amsterdam, The Netherlands.